ECOLOGY
OF
HALOPHYTES

ACADEMIC PRESS RAPID MANUSCRIPT REPRODUCTION

ECOLOGY OF HALOPHYTES

edited by

ROBERT J. REIMOLD
University of Georgia Marine Institute
Sapelo Island, Georgia

WILLIAM H. QUEEN
Chesapeake Research Consortium
University of Maryland
College Park, Maryland

89408

ACADEMIC PRESS, INC.   New York and London   1974
A Subsidiary of Harcourt Brace Jovanovich, Publishers

COPYRIGHT © 1974, BY ACADEMIC PRESS, INC.
ALL RIGHTS RESERVED.
NO PART OF THIS PUBLICATION MAY BE REPRODUCED OR
TRANSMITTED IN ANY FORM OR BY ANY MEANS, ELECTRONIC
OR MECHANICAL, INCLUDING PHOTOCOPY, RECORDING, OR ANY
INFORMATION STORAGE AND RETRIEVAL SYSTEM, WITHOUT
PERMISSION IN WRITING FROM THE PUBLISHER.

ACADEMIC PRESS, INC.
111 Fifth Avenue, New York, New York 10003

United Kingdom Edition published by
ACADEMIC PRESS, INC. (LONDON) LTD.
24/28 Oval Road, London NW1

Library of Congress Cataloging in Publication Data
Main entry under title:

Ecology of halophytes.

    1. Halophytes.   I.   Reimold, Robert J., ed.
II. Queen, William, ed.
QK922.E24   581.5'265   72-88347
ISBN 0–12–586450–7

PRINTED IN THE UNITED STATES OF AMERICA

CONTENTS

Contributors ix
Preface xi
Foreword by Alfred C. Redfield xiii

Part I
Halophytes: An Overview

Salt marshes and salt deserts of the world
    V. J. Chapman 3

Part II
Halophytes of the United States:
Distribution, Ecology, Anatomy, and Physiology

Vascular halophytes of the Atlantic and Gulf Coasts
of North America north of Mexico
    Wilbur H. Duncan 23

Mangroves: a review
    Gerald E. Walsh 51

Beach and salt marsh vegetation of the North American Pacific Coast
    Keith B. Macdonald and Michael G. Barbour 175

Inland halophytes of the United States
    Irwin A. Ungar 235

A review of structure in several North Carolina salt marsh plants
    Charles E. Anderson 307

Physiology of coastal halophytes
    William H. Queen 345

Physiology of desert halophytes
  Martyn M. Caldwell 355

Salt tolerance of mangroves and submerged aquatic plants
  Calvin McMillan 379

Part III
Habitat Associations of Halophytes

Mathematical modeling – *Spartina*
  Robert J. Reimold 393

The role of overwash and inlet dynamics in the formation of
salt marshes on North Carolina barrier islands
  Paul J. Godfrey and Melinda M. Godfrey 407

Probable agents for the formation of detritus from the
halophyte, *Spartina alterniflora*
  Mallory S. May, III 429

Marsh soils of the Atlantic Coast
  Leo J. Cotnoir 441

The relationship of marine macroinvertebrates to salt marsh plants
  John N. Kraeuter and Paul L. Wolf 449

Relationship of vertebrates to salt marsh plants
  G. Frederick Shanholtzer 463

Salt marsh plants and future coastal salt marshes in relation to animals
  Franklin C. Daiber 475

Part IV
Applied Research Related to Halophytes

Remote sensing as a tool for studying the ecology of halophytes
  John L. Gallagher 511

CONTENTS

Stabilization of coastal dredge spoil with *Spartina alterniflora*
    Ernest D. Seneca 525

Effects of herbicides on the *Spartina* salt marsh
    Andrew C. Edwards and Donald E. Davis 531

Nutrient limitation in salt marsh vegetation
    Ivan Valiela and John M. Teal 547

The potential economic uses of halophytes
    Peta J. Mudie 565

Halophytes, energetics and ecosystems
    Eugene P. Odum 599

Index 603

CONTRIBUTORS

Anderson, Charles E., Department of Botany, North Carolina State University, Raleigh, North Carolina 27607

Barbour, Michael G., Botany Department, University of California, Davis, California 95616

Caldwell, Martyn M., Department of Range Sciences, Utah State University, Logan, Utah 84321

Chapman, V. J., Department of Botany, Auckland University, Auckland, New Zealand

Cotnoir, Leo J., Plant Science Department, College of Agriculture Science, University of Delaware, Newark, Delaware 19711

Daiber, Franklin C., College of Marine Studies and Department of Biological Sciences, University of Delaware, Newark, Delaware 19711

Davis, Donald E., Department of Botany and Microbiology, Auburn University Agricultural Experiment Station, Auburn, Alabama 36830

Duncan, Wilbur H., Department of Botany, University of Georgia, Athens, Georgia 30602

Edwards, Andrew C., Department of Botany and Microbiology, Auburn University Agricultural Experiment Station, Auburn, Alabama 36830

Gallagher, John L., The University of Georgia Marine Institute, Sapelo Island, Georgia 31327

Godfrey, Melinda M., Department of Botany/U.S. National Park Service, University of Massachusetts, Amherst, Massachusetts 01002

Godfrey, Paul J., Department of Botany/U.S. National Park Service, University of Massachusetts, Amherst, Massachusetts 01002

Kraeuter, John N., The University of Georgia Marine Institute, Sapelo Island, Georgia 31327

CONTRIBUTORS

Macdonald, Keith B., Department of Geological Sciences, University of California, Santa Barbara, Santa Barbara, California 93106

May, Mallory S., III, Biology Department, Brunswick Junior College, Brunswick, Georgia 31520

McMillan, Calvin, The Department of Botany and the Plant Ecology Research Laboratory, The University of Texas at Austin, Austin, Texas 78712

Mudie, Peta J., Foundation for Ocean Research, Scripps Institute of Oceanography, La Jolla, California 92037

Odum, Eugene P., Institute of Ecology, The University of Georgia, Athens, Georgia 30602

Queen, William H., Chesapeake Research Consortium, University of Maryland, College Park, Maryland 20742

Reimold, Robert J., The University of Georgia Marine Institute, Sapelo Island, Georgia 31327

Seneca, Ernest, D., Departments of Botany and Soil Science, North Carolina State University, Raleigh, North Carolina 27607

Shanholtzer, G. Frederick, The University of Georgia Marine Insitute, Sapelo Island, Georgia 31327

Teal, John M., Woods Hole Oceanographic Institution, Woods Hole, Massachusetts 02543

Ungar, Irwin A., Department of Botany, Ohio University, Athens, Ohio, 45701

Valiela, Ivan, Boston University Marine Program, Marine Biological Laboratory, Woods Hole, Massachusetts 02543

Walsh, Gerald E. Environmental Protection Agency, Gulf Breeze Laboratory, Sabine Island, Gulf Breeze, Florida 32561

Wolf, Paul L., Tyrone Biological Laboratory, Lebanon Valley College, Annville, Pennsylvania 17003

PREFACE

This publication arose as a result of a symposium on the ecology of halophytes sponsored by the Physiological Ecology section of the Ecological Society of America and held as a portion of the American Institute of Biological Sciences meetings, August 1972. The interest generated in preparing and presenting this symposium all pointed to the need for a review volume on the salt marsh ecosystem, the saline soil ecosystem, and what was known about these systems. The diversity of interests of the contributors demonstrates the comprehensive nature of the *Ecology of Halophytes*.

The *Ecology of Halophytes* considers the fundamentals of distribution, anatomy, and physiology of halophytes. It also provides an overview of the role of the halophyte in ecosystems in various parts of the world. A section on habitat associations of halophytes considers the relation of the plants to other fauna and flora in natural systems. A final section deals with recent applied research related to halophytes and quantification of the impact of man on the ecology of halophytes.

It is hoped that this publication will be of use for various disciplines working in saline wetlands ecosystems. It is intended to serve land use planners, federal and state natural resources and transportation interests, and real estate developers in providing a comprehensive summary of the "state of the art" in understanding halophytic ecosystems. With a better fundamental knowledge of the system, the above mentioned professionals should be better able to plan activities and uses compatible with the natural halophytic ecosystem and hopefully avoid some of the past errors man has made.

We acknowledge the assistance of Charles R. Malone, program chairman for the Ecological Society of America, and F. John Vernberg, chairman of the Physiological Ecology Section of ESA, for arranging the symposium. Thanks also go to Winona B. Vernberg, Lee N. Miller, and James G. Gosselink for their assistance as symposium session chairman.

We extend our sincere appreciation to the many scientists who have served as editorial reviewers for each chapter. In particular we recognize the valuable assistance of Rick A. Linthurst for his dedicated efforts of manuscript, proof and camera ready copy production. The untiring motivation from our families is also acknowledged.

Robert J. Reimold
William H. Queen

FOREWORD

Salt marshes are many things to many men. To those who live on the east coast they are the one remaining accessible bit of undisturbed wilderness. Unfortunately as unoccupied space they offer great opportunity to those who would destroy them by development. On the other hand, being the favored habitat of water fowl, substantial areas of marshland are being set aside as wildlife reservations which require understanding of their ecology for their maintenance.

To the scientist salt marshes provide for study an ecological system with well-defined boundary conditions within which physical and biological factors interact. The physical environment is hostile to most of the organisms of the land, of the fresh water wetlands, and of the open sea. The result is a biota limited to the relatively few species of plants and animals adapted to periodic inundation and to life in waters of varied salinity.

The salt marshes have been formed in sheltered places by waterborne sediments collected by those halophytes which can populate the intertidal zone. Of these *Spartina alterniflora* is the principal species throughout its range because it can survive longer periods of submergence than others and hence occupy the foreshore at lower tide levels. The marshes, as we know them today, have developed during a period of rising sea level. The peats formed each year have buried and often left undisturbed those formed in earlier years. A record is preserved in the depths of the peat deposits from which the history of the marsh may be reconstructed and which provides a chronology of the recent rise in sea level. Also preserved are artifacts which give evidence of man's occupation of marshland areas and of his culture during earlier times.

The interaction of tidal flow, the resulting sedimentation and erosion, and the physiology of the halophytes has produced a unique land form for the attention of the geomorphologist. The creeks which drain the marshes have an hydrology which differs in important respects from terrestrial streams. The soil of the marshlands is distinctive in its saline character, its prevalent saturation with water, its high organic content, and its proverty in oxygen.

In the study of salt marshes, the botanists have done a more thorough job than others. In the present volume they have continued to lead the way by the investigation of the halophytes and their ecology. The collection of papers pre-

FOREWORD

sented should be useful to those with diverse interests as an account of what is being done currently toward understanding the ecology of the marshes.

Alfred C. Redfield
Woods Hole, Massachusetts

PART I. HALOPHYTES: AN OVERVIEW

SALT MARSHES AND SALT DESERTS OF THE WORLD

V. J. Chapman

Department of Botany
Auckland University
Auckland, New Zealand

Introduction

In this contribution I do not intend to say anything about the physiology and morphology of salt marsh plants, nor anything about salt marsh soils and animals because these topics are covered in other articles in this volume. Since many of you are familiar with my book on salt marshes and salt deserts of the world, I will consider intensively only those areas in which additional information has come to hand since the book was written.

In the past there has been considerable debate over what are halophytes. Generally, species capable of tolerating 0.5% or more NaCl are regarded as halophytes. Some evidence has been published in the past to indicate that some species are obligate halophytes and reach their optimum growth at moderate to high salinities. Barbour (1970) has discussed this aspect in some detail and concluded that very few species are restricted above 0.5% NaCl.

One characteristic of salt marshes and salt deserts is provided by the fact that only a relatively small number of plant species are capable of tolerating the degrees of salinity that occur. As a result, there are broad geographical areas in which there is a substantial uniformity in the vegetation. In some cases, subdivision can be based on temperatures or upon soil type. With this as background it seems to me that the major groups of maritime salt marshes and salt deserts (Fig. 1) are as follows:

Maritime Marshes

Arctic Group

The Arctic Group is represented by the salt marshes of the Canadian and American Arctic, Greenland, Iceland, northernmost Scandinavia, and arctic Russia. The environment is extreme, especially in winter, and the marshes appear to be somewhat fragmentary as a result. Succession is very simple. The vegetation is dominated by the grass *Puccinellia phryganodes*, though species of *Carex*, especially *C. subspathacea* and *C. maritima*, play an important part at higher levels.

Northern European Group

This group includes marshes from the Iberian Peninsula northward around the English Channel, North Sea Coasts, and the Baltic Sea, as well as those on

Fig. 1.
World distribution of salt-marshes, salt deserts (solid black) and mangrove swamps (latched)

the Atlantic coasts of Eire and Great Britain. Whilst there is a fundamental similarity in physiognomy and floristic composition, nevertheless, different communities can be recognized in specific areas. These sub-areas are related either to soil differences or, in the case of the Baltic, to salinity differences. Throughout the region there is a dominance of annual *Salicornia* species, *Puccinellia maritima, Juncus gerardi,* and the general salt marsh community. The North European Group can be divided into the following five subgroups:

Scandinavian Subgroup — Scandinavian marshes are characterized by a high proportion of sand in the soil and have developed on a rising coastline where newly exposed shore sand can be blown inland. They are found in Scandinavia, Schleswig-Holstein, the west coast of Britain from the Severn to Northern Scotland, on the east coast of Eire, and from Kincardine to Inverness on the east coast of Scotland. These marshes are dominated by grasses, especially *Puccinellia maritima*, *Festuca rubra*, and *Agrostis stolonifera*, and for this reason are grazed by domestic animals. Grazing probably results in the elimination, or control, of dicotyledonous herbs such as *Aster, Limonium, Triglochin,* etc. Since the publication of *Salt Marshes and Salt Deserts of the World* a number of contributions have been made to the literature of the Scandinavian Subgroup. Foremost among these have been the papers by Dalby (1970), South Wales marshes; Gillner (1960), Swedish marshes; Gimingham (1964), East Scottish marshes; Packham and Liddle (1970), West Wales marshes; Round (1960), Dee marshes; and Taylor and Burrows (1968), Dee marshes.

North Sea Subgroup — North Sea marshes have much more clay and silt in the soil and there tends to be a wider range of communities. Grasses are not so dominant and the general salt marsh community plays a greater role in the successions. These marshes are generally associated with a subsiding coastline, but their character is currently changing through the introduction or natural spread of *Spartina townsendii* or *S. anglica*. North Sea marshes are represented by the marshes of eastern England, southeast Scotland, northern Germany, and the low countries. Recent significant publications on the marshes of the North Sea Subgroup have been those by Koster (1961) and Beeftink (1965, 1966).

Baltic Subgroup — The Baltic Subgroup differs from the Scandinavian and North Sea Subgroups by the presence of some species, e.g. *Carex paleacea, Juncus bufonius,* and *Desmoschoenus bottnica,* that occupy dominant places in the succession. The brackish nature of the Baltic also permits *Scirpus* to be the primary colonist.

English Channel Subgroup — Marshes in this subgroup were probably comparable with the North Sea Subgroup, but since the appearance of *Spartina townsendii* and *S. anglica* towards the end of the last century and their subsequent spread, these marshes physiognomically now look more like those of the U.S.A. In Poole Harbour, the origianl home of *S. townsendii*, aerial photography has been employed to see whether there have been any recent changes in the vegetation (Ranwell et al. 1964, Bird and Ranwell

1964). The Total area of *Spartina* has not altered materially though there has been some seaward spread of *Scirpus* (2 ft. per annum), *Glaux*, *Phragmites* (4 ft. per annum), *Puccinellia*, and *Aster tripolium*.

Southwest Eire Subgroup — This subgroup is still regarded as a separate entity because of the peaty nature of the soil and the simplicity of the succession. It is a group that requires further study.

Mediterranean Group

Whilst Mediterranean marshes show some affinity with those of northern Europe, there are species, e.g. of *Arthrocnemum* and *Limonium*, which are characteristic. The eco-climax is dominated by *Juncus acutus* and the Mediterranean is the only region where this species attains such prominence. It is perhaps convenient to recognize western and eastern subgroups though obviously there is a gradual transition from one to the other. The marshes of the Caspian would appear to belong to a separate subgroup.

Western Mediterranean Subgroup — The marshes of southern France, particularly of the Rhone Delta (Knierr 1960, Corre 1962, Molinier et al. 1964) are typical of this subgroup. The primary colonists are *Salicornia herbacea* agg. or *Salsola soda* plus *Suaeda* or a *Kochio-Suaedetum*. *Salicornia fruticosa* is also prominent. Other characteristic communities are the Limonieto-Artemisietum, Limonieto-Limoniastretum, Crithmo-Staticetum, and Halimionetum portulacoidis.

Eastern Mediterranean Subgroup — Whilst many of the dominants are the same as in the western subregion, there are eastern species present, such as *Halocnemum strobilaceum*, *Petrosimonia crassiflora*, *Bupleurum gracile*, and *Suaeda altissima*.

Caspian Salt Marshes — Annual *Salicornia* or *Halocnemum* (Tagunova 1960) is the primary colonist of the Caspian marshes with thickets of *Salsola sola*, *Suaeda*, or *Puccinellia gigantea* behind (Agadzhanoz 1962). *Kalidium caspicum* and *Anabasis aphulla* are other typical species.

Western Atlantic Group

Marshes in this group extend from the St. Lawrence down to Florida where there is a transition to mangrove. They are subdivided essentially on differences in mode of formation, though there are floristic differences as well in respect to the southern subregion. The halophyte flora of the Atlantic coast of North America is discussed in detail by Duncan in another article of this volume.

Bay of Fundy Subgroup — This subgroup is characterized by the significant role of *Puccinellia americana*, and *Juncus balticus* var. *littoralis* at the highest levels, and by a transition to bog rather than fresh-water swamp. These marshes are formed in front of a weak rock upland.

New England Subgroup — New England marshes are formed in front of a hard rock upland, and the soil is a peat rather than a muddy clay, as in the other two subgroups. *Puccinellia americana* is not a dominant, and the transition is to reed swamp.

Coastal Plain Subgroup — Recent descriptions of coastal plain marshes have been given by Adams (1963) and Davis and Gray (1966), whilst Blum (1963, 1968) has described the algal communities. The marshes have developed in front of a soft rock upland.

Pacific American Group

The marsh communities of the American Pacific Coast are entirely different from those elsewhere and the successions appear to be very simple. The marshes extend from California to southern Alaska. They are discussed by MacDonald and Barbour in another paper in this volume.

Sino-Japanese Group

Recent accounts (Ito 1961, 1963; Ito and Leu 1962; Umezugi 1964; Miyawaki and Ohba 1965) have added much valuable information about this region. Primary communities are dominated by *Triglochin maritima*, *Salicornia brachystachya*, or *Limonium japonicum*, though in the far north *Suaeda japonica* with *Atriplex gmelini* occupy this position with *Limonium tetragonum* behind. Southwards the equivalent zone is occupied by *Zoysia macrostachya* or *Puccinellia kurilensis*. In the northern part a relationship to the arctic region is evidenced by the presence of *Carex ramenskii*. It is probable that this region should be subdivided into northern and southern subgroups.

Australasian Group

The Australasian Group is characterized by some specifically southern hemisphere species such as *Salicornia australis*, *Suaeda nova-zelandiae*, *Triglochin striata*, *Samolus repens*, and *Arthrocnemum* species. Some recent descriptions are those by Sauer (1965) and Clarke and Hannon (1967). Floristic similarities justify regarding them as one region, but there are sufficient floral differences to render it desirable to subdivide the group.

Australian Subgroup — This subgroup is characterized by *Arthrocnemum arbuscula*, *A. halocnemoides*, *Hemichroa pentandra*, and *Suaeda australis*. At higher levels *Atriplex paludosa*, *Limonium australis*, and *Frankenia pauciflora* are important.

New Zealand Subgroup — Nearly all species of the New Zealand subgroup are found on Australian salt marshes with the exception of those listed above.

South American Group

This group is characterized by species of *Spartina (S. brasiliensis, S. montevidensis)*, *Distichlis*, *Heterostachys*, and *Allenrolfea* which are not found elsewhere. These marshes are in need of further study.

Tropical Group

Tropical marshes generally occur at high levels behind mangrove swamps and are flooded only by extreme tides. Typical species are *Sesuvium portulacastrum* and *Batis maritima*.

Inland Marshes and Alkali Regions

Inland European Group
Whilst there is a general similarity throughout, biogeographically this group seems subdivisible into northern and southeastern subgroups.

Northern European Subgroup — Floristically the marshes in this subgroup are closely related to the northern European maritime marshes. They possess the coastal species *Salicornia* and *Plantago maritima*, whilst *Puccinellia distans* and *Halimione pedunculata* are characteristic species.

Southeastern European Subgroup — These marshes contain both northern and eastern species. In Italy, *Crypsis aculeata* and *Aster tripolium* var. *pannoricus* are of eastern affinities. A clearer picture of this vegetation is found in recent accounts by Bodrogkozy (1965, 1966). Species that could be regarded as representative of this subgroup include the two above together with *Cyperus pannonicus, Puccinellia limosa, Lepidium cartilagineum, Frankenia* spp,. *Halimione verrucifera, Salsola soda*, and *Suaeda splendens*.

Inland Asian Group
It would appear that this group can be subdivided into at least three subgroups. There may prove to be more, and one or more may need elevation to group status.

Aralo-Caspian Subgroup — Within this subgroup there may be three minor regions, namely, Aralo-Caspian, Turkestan, and Trans-caucasian. Birand (1960), Kutnetsov (1960), Rusyaeva (1961), Rustamov (1962), Vinogradov (1960), and Momotov (1963), among others, have provided valuable accounts. Whilst there are floristic affinities with the Southeastern European area, there are species such as *Kalidium foliatum, K. caspicum, Halostachys caspia*, and *Halocnemum crassifolia* that are typical of these salt deserts.

Irak-Central Asian Subgroup — This subgroup has affinities with the North African Subgroup (see below), but there are some characteristic species such as *Suaeda vermiculata, S. palestina, Seidlitzia rosmarinus*, and *Anabasis articulata*. For a recent general account, see papers by Zohary (1963) and Habib et al. (1971).

East Asian Subgroup — The information on this subgroup is very scanty but it seems to be characterized by species of *Atriplex* and *Artemisia*.

African Group
The African Group is conveniently divided into northern, eastern, and southwestern subgroups though further subdivisions may be desirable.

North African Subgroup — Apart from widespread species, this subgroup contains western species in the Moroccan area, and there is a gradual infiltration of eastern species as one travels towards Palestine. The most widespread community is dominated by *Salicornia fruticosa* and *Sphenopus divaricatus*. Other typical species include *Arthrocnemum glaucum*, *A. macrostachya*, and *Suaeda fruticosa*.

Southwest African Subgroup — This subgroup is found around the shores of saline lakes. *Scirpus robustus* and *S. spicatus* are primary colonists.

East African Subgroup — Whilst the East African Subgroup has some affinities with the North African Subgroup (e.g. *Arthrocnemum glaucum*), there are some typical species, e.g. *Halopeplis perfoliata* and *Aeluropus lagopoides*.

North American Group

Whilst some species of the North American Group are found also in salt marshes of either the Pacific or Atlantic coasts (Branson 1967), there are other species, such as *Salicornia utahensis*, *Allenrolfea occidentalis*, *Sarcobatis vermiculata*, and *Atriplex confertifolia* that are characteristic of the group. Arising from recent contributions (Dodd et al. 1964; Dodd and Coupland 1966; Hunt and Durrell 1966; Branson, Miller, and McQueen 1967; Cusick 1970; Ungar 1965, 1967, 1968, 1970; Ungar et al. 1969; Bradley 1970) it seems that the region can be subdivided into two subgroups.

North Western Subgroup — This subgroup is characterized by the presence of *Salicornia rubra* and *Puccinellia nuttallii*.

South Western Subgroup — This subgroup is characterized by *Sporobolus airoides* and *Tamarix pentandra*.

South American Group

Characteristic species of the inland saline areas of the South American Group are *Heterostachys ritteriana*, *Allenrolfea patagonica*, *Spartina montevidensis*, and *Salicornia gaudichaudiana*.

Australian Group

This Australian Group is mainly characterized by species of *Atriplex* and *Arthrocnemum halocnemoides*.

Discussion

The dominant vegetation of maritime salt marshes is essentially phanerogamic herbs, though some shrubs also occur. Inland salt marshes, and especially salt deserts, are typified by shrubs. Apart from the phanerogams, maritime salt marshes also carry an extensive algal vegetation which may, in places, be as important as the phanerogams. This is particularly the case where salt marsh fucoids are concerned (European and U.S.A. Atlantic marshes). As may be expected, Cyanophyceae are abundant and they can occur also on the soils of inland areas. There are also characteristic members of the Rhodophyceae, especially *Catenella* and *Bostrychia*, which are to be found at the base of plants at the higher levels. These two genera are interesting because they occupy a rather similar niche in mangrove swamps on pneumatophores and trunk bases. Chlorophyceae are ubiquitous, especially species of *Lola, Rhizoclonium, Chaetomorpha, Ulothrix,* and *Enteromorpha*. Species of the Xanthophycean *Vaucheria* are not uncommon as pioneers (Nienhuis and Simons 1971), especially on the mud banks of creeks.

On maritime salt marshes, continual accumulation of silt steadily raises the soil level. This reduces the number of inundations which has an effect on the soil salinity. The overall result is successive changes in vegetation giving rise to a succession. Provided the land/sea level remains constant, vertical growth of a marsh will cease and the final vegetation is regarded as the sere climax. In the past, some interest has centered around the time taken to develop to the sere climax from the arrival of the first colonists. Apart from the evidence that can be obtained from old maps, calculations can also be based upon experimentally determined accretion rates coupled with mean vertical height determinations of the various stages in the succession. More recently radioactive carbon techniques have also been employed (Lyon and Harrison 1960, Redfield and Rubin 1962, Sears 1963, Bloom and Ellis 1965, Bloom 1967), especially with respect to the New England (U.S.A.) marshes. This is currently the most accurate method, and has resulted in amendments to rates determined earlier by the older methods. Some of the rates, e.g. those for New Hampshire (3 ft. per century) and Nova Scotia (1.4 - 2 ft. per century), are widely removed from the general rate for that coast of 0.5 ft. per century and further measurements are required. At present it would seem that on this coast it must take about 600-700 years for a marsh to pass through all its phases up to the *Juncus gerardi* zone.

Apart from some of the earlier work on the movement of the water table of salt marshes and the effect of tidal height upon it, no real new information has appeared though Tyler (1971) has provided some data for Baltic Sea meadows. Whilst the existing information has enabled a generalized picture to be established, there may be marshes that do not conform. Early studies by the present author led to the hypothesis that on maritime salt marshes the water table never rises completely to the surface, even during flooding tides, so that an aerated layer existed in which most of the roots of salt marsh plants are to be found. The earlier work has been extended by the studies of Teal and Kanwisher (1961) on Georgia marshes and by Clarke and Hannon (1969) on Australian marshes. The former author found that only the upper few centimetres contained any oxygen and that, in general, reduction intensity was high except in well-drained areas. In Great Britain, 'die-back' in *Spartina townsendii* has been related to a toxic reduced inorganic ion - possibly sulphide - in the soil (Goodman and Williams 1961). These sulphides must lower the soil oxygen content, but how the oxygen deficiency operates upon the plant's metabolism has still to be solved (Lambert 1964). Clarke and Hammon (loc. cit.) in their study of Sydney salt marshes pointed out that the aerated layer is not completely continuous. It is clearly related to soil structure, abundance of crab, worm, and mollusc burrows, and to the existence of old decayed roots.

Critical work has still to be carried out on maritime salt marshes in order to relate their zonation to the zonation characteristic of rocky shores. It would seem that the primary colonists generally occupy the upper part of the eu-littoral belt and that most of the communities are located in the supra-littoral fringe.

The author is still of the opinion that the Montpellier system represents the best phyto-sociological system so far as salt marsh vegetation is concerned. Beeftink (1968) has discussed the classification of European salt marsh communities on this basis but I believe that because of the general uniformity of the vegetation, any such classification should be on a world-wide basis. The major classes and orders would be as follows:

1. Class Zosteretea
2. Class Salicornietea

3. Class Juncetea
4. Class Phragmitetea
5. Class Caricetea

Order 1.   Halobenthalia
Order 2.   Coeno-Salicornietalia
Order 3.   Coeno-Puccinellietalia
Order 4.   Coeno-Spartinetalia
Order 5.   Limonietalia
Order 6.   Coeno-Festucetalia
Order 7.   Halostachyetalia
Order 8.   Coeno-Juncetalia
Order 9.   Halo-Phragmitetalia
Order 10. Halo-Caricetalia

Because of the economic importance attached to *Spartina*, continuing studies of its ecological requirements have proceeded. These are discussed in other articles of this volume. A study of so-called *S. townsendii* has shown that there is the original sterile F_1 hybrid and a subsequent developed fertile autotetraploid (Hubbard 1967; Marchant 1963, 1966, 1967, 1968, 1970). This latter has been termed *S. anglica* (Hubbard 1968). The two species, collectively, occupy a world total of some 70 sq. miles (Ranwell 1967), the fertile hybrid being the more successful (Goodman et al. 1969). Both species have shown themselves to be very successful colonists of mudflats so long as daily tidal submergence is not more than six hours (Hubbard 1965a, b; Chalter 1965; Taylor and Burrows 1968; Lambert 1964; Bascand 1969). Both species appear to require low winter temperatures because neither do well, even in warm temperate areas. In the north island of New Zealand, *S. alterniflora* and *S. gracilis* are both better colonists of mudflats (Bascand 1968, 1970). In places where the plants have become too successful and tend to block channels, control can be achieved by two successive applications of Dalapon (Ranwell and Bowning 1960; Bascand 1968). Hubbard and Ranwell (1966) have reported on the value of cropping *Spartina* marsh for silage. A parasitic pyrenomycete *Haligena spartinae* has been recorded from the species (Jones 1962), but whether this can affect the success of the plant is not known.

It is difficult to determine the extent of inland saline lands because there is no generally accepted criterion as to when a soil is to be regarded as saline. Very considerable areas formerly existed in Eruope [300,000 hectares in Rumania but now reduced by reclamation (Oprea 1965)], particularly in Austria, Hungary, and Bulgaria (Ovdenicharov 1959). At least 75 million hectares or 3.4% of the available land in the U.S.S.R., China, and neighbouring states are saline or alkaline (Boumans and Husbos 1960, Muratova 1959, Agaer 1965). The exact extent is unknown and only a small percentage has so far been put to any use. In India and Pakistan there are about 20 million acres of saline soils (Raheju *in* Bayko 1966), and the same writer reports the development of 300,000 acres in Egypt as a result of irrigation from the Nile bringing the subsurface salts to the surface. The full extent of the vast saline areas on the African continent is just not known. I can find no published data on the area of saline soils in Australia where they are mainly confined to the southern part of the continent. Figures are probably available for Canada and the U.S.A. because Cairns (1969a) has stated that 8-12 million hectares of alkali soils exist in Alberta and Saskatchewan.

With the world at present facing an over population problem and a food shortage, maritime salt marshes and inland salt deserts both represent potential land that could and should be reclaimed for arable purposes. Successful reclamation of these soils is possible and indeed has been carried out in many parts of the world. The method to be used in any one case depends upon an understanding of the base-exchange conditions in the soil, especially that portion of the exchangeable sodium bound to the finer particles (Sambin 1963). Boyko (1966) has pointed out that new views, such as a better understanding of plant reaction to osmotic pressures, and the use of seawater which can be more successful in irrigation than other alkaline waters, especially in promoting increased resistance to drought, give great hopes for further successful reclamation. Raheju (1966) has listed the various reclamation methods used in different countries. It has to be noted that much salinization has arisen as a result of man's disturbance of the ecological balance. We have, however, now become very sensitive to this kind of effect and it should not occur in the future. Most recent reclamation work has centered on 1) the use of gypsum (Sambin 1963; Raikov and Kavadgiev 1965; Ramos et al. 1961; Padhi et al. 1965; Boumans and Ausbos 1960; Agaer 1965; Szabolcs and Abrakham 1964; Prettenhoffer 1964, 1965; Saini 1971) though Szabolcs (1965) found calcium nitrate equally satisfactory; 2) deep ploughing (Cairns 1961, 1969a, 1970; Bowser and Cairns 1967; Cairns and Bowser 1969; Cairns 1971; Prettenhoffer 1964a, 1964b; Zvereva 1960; Bolshakov 1960; Raikov and Kavadgiev 1965; Bodrogkozy 1961-62); and 3) irrigation (Ebert 1971). It is likely, therefore, that economic factors will largely determine the choice of method.

With the reclamation of saline soils, there is also the problem of salt tolerance of crop plants because this determines which crop can first be successfully planted. Most of the recent work here has been summarized by Strogonov (1964) and Boyko (1966).

In conclusion, one may draw attention to the enormous amount of literature that has appeared in the last ten years, and which is indicative of the great importance attached to the reclamation of saline soils. It is inevitable that, as natural phenomena, salt marshes and salt deserts will be greatly reduced in area in future years.

Literature Cited

Adams, D. A. 1963. Factors influencing vascular plant zonation in North Carolina Salt marshes. Ecol. 44:445-456.

Agadzhanov, S. D. 1962. Solonchakovaya vastite L'nost'primorskikh peskov Azerbaidzhansk (the saline vegetation of the coastal sandplain of Azerbaidzhan). Uch. Zap. Azerbaidzhansk Univ. Ser. Biol. Nauka. 2:3-13.

Agaev, B. M. 1965. Preliminary results of the reclamation of soda-saline soils in the Karabah Plain, Azerbaidzhan. Agrokem. Talajtan 14 (suppl.):189-194 (Engl. summ.).

Baker, J. M. 1971. Seasonal effects of oil pollution on salt marsh vegetation. Oikos 22(1):106-110.

Barbour, M. G. 1970. Is any Angiosperm an obligate halophyte? Amer. Mid. Nat. 84(1):105-120.

Bascand, L. D. 1968. Some aspects of the autecology of *Spartina* species in New Zealand. M.Sc. Thesis, Auckland Univ., N. Z.

Bascand, L. D. 1970. The roles of *Spartina* species in New Zealand. Proc. N. Z. Ecol. Soc. 17:33-40.

Beeftink, W. G. 1965. De Zontvegetatie van ZW-Nederland beschond in Europees verband (salt marsh communities of the S. W. Netherlands in relation to the European halophytic vegetation. Middel. Land-Bouwhogeschwageningen 65(1):1-167 (Engl. summ.).

Beeftink, W. G. 1966. Vegetation and habitat of the salt marshes and beach plains in the S.W. part of the Netherlands. Wentig 15:83-108.

Beeftink, W. G. 1968. Die Systematik der Europaischen Salzpflanzengesellschaften in Pflanzensoziologische Systematik. Edited by R. Tuxen. 239-63 pp.

Birand, H. 1960. Erste Ergebnisse vegetations — Untersuchungen in der Zentralanatolischen steppe. I. Halophytengesellschaften des Tuzgolii. Bot. Jahrb. 79(3):255-296.

Bird, E. C. F. and D. S. Ranwell. 1964. *Spartina* salt marshes in Southern England. IV. J. Ecol. 52:255-66.

Bloom, A. L. 1967. Pleistocene shorelines: a new test of Isostacy. Geol. Soc. Amer. Bull. 78:1477.

Bloom, A. L. and C. W. Ellis. 1965. Post-glacial stratigraphy and morphology

of Coastal Connecticut. Connecticut Nat. Hist. Survey Guidebook No. 1.
Blum, J. L. 1963. Interactions between certain salt marsh algae and angiosperms. Bull. Ecol. Soc. Amer. 44:92.
Blum, J. L. 1968. Salt marsh *Spartinas* and associated algae. Ecol. Monog. 38:199-221.
Bodrogkozy, G. 1961-62. Coenological evaluation of Grass-Clover combinations planted after sod-ploughing on alkali (Szik) soils of the Hortibagy Steppe. Act. Agron. 11(¾):345—67.
Bodrogkozy, G. 1965. Ecology of the halophilic vegetation of the Pannonicum. III, IV. Act. Biol. Szeged. 11:3-25, 207-227.
Bodrogkozy, G. 1966. Ecology of the halophilic vegetation of the Pannonicum. V. Act. Bot. Acad. Sci. Hung. 12:9-26.
Bol'shakov, A. F. 1960. An experiment on improvement of Solonchakic solonetzes and the means of reclaiming soils of the solonetz complex. Tr. Inst. Lesa. Akad. Nauk. S.S.R. 38:12-20, 1958. [Ref. Zhur. Biol. 1960. No. 54863 (trans.).]
Boumans, J. H. and W. C. Husbos. 1960. The alkali aspects of the reclamation of saline soils in Irak. Netherl. Journ. Agric. Sci. 8(3):225-35.
Bowser, W. E., and R. R. Cairns. 1967. Some effects of deep ploughing a solonetz soil. Can. J. Soil. Sci. 47:239-44.
Boyko, H. 1966. Ed. *Salinity and Aridity*. Monog. Biol. Vol. 16. The Hague VIII. 408 pp.
Bradley, W. G. 1970. The vegetation of Saratoga Springs, Death Valley, National Monument, California. Southwest Nat. 15(1):111-129.
Branson, F. A., R. A. Miller, and I. S. McQueen. 1967. Plant communities and associated soil and water factors on shale derived soils in north eastern Montana. Ecol. 51(3):391-407.
Cairns, R. R. 1961. Some chemical characteristics of a solonetzic soil sequence at Vegreville, Alberta, with regard to possible amelioration. Can. J. Soil Sci. 41:24-34.
Cairns, R. R. 1969. A Canadian solonetz and its field associated eluviated black soil and then ground waters. Agrokem. Talajtan 18:159-66 (Engl. summ.)
Cairns, R. R. 1969a. Canadian solonetz soils and their reclamation. Ibid. 18:233-37.
Cairns, R. R. 1970. Effect of solonetz soil horizon mixing on alfalfa growth. Can. J. Soil. Sci. 50:367-371.
Cairns, R. R. 1971. Effect of deep ploughing on the fertility status of black solonetz soils. Can. J. Soil Sci. 51:411-14.

Cairns, R. R., and W. E. Bowser. 1969. Solonetzic soils and their management. Canad. Dept. Agric. Publ. 1391:1-23.
Chapman, V. J. 1959. Studies in Salt Marsh Ecology. IX. J. Ecol. 47:619-39.
Chater, E. H. 1965. Ecological aspects of the dwarf brown form of *Spartina* in the Dovey estuary. J. Ecol. 53:789-97.

Clarke, L. D., and N. J. Hannon. 1967. The mangrove swamp and salt marsh communities of the Sydney district. I. Vegetation, Soils, and Climate. J. Ecol. 55:753-81.

Clarke, L. D. and N. J. Hannon. 1969. The mangrove swamp and salt marsh communities of the Sydney district. II. The Holocenotic complex with particular reference to physiography. J. Ecol. 57:213-34.

Corre, J. J. 1962. Une zone de terraine sales en bordure de L'etang de Mangio. Bull. Serv. Cant. Phytogeog., Ser. B. 6(2):1-105; 7(1):9-48.

Cusick, A. W. 1970. An assemblage of halophytes in northern Ohio. Rhodora 72:285-286.

Davis, L. V., and I. E. Gray. 1966. Zonal and seasonal distribution of insects in North Carolina salt marshes. Ecol. Mon. 36(3):275-95.

Dalby, D. H. 1970. The salt marshes of Milford Haven, Pembrokeshire. Field Stud. 3(2):297-330.

Dodd, J. D., D. A. Rennie, and R. J. Coupland. 1964. The nature and distribution of salts in uncultivated saline soils in Saskatchewan. Can. J. Soil. Sci. 44:165.

Dodd, J. D., and R. T. Coupland. 1966. Vegetation of saline areas in Saskatchewan. Ecol. 47(6):958.

Ebert, C. H. 1971. Irrigation and salt problems in Renmark, S. Australia. Geog. Rev. 61(3):355-69.

Gillner, V. 1960. Vegetations – und Standorts – Untersuchungen in der Stradweisen der Schwedischen West Kuste. Act. Phyt. Suec. 43:1-198.

Gimingham, C. H. 1964. Maritime and sub-maritime communities. Pages 67-143 in The Vegetation of Scotland. Edited by J. H. Burnett. Oliver & Boyd. Edinb.

Goodman, P. J., and W. T. Williams. 1961. Investigations into 'dieback' in *Spartina townsendii* agg. III. Physiological correlates of 'dieback'. J. Ecol. 49:391-8.

Goodman, P. J., E. M. Braybrooks, J. M. Lambert, and E. J. Marchant. 1969. *Spartina* Schreb. J. Ecol. 57:285-314.

Habib, I. M., J. A. Al-Ani, M. M. Al-Mufti, B. H. Al-Tawil, and B. A. Takessian. 1971. Plant indicators in Irak. I. Native vegetation as indicators of soil salinity and water-logging. Plant Soil 34(2):405-15.

Hubbard, J. C. E. 1965a. *Spartina* marshes in southern England. VI. Pattern of invasion in Poole Harbour. J. Ecol. 53:799-813.

Hubbard, J. C. E. 1965b. The earliest record of *Spartina maritima* in Britain. Proc. Bot. Soc. Br. Isl. 6:119.

Hubbard, J. C. E. 1967. Cleistogamy in *Spartina*. Watsonia 6:290-1.

Hubbard, J. C. E. 1968. Grasses. London (2nd edition).

Hubbard, J. C. E., and D. S. Ranwell. 1966. Cropping *Spartina* marsh for silage. J. Br. Grassld. Soc. 21:214-7.

Ito, K. 1961. On the salt marsh communities of Notsuke Zaki (Notsuke sand beach), Prov. Nemuro, Hokkaido, Japan – ecological studies on the salt marsh vegetation in Hokkaido, Japan. 4. Japan J. Ecol. 11(4):154-9.

Ito, K. 1963. Study on the vegetation of the salt marshes in Eastern Hokkaido, Japan. Sapporo Bull. Bot. Gard. Hokkaido Univ. 1:1-102.

Ito, Kard, and T. Leu. 1962. Ecological studies on the salt marsh vegetation in Hokkaido, Japan. 5. Japan J. Ecol. 12(1):17-20.

Jones, G. O. 1962. Reclamation of 'salty' land in the Kerang district. J. Agric. 60(10):453-7.

Knoerr, A. 1960. Le milieu, la flore, la vegetation, la biologie des halophytes dans l'Archipel de Rion et sur la cote sud de Marseille. Bull. Mus. D'Hist. Nat. Marseille 20:89-173.

Kotter, F. 1961. Die Pflanzenegesellschaften in Tidgebiet der Unterelbe. Arch. Hydrobiol. (suppl.) 26:106.

Kuznetsoy, L. A. 1960. Data on geobotanical characteristics of the Ariabaseta Salsae formation. Uch. Zap. Leningradsk Gos Pedagog. Inst. Im. A. I. Gertsena 178:119-144, 1959. Ref. Zhur. Biol. 1960, No. 80271 (transl.).

Lambert, J. M. 1964. The *Spartina* story. Nature (Lond.) 204:1136-1138.

Lyon, C. J., and W. Harrison. 1960. Rates on submergence of coastal New England and Arcadia. Science 132:295.

Marchant, C. J. 1963. Corrected chromosome numbers for *Spartina townsendii* and its parent species. Nature (Lond.) 199:9129.

Marchant, C. J. 1966. The cytology of *Spartina* and the origin of *Spartina townsendii*. Pages 71-2 in Chromosomes Today. Edited by Darlington & Lewis. Oliver & Boyd.

Marchant, C. J. 1967. Evolution in *Spartina* (Gramineae). 1. The history and morphology of the genus in Britain. J. Linn. Soc. (Bot.) 60:1-24.

Marchant, C. J. 1968. Ibid. II. Chromosomes, basis relationships and the problem of *S. townsendii* agg. III. Species chromosome numbers and their taxonomic significance. Ibid. 60:381-409, 411-417.

Marchant, C. J. 1970. Ibid. IV. The cytology of *S. alterniflora* Loisel. Ibid. 63:321-6.

Miyawaki, A., and T. Ohba. 1965. Studien-uber Strand-Salzwiesengesellschaften auf Ost-Hokkaido (Japan). Sci. Rept. Yokohama Nat. Univ. Sec. II 12:1-25.

Molinier, R., Viano Mme Le Forestier Mme, and J. P. Devaux. 1964. Edudes phytosociologigus et Ecologigues en Camargue et sur le Planktu Bourg. Ann. Fac. Sci. Mrs. 36:3-100.

Momotov, I. F. 1963. The characteristic of the natural conditions and vegetation of the Kyzylkum desert station. Ref. Zh. Brd. No. 4247 (trans.).

Muratov, V. S. 1959. Solonetzes of the Mil'sk alluvial plain (Kura-Araks Lowland). Pochvovodenie 1959(9):1041-53 (transl.).

Nienhuis, P. N., and J. Simons. 1971. *Vaucheria* species and some other algae on a Dutch salt marsh, with ecological notes on their periodicity. Acta. Bot. Neerl. 20(1):107-118.

Oprea, C. V. 1965. The genesis, development and amelioration of alkali soils on the western lowland of the Rumanian People's Republic. Agrokem.

Talajtan 14(suppl.):183-88 (Engl.).

Ovidenicharov, I. N. 1959. Solonetz soils of the Frakia-depression in Bulgaria. Pochvovodenie 1959(9):1090-99 (transl.).

Packham, J. R., and M. J. Liddle. 1970. The Cefni salt marsh, Anglesey, and its recent development. Field Stud. 3(2):331-56.

Padhi, U. C., R. T. Odell, J. B. Fehrenbacker, and R. D. Seif. 1965. Effect of gypsum and starch on water movement and sodium removal from solonetzic soils in Illinois. Soil Sci. Soc. Amer. Proc. 29(2):227-9.

Prettenhoffer, I. 1964. Die Weiterentwicklung der Melioration der Karbonatfrein Alkaliboden (Weisen-Solonetzboden) durch Untergrundlockerung. Agrokem. Talajtan (suppl.), pp. 227-35.

Prettenhoffer, I. 1964a. Tizantuli szikes Gyepjavitasi Kiserletek eddigi Evred menyei [recent results of experiments with amelioration of alkali ('szik' solonetz) grasslands east of the River Tisza. V.]. Kiserl. Kozlem. 17A:131-43 (Engl. summ.).

Prettenhoffer, I. 1964b. Kiserleti adatok a nesztelen szikesek forditasos melymurelesi Kerdesehez. Agrokem. Talajtan 13½:51-72 (French summ.).

Prettenhoffer, I. 1965. Amelioration of sodic solonetz soils in the region east of the River Tisza. Agrokem. Talajtan 14(suppl.):323-28.

Raheju, P. C. 1966. Aridity and salinity (a survey of soils and land use). Pages 43-127 in Salinity and Aridity. Edited by Boyko, Junk.

Raikov, L., and J. Kavadgier. 1965. Experiments on the amelioration of sodic solonetz in Bulgaria. Agrokem. Talajtan 14(suppl.):225-8 (Engl. summ.).

Ramos, G. A., J. A. Bonet, and J. O. Velez. 1961. Degree of reclamation of a saline-sodic soil in Lajas Valley, Puerto Rico. J. Agric. Univ. Puerto Rico 45(3):157-71.

Ranwell, D. S. 1967. World resources of *Spartina townsendii* (sensu lato) and economic use of *Spartina* marshland. J. Appl. Ecol. 4:239-56.

Ranwell, D. S., and B. M. Downing. 1960. Dalapon and substituted urea herbicides for control of cord grass in inter-tidal zones of estuarine marsh. Weeds 8:78-88.

Ranwell, D. S., E. C. F. Bird, I. C. E. Hubbard, and R. E. Stebbings. 1964. *Spartina* salt marshes in Southern England. V. Tidal submergence and chlorinity in Poole Harbour. J. Ecol. 52:627-41.

Redfield, A. C. 1965. Ontogeny of a salt marsh estuary. Science 146:50-55.

Redfield, A. C., and M. Rubin. 1962. The age of a salt marsh peat and its relation to recent changes in sea level at Barnstaple, Mass. Proc. Nat. Acad. Sci. 48(10):1728-35.

Round, F. E. 1960. The diatom flora of a salt marsh on the River Dee. New Phyt. 59(3):332-48.

Rustamov, I. G. 1962. Kharakteristike rastitel' nosti Zapadnoga Usboya. (The nature of the vegetation of a Western Usboi.) Isvest Akad. Nauk. Turkmen. S.S.R. Ser. Biol. Nauk 4:31-9. Ref. Zhur. Biol. No. 18V193 (transl.).

Rusyaeva, G. G. 1961. Rastitel' nost' schebnistoi pustyni Ozhungarskikh Vorot. Ref. Zhur. Biol. No. 18V193 (transl.).

Saini, G. R. 1971. Comparative effect of gypsum and limestone on drainage and salt removal from coastal alluvial soils of New Brunswick. Plant Soil 34(1):159-64.

Sambur, G. N. 1963. Podvizhnost' obmennogo natriya i obosnovanie norm gipsa pri melioratsii solontosov. (Exchangeable sodium and gypsum norms for solonetz solid reclamation.) Pochvovodenie 11:35-46 (Engl. summ.).

Sauer, J. 1965. Geographic reconnaissance of Western Australian seashore vegetation. Aust. J. Bot. 13:39-70.

Sears, P. B. 1963. Vegetation, climate and coastal submergence in Connecticut. Science 140:59-60.

Stebbings, R. E. 1970. Recovery of salt marsh in Brittany sixteen months after heavy pollution by oil. Environ. Pollut. 1(2):163-7.

Strognov, B. P. 1964. The physiological basis of salt tolerance of plants. Isr. Res. Council Transl.

Szabolcs, J. 1965. Salt affected soils in Hungary. Agrokem. Talajtan 14(suppl.):275-90.

Szabolcs, J., and L. Abrakham. 1959. The use of small applications of meliorative substances on solonetz and solonetz-like soils of The Great Hungarian Plain. Pochvovodenie 3:311-18 (transl.).

Tagunova, L. N. 1960. O svyazyakh pochvenno-rasitel'nogo pokvovo severo-vostohnogo proberezh'ya Kaspiiskogo morya s usloviyami Zasoleniya i uvlazhneniya. (The relation between soil and plant cover of the north eastern shore of the Caspian Sea and salinity and moisture conditions.) Bull. Moskov. Obschchestva Ispytatele Privody Otdel Biol. 65:61-76 (Engl. summ.). Ref. Zhur Biol. 1961, No. 7V242 (transl.).

Taylor, M. C., and E. M. Burrows. 1968. Studies s on the biology of *Spartina* in the Dee Estuary, Cheshire. J. Ecol. 56(3):795-810.

Teal, J. M., and J. W. Kanwisher. 1961. Gas exchange in a Georgia salt marsh. Limnol. & Oceanogr. 6(4):388-99.

Teal, J. M., and J. W. Kanwisher. 1966. Gas exchange in marsh grass *Spartina alterniflora*. J. Expt. Bot. 17(51):355-61.

Tyler, G. 1971. Hydrology and salinity of Baltic seashore meadows. III. Oikos 22(1):1-20.

Umezu, Y. 1964. Ueber die salzwasser pflanzengesellschaften in der nahe von Yuki hasi, Nordkyuschu, Japan. Jap. J. Ecol. 14(4):153-60.

Ungar, I. A. 1965. An ecological study of the vegetation of the Big Salt Marsh, Stafford County, Kansas. Univ. Kan. Sci. Bull. 46(1):1-99.

Ungar, I. A. 1967. Vegetation and soil relationships in saline areas of northern Kansas. Amer. Mid. Nat. 78:98-120.

Ungar, I. A. 1968. Species-soil relationships on the Great Salt Plains, northern Oklahoma. Ibid. 80:392-406.

Ungar, I. A. 1970. Species-soil relationships on sulphate dominated soils of South Dakota. Ibid. 83(2):343-57.

Ungar, I. A., W. Hogan, and M. McClelland. 1969. Plant communities of saline soils at Lincoln, Nebraska. Amer. Mid. Nat. 82(2):564-77.

Vinogradov, B. V. 1960. Pochvenno-geobotanischeskaya Kharakeristika geomorfologisches Kikh elementor doliny Nizhnego Uzboya. (Soil geobotanical characters of the geomorphological elements of the valley of the lower Uzboi.) Bot. Zhur. 45:720-7. Ref. Bot. Zhur. 1961, No. 8V232 (transl.).

Zohary, M. 1963. On the geobotanical structure of Iran. Bull. Res. Counc. Isr. Sect. D. (suppl.) 11:1-112.

Zvereva, E. A. 1960. Differentiation of the methods of meliorative treatment of different solonetzes. Dokl. Vses. Akad. selskokhoz Nauk. Im. Lenina 10:35-41, 1958. Ref. Zhur. Biol. 1960, No. 54862 (transl.).

PART II. HALOPHYTES OF THE UNITED STATES: DISTRIBUTION, ECOLOGY, ANATOMY, AND PHYSIOLOGY

VASCULAR HALOPHYTES OF THE ATLANTIC AND GULF COASTS OF NORTH AMERICA NORTH OF MEXICO

Wilbur H. Duncan

Department of Botany, University of Georgia

Abstract

Difficulties in delimiting halophytes and halophytic habitats are discussed and an annotated list of halophytes presented. Data included for each species are: the halophytic habitats in which they are reported to occur, the authors reporting the habitats, page citations for these reports, and composites of the coastal geographic distributions compiled from numerous authors. The list of halophytes includes 347 species in 177 genera and 75 families. The possibilities that some species may be erroneously listed and others apparently omitted are discussed. Geographic distributions are analyzed by areas and summarized. The most species, 197, occurrs in the are from Georgia into Delaware. Thirty-eight occur only in Florida, 17 only in Canada, and 32 in every one of the coastal states from Texas into Canada. Habitat terminology lacks uniformity.

Introduction

The above title probably has a definite meaning to most persons, but their interpretations of the word "halophyte" will often differ. This is emphasized by the many definitions of halophytes that exist. Chapman (1960) describes halophytes as "salt-tolerant plants," Fernald (1950) says they are plants "growing in saline soils," and Dansereau (1957) that a halophyte is "a plant that grows exclusively on salt soil; e.g., all species of *Salicornia*." Other definitions include: plants of salty or alkaline soils (Correll and Johnston 1970); plants that can tolerate the concentrations of salts found in saline soils (Oosting 1956); and plants tolerant of various mineral salts in the soil solution, usually sodium chloride (Lawrence 1951). No matter which definition is used there is the problem of where the line should be drawn between "salty" habitats and "fresh" habitats. Decisions on various habitats have been and will continue to be a matter of individual judgment since salinity data are available for relatively few habitats. Even without salinity data many areas are obviously salty, saline, or brackish. But what about those parts of tidal creeks and rivers farther from the sea? What about successive zones back of the beaches? How far back does "salt" occur? Obviously, different conclusions will be reached concerning whether certain habitats are halophytic or not.

Because of the above situations, no literature survey can provide a completely reliable list of all vascular halophytes of the seacoasts of eastern North America north of Mexico. However, a useful list can be compiled from reports of species shown to be halophytic plus those suspected of being halophytic on the basis of the kind(s) of habitat(s) in which each is reported to grow. Admittedly, some species in such a list may eventually be shown not to be halophytic.

I have prepared a list of species known to be or suspected of being halophytic, i.e., those that tolerate seawater, pure or diluted. Salinity measurements provide the basis for including many species in the list. However, most are included on the basis of the kinds of habitats in which they are reported to grow. Words interpreted as indicating halophytic habitats, or probable ones, include salt, saline, brackish, tidal, sea, strand, beach, reef, and estuary. Finally, some species can be listed as halophytic because of their occurrence in certain habitat types such as *"Spartina* Marshes" or most "Mangrove Swamps."

Some authors give types of habitats that can be halophytic or not and do not indicate which. For example, the term "marshes" is widely used to indicate fresh-water situations but is not always confined to them. Some authors give the habitat for certain species as "marshes" when the species are abundant in salt marshes and absent or uncommon in fresh-water ones. This occurs uncommonly and so marshes not designated by the authors as halophytic are considered here to be fresh, and species reported in them as not being halophytic in those instances.

Another source of error in listing halophytes can occur when fresh ground water from above tide levels flows into soil subject to saline or brackish water of tidal areas. The observer can easily be unaware of such situations and incorrectly interpret plants growing in those places as being halophytic. Upper sections of tidal areas of creeks and rivers can easily be misinterpreted as being brackish; rises in water level is often due to piling up of fresh water against the tidal surges. Species growing in these areas may thus be considered incorrectly as being halophytes. *Eleocharis ovata, Heteranthera reniformis, Lepidium latifolium, Cardamine longii, Elatine americana, Bacopa stragula, Micranthemum micranthemoides,* and *Bidens hyperborea* may be examples of this type of misinterpretation. Salinity measurements appear to be lacking for habitats involving these species.

Misidentification is also a likely source of error. Few persons are able to identify with reasonable certainty all species that are likely to be encountered in halophytic habitats. Since this is true of plant taxonomists, it is obviously that those less experienced in identification (i.e., salt marsh ecologists) are likely to make even more errors. *Pluchea camphorata* may be an example of this type of error. Godfrey (1952), in a monograph of the genus, says that this species grows in fresh-water habitats but gives halophytic habitats for *P. purpurascens.*

Apparent omissions from a list of halophytes can be due to nomenclatural situations. Names long familiar as halophytes are sometimes replaced by legitimate names that might not be recognized. Examples in this compilation include *Syringodium filiformis* Kuetz for *Cymodocea manatorum* Ashers and *Cynanchum palustre* (Pursh) Heller for *Seutera maritima* Decne. Taxonomic decisions are also involved in apparent omissions. For example, studies sometimes show that what was previously thought to be two species should be combined into one. Only one of the two species names would then appear as a primary name on the list. In the compilation here, an attempt has been made to list as synonyms all names abandoned for whatever cause in the last two or three decades, using as primary names those I think to be the proper ones. These include especially those instances in which two or more authors have used different names for the same species. Common names are also given for most species.

In my list of halophytes to follow, the ecological distributions stated are composites of the halophytic habitats from the authors indicated, and are, therefore, not necessarily those listed by an individual author. Many of the species also occur in non-halophytic habitats but these places are not indicated. When using the ecological data, the reader should also be aware that a given species can occur in halophytic habitats in one region but, insofar as is known, does not in some other region of its distribution. This is especially true of north-south regions, and a study of this subject could be profitable. On the basis of the literature I have reviewed, examples include *Agrostis palustris, Zizaniopsis miliacea, Scirpus americanus, Dichromena colorata, Cladium mariscoides, Rhynchospora macrostachya, Aneilema keisak, Atriplex hastata, Iresene rhizomatosa, Nuphar advena, Cakile edentula, Ludwigia alata, Ptilimnium capillaceum, Sabatia dodocandra, Lobelia elongata, Solidago tenuifolia,* and *Baccharis halimifolia.*

Geographical distributions given in my list are composites from all authors but the authors involved for each species are not indicated. These distributions include the entire coastal range of each species regardless of whether halophytic habitats are involved throughout or not. They do not include distributions in the interior. They are presented by states, or provinces, beginning with the westernmost and/or southernmost and ending with the easternmost and/or northernmost. With these restrictions, the meaning of the abbreviations for the states or provinces should be evident. To illustrate, if a distribution is given as "Tex-Ga," the species does not necessarily occur in all of Florida but could very likely be absent from the tropical southern end of that state. A species could be reported as "Va-Que" and be in southern Newfoundland. Similarly, if the distribution is "Va-Nfld," the species could be in coastal Quebec, somewhere south of the northernmost

Newfoundland station. When the distribution is given as being in one state only, the species does not necessarily occur throughout the coastal part of the state. The same is true for those states or provinces indicating distributional limits. "Fla-NC," for example, could be a distribution from northwest Florida into northeast North Carolina, or, at the other extreme, from northeastern Florida into southeastern North Carolina. The composite geographical distributions are from sources too numerous to list here. All sources are included in the literature cited, the most important being Small (1933), Muenscher (1944), Fernald (1950), Hitchcock (1950), Radford et al. (1968), Correll and Johnston (1970), and Long and Lakela (1971).

The arrangement of the species follows that of the manuals indicated above with a few exceptions, the main exception being that the genera of the Poaceae follow the order presented by Gould (1968). The arrangement in Hitchcock (1950) and in most other manuals is outdated, having long been recognized as strongly artificial.

The sources for the ecological data presented for each species are indicated by letters and numbers. For each citation the letter(s) indicate(s) the author as follows: C - Correll and Johnston (1970), D - Davis (1942), F - Fernald (1950), H - Hitchcock (1950), K - Kurz and Wagner (1957), L - Long and Lakela (1971), M - Muenscher (1944), R - Radford, et al. (1968), S - Small (1933), and W - West and Arnold (1956). The number following each letter indicates the page for the citation. Other references are indicated by author's name, date of publication, and page.

PTERIDACEAE

Acrostichum danaeaefolium langsd. & Frisch. SALT MARSH, LEATHER FERN. Brackish swamps. s Fla. L86.
Onoclea sensibilis L. SENSITIVE FERN. Moist brackish meadows. Tex-Lab. Harshberger & Burns (1919)17.
Aspidium thelypteris (L.) Sw. WOOD FERN, MARSH FERN. Shaded brackish situations. Tex-Nfld. Harshburger & Burns (1919)17.

PINACEAE
Pinus elliotii Engelm. SLASH PINE. Peripheral flat woods. La-SC. K83.

CUPRESSACEAE
Juniperus silicicola (Small) Bailey. COASTAL CEDAR. Barrier beaches. Tex-NC. K15.

TYPHACEAE
Typha angustifolia L. CAT-TAIL. Brackish marshes, brackish habitats, tolerates some salinity. Miss-Va. Fassett & Calhoun (1952)377 Harshburger & Burns (1919)20 R44.
Typha domingensis Persoon. CAT-TAIL. Brackish marshes and pools, brackish coastal marshes. Tex-Del. C85 F61 L116 R44.

ECOLOGY OF HALOPHYTES

Typha glauca Godron. CAT-TAIL. Slightly brackish lakes, ponds, and rivers. Ala-Va. R43.

Typha latifolia L. COMMON CAT-TAIL. Tolerates some salinity. Tex-Nfld. Hotchkiss and Dozier (1949)242.

ZOSTERACEAE

Zostera marina L. EELGRASS, GRASS-WRACK. Shallow bays, estuaries, or lagoons; soft or muddy bottoms; intertidal zone; shallow sea water. NC-Lab. Davis (1910)625 F64 M64 R44 S18 Transeau (1909)273.

Potamogeton crispus L. PONDWEED. brackish ponds. NC-NS. F71 R46.

Potamogeton curtissii Morgon. Tidal creeks. w Fla. S17.

Potamogeton filiformis Pers. Brackish waters. NH-Nfld. F70.

Potamogeton foliosus Raf. Brackish waters of lakes, streams, or ponds, brackish waters. Tex-NY. F71 R47.

Potamogeton friesii Rupr. Brackish waters. Va-Nfld. F71.

Potamogeton pectinatus L. BRACKISH WATER PONDWEED. Brackish ponds, saline, or brackish waters, shallow brackish water. Tex-Nfld. F70 Nichols (1920)534 R46.

Potamogeton perfoliatus L. Brackish waters; brackish waters of lakes, streams, and estuaries. La-Fla; NC-Lab. F79 R47.

Potamogeton richardsonii (Ar. Benn.) Rydb. Brackish lakes and rivers. NY-Lab. F79.

Potamogeton vaginatus Turez. Deep brackish waters. Lab-Nfld. F70.

Ruppia maritima L. WIDGEON-GRASS, DITCH GRASS, SEA-GRASS. Brackish waters, shallow brackish ponds and estuaries, saline lagoons. Tex-Me. C94 L116 M59 R48 S15 Setchell (1924)286.

Zanchellia palustris L. HORNED PONDWORT, PONDWEED. Slightly brackish water of pools, ponds, and estuaries; brackish waters. Fla-Nfld. C93 F81 L118 M61 R48 S15.

CYMODOCEACEAE

Syringodium filiformis Kuetz. *(Cymodocea manatorum* Aschers.). MANATEE GRASS. Shallow salt water bays and brackish streams, coastal waters, mudflats. La; Fla. C94 K14 L121 M56.

Halodule beaudettei (Den Hartog) Den Hartog. (*H. wrightii* Aschers.) [*Diplanthera wrightii* (Ashers.) Aschers.]. HALODULE. Shallow waters mostly offshore, mudflats, salt water of bays. Tex; Fla; NC. C93 K15 L123 M57 R45 S19.

NAJADACEAE

Najas conferta A. Br. WATER-NYMPH, NAIAD. Tidal creeks. w Fla. S20.

Najas flexilis (Willd.) R. & S. NAIAD. Shallow brackish waters. Va-Nfld. F82 L119.

Najas guadalupensis (Spreng.) Magnus. NAIAD. Brackish waters. Tex-Mass. L119.

POSIDONIACEAE
Posidonia oceania Konig. Marine. Tex. C95.

JUNCAGINACEAE
Triglochin maritima L. ARROW-GRASS. Brackish or salt marshes along coasts, edge tide pools, salt marsh pools. Del-Lab. Conrad (1935)459 F83 Harshberger (1916)483 M76 Rich (1902)89.
Triglochin palustris L. ARROW-GRASS. Brackish places. Me-Lab. F83.
Triglochin striata R. & P. ARROW-GRASS. Brackish and salt marshes, tidal flats, inner marsh, coastal strands. La-Del. F83 Kearney (1900)45 L123 M76 R50 S20 Wells (1928)232.

ALISMATACEAE
Lophotocarpus calycinus (Engelm.) J. G. Smith. (*Sagittaria calycina* Engelm.). LOPHOTOCARPUS. Tidal mudflats in brackish bays and estuaries. NC-Me. M84.
Lophotocarpus spongiosus (Engelm.) J. G. Smith LOPHOTOCARPUS. Tidal mud of brackish estuaries. Va-NB. F85.
Sagittaria eatoni J. G. Smith. GRASS-LEAVED SAGITTARIA. Tidal muds and sands. Va-NH. F90.
Sagittaria calycina. See *Lophotocarpus calycinus.*
Sagittaria falcata Pursh. LANCE-LEAVED SAGITTARIA. Brackish shores, tidal marshes along streams and ditches. Tex-Md. F89 R54.
Sagittaria graminea Michx. GRASS-LEAVED SAGITTARIA. Brackish water and marshes. Tex-Lab. R52.
Sagittaria lancifolia L. LANCE-LEAVED SAGITTARIA. Brackish tidal marshes. Tex-Del. C99.
Sagittaria rigida Pursh. SESSILE-FRUITING ARROW-HEAD. Brackish muds or waters. Va-Me. F88.
Sagittaria subulata (L.) Buch. SUBLATE SAGITTARIA. Brackish tidal muds or silts, tidal flats or waters. Miss-Mass. F88 L128 M93 R52.

HYDROCHARITACEAE
Halophila baillonis Aschers. SEA-GRASS. Sandy or marly reef bottoms, marine lagoons. Fla Keys. L130 M106 S27.
Halophila engelmannii Aschers. SEA-GRASS. Bays, open waters of coast, and reefs; marly sands in marine bays and reefs; mudflats, shallow salt water. Tex-Fla. C103 K15 L130 M106 S27.
Hydrilla verticillata Royle. HYDRILLA. Shallow brackish waters. s Fla. L130.
Thalassia testudinum Konig. TURTLEGRASS. Salt water in shallow bays and about reefs, mudflats. Tex-s Atl. coast Fla. C102 K16 L132 M107 S29.

POACEAE
Festuca ovina L. SHEEP FESCUE. Salt marsh. NJ-Que. Ganong (1903)363.

Festuca rubra L. RED FESCUE. Brackish meadows. NC-Lab. S113.

Puccinellia fasciculata (Torr.) Bicknell. GOOSEGRASS. Salt marshes or meadows, saline shores. Va-NS. F109 H79 Nichols (1920)530.

Puccinellia laurentiana Fern. & Weath. Gravelly seashores. NB-Que. F110.

Puccinellia lucida Fern. & Weath. Saline marshes, coastal sands. Que. F110.

Puccinellia maritima (Huds.) Parl. GOOSEGRASS. Saline or salt marshes, SALINE OR BRACKISH SHORES, RI-Que. F109. Ganong (1903)363 H79.

Puccinellia papercula (Holm) Fern. & Weath. Saline shores. Conn-Lab. F110.

Ammophila arenaria (L.) Link. EUROPEAN BEACHGRASS. Middle beach. Del-Mass. Snow (1902)291.

Agrostis palustris Huds. (*A. alba* var. *maritima* Mey.). CREEPING BENT. Saline shores, border of salt marsh, shallow saline waters. NC-Lab. F161 Transeau (1913)207.

Hierochloe nashii (Bichn.) Kaezmarek. Borders of brackish marshes. NJ-NY. F186.

Hierochloe odorata (L.) Beauv. HOLY GRASS, SWEET GRASS. Brackish shores. NJ-Lab. F186.

Hordeum jubatum L. BARLEY GRASS. High salt marsh. Del-Lab. Ganong (1903)364.

Catabrosa Aquatica (L.) Beauv. Water hairgrass. Brackish waters. NB-Lab. F127.

Parapholis incurva (L.) C. E. Hubb [*Pholiurus incurvus* (L.) Schriz. and Thell.]. HARD GRASS, THIN-TAIL, SICKLE GRASS. Borders of brackish marshes, salt marshes, tidal flats, brackish shores. NC-NJ. C118 F133 H279 M132.

Pholiurus incurvus. See *Parapholis incurva*.

Eriochloa michauxii (Poir.) Hitchc. CUPGRASS. Brackish meadows and marshes. Fla-Ga. H590 L167.

Paspalum caespitosum Flugge. PASPALUM. Strand beaches. s Fla. D136.

Paspalum distichum L. KNOTGRASS. Brackish places or marshes, brackish waters, high tide marsh, inner marsh. Tex-NJ. H603 Kearney (1900)17 L171 R133 Wells (1928)232.

Paspalum vaginatum Swartz. CREEPING PASPALUM, KNOTGRASS. Brackish marshes and sands; moist saline to brackish sands at edges of lagoons, bays, and river mouths; hard packed sand in salt marsh; high marshes. Tex-Ga; NC. Brown (1959)40 C163 H603 K15 L171 R134 S57.

Panicum amarulum H. & C. BITTER PANICGRASS. Strand beaches, beaches. Tex-NJ. C177 D136.

Panicum geminatum Forsk. [*Paspalidium geminatum* (Forsk.) Stapf]. PANICUM. Brackish waters. Tex-s Fla. L176.

Panicum virgatum L. SWITCHGRASS. Brackish marshes and shores, salt marshes, high marshes. Tex-NS. F207 H697 K15 S70 Wells (1928)232.

Echinochloa walteri (Pursh) Heller. WILD-MILLET. Brackish meadows, marshes, and waters. Tex-NH. H715 Harshberger & Burns (1919)18 L164.

Setaria geniculata (Lam.) Beauv. [*Chaetochloa geniculata* (Lam.) Millsp. & Chase, *C. imberbis perennis* Schribn. & Merr.]. KNOTROOT, BRISTLEGRASS. Salt marshes, inner marsh, borders of saline marshes. Tex-Mass. F226 H720 K14 Kearney (1900)17.

Setaria glauca (L.) Beauv. [*S. lutescens* (Weig.) F. T. Habb.]. FOXTAIL, PIGEON-GRASS. Less brackish parts of marsh. Tex-NB. Harshberger & Burns (1919)19.

Setaria magna Griseb. GIANT MILLET, CATTAIL MILLET. Marshes along coast, brackish marshes, brackish swales. Tex-NJ. F226 H724 R126.

Cenchrus pauciflorus Benth. SANDBUR. Strand beaches. Tex-Mass. D136.

Andropogon glomeratus (Walt.) BSP. BUSHY BEARDGRASS, BUSH BROOMSEDGE. High marshes, strand beaches. Tex-Mass. D136 K14.

Andropogon scoparius Michx. LITTLE BLUESTEM. Border of salt marsh. Tex-Que. Transeau (1913)207.

Andropogon virginicus L. VIRGINIA BROOMSEDGE. High marshes. Tex-Mass. K14.

Elyonurus tripsacoides Humb. BALSAMSCALE. High marshes. Tex-Ga. K14.

Muhlenbergia capillaris (Lam.) Trin. HAIR GRASS. High marshes. Tex-Mass. K15.

Sporobolus virginicus (L.) Kunth. VIRGINIA DROPSEED. Saline or brackish marshes, brackish sands, barrier beaches, strand beaches, packed loamy saline soil. Tex; Miss; Fla-NC. C222 D136 H418 K16 L141 R105.

Leptochloa fascicularis (Lam.) A. Gray [*Diplachne fascicularis* (Lam.) Beauv.; *D. maritima* Bickn.]. SALT MEADOW GRASS. Brackish marshes, soils, or shores. Tex-Miss; Fla; NC-Me. F127 H493 S119.

Chloris glauca (Chapm.) Wood. FINGERGRASS. Brackish marshes. Fla; SC-NC. H520 R117 S115.

Chloris neglecta Nash. FINGERGRASS. Barrier beaches, high marshes. Fla. K14.

Chloris petraea Swartz. FINGERGRASS. Strand beaches. Tex-NC. D136.

Spartina alterniflora Lois. (*S. strictaa glabra* Gray). SMOOTH CORDGRASS, SEDGE, SALT CANE. Salt marshes, outer marsh, saline shores or areas, brackish waters of coastal lagoons and marshes, beaches, tidal marshes, low marsh, mid-tide level, tidal shores or brackish to hypersaline bays and river mouths, muddy shores. Tex-Nfld. Adams (1963)455 C250 F179 Ganong (1903)351 H511 K15 Kearney (1900)16 L153 M135 Newman and Munsart (1968)86 Penfound (1952)432 R120 Reed (1947)607 S113.

Spartina bakeri Merr. BUNCH CORDGRASS. Edge of salt marshes, salt water sites, margins tidal streams, high marshes. Fla-SC. K15 Mobberly (1956)518.

Spartina cynosuroides (L.) Roth. BIG CORDGRASS, SALT REED-GRASS. Salt or brackish water, margins tidal streams, muck of tidal shores of brackish bays and river mouths. Tex-Mass. C250 F179 H509 M136 R120 S113.

Spartina patens (Ait.) Muhl. (*S. juncea* Willd.). SALT MEADOW

CORDGRASS, FOX GRASS, COUCH GRASS, RUSH SALT GRASS. Salt marshes and meadows, high salt marsh, brackish marsh, saline shores and marshes, creek bank, salt marsh flats. Tex-Nfld. Blum (1968)205 F180 Ganong (1903)363 H514 Harshberger & Burns (1919)19 L153 M137 Penfound (1952)433 R119 S114 Statler and Batson (1969) 1087.

Spartina pectinata Link. PRAIRIE CORDGRASS. Brackish marshes. Ga; NC-Nfld. M138.

Spartina spartinae (Trin.) Merr. CORDGRASS. High marshes, saline poorly drained flats. Tex-Fla. C250 K16.

Distichlis spicata (L.) Greene. SEASHORE SALT-GRASS, SPIKE GRASS. Salt or brackish marshes and flats, high or upper marsh, mid marsh, low high marsh. Tex-NS. Adams (1963)455 C254 F131 H175 L145 M117 R62 Reed (1947)607 S128 Statler & Batson (1969)1087 Taylor (1938)47.

Monanthochloe littoralis Engelm. SALT-FLAT GRASS. Muddy seashores and tidal flats along seashores, coastal marshes, rocky or muddy saline shores. Tex-s tip Fla. H175 L144 M128.

Uniola paniculata L. SEA OATS. Strand beaches. Tex-Va. D136.

Zizania aquatica L. ANNUAL WILDRICE. Brackish marshes, quiet brackish waters. La-NB. F188 R125 S92.

Zizaniopsis miliacea (Michx.) Doell & Aschers. SOUTHERN WILDRICE, WATER-MILLET. Brackish marshes, tidal swamps, margins of tidal streams. Tex-Md. F188 R125.

Phragmites communis Trin. REED. Brackish marshes. Tex-Miss; Fla; NC-NS. Conrad (1935)469.

Chasmanthium laxum (L.) Yates. [*Uniola laxa* (L.) BSP.]. SPANGLEGRASS. High marshes. Tex-NY. K16.

CYPERACEAE

Cyperus brummeus Sw. SWEET-RUSH. Strand beaches, sea beaches. Fla. D136. S153.

Cyperus erythrorhizos Muhl. SWEET-RUSH. High marshes, inner salt marsh. Tex-Mass. Harshberger (1909)389 K14.

Cyperus filicinus Vahl. (*C. nuttallii* Eddy). SLENDER CYPERUS, SWEET-RUSH. Brackish marshes or soil, border of saline marshes, brackish sands. La-Me. Harshberger & Burns (1919)19 F241 R171.

Cyperus odoratus L. SWEET-RUSH. Brackish or saline shores. Tex-Mass. F243 L208.

Cyperus planifolius Richard. CYPERUS. Brackish sands on beaches. Fla. L210.

Cyperus tetragonus E11. CYPERUS. Brackish marshes. Fla-NC. R180.

Scirpus americanus Pers. (*S. pungens* Vahl.). THREE SQUARE, SWORD-GRASS, CHAIR-MAKERS' RUSH. Inner salt marsh, mid-marsh, brackish or saline waters or shores. Tex-Nfld. Brown (1959)40 F269 Harshberger (1900)380 Johnson & York (1915)42 M167.

Scirpus cyperinus (L.) Kunth. WOOL-GRASS. Edge of salt marsh streams.

Tex-Nfld. Taylor (1938)59.

Scirpus etuberculatus (Steud.) Kuntze. Brackish marshes. La-Del. Cappel (1954)89 F271.

Scirpus olneyi Gray. BULRUSH. Saline to brackish marshes, saline ponds, salt marshes. Tex-NS. Cappel (1954)88 F270 M170 R196 S170.

Scirpus paludosus Nels. BAYONET-GRASS. Brackish ponds or marshes, saline marshes and shores. NJ-NB. F272. M170.

Scirpus maritimus L. Saline to brackish marshes, brackish tidal shores. SC-NS. Cappel (1954)80 F272.

Scirpus nanus. See *Eleocharis parvula.*

Scirpus robustus Pursh. SALTMARSH BULRUSH. Brackish or saline marshes or ponds, brackish waters. Tex-NS. Cappel (1954)81 F271 L221 M170 R197 S170.

Scirpus validus Vahl. GREAT BULRUSH. Brackish water and marshes. Upper edge of salt marsh. Tex-Nfld. F270 Taylor (1938)60.

Eleocharis albida Torr. WHITE SPIKE RUSH. Brackish soil and pools, damp to wet brackish soils and sand. Tex-Md. C278 F255 R188 S165.

Eleocharis cellulosa Torr. SPIKE RUSH. Brackish marshes. Tex-Ga; NC. M151 R183.

Eleocharis fallax Weatherby. SPIKE RUSH. Brackish marshes or shores. NC-Va. F258 R184.

Eleocharis halophila Fern. & Brack. SPIKE RUSH. Saline or brackish shores, brackish marshes. NC-Nfld. F257 R186.

Eleocharis ovata R. & S. OVOID SPIKE RUSH. Tidal flats of lakes, rivers and estuaries. NJ-Nfld. M156.

Eleocharis parvula (R. & S.) Link. (*Scirpus nanus* Spreng.) DWARF CLUB-RUSH. Brackish or salt marshes, wet saline or brackish shores, mid marsh. La-Nfld. Conrad (1935)455 Conrad & Gallegar (1929) F252 Johnson and York (1915)43 R186 S169.

Eleocharis rostellata Torr. SPIKE RUSH. Salt or brackish marshes, brackish meadows. Fla-NS. F253 M156 R186 S166 Transeau (1913)205.

Eleocharis uniglumis (Link) Schult. SPIKE RUSH. Brackish marshes and ponds. Del-Me. M156.

Fimbristylis caroliniana (Lam.) Fern. Brackish sands. Tex-NJ. Kral (1971)112.

Fimbristylis castanea (Michx.) Vahl. [*F. spacidea* (L.) Vahl. in part]. MARSH FIMBRISTYLIS. Salt meadows, brackish marshes and shores, brackish sands, high marshes, mid marsh. Tex-NY. Adams (1963)455 F261 K15 L216 R193 S157 Wells (1928)232.

Fimbristylis spadicea (L.) Vahl. [*F. caroliniana* (Lam.) Fern.]. STIFF FIMBRISTYLIS. Brackish marshes, sands, and shores; saline sands; sandy peats of beaches. Tex-NJ. C281 F262 L261.

Fimbristylis spathacea Roth. SOUTHERN FIMBRISTYLIS. Brackish soils along canals, mangrove borders, brackish marshes. s Fla. L216.

Dichromena colorata (L.) Hitchc. WHITE-TOP Brackish swamps. Tex-Va. F260.
Cladium jamaicensis Crantz. [*Mariscus jamaicensis* (Crantz) Britt.]. SAW-GRASS. Brackish waters and marshes, saline swamps. Tex-Va. F290 K15 M147 R214 S187.
Cladium mariscoides (Muhl.) Torr. [*Mariscus mariscoides* (Muhl.) Kuntze]. TWIG-RUSH, POND-RUSH. Brackish waters. Ala-Nfld. F290 M147.
Rhynchospora macrostachya Torr. HORNED-RUSH. Salt marshes. Tex-Me. S179.
Carex bipartita Bellardi. SEDGE. Saline or brackish shores. NB-Nfld. F313.
Carex hormathodes Fern. SEDGE. Brackish sands, rocks, and marshes. Va-Nfld. F327.
Carex hyalinolepis Steud. SEDGE. Brackish swamps, swales, and shores. Tex-Md. F372.
Carex mackenziei Krecz. SEDGE. Saline or brackish marshes. Me-Lab. F313.
Carex paleacea Wahlenb. SEDGE. Saline or brackish marshes or shores. Mass-Lab. F338.
Carex salina Wahlenb. SALTMARSH SEDGE. Saline or brackish shores, marshes, and swales. Mass-Lab. F339.

ARECACEAE
Sabal palmetto Lodd. ex Schultes. CABBAGE PALMETTO. Brackish marshes, salt waters, high marsh, barrier beaches, margin salt marsh. Fla-NC. Harshberger (1914)82 K15 R255 S240.
Serenoa repens (Bartr.) Small. SAW PALMETTO. Barrier beaches. La-SC. K15.
Cocos nucifera L. COCONUT. Sea shores. s Fla. S237.

ERIOCAULACEAE
Eriocaulon parkeri B. L. Robinson. PIPEWORT. Tidal muds and estuaries, muddy tidewater riverbanks. NC-Que. F391 Kral (1966).

COMMELINACEAE
Aneilema keisak Hassk. ANEILEMA. Fresh tidal marshes and shores. Ga-Va. F393.

PONTEDERIACEAE
Heteranthera reniformis R. & P. MUD-PLANTAIN. Tidal mudflats of rivers and creeks. Tex; NC-Conn. M202.

JUNCACEAE
Juncus balticus Willd. RUSH. Sandy brackish shores, salt marshes. NY-Lab. F408 Transeau (1909)274.
Juncus canadensis J. Gay RUSH. Borders of brackish or saline marshes, brackish marsh. NC-Nfld. F412 Harshberger & Burns (1919)20.

Juncus gerardi Loisel. BLACK-GRASS. Salt or saline marshes, saline spots, brackish shores, salt meadows, upper marsh, mid marsh, salt marsh flats. Fla; Va; NJ-Nfld. Blum (1968)202, F403 Harshberger (1900)653 M210 Reed (1947)607 S284 Taylor (1938)47.
Juncus megacephalus M. A. Curtis. RUSH. Brackish marshes, coastal strands. Tex-Va. F411 L280 R278.
Juncus roemerianus Scheele. MARSH RUSH, BLACK RUSH. Brackish marshes and ditches, salt flats, low marsh, barrier beaches, coastal marshes, inner margin. Tex-Md. Adams (1963)455 C373 K15 L278 R275 S283 Wells (1932)19.
Juncus scirpoides Lam. RUSH. Saline marsh. Tex-NY. Harshberger (1900)653.

AMARYLLIDACEAE
Hymenocallis collieri Small. SPIDER-LILY. Sandy shores and banks of estuaries. s Fla. S322.
Hymenocallis crassifolia Herbert. SPIDER-LILY. Brackish marshes. Ala-NC. R320.
Hymenocallis keyensis Small. SPIDER-LILY. Mangrove swamps. s Fla. S322.

IRIDACEAE
Iris hookeri Penny. BLUE FLAG. Turfy crests of headlands, rocky slopes, upper borders of beaches, dunes, etc. within reach of ocean sprays. Me-Lab. F460.
Iris prismatica Pursh. SLENDER BLUE FLAG. Brackish or saline marshes, sandy shores, or meadows. NC-NS. F460.
Sisyrinchium atlanticum Bicknell. BLUE-EYED-GRASS. Edges of salt marshes. Tex-NS. S329.

ORCHIDACEAE
Spiranthes praecox (Walt.) S. Wats. LADIES'-TRESSES, SPIRAL ORCHID. Coastal salt marshes. Tex-NY. Correll (1950)218.
Spiranthes vernalis Engelm. & Gray. LADIES'-TRESSES, SPIRAL ORCHID. Coastal salt marshes. Tex-Mass. Correll (1950)277.

MYRICACEAE
Myrica cerifera L. WAX-MYRTLE. Slightly brackish banks, barrier beaches, peripheral flat woods. Tex-NJ. K15 + 83 Kurz and Godfrey (1962)32.

LEITNERIACEAE
Leitneria floridana Chapm. CORKWOOD. Brackish swamps. Tex; w Fla. C457 S408.

FAGACEAE
Quercus virginiana Mill. LIVE OAK. Barrier beaches. Tex-Va. K15 + 83.

POLYGONACEAE

Rumex maritimus var. *fueginus* (Phil.) Dusen. GOLDEN DOCK. Saline or brackish marshes and shores. NY-Que. F571.

Rumex orbiculatus Gray. (*R. britannica* L.). GREAT WATER-DOCK. Brackish flats. NJ-Nfld. Harshberger & Burns (1919)22.

Rumex pallidus Bigel. WHITE OR SEABEACH DOCK. Saline marshes, beaches, and rocks. NY-Nfld. F569.

Rumex verticellatus L. WATER DOCK. Inner edge of marsh. Tex-Que. Kearney (1901)364.

Polygonum achoreum Blake. KNOTWEED. Saline marshes. Nfld-Lab. F579.

Polygonum arifolium L. HALBERD-LEAVED TEARTHUMB. Tidal marshes, brackish habitats. Fla-Del. F588 R412.

Polygonum aviculare var. *littorale*. (Link) W. D. J. Koch. BIRD-KNOTGRASS. Salt marshes. SC-Lab. F580.

Polygonum exertum Small. KNOTWEED. Saline or brackish soils. NJ-Que. F578.

Polygonum glaucum Nutt. SEABEACH KNOTWEED. Saline pond shores. Ga-Mass. F577.

Polygonum hydropiperoides Michx. MILD WATER-PEPPER. Brackish habitats. Tex-NS. Harshberger & Burns (1919)22.

Polygonum pensylvanicum var. *rosaeflorum* J. B. S. Norton. PINKWEED. Brackish clearings. Va-Del. F583.

Polygonum prolificum (Small) Robins. KNOTWEED. Saline or brackish marshes or shores. Va-Me. F578.

Polygonum punctatum var. *parvum* Vict. & Rousseau. WATER-SMARTWEED. Brackish tidal shores. NJ-Que. F586.

Polygonum sagittatum L. ARROW-LEAVED TEARTHUMB. Brackish tidal marshes, brackish areas. Tex-Nfld. F578 R412.

Coccoloba uvifera (L.) L. SEA-GRAPE. Coastal beaches, salt tolerant. s Fla. W63.

CHENOPODIACEAE

Chenopodium ambrosioides L. MEXICAN-TEA. Brackish meadows. Tex-NY. Harshberger & Burns (1919)23.

Chenopodium humile Hook. GOOSEFOOT. Saline or brackish soils. Me-NB. F596.

Chenopodium rubrum L. COAST-BLITE. Salt marshes or saline soils. NB-Nfld. F595 Harshberger (1909)380.

Atriplex arenaria Nutt. BEACH ORACH. Borders of salt marshes, sea beaches, sandy seashores, strand beaches, middle beach. Tex-NH. C543 D136 F597 Nichols (1920)517 S468 Snow (1902)291.

Atriplex hastata L. SALT-BUSH. Saline marshes, salt marshes, saline or brackish soils. SC-Nfld. Ganong (1903)359 S467.

Atriplex patula L. ORACH, SALT-BUSH. Brackish marshes, salt marsh pools, saline or brackish soils, coastal salt marshes, high marsh, middle beach.

Tex; SC-Nfld. Blum (1968)216 C540 F597 Ganong (1903)359 Harshberger (1916)483 Nichols (1920)517 R420 S467.

Atriplex pentandra (Jacq.) Standl. Sandy seashores. Tex-Fla. C543.

Bassia hirsuta (L.) Aschers. Saline or brackish soils. Md-Mass. F591.

Salicornia bigelovii Torr. (*S. mucronata* Bigel.). GLASSWORT. Salt marshes and flats, brackish marshes or shores. Tex-Me. C548 F599 K15 L379 M219 R420 S468.

Salicornia europea L. (*S. herbacea* L.). SAMPHIRE, CHICKENCLAWS, CROWFOOT. Salt or saline marshes, mid marsh, low marsh, brackish marshes, tidal Flats, denuded parts salt marshes. Tex; ne Fla-NS. Adams (1963)455 Conrad (1935)458 F599 Ganong (1903)356 Harshberger (1900)653 M219 R420 Reed (1947)607 S468.

Salicornia virginica L. (*S. perennis* Mill., *S. Ambigua* Michx.). SAMPHIRE, PERENNIAL SALTWORT, WOODY GLASSWORT. Salt marshes, meadows, and flats; brackish marshes; mid marsh; low marsh; sea beaches; seacoasts. Tex; ne Fla-Mass. Adams (1963)455 C548 F599 K15 L379 M219 R420 Reed (1947)607 S468 Transeau (1913)205.

Suaeda americana (Pers.) Fern. SEA-BLITE. Salt marshes, sea strands. Me-Que. F600.

Suaeda conferta (Small) I. M. Johnst. Seacoast. Tex. C550.

Suaeda linearis (Ell.) Moq. [*Dondia linearis* (Ell) Millsp.]. SEA—BLITE. Salt or brackish marshes, coastal beaches, strand beaches, sandy coasts. Tex-Me. C550 D136 L381 M223 R420 S469.

Suaeda maritima (L.) Dumort. SEA-BLITE. Salt marshes, mid brackish bays, tidal mudflats, sea strands, margins of seashores. Fla; Va-e Que. F600 Ganong (1903)358 L381 M223 Reed (1947)607.

Suaeda richii Fern. SEA-BLITE. Salt marshes and sea strands. Mass-Nfld. F600.

Suaeda tampicensis (Standl.) Standl. SEA-BLITE. Coastal sand. Tex. C550.

Salsola kali L. SALTWORT. Sea beaches, ocean beaches, sandy seashores, middle beach. Tex-Nfld. Conrad (1935)462 Snow (1902)291 F600 R421 S469.

AMARANTHACEAE

Amaranthus greggii Wats. Coastal beaches. Tex. C556.

Acnida cuspidata Spreng. (*A. alabamensis* Standl.). CARELESS. Brackish marshes, salty places of coast. Tex-Ala. C562 S475.

Acnida cannabina L. [*Amaranthus cannabinus* (L.) J. D. Sauer]. WATER-HEMP. Salt or brackish marshes, tidal bays and marshes, brackish stations. ne Fla-Me. F604 Harshberger & Burns (1919)23 L383 M224 R427 S474

Acanthochiton wrightii Torr. Beaches. Tex. C562.

Alternanthera philoxeroides (Mart.) Griseb. [*Achyranthes philoxeroides* (Mart.) Standl.]. ALLIGATOR-WEED. Barrier beaches, high marshes. Tex-NC. K14.

Iresene rhizomatosa Standl. High salt marsh. Tex-Md. (I have seen this in habitat indicated.)

BATACEAE
Batis maritima L. SALTWORT, BEACHWORT. Open brackish marshes, salt marshes and flats, strand beaches, muddy flats of the seashore. Tex-NC. C600 D136 L351 R422 S486.

AIZOACEAE
Mollugo verticillata L. CARPET WEED. Edge of brackish area. Tex-Que. Snow (1913)47.

Sesuvium maritimum (Walt.) BSP. SEA-PURSLANE. Sea beaches, damp coastal sands, coastal shores, tidal flats. Tex-NY. F607 L397 Wells (1928)232.

Sesuvium portulascastrum (L.) L. SEA-PURSLANE.. Brackish marshes, coastal shores, sea beaches, sea strands. Miss-NC. C604 D136 K15 L397 S492.

PORTULACEAE
Portulaca oleracea L. PURSLANE. Strand beaches, salt marshes. Tex-NB. C607 D136.

Portulaca phaeosperma Urban. PURSLANE. Sea beaches. s Fla. L400.

CARYOPHYLLACEAE
Stellaria humifusa Rottb. CHICKWEED. Saline or brackish shores. Me-Nfld. F623.

Arenaria peploides L. (*Ammodenia peploides* Rupr.). SANDWORT. Beaches. Md-Lab. Conrad (1935)462.

Spergularia canadensis (Pers.) G. Don. (*S. borealis* Robinson). SAND SPURRY. Salt marshes, brackish tidal shores. NY-Que. Ganong (1903)356 F615 M227.

Spergularia echinosperma Celak. SAND SPURRY. Salt marshes and flats. Tex. C630.

Spergularia marina (L.) Griseb. [*Tissa marina* (L.) Britton]. SAND SPURRY, PINK MARSH SPURRY. Saline or brackish soils, saline places, tidal flats, saline marsh, salt marshes, salt marsh pools. Tex-Que. C630 Conrad (1935)460 F614 Harshberger (1900)653 Harshberger (1916)483 R436 S502 Wells (1928)232.

Spergularia media (L.) C. Presl. SAND SPURRY. Saline soils. NY. F615.

NYMPHAEACEAE
Nuphar advena (Ait.) Ait. F. YELLOW POND-LILY. Tidal waters. Tex-Me. F639.

RANUNCULACEAE
Ranunculus cymbalaria Pursh. SEASIDE BUTTERCUP. Brackish streams and

marshes, saline or brackish shores. NJ-Nfld. F647 M248.
Ranunculus hyperboreus Rottb. NORTHERN BUTTERCUP. Shallow brackish water. n Nfld. F648.
Ranunculus sceleratus L. DITCH CROWFOOT, CURSED CROWFOOT. Brackish habitats, saline soils. La-Nfld. F650 S521.
Ranunculus subrigidus W. B. Drew. WHITE WATER-CROWFOOT. Brackish waters. NH-Que. F647.
Ranunculus trichophyllus Chaix. WHITE WATER-CROWFOOT. Brackish waters. NJ-Lab. F647.

LAURACEAE
Persea palustris (Raf.) Sarg. [*P. pubescens* (Pursh) Sarg.]. RED BAY. Barrier beaches. Tex-Del. K15.

BRASSICACEAE
Lepidium latifolium L. PEPPERWORT. Tidal shores, beaches. NY-Mass. F702.
Cochlearia tridactylites Banks. SCURVY-GRASS. Brackish soils. Nfld-Lab. F705.
Cakile edentula (Bigel.) Hook (*C. americana* Nutt.). SEA-ROCKET. Beaches, sandy or gravelly beaches, coastal beaches. Fla-Lab. L431 S575.
Cakile geniculata (Robins.) Millsp. SEA-ROCKET. Beaches. Tex-nw Fla. C691.
Cakile fusiformis Greene. SEA-ROCKET. Sandy shores, coastal beaches. Miss-pen Fla. L431 S574.
Cakile harperi Small. SEA-ROCKET. Sandy beaches. La-NC. S575.
Cakile lanceolata (Willd.) O. E. Schulz. SEA-ROCKET. Coastal shores, strand beaches. s Fla. D136 S575.
Cardamine longii Fern. BITTER CRESS. Tidal estuaries. Va-Me. F723.

CRASSULACEAE
Tillaea aquatica L. PIGMYWEED. Brackish or tidal shores, tidal mudflats of rivers or estuaries. Tex-La; Md-Que. F731 M258.

ROSACEAE
Potentilla anserina L. SILVERWEED. Salt marsh. NY-Nfld. Transeau (1909)274.

CHRYSOBALANACEAE
Chrysobalanus icaco L. COCO-PLUM. Coastal beaches. Fla. L444.

FABACEAE
Aeschynomeme virginica (L.) BSP. SENSITIVE-JOINT-VETCH. Brackish tidal shores, tidal marshes. Va-NJ. F914 R602.
Strophostyles umbellata var. *paludigena* Fern. WILD BEAN. Brackish tidal marshes. Va-DC. F937.

RUTACEAE
Xanthoxylum clava-herculis L. TOOTHACHE TREE, PRICKLY-ASH, PEPPER-WOOD. Barrier beaches. Tex-Va. K16.

SURIANACEAE
Suriana maritima L. BAY-CEDAR. Coastal beaches, seashores, salt or brackish shores, strand beaches. s Fla. D136 L520 W104.

SIMAROUBACEAE
Picramnia pentandra Sw. BITTER-BUSH. Shores. s Fla. W102.

EUPHORBIACEAE
Chamaesyce ammannioides (HBK.) Small. (*Euphorbia ammannioides* HBK.). SEASIDE SPURGE. Saline beaches, upper borders of coastal beaches, beaches, Tex-Va. C972 F971 L551.
Chamaesyce buxifolium (Lam.) Small. SPURGE. Strand beaches. s Fla. D136.
Chamaesyce mesembryanthemifolia (Jacq.) Dugand. SPURGE. Coastal beaches. s Fla. L550.
Chamaesyce polygonifolia (L.) Small. (*Euphorbia polygonifolia* L.). SEASIDE SPURGE. Beaches. Ga-NB. Conrad (1935)462.
Poinsettia Pine Torum Small. SPURGE. Strand beaches. s Fla. D136.
Pedilanthus tithymaloides (L.) Point. [*Tithymalus smallii* (Millsp.) Small]. RED-BIRD FLOWER. Strand beaches. s Fla. D136.

CALLITRICHACEAE
Callitriche hermaphroditica L. WATER-STARWORT. Quiet brackish waters. Nfld-Lab. F973.

ANACARDIACEAE
Rhus radicans L. POISON IVY. Barrier beaches. Tex-Me. K15.

AQUIFOLIACEAE
Ilex vomitoria Ait. CASSENA, YAUPON. Barrier beaches, peripheral flat woods, inner marsh. Tex-Va. K15 + 83 Kearney (1900)17.

HIPPOCRATEACEAE
Hippocratea volubilis L. HIPPOCRATEA. Shores, mangrove swamps. s Fla. L570 S820.

MALVACEAE
Hibiscus moscheutos L. ROSE MALLOW, SWAMP ROSE MALLOW, SEA-HOLLYHOCK. Brackish marshes, salt marsh, inner edge marsh, brackish flats. Ala-Md. Harshberger (1902)654 Harshberger & Burns (1919)26 Kearney (1901)364 R705 S856 Taylor (1938)63.
Hibiscus palustris L. SEA HOLLYHOCK. Saline or brackish marshes.

NC-Mass. F1006.

Kosteletzkya virginica (L.) Presl. SALT-MARSH MALLOW, SEASHORE MALLOW. Brackish marshes, nearly fresh marshes, shores salt marsh. Tex-Del. C1029 F1005 Kearney (1900)47 R703.

ELATINACEAE

Elatine americana (Pursh) Arn. [*E. triandra* var. *americana* (Pursh) Fassett]. WATERWORT. Silty tidal flats, muddy tidal shores. NC-Que. F1015 Fassett (1939)373 M264.

Elatine minima (Nutt.) Fisch. & Meyer. WATERWORT. Muddy tidal shores. Va-Nfld. Fassett (1939)369.

TAMARICACEAE

Tamarix gallica L. TAMARISK. Saline marshes. Tex-Mass. F1016.

LYTHRACEAE

Ammania teres Raf. (*A. koehnei* Britt.). TOOTH-CUPS. Salt, tidal, or brackish marshes. Miss-NJ. F1046 Kearney (1900)47 R738.

Lythrum lineare L. LOOSESTRIFE. Brackish marshes and pools, saline marshes, brackish or saline waters. La-NY. F1047 L634 M268 R739.

RHIZOPHORACEAE

Rhizophora mangle L. RED MANGROVE. Salt or brackish waters along coast, shores of creeks and rivers. pen Fla. L636 S939.

COMBRETACEAE

Bucida bucerus L. BLACK-OLIVE. Salt loving. s Fla. W154.

Conocarpus erecta L. BUTTONWOOD. Brackish waters, sandy shores, muddy saline shores. pen Fla. L638 S933 W155.

Laguncularia racemosa Gaertn. f. WHITE MANGROVE. Rocky and sandy shores near brackish water. pen Fla. L639 S934.

ONAGRACEAE

Ludwigia alata Ell. FALSE LOOSESTRIFE. Brackish or tidal swales and marshes. La-Va. F1054.

Ludwigia lanceolata Ell. FALSE LOOSESTRIFE. Brackish marshes. Miss; Fla-Ga; NC. R749.

Oenothera fruticosa L. [*Kneiffia fruticosa* (L.) Raimann]. EVENING PRIMROSE, SUNDROPS. Brackish marshes, swales, and meadows; saline meadows. Miss-Conn. F1067.

HYDROCARYACEAE

Trapa natans L. WATER-NUT, WATER-CHESTNUT. Slow backwaters of bays and rivers. Va-Mass. M276.

HALORAGACEAE
Myriophyllum exalbescens Fern. WATER-MILFOIL. Brackish ponds and pools. Md-Lab. F1073.

HIPPURIDACEAE
Hippuris tetraphylla L. f. MARE'S-TAIL. Saline or brackish marshes. Que-Lab. F1076.

APIACEAE
Eryngium aquaticum L. WATER ERYNGO, CORN-SNAKEROOT. Brackish marshes. La-NJ. F1090 R769.
Sium suave Walt. (*S. cicutaefolium* Schrank.). WATER-PARSNIP. Brackish marshes, estuaries. La-Nfld. F1097 R783.
Ptilimnium capillaceum (Michx.) Raf. MOCK BISHOP'S-WEED. Brackish marshes, brackish marshland. Tex-Mass. F1098 Harshberger & Burns (1919)27.
Lilaeopsis chinensis (L.) Ktze. [*L. lineata* (Michx.) Greene]. In mud of brackish marshes and tidal shores, tidal flats of rivers, salt marshes, mid marsh. Miss-NS. Conrad (1935)455 F1099 Johnson & York (1915)42 M295 R783 S972 Wells (1928)232.

PRIMULACEAE
Glaux maritima L. SEA-MILKWORT. Salt marshes, brackish bays, tidal mudflats, saline or brackish shores, marshes and sands. Va-Que. F1143 Ganong (1903)364 M298 Transeau (1909)274.

PLUMBAGINACEAE
Limonium carolinianum (Walt.) Britt. (*L. obtusilobum* Blake). SEA-LAVENDER, MARSH-ROSEMARY. Salt or brackish marshes, mid marsh, salt meadows, low high salt marsh, barrier beaches. Miss-NH. F1144 K15 R826 Reed (1947)607 S1021 Statler & Batson (1969)1087.
Limonium nashii Small. [*L. angustatum* (A. Gray) Small]. SEA-LAVENDER, MARSH-ROSEMARY. Brackish or salt marshes, beaches, salt flats. Tex-Nfld. C1186 F1144 R826 S1021.

SAPOTACEAE
Bumelia lanuginosa (Michx.) Pers. GUM BUMELIA, GUM-ELASTIC. Barrier beaches. Tex-Ga. K14.

GENTIANACEAE
Sabatia arenicola Greenm. SEA-PINK, MARSH PINK. Beaches, salt flats. Tex-La. C1206 Wilbur (1955)47.
Sabatia calycina (Lam.) Heller. SABATIA. Transition zone. Tex-Va. Wells (1932)20.
Sabatia campanulata (L.) Torr. SABATIA. Salt marsh meadows. La-Mass.

Conrad (1935)460.

Sabatia dodocandra (L.) BSP. SEA-PINK, MARSH-PINK. Salt, saline, or brackish marshes; brackish habitats. Tex-Conn. C1206 F1156 Wilbur (1955)82.

Sabatia stellaris Pursh. SEA-PINK; MARSH-PINK. Saline or brackish marshes and meadows, grassy marsh. La-Mass. F1156 Harshberger (1900)653 Kearney (1901)363 R838 Wilbur (1955)58.

Centaurium calycosum (Buckl.) Fern. CENTAURY. Salt marshes. Tex. C1207.

Centaurium spicatum (L.) Fern. CENTAURY. Borders of sandy salt marshes. Va-Md; Mass. F1157.

Gentiana gaspensis Vict. GENTIAN. Brackish marshes. Que. F1160.

Gentiana victorinii Fern. GENTIAN. Brackish tidal estuaries. Que. F1160.

APOCYNACEAE

Rhabdadenia biflora (Jacq.) Muell.-Arg. RUBBER-VINE. Salt marshes. s Fla. Davis (1940)344.

ASCLEPIADACEAE

Asclepias lanceolata Walt. MILKWEED. Brackish marshes. Tex-NJ. C1230 F1172 R850.

Cynanchium angustifolium Pers. [*C. palustre* (Pursh) Heller, *Lyonia palustris* (Pursh) Small, *Seutera maritima* Decne]. SALTMARSH MILKWEED. Salt marshes, high marshes. Tex-NC. C1235 K15 R856 S1074.

CONVOLVULACEAE

Ipomea pes-caprae (L.) R. Brown. RAILROAD VINE. Sandy beaches, strand beaches. Tex-Ga. C1252 D136 L722.

Ipomea stolonifera (Cyr.) Gmel. BEACH MORNING GLORY. Beaches. Tex-NC. C1252.

Ipomea sagittata Cav. MORNING-GLORY. Marings of brackish marshes, beaches. Tex-NC. C1253 R868.

BORAGINACEAE

Tournefortia gnaphalodes R. Br. [*Mallotonia gnaphalodes* (Jacq.) Britt.]. SEA LAVENDER. Strand beaches, beaches. s Fla. D136 L731.

Heliotropium curassavicum L. SEASIDE HELIOTROPE. Sandy seashores, borders of saline marshes, beaches. Tex-Del. C1288 F1197 L729 S1132.

VERBENACEAE

Verbena scabra Vahl. VERBENA, SANDPAPER VERVAIN. Margins brackish marshes. Tex-Va. R889.

Verbena xutha Lehn. GULF VERVAIN. Sandy soil of beaches. Tex-Ala. C1320.

Lippia. See *Phyla.*

Phyla incisa Small. TEXAS FROG-FRUIT. Seashores. Tex. C1332.

Phyla lanceolata (Michx.) Greene. (*Lippia lanceolata* Michx.). NORTHERN FROG-FRUIT. Wet brackish sands, coastal marshes. Tex-NJ. C1332 F1212.

Phyla nodiflora (L.) Greene. [*Lippia nodiflora* (L.) Michx.]. COMMON FROG-FRUIT, CAPE-WEED. Beaches, high salt marsh. Fla-Md. C1333 Kearney (1900)48 Wells (1928)232.

AVICENNIACEAE

Avicennia germinans (L.) Stearn. (*A. nitida* Jacq.). BLACK MANGROVE. Sandy shores, coastal swamps, saline soil, tidal shores. Tex-La; Fla. C1311 L732 S1145.

LAMIACEAE

Teucrium canadense L. AMERICAN GERMANDER. Salt marsh. Tex-NB. Rich (1902)89.

Physostegia granulosa Fassett. [*P. virginiana* var. *granulosa* (Fassett) Fern.]. FALSE DRAGONHEAD. Tidal shores, estuaries. Md-Que. F1227 Fassett (1939)377.

SOLANACEAE

Lycium carolinianum Walt. CHRISTMAS BERRY. Salt marshes, barrier beaches. Tex-Fla; SC. K15 S1118.

Physalis angustifolium Nutt. GROUND-CHERRY. Sea beaches. La-Fla. S1112.

Physalis viscosa L. (*P. maritima* M. A. Curtis). GROUND CHERRY. Sandy beaches or shores. Tex-Va. C1389.

Solanum nigrescens Mart. & Gal. (*S. gracile* Link.). NIGHTSHADE. Margins brackish marshes. Ala-NC. R933.

SCROPHULARIACEAE

Bacopa monniera (L.) Pennell. [*Brama monnieri* (L.) Pennell, *Monniera monniera* (L.) Britt.]. BACOPA. Salt or brackish marshes, water pools, and shores; salt marshes; tidal areas. Tex-Va. Kearney (1900)49 M308 R939 S1189.

Bacopa simulans Fern. WATER-HYSSOP. Fresh tidal muds. Va. F1277.

Bacopa stragula Fern. WATER-HYSSOP. Fresh tidal muds. Va-Md. F1278.

Micranthemum micranthemoides (Nutt.) Wettst. Fresh tidal muds. Va-NY. F1278.

Limosella aquatica L. MUDWORT. Shores of brackish ponds, tidal shores of rivers. NJ-Lab. F1278 M316.

Limosella subulata Ives. MUDWORT. Brackish muds and sands, intertidal mudflats. NC-Nfld. F1278 R937.

Agalinis fasciculata (Ell.) Raf. GERARDIA. Tidal marshes. Tex-Md. C1438.

Agalinis maritima (Raf.) Raf. (*Gerardia maritima* Raf.). SALT MARSH GERARDIA. Salt marshes, saline marshes, upper slopes. Tex-NS. C1435

Conrad (1935)460 Miller & Egler (1950)158.

Euphrasia randii Robins. EYEBRIGHT. Brackish shores. Me-Lab. F1295.

PLANTAGINACEAE

Plantago eriopoda Torr. PLANTAIN. Saline marshes and shores. Que. F1315.

Plantago juncoides Lam. (*P. decipiens* Barneoud, *P. maritima* L.). SEASIDE PLANTAGO, GOOSETONGUE. Salt marshes or meadows, salt marsh pools, upper slope of marsh. NJ-Lab. Conrad (1935)460 Ganong (1903)363 Harshberger (1909)379 Harshberger (1916)483 Miller & Egler (1950)158 Nichols (1920)530.

Plantago major L. var. *scopulorum* Fries. & Brob. PLANTAIN. Brackish or saline shores. Del-NS. F1315.

Plantago oliganthos R. & S. SEASIDE PLANTAIN. Salt marshes, saline or brackish shores. NJ-Lab. F1316.

CAMPANULACEAE

Lobelia elongata Small. LOBELIA. Brackish marshes and swamps. La-Del. F1355.

GOODENIACEAE

Scaevola plumeri (L.) Vahl. Strand beaches, beaches. s Fla. L820 D136.

ASTERACEAE

Eupatorium mikanioides Chapm. SEMAPHORE EUPATORIUM. Salt marshes. Fla. L873.

Eupatorium perfoliatum var. *colpophilum* Fern. & Grisc. BONESET. Brackish tidal shores. Me-Que. F1369.

Mikania scandens (L.) Willd. CLIMBING HEMPWEED. Inner edge salt marsh. Tex-NH. Kearney (1901)364.

Solidago austrina Small. GOLDENROD. Brackish swamps. Ga-Va. S1354.

Solidago flavovirens Chapm. GOLDENROD. Brackish marshes. nw Fla. S1354.

Solidago mexicana L. GOLDENROD. Salt marshes, banks tidal rivers, beaches, high marshes, swales. Tex-Mass. C1589 K15 R1091 S1354.

Solidago sempervirens L. (*S. mexicana* L.). SEASIDE GOLDENROD. Saline or brackish habitats, banks tidal rivers, inner marsh, brackish marshes, beaches, brackish flats. Tex-Nfld. C1589 F1398. Harshberger & Burns (1919)31 K15 Kearney (1900)17 R1091 S1354.

Solidago tenuifolia Pursh. GOLDENROD. Brackish marshes. Ga-NS. R1098.

Boltonia asteroides var. *glastifolia* (Hill) Fern. BOLTONIA. Tidal marshes. La-NJ. F1415.

Aster exilis Ellis. ASTER. Brackish marshes. Miss; Fla; SC. R1081.

Aster laurentianus Fern. ASTER. Saline marshes, brackish sands or muds. NB-se Que. F1442.

Aster novi-belgii var. *litoreus* Gray. ASTER. Seashores, borders of saline marshes. Del-NS. F1438.

Aster puniceus var. *compactus* Fern. ASTER. Brackish or saline marshes. Va-Mass. F1429.

Aster racemosus Ell. ASTER. Brackish marshes. SC-NC. R1081.

Aster subulatus Michx. ASTER. Saline, tidal, or brackish marshes; salt marshes, mid marsh, inner marsh. Tex-NB. F1441 Kearney (1900)17 Penfound (1952)433 R1080 S1393 Wells (1928)232.

Aster tenuifolius L. MARSH ASTER. Saline or brackish marshes or shores, high marshes, inner marsh, low marsh, salt marshes, brackish mud. Tex-NH. Adams (1963)455 C1599 F1440 K14 Kearney (1900)17 L862 Penfound (1952)433 R1081 S1393.

Baccharis angustifolia Michx. FALSE-WILLOW, SILVERLING. Brackish marshes, salt marshes, brackish swamps. Fla-NC. L850 R1067 S1398.

Baccharis dioica Vahl. SILVERLING. High marshes. s Fla. K14.

Baccharis glomeruliflora Pers. SILVERLING, BUCKBRUSH. Salt marshes and swamps, salt water swamps. Fla-NC. L851 Penfound (1952)417 S1398.

Baccharis halimifolia L. SEA-MYRTLE, GROUNDSEL TREE, SILVERLING. Brackish marshes, borders of salt marshes, extreme upper edge of marsh, high high salt marsh. Tex-Mass. L851 S1398 Statler and Batson (1969)1087 Taylor (1938)60.

Pluchea camphorata (L.) DC. MARSH FLEABANE, CAMPHOR-WEED, STINKWEED. Brackish marshes, shores, and ditches; salt marshes; salt marsh pools. Tex-Del. F1449 Harshberger (1916)483 Harshberger & Burns (1919)31 S1399.

Pluchea purpurascens (Swartz) DC. SALT-MARSH FLEABANE. Saline or brackish marshes, salt marshes, salt flats or plains. Tex-Me. F1449 Godfrey (1952)255 L867 R1062.

Iva imbricata Walt. MARSH-ELDER. Coastal beaches, strand beaches, sandy beaches. Tex-Va. C1628 D136 L828.

Iva frutescens L. (*Iva oraria* Bartlett). MARSH-ELDER, HIGH-WATER SHRUB. Brackish marshes, saline marshes or shores, salt marshes, barrier beach, upper slope or reaches of marsh, mid marsh, inner marsh, high high salt marsh. Tex-NH; w NS. F1468 K15 Kearney (1900) 17 L828 Miller & Egler (1950)158 R1015 S1298 Statler & Batson (1969)1087 Taylor (1938)60 Wells (1928)232.

Ambrosia hispida Pursh. RAGWEED. Strand beaches, sea beaches, beaches. Fla. D136 L827 S1300.

Xanthium echinatum Murr. SEA-BURDOCK. Borders of saline marshes, beaches. Va-NS. Conrad (1935)462 F1474.

Borrichia arborescens (L.) DC. SEA OX-EYE. Sea beaches, strand beaches. s Fla. D136 L834.

Borrichia frutescens (L.) D.C. SEA OX-EYE. Saline or brackish marshes, salt marshes, low saline grounds, low high salt marsh, high marsh. Tex-DC. Adams (1963)455 F1487 L834 R1110 S1430 Statler & Batson (1969)1087.

Melanthera hastata Michx. MELANTHERA. Strand beaches. La-SC. D136.

Bidens bidentoides (Nutt.) Britt. BUR-MARIGOLD. Brackish tidal shores. Del-NJ. F1503.

Bidens cernua var. *oligodonta* Fern. & St. John. STICK-TIGHT. Brackish or saline shores. Mass-Que. F1501.

Bidens eatoni Fern. BEGGAR'S-TICKS. Brackish marshes, tidal shores. NY-Conn; Mass; Me; Que. F1504.

Bidens heterodoxa (Fern.) Fern. & St. John. BEGGAR'S-TICKS. Brackish or saline marshes. Conn; e Que. F1504.

Bidens hyperborea Greene. ESTUARY-BEGGAR-TICKS. Tidal muds. Mass-Que. F1502.

Bidens infirma Fern. BEGGAR'S-TICKS. Tidal flats. Que. F1505.

Bidens laevis (L.) BSP. BEGGAR'S-TICKS. Brackish marshes and pools, brackish water. Tex-NH. F1500 L832.

Bidens mariana Blake. BEGGAR'S-TICKS. Tidal shores. Md. F1504.

Bidens mitis (Michx.) Sherff. BEGGAR'S-TICKS. Brackish swamps or marshes. La-Ga; NC; Md. F1507 R1128 S1453.

Flavaria linearis Lag. FLAVARIA. Strand beaches, coastal sands. s Fla. D136 S1464.

Helenium autumnale var. *canaliculatum* (Lam.) T. & G. SNEEZEWEED. Tidal shores. Mass-Que. F1513.

Artemesia stelleriana Bess. WORMWOOD. Sandy beaches. Va-Que. Conrad (1935)462 F1522.

Erechtites hieracifolia (L.) Raf. FIREWEED, PILEWORT. Salt marsh. Tex-NS. Taylor (1938)63.

Hedypnois cretica (L.) Willd. Coquina beds at shoreline. Tex. C1730.

Analysis and Summary

1. Halophytes are interpreted here as those plants that can tolerate sea water, pure or diluted.
2. The annotated list of halophytes, which was derived almost exclusively from literature, contains 347 species in 177 genera and 75 families. Useful synonyms are given.
3. Families with the largest number of representatives are POACEAE — 50 species in 28 genera, ASTERACEAE — 43 species in 18 genera, CYPERACEAE - 36 species in 8 genera, CHENOPODIACEAE — 18 species in 6 genera, POLYGONACEAE — 15 species in 3 genera, and ZOSTERACEAE — 12 species in 4 genera. All other families have less than 10 representatives.

4. Of the 357 species, 38 occur in Florida only, 17 in Canada only, and 9 in Texas only.
5. Twenty-four species occur only along the Gulf Coast from Texas to Alabama whereas 137 species occur only along the Atlantic coast.
6. Thirty-two species occur in all coastal states from Texas into Canada.
7. Some indication of the relative richness of the halophytic vegetation in different regions of the coast is given by the total numbers of species reported for the regions; namely, 133 species along the Gulf coast from Texas into Alabama, 197 species from Georgia into Delaware, 161 species from New Jersey into Maine, and 125 species in Canada.
8. It is probable that there are misidentifications by the various authors, that they have misinterpreted some habitats as being halophytic, or that I have included some species on the basis of habitats I misinterpreted as being halophytic. More likely, however, is the inclusion of at least a few species on the list that are not halophytes.
9. There is little uniformity in the literature in the usage of terms to identify types of halophytic habitats. This situation is probably mostly due to the diversity of halophytic habitats and the different interpretations of specific habitats by people.

Literature Cited

Adams, D. A. 1963. Factors influencing vascular plant zonation in N. Carolina salt marshes. Ecol. 44:445-456.

Blum, J. L. 1968. Salt marsh Spartinas and associated algae. Ecol. Monogr. 38:199-221.

Brown, C. A. 1959. Vegetation of the outer banks of North Carolina. Coastal Studies, Series No. 4. Louisiana State Univ. Press., Baton Rouge, La. 179 pp.

Cappel, E. D. 1954. The genus *Scirpus* in North Carolina. Jour. Elisha Mitchell Soc. 70:75-91.

Chapman, V. J. 1960. Salt marshes and salt deserts of the world. Plant Science Mongraphs. Interscience Publishers, Inc., N. Y. 392 pp.

Conrad, H. S. 1935. The plant associations of central Long Island. Amer. Midland Natur. 16:433-516.

Conrad, H. S., and G. C. Galligar. 1929. Third survey of a Long Island salt marsh. Ecology 10:326-336.

Correll, D. S. 1950. Native orchids of North America. Chronica Botanica Company, Waltham, Mass. 399 pp.

Correll, D. S., and M. C. Johnston. 1970. Manual of the vascular plants of Texas. Texas Research Foundation, Renner, Tex. 1881 pp.

Dansereau, Pierre. 1957. Biogeography, an ecological perspective. Ronald Press, N. Y. 394 pp.

Davis, C. A. 1910. Salt marsh vegetation near Boston, and its geological significance. Ecol. Geol. 5:623-639.

Davis, J. H., Jr. 1940. The ecology and geologic role of mangroves in Florida. Carnegie Inst. of Washington, Tortugas Laboratory. Paper No. 16, vol. 32:305-412. 12 plates.

Davis, J. H., Jr. 1942. The ecology of the vegetation and topography of the Sand Keys of Florida. Carnegie Inst. of Washington, Tortugas Laboratory Paper No. 6, vol. 33:114-195. 7 plates.

Fernald, M. L. 1950. Gray's manual of botany. 8th ed. American Book Company, N. Y. 1632 pp.

Fassett, N. C. 1939. Notes from the Herbarium of the University of Wisconsin. XVII. Elatine and other aquatics. Rhodora 41:367-377.

Fassett, N. C., and B. Calhoun. 1952. Introgression between *Typha latifolia* and *T. aangustifolia*. Evolution 6:367-379.

Ganong, W. F. 1903. The vegetation of the Bay of Fundy salt and diked marshes: an ecological study. Bot. Gaz. 36:161-186, 280-302, 349-367, 429-455.

Godfrey, R. K. 1952. Pluchea, Section Stylimnus, in North America. Jour. Elisha Mitchell Soc. 68:238-271.

Gould, Frank W. 1968. Grass systematics. McGraw-Hill Book Co., N. Y. 382 pp.

Harshberger, J. W. 1900. An ecoligical study of the New Jersey strand flora. Acad. Nat. Sci. Phila. Proc. 1900:623-671.

Harshberger, J. W. 1902. Additional observations on the strand flora of New Jersey. Acad. Nat. Sci. Phila. Proc. 1902:642-669.

Harshberger, J. W. 1909. The vegetation of the salt marshes and of the salt and fresh water ponds of northern coastal New Jersey. Acad. Nat. Sci. Phila. Proc. 1909:373-400.

Harshberger, J. W. 1914. The vegetation of south Florida south of 27 so30' north, exclusive of the Florida Keys. Wagner Free Inst. Sci. Phila. Trans. 7:47-189.

Harshberger, J. W. 1916. The origin and vegetation of salt marsh pools. Amer. Phil. Soc. Proc. 55:481-484.

Harshberger, J. W., and V. G. Burns. 1919. The vegetation of the Hackensack Marsh: a typical American fen. Wagner Free Inst. Sci. Phila. Trans. 9:1-35.

Hitchcock, A. S. 1950. Manual of the grasses of the United States. U. S. Dept. of Agriculture, Msc. Pub. No. 200. U. S. Govt. Printing Office, Washington. 1050 pp.

Hotchkiss, N., and H. L. Dozier. 1949. Taxonomy and distribution of N. American cat-tails. Amer. Midland Nat. 41:237-254.

Johnson, D. S., and H. H. York. 1915. The relation of plants to tide levels. Carnegie Inst. Wash. Publ. 206. 162 pp.

Kearney, T. H. 1900. The plant covering of Ocracoke Island; a study in the ecology of the North Carolina strand vegetation. U. S. Nat. Museum,

Contr. U. S. Nat. Herbarium 5:5-63.

Kearney, T. H. 1901. Report on a botanical survey of the Dismal Swamp region. U. S. Natl. Museum, Contr. U. S. Natl. Herbarium 5:321-585.

Kral, R. 1966. Eriocaulaceae of continental North American north of Mexico. SIDA 2:285-332.

Kral, R. 1971. A treatment of *Abildgaardia, Bulbostylis* and *Fimbristylis* (CYPERACEAE) for North America. SIDA 4:57-227.

Kurz, H., and R. K. Godfrey. 1962. Trees of northern Florida. University of Florida Press, Gainesville. 311 pp.

Kurz, H., and K. Wagner. 1957. Tidal marshes of the Gulf and Atlantic coasts of northern Florida and Charleston, S. C. Fla. State Univ. Studies No.24. Florida State University, Tallahassee.

Lawrence, G. H. M. 1951. Taxonomy of vascular plants. MacMillan Co., N. Y. 823 pp.

Long, R. W., and O. Lakela. 1971. A flora of tropical Florida. University of Miami Press, Coral Gables, Fla. 962 pp.

Miller, W. R., and F. E. Egler. 1950. Vegetation of the Wequetequock-Pawcatuck tide-marshes, Connecticut. Ecol. Monogr. 20:143-172.

Mobberley, D. G. 1956. Taxonomy and distribution of the genus *Spartina*. Iowa State College Jour. of Sci. 30:471-574.

Muenscher, W. C. 1944. Aquatic plants of the United States. Comstock Publishing Co., Inc., Ithaca, N. Y. 374 pp.

Newman, W. S., and C. A. Munsart. 1968. Holocene geology of the Wachapreague Lagoon, Eastern Shore Peninsula, Virginia. Marine Geology 6:81-105.

Nichols, G. E. 1920. The vegetation of Connecticut. VII. The associations of depositing areas along the seacoast. Torrey Bot. Club Bull. 47:511-548.

Oosting, H. J. 1956. The study of plant communities. W. H. Freeman and Co., San Francisco. 440 pp.

Penfound, W. T. 1952. Southern swamps and marshes. Botanical Rev. 18:413-446.

Radford, A. E., H. E. Ahles, and C. R. Bell. 1964. Manual of the vascular flora of the Carolinas. University of North Carolina Press, Chapel Hill, N. C. 1183 pp.

Reed, J. F. 1947. The relation of the *Spartinetum glabrae* near Beaufort, North Carolina to certain edaphic factors. Amer. Midl. Nat. 38:605-614.

Rich, W. P. 1902. Oak Island and its flora. Rhodora 4:87-94.

Setchell, W. A. 1924. *Ruppia* and its environmental factors. Proc. Nat. Acad. Sci. 10:286-288.

Small, J. K. 1933. Manual of the southeastern flora. The University of North Carolina Press, Chapel Hill. 1554 pp.

Snow, L. M. 1902. Some notes on the ecology of the Delaware coast. Bot. Gaz. 34:284-306.

Snow, L. M. 1913. Progressive and retrogressive changes in the plant associations of the Delaware coast. Bot. Gaz. 55:45-55.

Statler R., and W. T. Batson. 1969. Transplantation of salt marsh vegetation. Georgetown, South Carolina. Ecology 50:1087-1089.

Taylor, N. 1938. A preliminary report on the salt marsh vegetation of Long Island, New York. Bull. N. Y. State Mus. 316:21-84.

Transeau, E. N. 1909. Successional relations of the vegetation about Yarmouth, Nova Scotia. Plant World 12:271-281.

Transeau, E. N. 1913. The vegetation of Cold Spring Harbor, Long Island. I. The littoral successions. Plant World 16:189-209.

Wells, B. W. 1928. Plant communities of the Coastal Plant of North Carolina and their successional relation. Ecology 9:230-242.

Wells, B. W. 1932. The natural gardens of North Carolina. University of North Carolina Press, Chapel Hill. 458 pp.

West, E., and L. E. Arnold. 1956. The native trees of Florida. University of Florida Press, Gainesville. 218 pp.

Wilbur, R. L. 1955. A revision of the North American genus *Sabatia* (Gentianaceae). Rhodora 57:1-33, 43-71, 78-104.

MANGROVES: A REVIEW[1]

Gerald E. Walsh

Environmental Protection Agency

Gulf Breeze Laboratory

Sabine Island

Gulf Breeze, Florida 32561

Associate Laboratory of the National Environmental Research Center,

Corvallis

"The beaches on that coast I had come to visit are treacherous and sandy and the tides are always shifting things about among the mangrove roots...A world like that is not really natural...Parts of it are neither land nor sea and so everything is moving from one element to another, wearing uneasily the queer transitional bodies that life adopts in such places. Fish, some of them, come out and breathe air and sit about watching you. Plants take to eating insects, mammals go back to the water and grow elongate like fish, crabs climb trees. Nothing stays put where it began because everything is constantly climbing in, or climbing out, of its unstable environment."

INTRODUCTION

The quotation above from Loren Eisley's eloquent book, "The Night Country," portrays in poetic terms the fascination of the tropical mangrove forest for those who have studied and researched that "not really natural"

---

[1] Publication No. 154 from the Gulf Breeze Laboratory, Environmental Protection Agency, Gulf Breeze, Florida 32561 Associate Laboratory of the National Environmental Research Center, Corvallis.

world. In the mangrove ecosystem, where tides and coastal currents bring unremitting variation to the forest, plants, and animals adapt continuously to changing chemical, physical, and biological characteristics of their environment. Many species use the environment dominated by mangrove trees for food and shelter during part or all of their life cycles. There is constant movement of living and non-living matter into and out of the mangrove swamp, and the effects of such movement may be felt miles away (Heald 1971, Odum 1971). Of course, not all tropical coasts are lined with mangrove forests; often a mangrove stand is small, or only an occasional tree dots the shoreline.

The factors which determine development of coastal forests, the ecological roles of mangroves in estuaries, and their utilization by man have been studied at length. The references at the end of this review give over 1,200 published accounts on mangroves. I am certain to have missed many publications in my search, but the number gives testimony to the importance of mangroves in estuaries. For an historical sketch of published works on mangrove, see Bowman (1917), who traced the mangrove literature back to 325 B.C. and the chronicle of Nearchus, commander of the fleet of Alexander the Great. Additional information is given in the reports of Walter and Steiner (1936) on East African mangroves, Davis (1940b) on the ecology and geologic roles of mangroves in Florida, and Macnae (1968) on the flora and fauna of mangrove swamps in the Indo-West-Pacific region. See also the excellent discussion of ecology of the Rhizophoraceae by van Steenis in Ding Hou (1958).

Davis (1940b) described "mangrove" as a general term applied to plants which live in muddy, loose, wet soils in tropical tide waters. According to Macnae (1968), mangroves are trees or shrubs that grow between the high water mark of spring tides and a level close to but above mean sea level. They are circumtropical on sheltered shores and often grow along the banks of rivers as far inland as the tide penetrates. Chapman (1939, 1940, 1944a) described silt, sand, peat, and coral reefs as mangrove habitats. On the reef, seedlings develop in holes and crevices in the porous coral rock, but the trees are usually stunted and the area occupied by the stand is not large. The reef may be a habitat only in those areas where tidal height is not great, because total inundation for extended periods of time can be fatal to seedlings (Rosevear 1947). Another mangrove habitat, the sand beach, described by Chapman (1940) supports *Rhizophora mangle* L. Later, van Steenis (1962) stated that *R. stylosa* Griffith is often found in sand in the Indo-Pacific region. Hathaway (1953) and Moul (1957) reported stands of *R. mucronata* Lamk., *Sonneratia caseolaris* (L.) Engler, and *Bruguiera conjugata (gymnorhiza?)* Lamk is sand on several atolls in the Pacific Ocean. I saw *R. mangle* growing in sand in Hawaii.

Boughey (1957) described mangroves which grew in two types of lagoons on the west coast of Africa. In open lagoons, some of the mud around the margins was exposed daily at low tide. *Rhizophora racemosa* G.F.W. Meyer and *R. harrisonii* Leechman grew on the exposed mud. *Rhizophora* species grew only in open lagoons which were flooded daily. In closed lagoons, *Avicennia nitida* Jacq. was the dominant form in association with *Conocarpus erectus* L., *Laguncularia racemosa* Gaertn., and *Dodonea viscosa* L.

Burtt Davy (1938) classified tropical woody vegetation types according to "mature" or "apparently stable" communities. Two of his classes apply to mangrove vegetation and are given here because much of the nomenclature is in common usage today.

1. Tropical Mangrove Woodland

Name Suggested for Adoption: Mangrove woodland.
Synonyms: Mangrove, Mangrove swamp, Tidal forest.
Brief Definition: Woodland formation below high-tide mark; sometimes forest-like. Nearest in form to dry evergreen forest. A subformation of the littoral swamp forest.
Habitat: Soil flooded with water either permanently or at high tide; water usually more or less brackish; on estuarine mud.

2. Tropical Littoral Woodland

Name Suggested for Adoption: Littoral woodland.
Synonyms: Strand vegetation, Beach forest, Dune forest.
Brief Definition: Woodland formations in situations mentioned below; somewhat resembling semi-evergreen forest; open herbaceous vegetation.
General Description: The most characteristic species of this formation in India and Burma is the evergreen but very light-foliaged *Casuarina*, which often forms an almost pure fringe on sandy beaches and dunes along the sea face. Scattered smaller evergreen trees occur, with fewer deciduous trees, and these, in the absence of *Casuarine*, form the dominant canopy. On the east coast of Tropical Africa are such species as *Heritiera littoralis* Dryand, *Barringtonia racemosa* L., *Terminalia catappa* L., *Phoenix reclivata*, and *Diospyros vaughaniae*; species of *Pandanus* and *Cocos nucifera* L. are characteristic of this formation, which naturally includes several species whose seeds or fruits are current-borne. *Ipomea pescaprae* commonly occurs as a surface creeper on exposed sand dunes. Xerophytic herbs such as *Sansevieria, Opuntia, Kalanchoe*, and *Euphorbia* are common.
Habitat: Sandy and gravelly seashores; not subject to immersion, but under constant maritime influence. All around the coast wherever a fair width of sandy beach occurs, including sandy bars on the sea face of river deltas.

In this discussion, I shall follow Macnae (1968) and use the word "mangrove" with reference to individual kinds of trees, and the word "mangal" with reference to the swamp forest community.

It has been estimated that between 60% and 75% of the tropical coastline is lined with mangrove trees (McGill 1958) though some stands are more extensive than others. There seem to be five basic requirements for extensive mangal development. They are:

1. **Tropical temperatures.** Well developed mangals are found only along coastal areas where the average temperatures of the coldest month is higher

than 20°C and the seasonal temperature range does not exceed 5°C (West 1956, van Steenis 1962).

2. **Fine-grained alluvium.** Mangrove stands are best developed along deltaic coasts or in estuaries where soft mud comprised of fine silt and clay and rich in organic matter, is available for growth of seedlings. Quartzitic and granitic alluvia are generally poor substrata, whereas volcanic soils are highly productive of mangroves (Schuster 1952, West 1956, Haden-Guest et al. 1956, Macnae and Kalk 1962, Macnae 1968).

3. **Shores free of strong wave and tidal action.** Mangroves develop best along protected shores of estuaries because strong wave and tidal actions uproot seedlings and carry away soft mud (Young 1930, Cockayne 1958).

4. **Salt water.** Salt water per se is not a physical requirement of mangroves (Bowman 1917, Warming and Vahl 1925, Rosevear 1947, Egler 1948, Daiber 1960). Mangroves are faculative halophytes that occupy tidal areas where fresh-water plants, which are intolerant to salt, cannot live (West 1956).

5. **Large tidal range.** A wide, horizontal tidal range has been cited as requisite for extensive growth of mangrove (Foxworthy 1910, West 1956) and Chapman and Trevarthen (1953) stated that a universal scheme for comparison of different shores can be based only on the tides as a universal controlling factor. Although the tide per se is probably of little importance in determining the extent of mangal development, on a shore of gentle gradient and large tidal range, a wide belt of alluvium will be formed, and with it, a wide belt of mangrove. Deep tidal penetration would also cause saline water to be distributed far inland. Davis (1940) described the action of wind in driving salt water inland in Florida.

These five factors can determine the occurrence of mangroves, the species present, and the area occupied by a mangal. Once established, mangals throughout the tropics have many ecological similarities. In the following pages I attempt to summarize from accounts available to me, what is known about mangroves and mangals.

GEOGRAPHICAL DISTRIBUTION

Geographical distribution of mangroves is similar in many ways to that of sea grasses (Den Hartog 1957) and marine angiosperms in general (Good 1953). The main difference is that some mangrove species occur on both sides of the Atlantic Ocean and on the Atlantic and Pacific coasts of the Americas.

Fig. 1 shows the general geographic distribution of mangroves. Among individual genera and species, distribution is undoubtedly influenced by whether or not the plant is viviparous, and the ability of the seedling to survive in sea water for an extended period of time. Dispersal of resting seedlings by drift in the open ocean and by alongshore surface currents permits wide geographic range, and temperature and geomorphological characteristics determine distribution along individual coasts.

Geographical distribution is restricted, in general, to the tropics, but Oyama (1950) reported a small stand of mangrove on the southern tip of Kyushu Island at 35°N latitude. Later, van Steenis (1962a) identified the species there as *Kandelia kandel* (L.) Druce. Vu Van Cuong (1964) reported the Ryukyu Islands (about 27°N latitude) to be the northern limit of *R. mucronata, B. gymnorhiza, Avicennia marina* (Forsk.) Vierth., *Xylocarpus moluccensis* (Lamk.) Roem., *X. granatum* Koenig, *Lumnitzera littorea* (Jack) Voigt, and *Lumnitzera racemosa* Willd. In the southern hemisphere van Steenis (1962a) reported the southernmost stand to be on North Island of New Zealand at "less than 40° south." Chapman and Ronaldson (1958) reported that dwarfed *A. marina* grew in abundance in Auckland Harbor at 37°S latitude.

Mangroves are present on the Pacific coast of South America only to about 4°S latitude due to lack of sedimentation below that point. It was once thought that atmospheric drought caused absence of mangroves from that and other areas. Van Steenis (1962a) pointed out, however, that drought is not a factor in distribution as mangroves grow in the Arabian Gulf, the delta of the Indus River, southern Timor, and western Australia, where large silt deposits are found on arid coasts. The major differences between mangals on arid coasts and those on humid coasts is the paucity of the epiflora in the former.

Every mangal is composed of two classes of plants: (a) genera and higher taxa which are found only in the mangrove habitat and (b) species that belong to genera of inland plants but which are adapted for life in the swamp forest. World distribution of genera that occur in mangrove swamps is given in Table 1. For a detailed listing of many forms in class "b" above, see Vu van Cuong (1964). The fern *Acrostichium aureum* appears to be a circumtropical associate of mangroves since it has been reported in mangles of Ceylon (Abeywickrama 1964), India (Biswas 1927), Africa (Bews 1916, Boughey 1957), and the West Indies (Borgesen 1909).

Geographically, mangrove vegetation may be divided into two groups: that of the Indo-Pacific region and that of western Africa and the Americas. The Indo-Pacific region is comprised of East Africa, the Red Sea, India, Southeast Asia, southern Japan, the Philippines, Australia, New Zealand, and the southeastern Pacific archipelago as far east as Samoa. The West Africa-Americas region includes the Atlantic coasts of Africa and the Americas, the Pacific coast of tropical America, and the Galapagos Islands. Mangroves are not native to Hawaii, but *R. mangle, B. sexangula* (Lour.) Poir., *S. caseolaris*, and *Conocarpus erectus* have been introduced.

Distributions of several species found only in mangrove swamps are shown in Fig. 2-8. These figures, with Table 1, show that (a) the greatest number of genera and species occur along the shores on the Indian and western Pacific oceans, (b) there are no species common to East and West Africa, and (c) the species of the Americas and West Africa are related taxonomically. Species found on both the eastern shores of the Americas and the western shore of

Table 1. Distribution of plant genera that occur only in mangrove swamps (Chapman 1970).

Families and Genera	Total species	Indian Ocean W. Pacific	Pacific America	Atlantic America	West Africa
Rhizophoraceae					
Rhizophora	7	5	2	3	3
Bruguiera	6	6	0	0	0
Ceriops	2	2	0	0	0
Kandelia	1	1	0	0	0
Avicenniaceae					
Avicennia	11	6	3	2	1
Myrsinaceae					
Aegiceras	2	2	0	0	0
Meliaceae					
Xylocarpus	?10	?8	?	2	1
Combretaceae					
Laguncularia	1	0	1	1	1
Conocarpus	1	0	1	1	1
Lumnitzera	2	2	0	0	0
Bombacaceae					
Camptostemon	2	2	0	0	0
Plumbaginaceae					
Aegiatilis	2	2	0	0	0

Con't on next page

Table 1 con't

Palmae					
Nypa	1	1	0	0	0
Myrtaceae					
Osbornia	1	1	0	0	0
Sonneratiaceae					
Sonneratia	5	5	0	0	0
Rubiaceae					
Scyphiphora	1	1	0	0	0
	55	44	7	9	7

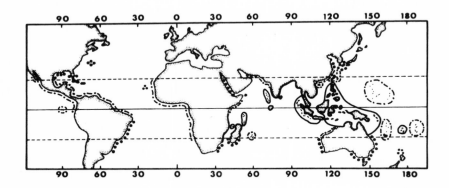

Figure 1. World distribution of mangroves (after Chapman 1970). • • • • less than five species present; – • – • – • five to twenty species present; ─────── more than 20 species present.

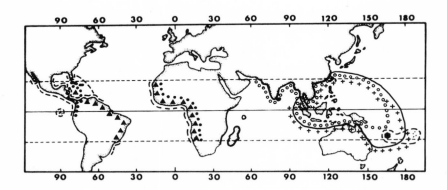

Figure 2. World distribution of Rhizophora species (after Ding Hou 1960, Vu Van Cuong 1964, and Chapman 1970). – – – R. mangle L.; ▲▲▲ R. racemosa F.F.W. Meyer; • • • R. harrisonii Leechm.; ─────── R. mucronata Lamk.; ○○○ R. apiculata Blume; + + + R. stylosa Griffith; ⬢ ⬢ R. lamarckii Montr.

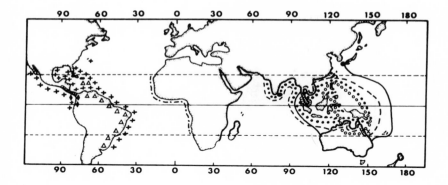

Figure 3. World distribution of <u>Avicennia</u> species (after Vu Van Couong 1964 and Chapman 1970). + + + <u>A. nitida (germinans?)</u> Jacq.; △△△ <u>A. schauerana</u> Stapft; ⊠⊠⊠ <u>A. bicolor</u> Standl.; ◆◆◆ <u>A. tonduzii</u> Moldenke; – • – •– <u>A. africana</u> P. Beauv.; ─────── <u>A. marina</u> (Forsk.) Vierh.; •••• <u>A. officinalis</u> L.; – – – <u>A. alba</u> Blume; X X <u>A. balanophora</u> Stapft and Modlenke; ○○○ <u>A. eucalyptifolia</u> Zipp; ═════ <u>A. lanata</u> Ridly.

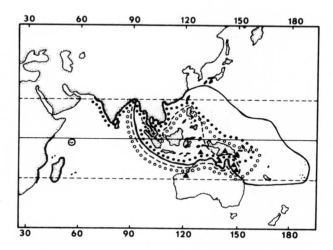

Figure 4. Distribution of <u>Bruguiera</u> species (after Vu Van Cuong 1964). ─────── <u>B. gymnorhoza</u> Lamk.; •••• <u>B. sexangula</u> (Lour.) Poir.; – – – <u>B. Cylindrica</u> (L.) Blume; ○○○ <u>B. parviflora</u> (Roxb) W. and A; + + + <u>B. hainesi</u> C. G. Robers; ▲▲▲ <u>B. exaristata</u> Ding Hou.

59

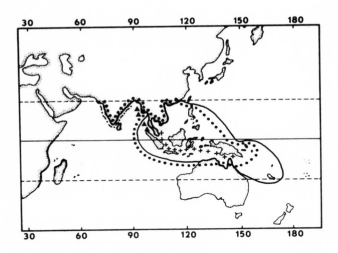

Figure 5.   Distribution of <u>Sonneratia</u> species (after Vu Van Cuong 1964 and Chapman 1970). ———— <u>S. alba</u> J. Smith; • • • • <u>S. caseolaris</u> (L.) Engler + + + <u>S. ovata</u> Backer; ▲▲▲ <u>S. griffithii</u> Kurz.; – • – • – <u>S. apetala</u> Buch.-Ham.

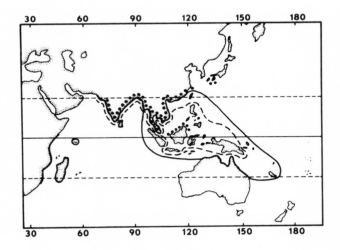

Figure 6.   Distribution of ———— <u>Ceriops tagal</u> (Perr.) C. B. Rob.; – – – <u>C. decandra</u> (Griffith) Ding Hou; • • • • <u>Kandelia Kandel</u> (L.) Druce (after Vu Van Cuong 1964).

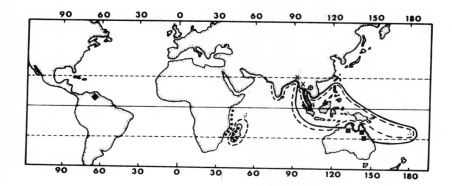

Figure 7. World distribution of Xylocarpus species (after Vu Van Cuong 1964). ──── X. granatum Koenig; – – – X. moluccensis (Lamk.) Roem.; X X – X. gangeticus Park; ▲▲▲ X. minor Ridley; ⊗⊗ X. parvifolius Ridley; ■■■ X. australasicum Ridley; ◆◆ X. guianensis; ●●●● X. benadirensis Moll.

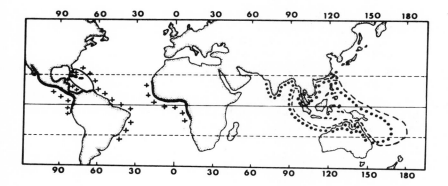

Figure 8. Distribution of ──── Conocarpus erectus L.; + + + Laguncularis racemosa Gaertn.; ●●●● Lumnitzera racemosa Willd.; and – – – Lumnitzera littorea (Jack.) Voigt.

Africa are *R. mangle, R. racemosa, R. harrisonii, Laguncularia racemosa,* and *C. erectus. Pelliciera rhizophorae* Planchon and Triana, a member of the tea family (Theaceae), is found only on the Pacific coast of tropical America. It occurs in small communities on exposed areas such as the seaward tips of point bars in estuaries or in spots having hard, clay soils (West 1956). Fuchs (1970) reported pure stands of *P. rhizophorae* on firm, sandy ground of low salinity. *Pelliciera* was associated with *Rhizophora* on low, muddy ground but in this habitat, trees of both genera were small.

There is confusion concerning taxonomy and distribution of *Avicennia* species on the shores of the eastern and western Atlantic Ocean. It was commonly held that a single species, *A. nitida,* occurred on both sides of the Atlantic Ocean. Moldenke (1960), however, recognized *A. nitida* in the Americas and *A. africana* Moldenke in West Africa. Vu van Cuong (1964) discarded the species *nitida* and recognized *A. germinans* in the Americas and *A. africana* in West Africa. Both Chapman (1970) and Vu van Cuong (1964) recognized four species in the Americas and only one, *A. africana,* in West Africa, and toxonomists at this time seem to agree that the American and West African species are closely related. Chapman (1970) speculated that the reason for confusion is that speciation is now occurring within the genus on both sides of the Atlantic.

There is also confusion in the common names of some mangroves. Those of the genus *Rhizophora* are called "red" mangrove in both the Americas and Africa. *Avicennia,* called the "black" or "honey" mangrove in the Americas, is known as the "white" mangrove in West Africa. *Laguncularia* is called the "white" mangrove in America.

From taxonomic and distributional considerations, Ding Hou (1960) and van Steenis (1962) concluded that *Rhizophora, Avicennia, Xylocarpus, Lumnitzera,* and *Laguncularia* arose in the Indo-Malaysian region and spread westward to East Africa and (except *Laguncularia*) eastward to the Pacific coast of the Americas. The genera reached the Caribbean Sea sometime between the Upper Cretaceous Period and the Lower Miocene Epoch, when the Isthmus of Panama was an open seaway. After establishment on eastern American shores, the trees reached West Africa when seedlings were carried across the ocean by surface currents.

To explain why the mangrove floras of East and West Africa are separate, van Steenis (1962a) postulated that the climate of South Africa during the Upper Cretaceous Period was not tropical and that mangroves could not have been distributed from east to west around the Cape of Good Hope.

As an interesting sidelight to the problem of distribution, Ding Hou (1960) pointed out that broad areas of the Pacific Ocean, to which favorable currents flow, do not contain mangrove. He attributed this to lack of suitable coasts for successful implanting of seedlings. He also described *R. mangle,* a native of the Americas, as present in New Caldonia, Fiji, and Tonga (see Fig. 2). Chapman (1970) speculated that early man carried seedlings from Pacific

America to those islands for growing trees to serve as a source of tannin.

ECOLOGY

Mangrove swamp forests are complex ecosystems that occur along intertidal accretive shores in the tropics. Dominated by estuarine trees, they draw many of their physical, chemical, and biological characteristics from the sea, inflowing fresh water, and upland forests. Mangrove swamps serve as ecotones between land and sea, and elements from each are stratified both horizontally and vertically between the forest canopy and subsurface soil.

The canopy is inhabited by floristic and faunistic elements from the tropical rain forest, including epiphytes, insects, reptiles, birds and mammals. Phytotelmata, filled with rain water, support a variety of algae, protozoa, and immature insects. Below the canopy, portions of tree stems are immersed, in relation to the tidal cycle, for various periods throughout the day. An extreme example is Inhaca Island in southeastern Mozambique where stems are often immersed for 8 to 12 hours per day at depths up to 2 m (Mogg 1963). The surface soil of swamps is alternately inundated and drained. It supports animals such as crabs, amphibians, reptiles, air-breathing fishes, and mammals, whose distributions are governed by degree of tidal penetration and by the nature of the substratum. At the landward edge of the swamp, typically fresh-water forms such as frogs, monitors, and crocodiles may be found (Macnae 1968), and I have observed the toad *Bufo marinus* in salinities up to 10 ppt (parts per thousand) in a mangal in Hawaii.

At the seaward edge, the mud surface is often a truly marine mid-littoral soft-bottom environment (Rutzler 1969) and supports crabs, shrimp, shellfish, etc. Numerous permanent and semi-permanent pools contain insects, shellfish, amphibians, and fish. Throughout the mangal is a network of rivulets, creeks, channels, and often rivers which change in depth with tidal ebb and flow. These contain numerous sessile forms such as algae, fungi, tunicates, sponges, and shellfish which live on mangrove prop- and aerial-roots. Mobile forms such as worms, crabs, shrimp, and fish migrate within the waterways in relation to the tidal cycle and nature of the substratum.

Jennings and Bird (1967) gave six environmental factors which affect geomorphological characteristics in estuaries and, therefore, the flora and fauna. The characteristics were: (1) aridity, (2) wave energy, (3) tidal conditions, (4) sedimentation, (5) mineralogy, and (6) neotectonic effects. All have been cited as factors in mangrove establishment. Troll and Dragendorff (1931) stated that water, salt, and oxygen contents of the soil are also important. On a short-term basis, tropical storms are very disruptive to mangals (West 1956, Alexander 1967) and are the greatest single sources of repeated set backs to the vegetation (Exell 1954). On the other hand, storms may carry propagules further inland than would normal tides (Egler 1952), and Mullan (1933) stated that seeds of mangroves are dispersed widely during the monsoon in Malaya.

Tides and type of substratum are probably the most important factors that govern the nature of intertidal communities (Chapman and Trevarthen 1953). In the case of mangroves, salinity of the surface and soil waters are also very important (Davis 1940), as are temperature, rainfall, rate of evaporation, topography, and geomorphology.

Surface Water

One of the distinctive features of mangrove vegetation is the ability to live in salt water as facultative halophytes. In such situations as reefs, lagoons, and the Florida Everglades, the surface water environment of mangroves is fairly stable in terms of physical and chemical composition. In mangals, salt and nutrient concentrations of surface water, whether in waterways or covering the swamp floor at high tide, is regulated by (1) inflow of fresh water from upland areas, (2) inflow and outflow of seawater with each tidal cycle, (3) precipitation, and (4) humidity.

Chemical and physical data on the surface waters of mangrove swamps have been reported from the Great Barrier Reef (Orr and Moorehouse 1933); Inhaca Island, Mozambique (Macnae and Kalk 1962); Cananeira, Brazil (Teixeria and Kutner 1963, Teixeria et al. 1965, Okuda et al. 1965); Tabasco, Mexico (Thom 1967); Hawaii (Walsh 1967); and Trinidad (Bacon 1968, 1971). Davis (1966) reported salinities up to 43 ppt and temperatures to $39.5^{\circ}C$ in a mangrove salt-water pool in Jamaica. Examples of extreme conditions in a single swamp were given by Walsh (1967) who analyzed the water at six stations located between the landward and seaward edges of a swamp in Hawaii. At the landward edge, tidal effect was minimal or non-existant between August 1961 and November 1962, and the water was always fresh. Oxygen content of the water was that of a dystrophic body, averaging 0.67 ml/L throughout the sampling period. None of the factors measured were subject to large diel, monthly, or annual changes. Proceeding from the landward to the seaward stations in the swamp, diel changes became greater. At the seaward edge, water chemistry at high tide was similar to that of open bay water with great variations in relation to the tidal cycle. At low tide the water was fresh, whereas at high tide salinity was always greater than 25 ppt. In spite of such great differences in physical and chemical properties of the surface water, *R. mangle* grew in a dense stand between the landward and seaward edges of the swamp.

Bacon (1968) gave similar data for a mangrove swamp in Trinidad. In addition, he reported diel variations in concentrations of dissolved nitrate, phosphate, silicate, and suspended solids. Increased concentrations of nitrate and silicate occurred at low tide. Bacon suggested that either the inflowing fresh water was richer in nutrients than tidal water or that nutrients were released from the mud at low water. Walsh (1967) found a similar phenomenon for both nitrate and phosphate, and ascribed this to greater solubility of the substances in fresh water than salt water. He also demonstrated the affinities of the various types of swamp substrata for nitrate and phosphate. In addition, nitrate and phosphate contents of swamp waters at low tide were greater than those of inflowing water. All of this indicates a dynamic system in which nitrate and phosphate are released from or taken up by sediments covered by surface water.

Watson (1928), DeHaan (1931), Walter and Steiner (1936), Macnae and Kalk (1962), and Macnae (1966) proposed schemes for classification of zonation of vegetation within mangals based upon tidal inundation and salinity. The details of these proposals were reviewed by Macnae (1968). In each scheme, vegetation of southeastern Asia was related to salinity of the water and Macnae (1968), using the data of Watson (1928), showed the general preference of *Avicennia intermedia* for coastal seawater and of *A. alba* for less saline water around the mouths of rivers. *Rhizophora mucronata* lived in water of greater salinity than *R. apiculata*. *Bruguiera sexangula* occurred in water of greater salinity than *B. cylindrica*, *B. parviflora*, and *B. gymnorhiza*. The three species of *Xylocarpus* lived in the less saline areas of the swamp.

Davis (1940) related salinity of surface water to distribution of trees in a swamp in southern Florida. American swamps are simpler floristically than those of Malaya and Davis demonstrated the relationship between tidal penetration and salinity to horizontal zonation. In Florida, *R. mangle* and *Luguncularia racemosa* are mixohaline, but the former is the pioneer species on seacoasts. *Rhizophora mangle* was found by Davis to grow in salinities that ranged from fresh water in the Everglades to 34.9 ppt along the seashore. *Laguncularia* grew in "nearly fresh water" to water of 45.8 ppt, and was usually found in association with the other mangrove species. Davis stated "no particular habitat is definitely most suitable for *Laguncularia*." This concept was extended by Thom (1967), who observed that *Laguncularia racemosa* in Tabasco formed communities with other mangroves and had less stringent habitat requirements than they.

Davis (1940) reported *A. nitida* growing in the field in salinities between 36.8 ppt and 38.6 ppt although it can grow in fresh water in the laboratory. This species seems to be adapted for survival in swamp areas with great salinity fluctuations. The community is not flooded deeply by tidal water and salt is concentrated by evaporation during dry periods. During periods of rain, the surface water is diluted greatly so that the *Avicennia* zone has a greater range of salinity than any other. Conversely, *Conocarpus erectus* grew only where salinity was low and the ground covered only occasionally by tidal water. Many *Conocarpus* localities had no surface water; where there was surface water, salinity averaged less than 2 ppt. An important factor for survival of *C. erectus* seemed to be high salinity of the soil water. This will be discussed in the next section.

Sedimentation and Soil

According to the nature of the substratum, mangroves may be classified as reef, sand, mud, and peat types (Chapman 1940, Rutzler 1969). Also, some are found occasionally among boulders, having roots within cracks or other niches, and use tidal water as the source of nutrients. The typical sediments of swamps are composed of peaty, soft, sandy or clayey mud. They are similar to the sediments of salt marshes, which occupy the same sedimentological position at higher latitudes. Mineralogy of mangrove sediments is concerned mainly with clastic detritus from rivers and calcareous debris formed either biologically by shelled organisms or by inorganic precipitation. Algae and bacteria can also function in precipitation and pyrite is often abundant in swamps, usually embedded within or attached to plant remains. Along some shores, where tidal and alongshore currents control the character of the sediments, siliceous and quartzitic sand may predominate.

Mangroves advance seaward only where sedimentary processes prepare shallow water areas for growth of seedlings. Mangals often are associated with lateral accretion of sediment along tropical shores, and location, size, and shape of swamps are influenced strongly by the pattern of coastal sedimentation. Hagen (1890) and van Steenis (1941) stated that natural coastal accretion by mud-silting is the major factor responsible for development of large mangals. Although the trees do not aid appreciably in lateral extension of shores, they do aid in accumulation of sediment with subsequent build-up of soils (Curtiss 1888, Vaughan 1909, Watson 1928, Holdridge 1940, Egler 1952, West 1956, Boughey 1957, Vann 1959, Stoddart 1962, Scholl 1963, Thom 1967, Macnae 1968). During high tide, brackish, sediment-laden water overflows the numerous creeks and channels of the swamp. Alluvium is deposited on the swamp floor and, with autochthonous organic and inorganic detritus, aids in land elevation. Freise (1938) stated that the black color of mangrove mud in Brazil was due to the presence of iron sulphide. The black mud was often covered by a 1-5 cm deep layer of grey-brown mud which was either deposited during tidal inundation or affected chemically by oxygen in tidal waters.

Accretion of sediment along alluvial coasts is regulated mainly by physiographic-geomorphic processes such as (a) the rate at which sediment is brought into an area by rivers and tides, (b) the angle of slope of the shore, (c) sedimentary distributional patterns, (d) subsidence or emergence of the coast, (e) other factors associated with changes in sea level, and (f) tidal-river channel development.. As in other estuaries, the coarser sediments of mangrove swamps are generally in the channels, the finer sediments along the shores of the channels (Walsh 1967). Also, when a river reaches the estuary, the heavier particulate elements have been sorted out above the mangal, so that the predominent sediments within the swamp are of fine-grained alluvium. Near the mouth of the estuary, coarser sediments may again be found. These originate from tidal and alongshore currents which have enough energy for deposition of small sand particles and calcareous detritus. The contribution of inorganic detritus from seawater is usually small, however, because strong currents do not allow seedling development. River-borne sediment is the greatest source of allochthonous material in most waamps and appears to be especially important in the Indo-Malayan region (Watson 1928, Schuster 1952, Macnae 1968).

Sedimentation of autochthonous matter is an important factor for mangroves which are not influenced greatly by fresh-water inflow. Davis (1940) stated that the authochthonous mangals of Florida develop over three primary soil types; namely, (1) siliceous sands, (2) calcareous sands, and (3) calcareous mud marls. Mature *R. mangle* trees are thus sometimes found on nearly bare rock with only small pockets for rooting, but more often grow on deep peat soils. Although the general physical and morphological features of soils vary greatly between mangals, the halotropic peats so often found with mangroves are composed mainly of calcium compounds from shells, biologically precipitated calcite and aragonite, and organic matter of floral and faunal origin.

Extensive autochthonous mangrove swamps have developed along the western side of Andros Island in the Bahamas where the rate of carbonate mud precipitation is great. Burkholder and Burkholder (1958) described the autochthonous sediments of Bahia Fosforescente in Puerto Rico. The sediments contained large amounts of mangrove roots, stems, and leaves, and the authors stressed the important influence of mangrove detritus on the chemistry and biology of the bay. At the present time, autochthonous peat swamps are developing along the southwestern coast of Florida because of the paucity of sediments from rivers and streams.

In southeast Asia, where large numbers of rivers drain uplands of volcanic origin, large allochthonous swamps form in deltas, estuaries, lagoons, and along sheltered open coasts. These allochthonous swamps are the most highly developed mangals in the world (Watson 1928, Macnae 1968).

Mixed authochonous-allochthonous swamps occur along the Pacific coast of Colombia where there is low to intermediate supply of river-borne sediment (West 1956).

Several systems have been proposed for the classification of mangrove swamp soils. Aubert (1954) and Dubois (1954) classified mangrove soils in relation to hydromorphic characteristics and salinity. Bonfils and Faure (1961) related halomorphic soil types to the degree of salt-and fresh-water flooding and to the relative concentrations of chlorides and sulphates. D'Hoore (1963) typed mangrove swamp soils as "juvenile soils on marine alluvium" and called them "weakly developed soils." This general classification was accepted by Giglioli and Thornton (1965), who suggested further subdivision for agricultural purposes according to soil texture, water regime, degree of gleying and/or mottling, amount of oxidizable sulfur, and relative amounts of chlorides and sulphates.

Grant (1938), Davis (1940b), Chapman (1940), Thom (1967), Walsh (1967), and Giglioli and King (1965) discussed the evolution of mangrove swamps in relation to silting and plant succession. Davis (1940b) listed three main factors which promote soil accretion: (1)molar, (2) physicochemical, and (3) biotic.

Molar factors are mainly tides, littoral currents, and winds. The first stages of accretion consist of marine and estuarial sedimentation, resulting in formation of shoals, bars, and flats in the shallow water. At the same time, deposition of sediments and physico-chemical precipitation of dissolved substances occurs when fresh and salt water mix (Jackson 1958). This adds carbonates, phosphates, nitrates, and other substances to the developing soil.

Hesse (1961b) reported that *R. racemosa* swamps in Sierra Leone were comprised of fibrous mud, whereas *A. germinans* swamp soils were non-fibrous. Also, *Rhizophora* swamps had higher pH values, C/N ratios, and contents of oxidizable sulfur, nitrogen, phosphorus, and carbon. Giglioli and Thornton (1965) described the early phases of swamp evolution in the Gambia, West Africa, where *R. racemosa* pioneers in virgin alluvia composed of soft, silty soil. Proliferation of fibrous roots at the soil surface produces a "felt-like" layer which entraps sediment and increases the rate of deposition of both alluvium and leaf litter. Rosevear (1947) suggested that the fibrous mat formed by *R. racemosa* prevents further establishment of that species and conditions the soil for colonization by *Avicennia*, which requires a more consolidated and elevated substratum (Jordan 1964). As the soil surface becomes elevated, the rhizophoretum dies and *A. germinans* replaces *R. racemosa* (Giglioli and Thornton 1965). The many pneumatophores of *Avicennia* further accelerate deposition and the forest floor becomes even more elevated. According to Giglioli and Thornton, if the amount of drainage from higher ground is large and the swamp near the main river or a tributary, a balance occurs between erosion and drainage of the swamp and land elevation.

Zieman (in press) found that in Biscayne Bay, Florida, circular beds of *Thalassia testudinum* Koenig and Sims laid over depressions in the bedrock. The depressions were over 5 m deep and filled with mangrove peat dated to be 3,680 years old. Wharton (1883) suggested that living mangroves and their peat produce organic acids that dissolve bedrock. Zieman suggested that bedrock was dissolved under mangrove hammocks and hypothesized that as the mangrove shoreline receded and sea level rose, *Thalassia* colonized the old mangrove areas. Dodd and Siemers (1971) described a very similar situation on Bahia Honda and Big Pine Keys in the lower Florida Keys. They stated that the topography developed during the lowered sea level of the Pleistocene strongly controls Holocene sediment thickness and present biotic distribution. They said, however, that the depressions were sinkholes and thick sediment in underwater sinkholes promoted growth of *Thalassia*, whereas depressions in shallow water or in the tidal zone supported growth of *R. mangle* and *A. nitida*.

Schuster (1952) reported deposition of sediment on the mangrove forest floor at every spring tide in Java, the processes of land elevation and soil formation being accelerated by growth of beach thistles (*Acanthus* sp.) which produced large quantities of organic matter.

Soil derived by sedimentation from river water is often poor in calcium and potassium and, in mangrove swamps, tidal water is the main source of salt. In the early stages of mangal development, the clay particles absorb calcium and potassium salts from seawater and a fine-grained soil, rich in minerals, results. As evolution of the swamp continues, shelled animals invade and grow on the trees and substratum. The organic content of the soil increases and in the moist environment, decay processes are rapid and calcareous particles are dissolved (Abel 1926). Wharton (1887) reported rapid corrosion of $CaCo_3$ in mangrove swamps at Aldabra. Fairbridge and Teichert (1947) concluded that pools on reef flats at Low Isles (Marshall and Orr 1931) were caused by solution of $CaCO_3$ by mangrove swamp acids. Reville and Fairbridge (1957) suggested that the principal agent for destruction of $CaCO_3$ in mangrove swamps is carbonic acid produced by decomposition of organic matter. They also suggested that tannic acid from mangrove bark and "humic acids" aid in decomposition.

Very little is known about the factors that form and condition mangrove mud, which may lie in an unconsolidated state to a depth of 1.5 m. Schuster (1952) discussed breakdown and modification of the substratum by bacteria, fungi, actinomycetes, and myxomycetes. He mentioned the occurrence of the bacteria *Clostridium* sp. and *Azobacter* sp. and the algae *Nostoc* sp. and *Anabena* sp. in mangrove swamps and speculated that those organisms are important in nitrogen fixation.

Most of the organic debris on and within mangrove soils is authochonous. Because of the saline water, relatively high pH of surface soil water (often as high as 7.8) and anaerobic conditions at low tide, plant detritus is only partially broken down by bacteria, fungi, and algae. This causes formation of peat, which is composed mainly of plant remains.

The role of birds in composition and fertility of mangrove soil has not been investigated adequately. Birds in mangrove have been described by Cawkell (1964), Haverschmidt (1965), Ffrench (1966), Parkes and Dickerman (1967), Nisbet (1968), Field (1968), Dickerman and Gavino T. (1969), Ffrench and Haverschmidt (1970), Dickerman and Juarez L. (1971), and Ricklefs (1971). Large numbers of birds (Ffrench reported 94 species in mangrove in Trinidad), including egrets, ibis, herons, spoonbills, anhingas, pelicans, storks, ospreys, and eagles, roost in mangrove trees but feed elsewhere. In this way, nutrients are brought into the swamps and the functions of such nutrients should be investigated.

Analyses of mangrove peat have been reported by several workers. Davis (1940) gave detailed accounts of soil profiles in the swamps of southern Florida and classified them according to their general composition, i.e. homogeneous, heterogeneous, or layered. He also classified them on the basis of the type of vegetation which covered the soil and the probable types of vegetation that formerly were present and contributed most to the accumulated materials. Davis reported various types of soil profiles, some of which indicated progressive soil accretion, while others did not. Giglioli and Thornton (1965) gave soil profiles from the Gambia River basin in West Africa. The profiles indicated typical alluvial soils in the process of silting.

The composition of mangrove swamp substratum is dependent upon its source, age, position in the swamp, organisms present, and scouring by water flow in the channels and over higher ground at high tide. Walsh (1967) reported that up to 74.6% of the substratum in the center of the channels of a swamp in Hawaii was composed of shells, pebbles, and gravel with diameter greater than 3.35 mm. All of the alluvial particles, however, were less than 0.23 mm in diameter, and most sediments in mangrove swamps are of small grain size.

Scholl (1963) compared grain-size distribution of clastic sediments in two mangrove swamps in southwestern Florida, where *R. mangle* was the dominant species. In the swamp of the Ten Thousand Islands area, the sediment was composed of fine to very fine calcareous-quartzitic sand and coarse silt. The quartz content was approximately 70%, carbonate mineral 10-20%, and organic matter usually less than 10%. Isopleths of grain size showed a zone of coarser-grained sediment (approximately 0.100-0.200 mm diameter) along the shore flanking a belt of finer-grained sediment (0.062-0.125 mm) inland. Another belt of coarser sediment (0.125-0.250 mm) lay landward of the finer-grained belt. In contrast, the sediments of the Whitewater Bay area swamp were composed mainly of mollusc shells and shell fragments. Less than 15% of the sediment was quartz and "little organic detritus" was present. The grain sizes of surface sediments fell between 0.054 and 0.540 mm in diameter. Distribution of grain size was variable throughout the swamp. Scholl attributed the differences in sediment characteristics between the Ten Thousand Islands and the Whitewater Bay areas to differences in patterns and strengths of the tidal currents. Strong tidal currents which washed the former were lacking in the latter. Tables 2 and 3 give grain-size distributions in several swamps.

There is a paucity of data on physical and chemical characteristics of mangrove soil. Values for physical and chemical factors from forests dominated by different trees with different substrata overlap (Doyne 1933; Davis 1940; Bharucha and Navalkar 1942; Chapman 1944a, b, c; Navalkar and Bharucha 1948, 1949; Schuster 1952; Wyel 1953; Hesse 1961a, b; Scholl 1963; Giglioli and Thornton 1965a; Giglioli and King 1966; Clarke and Hannon 1967; Kassas and Zahran 1967; Walsh 1967; Lee and Baker 1972a). The ranges of some factors reported are: pH 5.0-9.0, Chloride 1.9-87.0 ppt, carbon 0.05-11.9%, loss on ignition 3.0-72.8%, and C/N ratio 0.4-36.0. Diel and seasonal variations occur in relation to the tide, rainfall, and rate of evaporation (Navalkar 1941; Navalkar and Bharucha 1948, 1949, 1950; Barucha and Navalkar 1942; Clarke and Hannon 1967, 1969; Giglioli and Thornton 1965a, b) and at this time, it is impossible to relate plant distributional patterns to specific physical or chemical properties of the soil. In a study of the plant communities of the Sydney District, Australia, Clarke and Hannon (1967, 1969) found that variations in the physical and chemical properties of mangrove soils were similar to those of other plant stands.

Table 2. Percentage grain-size distribution (mm diameter) in mangrove swamp surface sediments.

El Salavador (Wyel 1953)		Brazil (Friese 1937)	
>1.0 mm	0.0-5.2%	>0.2 mm	4.7-7.0%
1.0-0.5	0.0-4.2	0.2-0.06	7.2-11.8
0.5-0.4	0.0-8.0	0.06-0.03	6.7-10.1
0.4-0.3	0.0-6.0	0.03-0.006	6.6-10.4
0.3-0.2	0.0-18.0	0.006-0.003	22.1-24.5
0.2-0.1	0.0-18.0	<0.003	43.6-46.6
0.1-0.06	0.0-16.6		
0.06-0.03	22.0-43.8	India (Navalkar 1941)	
0.03-0.017	7.0-32.0	2.0	3.9-4.0%
0.017-0.007	3.0-27.0	0.2	38.2-38.6
0.007-0.003	0.5-8.0	0.02	29.5-33.1
<0.003	0.5-5.0	0.002	3.9-4.3

Java (Schuster 1952)		Florida (Scholl 1963)
2 mm	0%	Median grain-size from 10
2.0-0.1	2	stations varied between 0.006
0.1-0.05	5	and 0.700 mm.
0.05-0.01	30	
0.01	63	

Jamaica (Chapman 1944)

Coarse sand	39.9%
Fine sand	26.5
Clay	5.1
Silt	15.7

Table 3. Grain-size distribution in surface sediments of swamps dominated by *Avicennia alba* (Navalkar 1941) and *A. marina* (Clarke and Hannon 1967). Coarse sand = 2.0 mm diameter, fine sand = 0.2 mm, silt = 0.02 mm, clay = 0.002 mm.

Species	Percentage of Dry Soil			
	Coarse sand	Fine sand	Silt	Clay
A. alba	4.0	38.4	31.3	4.1
A. marina	75.2	3.8	1.7	4.8

However, *Avicennia marina* and *Arthrocnemum australasicum* occurred only where high salinity of the soil water endured for long periods of time or where there were wide variations in salinity. Giglioli and King (1966) pointed out that *A. germinans* grew in old soils of high salinity and that this high salt content was a function of time. *Avicennia* was apparently able to exclude *Rhizophora racemosa* because it was better adapted to high concentrations of salt. *Avicennia*, unlike *Rhizophora*, absorbs large quantities of salt through its roots and excretes them through the leaves (Scholander et al 1962). As shown above, the fibrous nature of the substratum also appears to be important in colonization, and it is most likely that combinations of factors, including soil salinity, regulate species distribution.

Clarke and Hannon (1969) concluded that tidal action, as modified by microtopography, was the major factor which affected soil salinity over long periods of time, and that succession was mainly allogenic rather than autogenic. This concept was shown to be true by Thom (1967), who demonstrated that although biotic and geomorphic processes are effective in short-term changes on actively accreting shores, physiographic processes of sedimentation and subsidence are more important over a long period of time. Physiographic changes influence salinity of the soil, degree of water saturation, soil type, and drainage, and therefore greatly influence the species present.

Zonation and Succession

In general, in areas of large mangals, five geographic belts can be distinguished (West 1956): (1) a belt of shore water and mudflats along the coast, (2) a series of discontinuous sand beaches, variable in size, which are interrupted by tidal inlets and mudflats, (3) a zone of mangrove forest, usually one-half to three miles wide, (4) a fresh-water swamp, and (5) equatorial rain forest. Although the beach zone is frequently absent, this zonation was described in the Malay Peninsula (Watson 1928), western Africa (Grew 1941), the Congo (Pynaert 1933), and in Guiana (Martyn 1934).

Within the mangrove belt, there is usually a seral succession of vegetation in relation to hydrological and climatic conditions. Day et al. (1953) held the salinity gradient to be of great importance to distribution in South Africa, and described correlations between rainfall, evaporation, upflow of salt water from the sea, and seral change.

There have been several attempts to classify mangrove vegetation according to physical characteristics of the environment. Watson (1928) described two general classifications: (1) mangroves that grow on accretive shores and (2) those that grow on sand. Watson also related species to the tidal cycle and described five classes: (1) inundated by all high tides, (2) inundated by medium high tides, (3) inundated by normal high tides, (4) inundated by spring tides, and (5) occasionally inundated by exceptional or equinoctial tides.

Stevenson and Tandy (1931), working at Low Isles in Australia, described the mangrove habitat as (1) dense woodland, (2) muddy glades, and (3) shingle tongues. At present, mangrove types are sometimes considered to be related to the type of soil present. Troll and Dragendorff (1931) and Walter and Steiner (1937) described mud and reef mangroves; Chapman (1944a) added the categories of sand and peat mangroves. These four types are generally recognized today.

Tansley, Watt, and Richards (1939) suggested that mangrove vegetation be considered as a formation type on a world-wide basis. They recognized two subformations: (1) the New World subformation, including western Africa, and (2) the Old World subformation. Chapman (1944a) recommended that a third subformation, the Australasian, be recognized because the species of *Avicennia* are very distinct in their distribution and segregate into these three geographical regions.

Davis (1940) said that the mangrove formation is composed of seral communities. Although reef and sand communities appear to be climax stages, the statement of Davis is generally true. The order of zonation varies considerably even in geographically related areas. For example, in the Old World subformation, *Rhizophora* is the pioneer species along river banks and in the more protected regions along oceanic shores, whereas *Avicennia* or *Sonneratia* pioneer on shores of greater wave and tidal action. In Jamaica, *R. mangle* pioneers along protected shores, while *Laguncularia racemosa* pioneers on sand spits where wave action is greater (Chapman 1944a).

Chapman (1940) pointed out that the presence of sea grass in submerged areas accelerates the seaward extention of mangrove because it raises the height of the sea bed, allowing *R. mangle* seedlings to grow.

The seral nature of mangrove vegetation in the Indo-West-Pacific region was described in detail by Macnae (1968), who recognized succession in every mangal he visited. Macnae stated that variation in development was often found, succession being complete only where the amount of available fresh water exceeded that lost through evaporation and transpiration. When losses through evaporation and transpiration exceed income from rain and rivers, the soil becomes hypersaline and zonation is interrupted.

Macnae described the effects of fresh-water imbalance in a zone of *Ceriops tagal* (Perri) C. B. Rob. This zone was located between a seaward fringe of *Bruguiera gymnorhiza* and a landward fringe of *A. marina*. Where fresh-water loss exceeded gain, *C. tagal* became stunted. With increasing excess of evaporation, the *Ceriops* bushes died, forming a bare area which expanded both landward and seaward until only a few bushes grew near the *Avicennia* and *Bruguiera* fringes.

In Florida, Davis (1940b) was the first investigator to give a detailed account of succession in a mangal. He recognized seven principal communities: (1) The pioneer *Rhizophora mangle* zone. This seaward stage was composed of mangrove seedlings of various age growing in marl soil below the level of low tide in shallow undisturbed water. *Thalassia testudinum* Koenig and Sims and *Cymadocea manatorum* Aschers grew in shoal areas near this zone, and *Spartina alterniflora* Loisel was present in some parts of Florida. (2) Mature *Rhizophora* consocies. This stage was composed of mature *R. mangle* with well-developed prop roots growing in mangrove peat. (3) *Avicennia*-salt marsh consocies. This stage was composed of the tree *A. nitida* and the salt marsh plants *Batis maritima* L., *Salicornia perennis* Mill, *Spartina alterniflora, S. spartinae* (Trin.) Merv, *Monanthochloe littoralis* Englem., and *Sporobolus virginicus* (L.) Kunth. This mangrove salt marsh consocies grew on peaty soil and accumulated large amounts of organic and inorganic detritus. During dry periods, soil salinity was very high, whereas salinity was very low during rainy periods. In some places, *Avicennia* was more than 30 cm in diameter, but in other places was a small gnarled bush. (4) Mature mangrove associes. This stage consisted of large trees of *R. mangle* and *A. nitida* growing together on peat soils in water of low salinity at approximately the mean high tide mark. (5) *Laguncularia racemosa* consocies. This stage did not occupy a specific habitat, but was found with both the mature mangrove associes or between the *Avicennia*-salt marsh associes and a *Conocarpus* associes when the natural mangrove associes was not present. (6) *Conocarpus erectus* transition associes. This was the final stage of mangrove succession in Florida, as it was bordered on its landward edge by sand dunes, upland tropical forest, or fresh-water marsh. Davis considered the *Conocarpus* associes to be an ecotone, not a definite seral community. This was disputed by Chapman (1944a) who reviewed the work of Borgeson (1909) and concluded that *Conocarpus* formed a true seral stage. Chapman later (1970) considered the *Conocarpus* community to be an ecotone between saline and fresh-water communities. West (1956) described the final stage in seral succession in Colombia as dominated by *C. erectus* in the drier and less saline areas. (7) Dwarf-form mangroves. Davis recognized a scrub-mangrove facies of dwarfed *Rhizophora, Avicennia*, and *Laguncularia*, which grew above the high tide mark in fresh water. This dwarfed form was common in the Everglades region.

ECOLOGY OF HALOPHYTES

Thorne (1954) listed many other plants in the mangals of Florida. Holdridge (1940) gave an extensive description of the vegetative characters and general characteristics of *R. mangle, Languculuria racemosa, C. erectus*, and *A. nitida*, and reported *Petrocarpus officinale* Jacq., *Anona glabra* L., *Bucida buceras* L., and *Drepanocarpus lunatus* G.F.W. Meyer living in the mangals of Puerto Rico.

Chapman (1944a) compared succession in the swamps of Jamaica with that in Florida (Figs. 9 and 10). In both cases, *Rhizophora* was the pioneer form, with *Avicennia, Laguncularia*, and *Conocarpus* inland. Asprey and Robbins (1953) stated that there were few associates with mangrove in Jamaica, a pattern similar to mangals in other parts of the world. *Batis maritima, Salicornia ambigua* Michx., *Acrostichium aureum* L., *Alternanthera ficoides*, and *S. virginicus* occurred in the swamps of Jamaica. In other parts of the world, other genera and species occupy similar positions. For example, Taylor (1959) described the mangals of Papua, New Guinea, as follows:

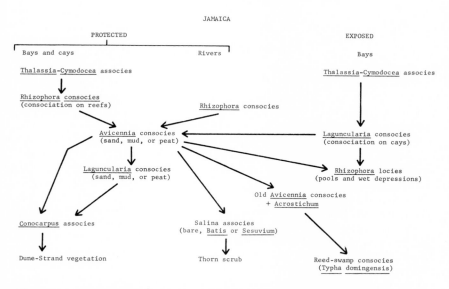

Figure 9. Succession in the mangrove swamps of Jamaica (Chapman 1944a).

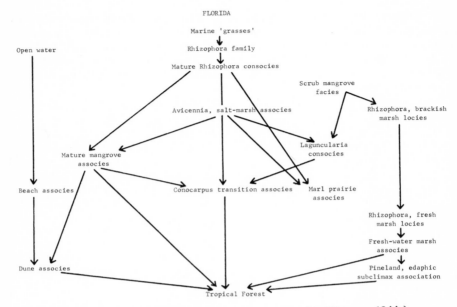

Figure 10. Succession in the mangrove swamps of Florida (Chapman 1944a).

1. Salt water swamps
   a. Tidal mangrove sequence
   b. Mangrove marsh sequence
2. Brackish water swamps
   a. Brackish swamp sequence
   b. Estuarine sequence.

In the tidal mangrove sequence, each succeeding community occurred at sites with successively longer periods of tidal inundation. The shore pioneer species was *C. tagal*, followed by a broad zone of *R. mucronata* and *B. gymnorhiza*. The final stage was dominated by *Heritiera littoralis*. The boundary between the *H. littoralis* zone and the rain forest was dominated by *Intsia bijuga* (Colebr.) O. Kunz. and was very sharp. All of the mangrove species occurred as scattered individuals in all of the communities, although the dominant species comprised over 50% of the number of species, and there were sharp transitions between the zones.

In the mangrove marsh sequence, there was a fringing area of *A. alba* trees up to 12 m high, which changed gradually to a thicket of the same species up to 6 m high. The thicket was bordered by swampy soil devoid of vegetation and Taylor suggested that the sequence from tall trees to bare swampy ground was regulated by soil salinity.

Brackish water swamps occurred where mangrove forest was bordered inland by fresh-water swamps. There were only two sharply-defined zones: (1) a zone composed predominantly of *A. alba* and *B. gymnorhiza*, with a dense ground cover of the fern *Acrostichium speciosum* Thunb.; and (2) a zone dominated by *Parinari corymbosum* (Bl.) Mig. *Hibiscus tiliaceus* L. was present in small numbers. *Acrostichium speciosum* was generally abundant in this zone, but *Acanthus ilicifolius* L. sometimes made up 50% of the ground cover.

The estuarine sequence was similar to the tidal mangrove sequence, except the palm *Nypa fruticans* Wurmb. dominated the zone which bordered the fresh-water swamp.

In West Africa (Nigeria), Jackson (1964) recognized six groups of mangrove on the dual basis of range of habitat and dispersal of seeds.

Group 1. Species restricted to the tidal areas and with specialized seed habits. This group included *R. racemosa*, the pioneer species at the water's edge and on the storm beach, and *A. nitida* dominant along the inner edges of closed lagoons. Boughey (1957) found *Rhizophora* only in open lagoons and *Avicennia* only in closed lagoons in West Africa. Bews (1912) reported *B. gymnorhiza* from lagoons in Natal. In the case of *Rhizophora* species, Keay (1953) considered *R. racemosa* to be the pioneer species, with *R. mangle* following on drier ground and *R. harrisonii* on wet ground. These were followed inland by *A. nitida, Laguncularia racemosa*, and *C. erectus*. Gledhill (1963) pointed out that the propagules of *R. racemosa* are between 30 and 65 cm long and, by virtue of their length, are suited for establishment in flooded mud. The seedlings of *R. mangle* are approximately 20 cm long and those of *R. harrisonii* 30 cm long. Gledhill felt these were adapted for establishment on less heavily silted soil under more vigorous water current conditions.

Group II. Species found normally in tidal areas and with specialized fruits but normal seeds, or with buoyant seeds. Genera in this group were *Drepanocarpus, Dalbergia, Ormocarpum,* and *Hibiscus.* The seeds of these groups germinate in the water.

Group III. Species widely distributed along water courses and in swampy areas, usually with unspecialized fruits. The genera here were *Pterocarpus, Cynometra, Lonchocarpus, Phyllanthus,* and *Phoenix*. Seeds of the first four are buoyant and are found floating in water. Seeds of *Phoenix* have never been found in the drift.

Group IV. Species not restricted to water courses in the forest areas and with no marked specializations. These species are associated with high or well-distributed rainfall. The fruits and seeds are not associated with dispersal by water. This group included *Anthocleista, Elaies, Combretum, Alchornea,* and *Paullinia.*

Group V. Cultivated plants. These are plants which occur in strand vegetation and have floating fruits which are dispersed by water. Genera include *Cocos, Terminalia*, and *Anacardium*.

Group VI. Species whose seeds and seedlings are found in the swamp but with few or no mature individuals present. These include *Lonchocarpus, Halomosia, Spondias, Cleistopholis, Dioclea*, and *Entada*.

It is clear that mangrove swamps are not the simple communities some writers thought them to be. Macnae (1966) described in detail a complicated succession in the swamps of Queensland, Australia. There, the pioneer tree was *A. marina* where there was a large amount of fresh-water inflow, or *Sonneratia alba* where the influence of saline water was strong. The *Avicennia* zone was composed of a row of mature trees, two or three trees deep, with thickets of seedlings and saplings extending out onto a beach. Often, where the influence of fresh water was strong, *Aegiceras corniculatum* Blanco occurred in large numbers in the seedling and sapling thickets. On the other hand, where *S. alba* fringed the shore, the pioneer belt was well-developed. The alga *Catenella nipa* Zanard colonized the pneumatophores of both *Avicennia* and *Sonneratia*. Macnae found no other algae there. This was exceptional as large numbers of algae are present on the pneumatophores of both genera in southern Australia and eastern Africa. The substrata of shoreline fringes were considerably firmer than either the foreshore in front or the *Rhizophora* forest behind because both *Avicennis* and *Sonneratia* have a mass of intertwining absorptive roots which lie 20 to 40 cm below the surface.

Behind the ocean fringe, there occurred an ocean shore sub-fringe composed mainly of *R. stylosa* and occasionally of *R. mucronata*. *Rhizophora* formed the fringing zone along creeks. Ding Hou (1958) and van Steenis (1962) held that *R. stylosa* was found only on sandy shores and coral terraces. Macnae stated that both *R. styloas* and *R. mucronata* grew in mud, sand, and on coral debris, and speculated that the two forms are actually variants of a single species. Whatever the taxonomic position may be, it is clear that in contrast to the New World genera the southeastern-asiatic and eastern Africa forms of *Avicennia* are better adapted for pioneering than Rhizophorous forms (see Watson 1928; Macnae 1963, 1968).

In contrast to the pioneer fringe, the substratum in the well-developed *Rhizophora* forest was always very soft and muddy due to entrapment of sediments between the prop roots.

Landward of the *Rhizophora* forest lay broad areas of either (1) thickets dominated by *C. tagal*, where the amount of rainfall was intermediate, or (2) forests dominated by *Bruguiera*, where the amount of rainfall was large. In the thickets, *C. tagal* was ordinarily the only species present, but *Bruguiera exaristata* Ding Hou was sometimes subdominant. Occasionally, *A. marina, B. gymnorhiza, R. apiculata, R. stylosa, Xylocarpus granatum* Konnig, and *X.*

australasicum Ridley were present. In areas of much rainfall, the *Ceriops* thickets were narrow and bordered by very dense forests dominated by either *B. parviflora* or *B. gymnorhiza*. In *B. gymnorhiza* forests, scattered specimens of *X. australasicum* occurred and the fern *A. speciosum* grew between the trees. The *Bruguiera* forests were the tallest of the Australian mangrove. Height of the trees appeared to depend upon the amount of fresh water available, with tallest trees in areas of highest rainfall.

The landward fringe of the Queensland mangals was the most diverse of all seral stages. *Avicennia marina* was the most abundant tree, but *B. exaristata* Ding Hou was common, and *C. tagal, C. decandra, Lumnitizera agallocha, L. littorea, R. apiculata,* and *Exocoecaria agallocha* L. were present. *Xylocarpus granatum* and *X. australasicum* were present occassionally. Where the landward fringe bordered a rain forest, many of the forest epiphytes grew on the mangrove trees. These included the orchids *Dischidia nummularia* R. Br. and *Dendrobium* sp., the ant plant *Myrmecodia (beccari* Hook.?), and the ferns *Drynaria rigidula* Bedd., *Platycerium* sp., *Polypodium acrostichoides* Forst., and *P. quercifolium* L.

The above description by Macnae of seral succession in a mangal is, in general, typical of the large swamps of the Indo-Pacific region. This author (1966, 1968) described characteristics of large mangals in detail, and concluded that zonation of mangrove trees was due to the interaction of (1) frequency of tidal flooding, (2) salinity of the soil water, and (3) water logging of the soil. All three are modified by the presence of creeks, gullies, channels, and rivers. The second and third depend upon rainfall and/or the supply of fresh water, evaporation, transpiration, and the nature and quality of the soil. Chapman and Trevarthen (1953) stated that on muddy or sandy shores, distribution of organisms is related to the nature of the substratum which controls drainage, aeration, and penetrability.

The possible role of tidal flooding in relation to succession of Jamaican mangroves was shown by Chapman (1944b), who related vegetation types to the number of tidal submergences per year. His data showed decreasing tidal influence between the seaward *R. mangle* stand and the landward *C. erectus* stand (Table 4).

Tidal flooding alone does not determine species composition, zonation, or succession in mangals. Clarke and Hannon (1967, 1969, 1970, 1971) studied the physical habitat of mangroves in Australia in great detail. They concluded that soil did not play a major role in control of plant distributional patterns and that plant reaction on the soil did not regulate seral change. Microclimate was important in providing conditions necessary for seedling development, determining soil characteristics, and influencing competition between species. Clark and Hannon showed that the holocoenotic complex (Fig. 11) was intricate and that variation in any of the components affected species distribution. The main factors were degree of tidal flooding, elevation of the land, and salinity of the soil water. Plant zonation was associated closely with elevation above mean sea level, seasonal patterns of soil salinity, and small

Table 4. Number of tidal inundations in a mangrove swamp in Jamaica (Chapman, 1944b).

Vegetation	Inundations per year
Rhizophora swamp	700+
Rhizophora/Avicennia boundary	524
Avicennia swamp	432
Avicennia/Laguncularia boundary	213
Center of salina	150
Laguncularia/Conocarpus boundary	4

differences in microtopography. Also, light and water-logging of the soil were important to distribution, and a comparison of environmental requirements of coastal halophytes was made (Fig. 12). The authors stated that the sharpness of zonation depended upon the intensity of species interaction at ecotones. Slight environmental changes related to topography produced intense competition which made significant factors that were normally of secondary importance. Generally, the severity of the environment, including covering by mud and tidal scouring, determined the success of a species in advancing seaward, whereas landward extension was governed by ability to compete with other species in relation to salinity, availability of fresh water, temperature, light, and humidity.

Macnae (1966) criticized the Watson (1928) scheme of classification based on frequency of tidal flooding (described above) because it applied only to ever-wet forests in Malaya. Instead, Macnae proposed that zonation be based on the dominant tree. Dansereau (1947), in a phytosociological study of mangrove in Brazil, described three natural associations: *Rhizophoretum manglei*, *Avicennietum tomentosae*, and *Lagunularietum racemosa*. Cuatresasas (1958) described the mangrove associations of South America as *Rhizophoretum brevistylae*, *Rhizophoretum mangleae*, *Brugieretum gymnorrhizae*, *Sonneratietum albae*, and *Avicennietum nitidae*. Schnell (1952) described five "edaphic" associations in West Africa: *Rhizophoretum racemosae*, *Avicennietum nitidae*, *Drepanocarpeta-Rhizophoretum*, *Ecastophylletum (Dalbergiaetum) brownei*, and *Cyperteum articulati*. Chapman (1970), in a very important paper on mangrove phytosociology, compared succession in mangals throughout the tropics and gave eight schemata that depicted zonation (Figs. 13-21). Chapman concluded that there is great similarity in the vegetational communities and suggested an extensive classification of natural associations according to the Braun-Blanquet system. The classification consisted of 8 alliances, 15 orders, and 40 associations, but must be considered tenuous at this time because of lack of taxonomic and systematic data from many localities.

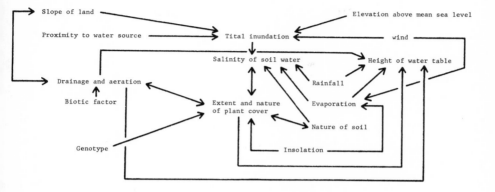

Figure 11. The holocoenotic complex in mangrove swamps and salt marshes of the Sydney District, Australia (Clarke and Hannon 1969).

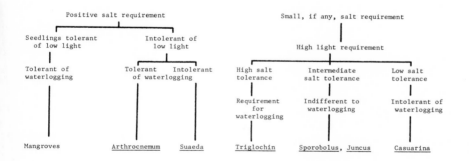

Figure 12. Environmental requirements of coastal halophytes of the Sydney District, Australia (Clarke and Hannon 1971).

Figure 13. Succession in the mangrove swamps of the Gulf of Mexico and the Caribbean Sea (Chapman 1970, Schema 1).

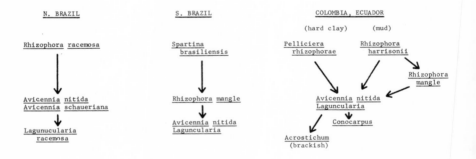

Figure 14. Succession in the mangrove swamps of South America (Chapman 1970, Schema 2).

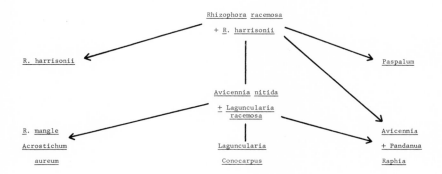

Figure 15. Succession in the mangrove swamps of West Africa (Chapman 1970, Schema 3).

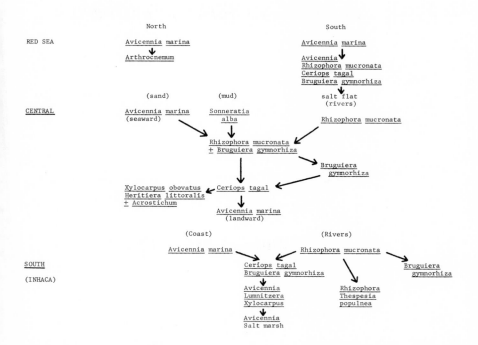

Figure 16. Succession in the mangrove swamps of East Africa (Chapman 1970, Schema 4).

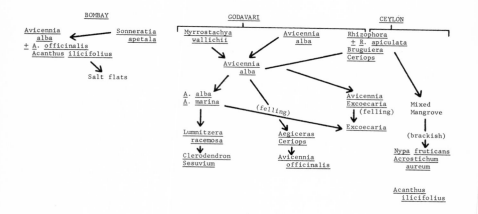

Figure 17. Succession in the mangrove swamps of India (Chapman 1970, Schema 5).

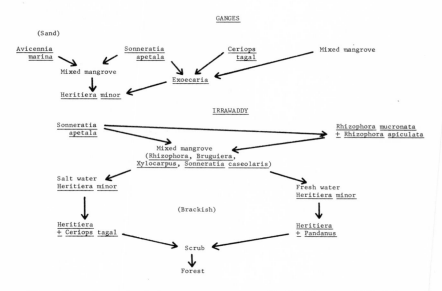

Figure 18. Succession in the mangrove swamps of India (Chapman 1970, Schema 6).

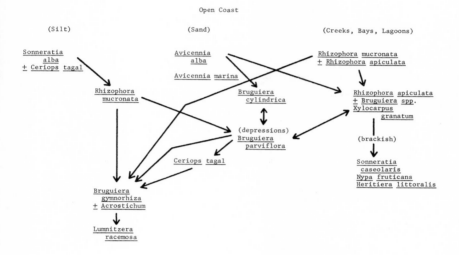

Figure 19. Succession in the mangrove swamps of Malaysia, Indonesia, and Borneo (Chapman 1970, Schema 7).

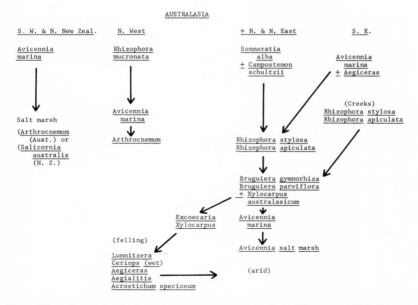

Figure 20. Succession in the mangrove swamps of Australia (Chapman 1970, Schema 8).

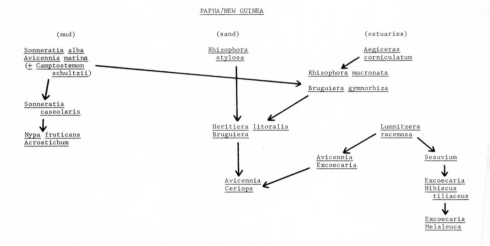

Figure 21. Succession in the mangrove swamps of Papua, New Guinea, the Philippines, and Oceania (Chapman 1970, Schema 9, part 1).

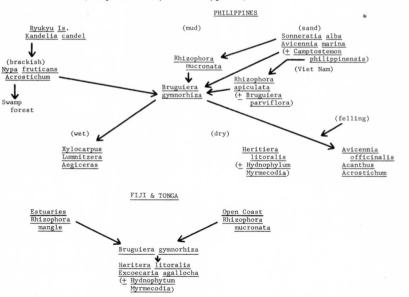

Figure 21. (Continued). Succession in the mangrove swamps of Papua, New Guinea, the Philippines, and Oceania (Chapman 1970, Schema 9, part 2).

ADAPTATIONS

Warming (1883) stated that mangroves have adapted to their environment through (1) mechanical fixation in loose soil, (2) respiratory roots and aerating devices, (3) viviparity, (4) specialized means of dispersal, and (5) development of xerophytic structures in relation to soil salinity. Walter (1931a, b; 1936a, b) and Walter and Steiner (1934) concluded from studies in East Africa that zonation was related to the capacity of mangroves to compete and survive in saline soils. Thus, they distinguished zones of *Rhizophora*, *Avicennia*, and *Sonneratia* and stated that *Rhizophora* and *Avicennia* bore great fluctuations in soil salinity, whereas *Sonneratia* required a constant chloride content. It is clear, however, that zonation depends also upon morphological and physiological adaptations. Wenzel (1925) gave detailed descriptions of the anatomy of *R. mucronata*, *R. mangle*, *B. gymnorhiza*, *C. candolleana*, *A. officinalis*, and *X. granatum*. Chapman (1944c) described functional morphology of *A. nitida* in detail and presented data on physiology of the pneumatophores. The gross morphology of most species is described in many manuals of tropical trees.

Anatomical

Marco (1935) described the anatomy of the woods of rhizophoraceous species from both mangals and upland forests of the Indo-West-Pacific region. He divided the family into three groups and stated that the mangrove genera formed a well-defined, natural aggregation that was readily separable from all other members of the family. He suggested that *Rhizophora*, *Bruguiera*, *Ceriops*, and *Kandelia* be placed in an independent family. Marco placed the four genera in the anatomical division Rhizophoreae, characterized by (1) heavily barred, exclusively scalariform perforation plates, (2) characteristic scalariform intervascular pitting, (3) little vasicentric parenchyma, (4) numerous fine-celled multiseriate rays and very few uniseriate rays, (5) libriform fibers with inconspicuous pits, and unilaterally and bilaterally compound pitting between rays and vessels. These features segregated the four genera from all other groups of the Rhizophoraceae, but their significance as adaptive features has not been determined.

Reinders-Rouwentak (1953) stated that, in the Sonneratiaceae, the mature wood of species from more saline environments contained a larger number of smaller vessels than species from less saline areas. For example, *S. griffithii* from the seashore of Bengal had 34-50 vessels/mm^2 and the diameter range was 85-100 u. *Sonneratia apetala* from the river had 18-32 vessels/mm^2 with diameter range of 135-150 u. Heiden (1893) gave a detailed account of anatomy of the Combretaceae, including the genera *Conocarpus*, *Lumnitzera*, and *Laguncularia*.

Macnae (1968) reviewed adaptations of mangroves with regard to growth in ill-consolidated mud, specializations of stems and leaves, relationships between root and shoot systems, and vivipary. Robyns (1971) considered mangroves to be the only truly viviparous plants. He defined vivipary as the process in which the seed remains attached in the fruit to the mother tree, germinates into a protruding embryo with a long hypocotyl, and finally falls from the tree. Genkel' (1962) speculated that mangroves evolved in the ancient tropical forest from xerotic plants in which the seeds had no dormant period and loss of fruit from the tree was delayed by chloride in the soil water.

The roots of mangroves are not deep and tap roots are not present. For descriptions of mangrove root systems see Goebel (1886), Schimper (1891), Troll (1930), Troll and Dragendorff (1931), Uphoff (1941), and Macnae (1968). Zieman (in press) found that height of *R. mangle* was related to root length. In *Rhizophora*, the primary roots of the hypocotyl function for only a short period of time and root functions are assumed by secondary roots which extend from the main trunk. The cause of cessation of growth of primary roots is not known, but Warming (1877), Johow (1884), and Schimper (1891) suggested that they are injured mechanically by crabs and snails. There are two kinds of roots in *Rhizophora:* (1) aerial roots that arise from the main trunk and form arched stilts which penetrate the ground (prop roots) and (2) subterranean roots that arise from the prop roots. Aerial roots also arise adventitiously from the lower branches of trees. The prop and aerial roots function in aeration and ventilation of the tree in general and of the subterranean roots in particular. Most mangroves have schizogenous lacunae in the cortex of the roots. The main function of the subterranean roots is absorption of water and nutrients.

The anatomy of aerial roots has been described by Warming (1883), Schenck (1889), Karsten (1891), Leibau (1914), Bowman (1917), Mullan (1931, 1932, 1933), and Gill and Tomlinson (1971). According to Gill and Tomlinson, aerial roots first appear on the hypocotyl or lower internodes of seedlings after 1 to 3 first-order branches have been produced. Later, they arise on higher internodes and lower branches. Aerial roots also develop on the high branches of mature trees. In general, the aerial roots originate on the shoot in acropetal sequence. Gill and Tomlinson (1971) gave a detailed account of root growth and anatomy.

When aerial roots reach and penetrate the ground, they undergo marked changes which relate to subterranean function. According to Bowman (1917), the absorptive subterranean roots are thick, spongy, and gas-filled due to great development of the primary cortex. The primary cortex of absorptive roots is composed of large cells and very large intercellular spaces in which idioblasts, trichoblasts, and root hairs are lacking (Bowman 1921). The periderm of the absorptive root consists only of cork cells, whereas that of the aerial root consists of both cork and "parenchymatic" tissue (Bowman 1921). Bowman also reported stone cells and idioblasts in all parts of *R.*

mangle except the flower. These were frequently associated with tannin cells. Sclerenchymatous tissue occupies a large portion of the stem and hypocotyl of mangrove and makes anatomical study very difficult.

Two other rhizophoraceous genera, *Bruguiera* and *Ceriops*, do not have aerial roots. Instead, they have subterranean cable roots which differentiate into knee roots that penetrate the soil surface, and absorptive roots (Marco 1935).

Troll and Dragendorff (1931) gave an extensive account of the cable root system of *Sonneratia*, and similar roots systems are present in some species of *Avicennia, Lumnitizera racemosa, X. australasicum,* and *X. moluccensis* Roem. For an extensive study of anatomy of respiratory roots of mangroves, see Ernould (1921).

Chapman (1944c) showed that the composition of gas in the roots of *A. nitida* was similar to air and that there was no fundamental difference between composition of gas in the pneumatophores and in the horizontal roots. He stated that the large cortical air spaces allowed longitudinal gas flow between organs. Scholander et al. (1955) studied respiratory gas exchange in the roots of *A. nitida* and *R. mangle*. The radial roots of *A. nitida* send numerous pneumatophores up to 30 cm above the ground. There is a direct gas connection between the radial roots and the pneumatophores. When the tide covered the pneumatophores, there was a decrease in the oxygen content of the whole root system. At low tide, oxygen comprised between 15 and 18% of the gas content. At high tide, oxygen content was about 7%. At high tide, the oxygen content dropped until the pneumatophores were again exposed to air at low tide. There was little change in carbon dioxide content of the roots over the tidal cycle.

Gas in the subterranean roots oof *R. mangle* contained 15 to 18% oxygen and there was always a direct gas connection between these roots and lenticels on the prop roots. The high oxygen tensions in the roots were maintained by means of ventilation through the lenticels on the prop roots (Scholander 1955).

Macnae (1968) gave diagrams of the cross sections of leaves of *Rhizophora, Avicennia,* and *Sonneratia.* Schimper (1891, 1898) showed that the leaves of most mangroves contain water storage tissue. This is initially in the form of a hypodermis in *Rhizophora* and *Avicennia* and a centrally located layer of cells in *Sonneratia.* Stace (1966) made a detailed study of leaf anatomy of seven genera (Tables 5 and 6) and also the epidermal characteristics of *Bruguiera* spp. and *Avicennia* spp. He concluded that the leaf and epidermal characteristics of mangroves are similar to most xeromorphs. All species had common epidermal features, notably a thick cuticular membrane, straight epidermal cell walls, and the presence of water-storage tissue, hydathodes, cork warts, and water stomata. See Artz (1936) for descriptions of the cuticula of *S. alba, C. candolleana, R. mucronata, B. gymnorhiza, Lumnitizera racemosa, X. obovatus,* and *A. officinalis.* In the study of Stace, almost all of the species studied had sunken stomata or stomata surrounded

Table 5. Characteristics of the leaves of rhizophoraceous mangroves (Stace 1966).

	Rhizophora	Ceriops	Bruguiera and Kandelia
Venous system on upper epidermis	Midrib only, very broad and conspicuous; cells broader than long	Midrib only, narrow and inconspicuous; cells broader than long	Midrib only, narrow but conspicuous; cells broader than long
Venous system on lower epidermis	Midrib only, very broad and conspicuous, or lateral veins also discernible	Midrib only, very broad and conspicuous	Midrib only, very broad and conspicuous
Epidermal cells of non-venuous areas	Straight- or curved-walled, not second divided, mostly ca. 11-25 µ across	Straight- or curved-walled, not second divided, mostly ca. 15-35 µ across	Mostly straight- or curved-walled, not second divided, mostly ca. 15-40 µ across
Stomata	Sunken, ca. 30-55 x 20-35 µ; outer stomatal ledge conspicuous, single or with minute second lip	Sunken, ca. 36-46 x 22-34 µ; outer stomatal ledge conspicuous, conspicuously two-lipped	Sunken, ca. 30-44 x 16-28 µ; outer stomatal ledge conspicuous in some spp. conspicuously two-lipped
Subsidary cells	5-8, cyclocytic	6-8, cyclocytic	4-6(8), cyclocytic

Con't on next page

Table 5 con't

Water stomata, hydathodes and cork-warts	Large conspicuous cork-warts on lower epidermis, sometimes also on upper epidermis; water-like structures on both epidermides	Cork-warts ± absent; frequent water-stomata-like structures on both epidermides	All apparently absent
Hypodermis, including extra epidermal layers	Upper three-to five-layered, sparsely chloroplasted; lower usually absent	Upper two-layered, sparsely chloroplasted; lower usually absent	Upper and lower one-layered, densely chloroplasted
Mesophyll	One to three layers of palisade and ca. eight to ten layers of spongy below upper hypodermis	Usually one layer of palisade and ca. eight to ten layers of spongy below upper hypodermis	Usually one layer of palisade and ca. eight to ten layers of spongy below upper hypodermis
Water-storage tissue	Upper hypodermis ?	Upper hypodermis ?	Absent, or ? sometimes in spongy mesophyll

Table 6. Characteristics of leaves of combretaceous mangroves (Stace 1966).

	Lumnitzera	Laguncularia	Conocarpus
Venous system on upper epidermis	Absent, or midrib only very inconspicuous; cells longer than broad	Midrib only, broad and conspicuous to very inconspicuous; cells longer than broad	At least midrib and lateral veins distinct; cells longer than broad
Venous system on lower epidermis	Midrib only, broad and conspicuous	Midrib only, broad and conspicuous	Midrib, lateral, secondary and lesser veins distinct
Epidermal cells of non-venous areas	Straight-walled, not second divided, mostly ca. 25-40 µ across	Straight- or slightly curved-walled, many second divided, mostly ca. 15-30 µ across	Mostly curved or straight-walled, not second divided, mostly ca. 15-35 µ across
Margin	Of several regular rows of rectangular cells with angular lumina	Of small cells with rounded lumina not arranged in rows	Of several regular rows of rectangular cells with angular lumina
Stomata	Sunken or not, not protected by hairs, always more frequent on upper epidermis, absent only from margins, randomly	Scarcely sunken, not protected by hairs, usually more frequent on upper epidermis, absent only from margins, orien-	Not sunken, protected by dense hairs or not, usually slightly more frequent on lower epidermis,

Con't on next page

Table 6 con't

	oriented, ca. 24-32 x 19.5-24.5 µ; outer stomatal ledge conspicuous, single	tated at right angles to midrib on upper epidermis, ca. 25-35 x 20-26 µ; outer stomatal ledge conspicuous, single	absent from margin and lower epidermal midrib, randomly orientated, ca. 25-30 x 17-25 µ; outer stomatal ledge fairly conspicuous, single
Subsidiary cells	(3)4-5(6), cyclocytic	(3-)4(-5), cyclocytic	3-6, not differentiated
Trichomes	Compartmented hairs only, often extremely sparse to absent on both epidermides	Compartmented hairs and apparently sessile deeply sunken glands on both epidermides	Compartmented hairs and stalked superficial glands on both epidermides
Water stomata, hydathodes and cork-warts	Large water-stomata present on both epid.; hydathode-like areas present, mostly on midrib of lower epidermis and on margin	As in *Lumnitzera* but very sparse	Usually apparently absent, rarely a few water-stomata present
Domatia	Shallow pits along margins may be rudimentary domatia	Absent	Large, primary-axillary lebetiform domatia on lower epidermis

Con't next page

Table 6 con't

Mesophyll	Two layers of palisade below each epidermis; spongy absent	Two layers of palisade below upper epidermis; one to two layers of spongy, palisade or mixed above lower epidermis	One or two layers of palisade below upper epidermis, one layer of palisade above lower epidermis; spongy absent
Water-storage tissue	ca. six to twelve layers of centrally placed ± isodiametric cells, not chloroplasted	ca. six to twelve layers of centrally placed ± isodiametric cells, very sparsely chloroplasted	ca. four to six layers of centrally placed ± vertically elongated cells, sparsely chloroplasted

by dense trichomes. All genera, except *Avicennia* and *Conocarpus*, lacked lateral and lesser epidermal veins, a condition associated with development of water-storage tissue. Stack also gave a key to the genera based on epidermal characters of the leaves.

Bowman (1921) observed that the water-storing hypodermis and tannin-containing cells of *R. mangle* were much larger in trees that grew in seawater than in trees from brackish water. Possible reasons for this will be discussed under "Physiological" in this report.

Sidhu (1962, 1968) reported chromosome numbers of mangrove species from India (Table 7). He concluded that most species from the mangrove habitat possess higher chromosome numbers than other species of the same genera from mesic habitats. However, among the mangroves, the size and number of chromosomes did not show any correlation to habitat conditions.

Physiological

Most research on physiology of mangroves has stressed halophytic adaptations. Halophytes are ordinarily distinguished from other plants by their ability to grow in high concentrations of salt. They complete their entire life cycles and compete successfully with other plants in saline environments. Genkel' and Shakone (1946) classified halophytes as (1) Euhalophytes (salt accumulating), (2) Crynohalophytes (salt excreting), (3) Glycohalophytes

Table 7. Chromosome numbers (n) of mangroves and related species from India (Sidhu 1962, 1968).

Family and Species	Chromosome Number	Family and Species	Chromosome Number
Rhizophoraceae		Chenopodiaceae	
R. mucronata	18	Suaeda nudiflora	18
R. conjugata	18	S. monoica	9
B. parviflora	18	S. maritima	9
B. gymnorhiza	18	Euphorbiaceae	
C. candolleana	18	Exocoearia agallocha	65
C. roxburghiana	18	Palmae	
Sonneratiaceae		Nypa fruiticans	8
S. apetala	12	Sterculiaceae	
Duabanga sonneratioides	12	Heritiera littoralis	19
Myrsinaceae		Meliaceae	
Aegiceras corniculatum	24	Xylocarpus moluccensis	21
Salvadoraceae		Xylocarpus granatum	21
Salvadora persica	13	Papilionaceae	
Acanthaceae		Derris uliginosa	10
Acanthus ilicifolius	24	Rubiaceae	
Verbenaceae		Ixora parviflora	11
Avicennia alba	16		

(salt impermeable), and (4) those in which salt is localized in special structures. Depending on the species, mangroves may be placed in (1), (2), or (4). It is doubtful that mangroves are intolerant or obligate halophytes, although Stern and Voight (1959) and Connor (1969) have shown that *R. mangle* and *A. marina* grow best when salt is present in the soil water. Some mangroves (e.g. *R. mangle, C. tagal, N. fruiticans*) adapt to glycophytic conditions and may be considered to be facultative halophytes. Mangroves species have been reared in fresh water in the laboratory, and Stocker (1924, 1925) proposed the term "miohalophytes" for such plants.

Barbour (1970) suggested that ability to reproduce, rather than short-term growth, should be the ultimate criterion of salt tolerance, but this has not been studied with regard to mangroves. In the field, Bowman (1917), Davis (1940), and Stern and Voight (1959) in Florida, and Pannier (1959) in Venezuela reported that *R. mangle* grew and reproduced in fresh water, but height of trees and area covered were greatest in brackish water.

In the laboratory, Winkler (1931) reported that *Bruguiera eriopetala* and *R. mangle* grew and flowered in pots of sand watered only with fresh water. Davis (1940) grew *R. mangle* in fresh water in the laboratory. Pannier (1959) grew the same species in rain water and in salinities up to full strength seawater. Although seedlings grew in the rain water, root growth was optimal at 50% seawater and shoot growth was optimal in 25% seawater. Stern and Voight (1959) reported that height, dry weight, and survival of *R. mangal* increased with increasing salinity in the laboratory. They used artificial seawater, and plant dry weight was approximately three times as great in the highest salinity than in the lowest. Maximum growth occurred at salt concentrations equivalent to seawater. Patil (1964) grew *R. mucronata, K. candel, B. parviflora, C. tagal, A. ilicifolius, X. moluccensis, E. agallocha*, and *Heritiera fomes* in salt concentrations between 0.3 and 1.2%. All species grew at all salinities, but growth was best at 1.2%. Clarke and Hannon (1970) found that *A. marina* at the 0-2 leaf stage maintained optimal growth in 20% seawater. Concentrations above 60% seawater retarded growth. Seedlings at the 2.-4 leaf stage were more tolerant, and optimal growth occurred at 40% seawater. Connor (1969) reported that the optimal concentration for laboratory growth of *A. marina* from Australia was approximately 1.5%, or half the concentration of seawater. Connor reported that suppression of height by higher salt concentrations was more marked than suppression of dry weight production. An important aspect of Connor's work was that while growth appeared normal when sodium chloride was the main component of the salt mixture, potassium chloride and calcium choloride suppressed growth. Connor suggested that high concentrations of calcium caused nutrient imbalance leading to iron deficiency and speculated that the responses of mangroves to specific ions reflected the physiological ability of the plants to adapt to concentrations in the root environment.

Temperature as a factor in seedling establishment of *A. germinans* was shown by McMillan (1971). Exposure to temperatures of 39-40°C for 48 hours was lethal to stemless seedlings, but not to seedlings with stems and roots.

Bharucha and Navalkar (1942) reported the chloride content of leaf cell sap of *A. alba* in relation to that of seawater and soisalinity (Table 9). They concluded that seasonal variations in the chloride content of leaf cell sap were dependent directly upon climatic conditions of temperature, rainfall, and humidity. It will be shown later that such high sap concentrations are common in mangroves that possess glands for salt excretion.

Table 9. Chloride content of seawater, soil water, and leaf cell sap of *A. alba* (Bharucha and Navalkar 1942).

	Percent Chloride
Seawater	0.77-3.24
Soil Water	0.55-3.47
Leaf cell sap	1.59-5.05

Blum (1941) reported osmotic pressures in the leaves of several mangrove species from Java (Table 10). Avicennia had the highest osmotic pressure, whereas *Rhizophora, Bruguiera,* and *Sonneratia,* genera which possess mechanisms for salt exclusions and/or dilution, had relatively low osmotic pressures.

Bole and Bharucha (1954) reported data on osmotic relationships in leaves of *A. alba* (Table 11) and concluded that higher rates of transpiration brought about higher accumulation of osmotically active substances in the older leaves. The osmotic pressure did not vary directly with the water content, but older leaves always had higher water contents and higher osmotic pressures than younger leaves.

Chapman (1968) stressed that research on saline vegetation must emphasize the roles of different ions upon plant metabolism. He stated that the interrelations of sodium and potassium are particularly important because the amount of potassium absorbed is influenced greatly by the presence of sodium. There is evidence that temperature and light affect the responses of halophytes to salinity (Tsopa 1939), but tolerance of plants depends mainly upon the type of soil salinity (Cl^-, $SO_4^=$, etc.), the species or variety of plant, and the stage of plant development (Chapman 1966).

Table 10. Osmotic pressures in the leaves and soil of mangroves from Java (Blum 1941).

Species	Osmotic Pressure, Atmos.		
	Leaf	Soil	Diff.
A. officinalis, High tide	45	23	22
A. officinalis, Low tide	40	18	22
Sonneratia acida, High tide	27	6	21
Sonneratia acida, Freshwater	20	0.3	20
Rhizophora conjugata, Seawater	31	23	8
Bruguiera gymnorhiza, Freshwater	25	0.3	25

Table 11. Osmotic relationships in young and old leaves of *A. alba* (Bole and Bharucha 1954).

	Leaf	
	Young	Old
Osmotic pressure, atmos.	38.8-47.7	51.5-57.4
Water content, percent	69.5-72.9	57.5-62.7
Total carbon, percent	37.1-40.8	37.9-42.2
Total nitrogen, percent	1.1- 1.4	1.0- 1.6
Water loss/m^2/hr, grams	0.56-0.74	0.92-1.00

Jennings (1968) demonstrated positive correlations between the sodium and water contents and the phosphorus and water contents of halophytes. Potassium had no appreciable relationship to succulence. Jennings stated that increased succulence produced by high light intensity, aridity, and sodium ions was brought about by essentially the same mechanism. Jennings also suggested three mechanisms used by halophytes to cope with toxic concentrations of ions. The first was export of the ions from the shoots and leaves. This could occur in either of two ways: (a) transportation of ions through the phloem to the roots and extrusion back to the soil or (b) extrusion through specialized glands in the leaves. The former has not been reported for mangrove, but salt excretion through epidermal glands occurrs in *A. alba* (Walter and Steiner 1936), *A. nitida* (Biebl and Kinzel 1965), *Aegiceras* (Areschog 1902a, b; Schmidt 1940a), and *Aegiatilis* (Ruhland 1915), Mullan (1931a, b, c) reported salt-excreting glands on the petioles and upper and lower epidermis of *A. alba, A. ilicifolius*, and *A. corniculatum*. The glands were most numerous on plants from hypersaline areas. They were absent from *A. ilicifolius* from fresh water.

The xylem sap of salt-excreting species is composed of approximately 0.2-0.5% sodium chloride. This concentration exceeds that of non-secreting mangrove species by 10 times, and that of ordinary land plants by about 100 times (Scholander et. al. 1962)

Scholander et. al. (1962) also reported salt excretion by *A marina, A. corniculatum,* and *Aegiatilis annulata*. In *Aegitalis*, secretions from salt glands contained between 1.8 and 4.9% sodium chloride, with highest values during the day. Diel variation in rate of secretion occurred in *Aegiceras*, the average sodium chloride content of salt gland secretions being 2.9% during the day and 0.9% at night. The sodium chloride content of the xylem sap of *Avicennia* was very high, ranging from 4 to 8 mg/ml; one sample of gland secretion contained 4.1% sodium chloride. Jennings (1968) reported that the secretion process selects sodium over potassium, the Na/K ratio being 13 in the exudate, but only 3 in the leaves of *Aegialitis*. Atkinson et. al. (1967) reported that in *Aegiatilis,* input of chloride to a mature leaf was approximately 100 u-equiv./day and this was balanced by secretion, mainly of sodium chloride, from the salt glands. Secretions from the salt glands contained 450 u-equiv/ml of chloride, 355 u-equiv/ml of sodium, and 27 u-equiv/ml of potassium. Rate of secretion varied between 93 p-equiv/cm^2/sec during the day and 3 p-equiv/cm^2/sec in darkness. Atkinson et. al. suggested that because the water potential of the secretion is similar to that of the leaf, the secretory process involves active transport of salt and movement of water by local osmosis. Atkinson also presented light and electron microscope studies of the salt glands.

Rains and Epstein (1967) studied preferential absorption of potassium by leaves of *A. marina* in the presence of high concentrations of sodium chloride. They demonstrated that *A. marina* could (1) absorb and concentrate potassium within its tissues in excess of the concentration in the substratum and (2) preferentially select potassium when in the presence of high concentrations of sodium, a closely-related ion. The ability to select potassium over sodium is an extremely important adaptive character in the marine environment.

Another significant adaptation is the ability to tolerate, without injury, high internal concentrations of salt. *Avicennia marina*, unlike other mangroves such as *Rhizophora, Laguncularia,* and *Sonneratia*, absorbs salt in substantial amounts. The concentrations of ions in leaves examined by Rains and Epstein (1967) were 30 mM potassium, 210 mM sodium, and 245 mM chloride. The authors concluded that the effect of preferential absorption of potassium was not to exclude sodium, but rather to raise the concentration ratio of potassium to sodium from the value in seawater (1/40) to 1/7 within the tissue. The tissue did contain a high concentration of sodium chloride (1.8 mM/g dry weight), but excretion by salt glands prevented higher and possibly deleterious concentrations from developing.

Genkel' (1962) suggested that viviparous species of mangroves utilize seedlings for exclusion of salt. He found that the chloride content of seedlings increased in proportion to size and were adapted to high salt content in the soil before dropping from the tree. This was shown to be true for *R. mangle* by Lotschert (1968). Excess chloride in soils delayed loss of seedlings from the tree and Genkel' concluded that vivipary is an adaptation to the salt regime in tidal areas.

The second mechanism suggested by Jennings (1968) was limitation of transfer of ions to the shoot by some mechanism located in the roots. Scholander et. al (1966) showed that *R. mucronata, Laguncularia racemosa,* and *S. alba* are efficient in salt exclusion. Atkinson et. al (1967) showed the same for *R. mucronata.* Scholander et. al. (1966) stated that the desalinization process in the root system produces a sap of fairly constant concentration that is independent of rate of transpiration. Salt glands, when present, eliminate salts left behind by transpiration.

Concentration of the soil solution, rainfall, tide, humidity, temperature, transpiration, nature of the organisms, leaf age, water content, nitrogen content, and carbon content have effects upon osmotic relationships of mangroves (Blatter 1909; Cooper and Pasha 1935; Navalkar 1940, 1942, 1948; Bharucha and Navalkar 1942; Bole and Bharucha 1954). Gessner (1967), however, found that water which passed from the stems to the leaves of *R. mangle* was nearly salt-free.

Scholander et. al. (1965) reported that halophytes such as *Rhizophora, Osbornia, Salicornia,* and *Batis* have strong negative sap pressures, ranging from -35 to -60 atmospheres, whereas the osmotic potential of seawater is approximately -25 atmospheres. The activity of water in the marine environment is always higher than that of water in the roots, xylem sap, and leaves. In *R. mangle, Laguncularia racemosa,* and *C. erectus,* the xylem sap content of sodium chloride was only 1.2-1.5 mg/ml. At night, when transpiration by *Rhizophora* and *Osbornia* was nil, sap tension was the same as the osmotic potential of seawater but the solute pressure in the leaves was 10 to 20 atmospheres higher than seawater (Scholander 1971). In *Rhizophora, Laguncularia,* and *Conocarpus* only about 50-70% of the freezing point depression in leaf cells was produced by sodium and chloride ions, most of the remaining solutes being organic (Scholander et. al 1966). Benecke and Arnold (1931) demonstrated that osmotic pressure of *S. alba* was lower under glycophytic than halophytic conditions. These pressure differences give mangrove such as *Rhizophora, Laguncularia, Sonneratia,* and *Conocarpus,* which do not possess salt-excreting organs, the ability to obtain fresh water osmotically from seawater by transpiration and by diffusion at the roots. Scholander (1968) concluded that *Rhizophora, Sonneratia, Avicennia, Osbornia, Bruguiera, Ceriops, Exocoecaria, Acrostichum, Aegiceras,* and *Aegialitis* separate fresh water from the sea by simple nonmetabolic ultrafiltration of the seawater combined with ion transport. The negative xylem pressure is produced by high salt concentration in the cells, resulting in a solute pressure which exceeds that of seawater.

The third mechanism proposed by Jennings (1968) for coping with toxic concentrations of ions was production of increased succulence. High concentrations of ions in the leaf may be prevented because of the dilution effect brought about by increased water content of cells. Bowman (1921) reported greater succulence in *R. mangle* from seawater than from fresh water, a phenomenon which gives support to the third mechanism. Reinders-Gouwentak (1953) found that succulent leaves were common in *Sonneratia* and were due to the presence of a distinct hypodermal aqueous tissue layer. In leaves that were immersed in tidal water, the hypodermal layer was three to five times as thick as leaves at higher levels of the same tree. The same author stated that the hypodermal layer was almost absent in trees grown in fresh water in botanical gardens. Reinders-Gouwentak believed that succulence in *Sonneratia* was related to the chloride content of the water.

Jennings (1968) related succulence in halophytes to sodium metabolism. He postulated that an outwardly-directed sodium pump exists in halophytes and that this pump is related to cation-activated ATPase in the cell wall. This same pump would drive potassium ions into the cell against an electrochemical potential gradient. Jennings admitted that the evidence for such a pump must be viewed with caution, but stated that there are no reasonable arguments against its existence and suggested a relationship to ATPases. In relation to succulence, Jennings proposed that sodium-activated ATPases might be involved in the synthesis of new wall material or in increasing cell wall extensibility. In a similar way, succulence is also induced by increased amount of light which increases the rate of photo-phosphorylation and production of ATP. Also, aridity causes succulence because increased rate of transpiration causes the ration of potassium to sodium in the shoot to change in favor of sodium. The concentration of sodium in the xylem sap reaches such a level that the sodium-activated ATPases in the plasmalemma bring about synthesis of ATP. It should be stated that these proposals of Jennings are highly theoretical and have yet to be tested.

Kylin and Gee (1970) presented evidence that the leaves of *A. nitida* possess ATPases that are dependent upon the ratio of sodium to potassium. Enzyme activities were directly related to ionic strength of the growth medium. Unlike animal systems in which synergistic effects of sodium and potassium yield a peak at only one ratio, *A. nitida* yielded three peaks. At 50 mM total concentration (NaCl + KC1), activity peaks occurred at Na:K ratios of 2:8, 5:5, and between 8:2 and 9:1. These results were interpreted as indicating that either several enzymes functioned in the membrane system, or else structural changes allowed more than one ion to activate a transport site. Whatever the mechanism, the report of Kylin and Gee gives credence to the hypothesis of Jennings that sodium-activated ATPases are present in halophytes. Their role in succulence has yet to be established.

Salt exclusion, salt excretion, and succulence are not the only physiological mechanisms whereby mangroves adapt to their saline environment. Metabolic processes of photosynthesis, growth, and respiration are also important in adaptation of mangrove, but little work has been reported. Chapman (1966) pointed out that little is known about respiration and photosynthesis by mature mangroves, but speculated that high concentration of salt in the soil would cause slower rate of water uptake, slower rate of upward water movement in the trunk, and slower transpirational loss as compared with many other tropical trees. He suggested that the net result would contribute toward a slow growth rate as compared with trees from mesophytic habitats.

Bharucha and Shirke (1947) stated that the respiratory activity of a plant is influenced by food reserves. In the case of *A. officinalis*, the intensity of respiration of seedlings increased from a minimal to a high rate and then gradually declined. As the seedlings grew, there was an increase in water content and fresh weight, but dry weight decreased, indicating that the growing plant utilized reserve material for growth. Also, the authors showed that in germinating seeds, respiration rate increased during the period of absorption of water and gain in fresh weight.

Bharucha and Shirke also studied the respiration of seedlings of *A. officinalis* from germination to the eighth day of growth in both air and under seawater. Their data, some of which are given in Table 12, showed that the rate of respiration increased with growth under both aerial and submerged conditions. However, respiration rate was much slower under water and this was ascribed to the limiting influence of oxygen in water. Chapman (1962b) reported minimal respiration rates in "medium sized" seedlings of *R. mangle*. He also reported that the cotyledonary body of *R. mangle* and the fruit wall of *A. marina* (both structures function in transport of food from parent to embryo) had very high respiration rates. On a dry weight basis, seedlings of *Avicennia* had a higher respiration rate than those of *Rhizophora*. Chapman suggested that this was related to differential development of aerenchyma.

Table 12. Respiration rates of *A. officinalis* in air and submerged in seawater (after Bharucha and Shirke 1947).

Stage of Growth	Respiratory Index	
	Air	Submerged
1-Day seedlings	2.79	1.21
2-Day seedlings	2.85	1.48
3-Day seedlings	2.94	1.45
4-Day seedlings	3.09	1.46
5-Day seedlings	3.34	1.50
6-Day seedlings	3.67	1.51
7-Day seedlings	3.84	1.62
8-Day seedlings	3.92	1.66

Bharucha and Shirke (1947) also reported increase in the respiration rate of seedlings when the fruit wall was removed. Chapman (1962a, b; 1966b) reported that the fruit wall caused a marked inhibition of respiration in *A. marina, A. nitida, R. mangle, R. apiculata,* and *B. gymnorhiza* and speculated that mangroves are capable of anaerobiosis, although the amount of energy released under anaerobic conditions was inadequate for growth. Anaerobiosis is apparently important during the periods when seedlings float in seawater and no growth occurs. Brown et al. (1969) demonstrated anaerobic respiration in *A. marina, B. gymnorhiza,* and *R. apiculata.*

Lotschert (1968) showed that chloride accumulates in the seedlings of *R. mangle* before they fall from the tree. Conversely, Chapman (1944c) reported that a salt exclusion mechanism operated in *A. nitida.* He stated that when the seedlings fell from the tree, there was an immediate uptake of salt and a sudden reduction in respiration rate. Successful colonization of *Avicennia* was related, therefore, to the capacity of seedlings to respond to sudden changes in internal salt content, whereas this was not necessary for *Rhixophora* seedlings.

Arnold (1955) showed that the transpiration rates of mangroves are very much lower than those of mesophytes. Because of this, Chapman (1962a) suggested that mangroves are lacking in dry tropical areas, such as the west coast of South America, because low humidity reduces respiration rate to the point where seedlings cannot grow.

Lewis and Naidoo (1970) reported that the apparent transpiration rate of *A. marina* in South Africa rose in the morning as light intensity increased and humidity decreased. Maximum transpiration occurred at mid-morning, after which the rate progressively decreased, regardless of atmospheric conditions. Tidal inundation after the mid-morning maximum caused increase in transpiration rate and a second maximum. The authors speculated that decrease in rate at mid-morning was caused by incipient wilting following excessive transpiration.

Chemical Composition

The chemical composition of mangrove trees has been studied by Sokoloff et. al. (1950), Sidhu (1963), Morton (1965), and Golley (1969), but little is known in relation to environmental factors and age of the trees. Sidhu (1963) stated that concentrations of ash and sodium in leaves of species of *Rhizophora, Avicennia,* and *Aegiceras* which grew near the sea were lower than from species which inland, but gave data only for *Avicennia* (Table 13).

Table 13. Ash and sodium contents of three species of *Avicennia* (Sidhu 1963).

	Species	Ash, percent	Sodium, percent
Sea	*A. officinalis*	14.8	2.3
	A. alba	15.8	3.3
	A. alba	19.4	3.7
Inland	*A. marina*	30.4	5.0

Sidhu (1963) also divided species into three catagories based on sodium content of the leaf:

A. Species with more than 5% sodium: *A. marina, Salvadora persica*.

B. Species with 3-5% sodium: *A. alba, Lumnitizera racemosa, Ceriops candolleana*, and *A. ilicifolius*.

C. Species with 1-3% sodium: *A. officinalis, R. mucronata, S. Apetala, B. Caryophylloides, A. Corniculatum, Eleopodendron inerme, Exocoecaria agallocha, D. uliginosa*.

Atkinson et. al (1967) reported concentrations of some ions in leaves of various ages in *R. mucronata* and *Aegiatilis annulata* R.Br. Measurements were made on successive leaf pairs of shoots (Tables 14 and 15). Concentrations of sodium and chloride in the leaves of *R. mucronata* increased with age, but in relation to amount of leaf water, chloride content was constant, sodium concentration increased, and potassium concentration deceased.

In contrast, there was a decrease in concentrations of sodium, potassium, and chloride with age of *A. annulata*. Atkinson et. al. ascribed this to excretion of salt by epidermal glands, an adaptation lacking in *Rhizophora*.

Table 14. Concentrations of Na^+, K^+, and Cl^- in the leaves of **R. mucronata** (after Atkinson et al. 1967). Leaves in Sample 1 were youngest in the sequence of increasing age.

	Sample Number					
	1	2	3	4	5	6
Dry weight (g)	0.16	0.50	0.50	0.61	0.57	0.63
Water (% fresh weight)	56	65	66	65	67	69
Na^+ (μ-equiv/leaf)	61	290	420	480	520	645
Na^+ (μ-equiv/ml H_2O)	305	313	431	435	461	461
K^+ (μ-equiv/leaf)	25	81	57	48	69	45
K^+ (μ-equiv/ml H_2O)	124	88	59	44	61	32
Cl^- (μ-equiv/leaf)	74	520	510	585	580	730
Cl^- (μ-equiv/ml H_2O)	370	562	522	530	515	522

Table 15. Concentrations of Na^+, K^+, Mg^{++}, and Cl^- in the leaves of A. annulata (after Atkinson et al. 1967). Leaves in Sample 1 were youngest in the sequence of increasing age.

	Sample Number				
	1	2	3	4	5
Dry weight (g)	0.32	0.45	0.45	0.50	0.44
Water (% fresh weight)	72	60	60	59	60
Na^+ (μ-equiv/leaf)	420	325	275	280	235
Na^+ (μ-equiv/ml H_2O)	518	480	411	388	356
K^+ (μ-equiv/leaf)	155	106	87	93	70
K^+ (μ-equiv/ml H_2O)	191	157	130	129	106
Mg^{++} (μ-equiv/leaf)	108	330	440	590	530
Mg^{++} (μ-equiv/ml H_2O)	133	488	659	819	802
Cl^- (μ-equiv/leaf)	415	290	270	361	386
Cl^- (μ-equiv/ml H_2O)	512	429	405	361	386

There is a paucity of literature data concerning elemental composition of mangroves. Most studies of which I am aware have not been published. Values given in Tables 16 and 17 were personal communications from F.B. Golley (Univeristy of Georgia, Athens), S. S. Sidhu (University of Western Ontario, London, Canada), and T. F. Hollister (U. S. Environmental Protection Agency, Gulf Breeze, Florida). The samples were collected in Panama (Golley), India (Sidhu), and Florida (Hollister).

There are wide differences in concentrations of each element between species. For example, *Avicennia* species contain relatively high concentrations of sodium and potassium in all organs. Also, the roots of *Laguncularia racemosa* contained very high concentrations of all elements, except magnesium.

Sidhu (personal communication) found no correlation between the mineral status of soils and the elemental content of 16 mangrove species. It is clear that research which relates species and habitat to elemental composition is needed.

Table 16. Concentrations of various elements in mangrove leaves (personal communications: + Siddhu, * Golley, # Hollister).

Species	Concentration, ppm dry weight									
	Ca	Co	Fe	K	Mg	Mn	Na	P	Si	Zn
Acanthus ilicifolius (+)	7800	-	-	8200	8600	-	30000	3200	30000	-
Aegiceras corniculatum (+)	6800	-	-	8400	10200	-	23100	2500	4600	-
Avicennia alba (+)	10400	-	-	12100	10000	-	35000	2300	4000	-
Avicennia marina (+)	11200	-	-	15200	13500	-	50600	2300	5000	-
Avicennia nitida (#)										
Overstory	12430	-	147	21000	3600	-	4590	-	-	154
Understory	6680	-	300	29400	2320	53	6500	-	-	24
Avicennia officinalis (#)	11600	-	-	19100	12200	-	23300	4000	20800	-
Bruguiera caryophylloides (+)	24800	-	-	6100	10000	-	27700	2400	17200	-
Ceriops candolleana (+)	20400	-	-	5300	11700	-	36600	2500	11000	-
Conocarpus erectus (#)										
Overstory	13350	22	305	4100	-	50	5500	-	-	160
Understory	7600	24	251	21600	-	88	4110	-	-	142
Derris ulginosa (+)	17600	-	-	25800	12400	-	12300	2600	18000	-
Exocaria agallocha (+)	14200	-	-	12900	14700	-	4800	4100	18800	-
Laguncularia racemosa (#)										
Overstory	6510	12	125	17700	4440	74	10520	-	-	92
Understory	7600	32	149	25000	4740	116	8620	-	-	84
Lumnitizera racemosa										
India (+)	24200	-	-	22400	15600	-	43100	2500	5800	-
Florida (#)	9100	-	169	19900	3940	-	9370	-	-	170
Rhizophora brevistyla (*)										
Overstory	12200	46	82	8400	4700	387	9800	9000	-	11
Understory	7800	56	672	8500	5000	125	8300	8000	-	15
Rhizophora mangle										
Hollister	10760	9	132	16400	4320	92	11130	-	-	146
Morton (1965)	13500	52	152	6500	8000	30	-	1400	-	43
Rhizophora mucronata (+)	19800	-	-	21600	14500	-	22800	3300	6000	-
Sonneratia apetala (+)	11200	-	-	3100	10500	-	14900	3100	4800	-

Table 17. Concentrations of various elements in organs, other than leaves, of mangroves (personal communications: * Golley, # Hollister).

Species	Concentration, ppm dry weight								
	Ca	Co	Cu	Fe	K	Mg	Mn	Na	Zn
Avicennia nitida (#)									
Overstory stems	7430	-	-	282	14700	3080		-	3730	227
Understory stems	7100	16	10	580	27800	3650		48	9340	28
Roots	3930	40	13	465	29200	3700		34	9000	121
Conocarpus erectus (#)									
Overstory stems	3930	15	13	329	4100	3400		47	412	180
Understory stems	5600	36	15	648	27300	2400		83	1260	112
Roots	21200	25	40	979	-	7160		75	3560	162
Laguncularia racemosa (#)									
Overstory stems	3100	18	9	49	10000	1880		92	4590	89
Understory stems	6850	16	-	516	9200	1220		46	4120	38
Fruit	8590	-	15	332	35000	2920		123	4120	79
Roots of seedlings	19390	54	-	1000	81000	1990		-	14100	514
Lumnitzera racemosa (#)									
Stems	11000	25	10	99	13200	3405		-	618	25
Rhizophora brevistyla (*)									
Overstory stems	12900	52	7	36	3000	1000		168	5500	11
Understory stems	5700	83	8	1000	3900	2900		255	9500	12
Overstory fruits	5900	81	6	82	10100	2900		191	9600	11
Understory fruits	6900	56	4	45	7100	3400		164	9700	7
Rhizophora mangle (#)									
Overstory stems	8936	-	-	252	2200	2820		74	8370	120
Fruit (2-10 cm long)	2350	58	12	155	4000	3050		121	8890	2
Fruit (11-20 cm long)	1260	-	15	113	2990	2460		69	10250	152
Roots of seedlings	600	36	12	422	13800	1070		71	12840	12
Prop roots	4850	-	11	253	13800	2540		-	10660	74
Rhizophora mucronata (#)									
Fruit	932	20	-	346	13860	2620		46	8620	6

Sokoloff et. al (1950) and Morton (1965) gave data from chemical analyses of leaves of *R. mangle* from Florida (Table 18.) Their values for vitamin content vary greatly but cannot be compared because methods of treatment of leaves and assay procedures were not given. Sokoloff et. al. suggested that leaf meal could replace alfalfa in chicken rations and Morton recommended its use as cattle feed.

Golley (1969) compared the caloric content of *R. brevistyla* in Panama with trees from tropical moist, premontane, and gallery forests. The energy values of the mangrove (Table 19) were generally greater than those of the other tropical trees.

Tannin is an important constituent of all parts of mangrove trees. Bowman (1921) showed that tannin is usually stored in solid masses in cells, but is frequently in solution in cytoplasm. He considered the tannin cells and water storage tissue of *R. mangle* to constitute a true hypodermis.

The function of tannin in mangrove is unknown. Tannins have a strong protein-binding capacity and, therefore, are able to inhibit enzymes. In the living plant it is possible that tannins aid in resistance to fungi, as fungi have been shown to cccur in large numbers in mangals (Swart 1958; Kohlmeyer 1965, 1966, 1968a, b, 1969a, b; Kohlmeyer and Kohlmeyer 1971; Ahern et al. 1968; Rai et al. 1969, and Ulken 1970). Lee and Baker (1972a, b) identified 52 species of soil fungi from a Hawaiian swamp. Swain (1965) suggested that the presence of tannin causes resistance of dead organic matter to attack by fungi and other decomposers. Crossland (1903) stated that the Arabs of Zanzibar used mangrove wood for houses and furniture because it was not attacked by termites, and suggested that the high tannin content repelled the termites. The ability of tannins to inhibit enzymes probably affects the rate of decay of plant detritus and, therefore, is important in relationships within the detritus based food web. In most plants, hydrolyzable tannins are usually present in leaves and fruit, whereas condensed tannins occur in the bark or heartwood. This implies that leaves and fruit of mangroves are less persistant as particulate detritus than woody parts.

Most research on mangrove tannin has been done on samples of bark. Drabble (1908) illustrated distribution of tannin in *R. mangle* and *Laguncularia racemosa*. Trimble (1892) reported the empirical formula a $C_{25}H_{25}O_{11}$ for tannin from the bark of *R. ,mangle*. Baillaud (1912) found that 30% of the dry weight of bark from *Rhizophora* and *Bruguiera* in Africa was comprised of tannin. Dry bark of *Xylocarpus* contained 26% tannin. Dried bark of *R. mucronata* from Africa contained between 41.3 and 42.8% tannin (Anon. 1904). Brown and Fisher (1918) pointed out that tannin content varied greatly between species from the Indo-West-Pacific region. In Malaya, Buckley (1929) reported the following percentages of tannin in fresh bark: *R. Mucronata* 20.7-30.8, *R. congugata* 7.9-17.6, *B. gymnorhiza* 14.5-25.6, *B. eriopetala* 17.3-23.0, *B. caryophylloides* 15.8, *B. parviflora* 4.7-7.6, *C. candolleana* 19.0-30.8, *Carapa obovata* 29.8-41.6. The fruits and

Table 18. Chemical analysis of dry leaves of *R. mangle* **from Florida (after Sokoloff et al. 1950 and Morton 1965).**

Total protein, percent	12.1-14.3	7.5
Crude fiber, percent	13.9	13.9
Crude fat, percent	2.9	3.6
Calcium, percent	1.6	1.4
Sulphur, percent	0.6	—
Ash, percent	6.7	10.1
Iodine, percent	0.8	0.5
Manganese, mg/kg	0.3	0.03
Thiamin, mg/kg	1.56-2.03	130
Riboflavin, mg/kg	4.5-5.6	190
Folic acid, mg/kg	0.60-0.67	320
Niacin, mg/kg	20.3-28.0	2,400
Pantothenic acid, mg/kg	4.0-4.5	53

Table 19. Mean caloric values and standard errors of *R. brevistyla* **from Panama (Golley 1969).**

	Energy, g cal/g dry weight	
Compartment	Mean	SE
Canopy leaves	4182	22
Canopy stems	4337	11
Understory leaves	4299	132
Understory stems	4204	12
Canopy fruit	4298	29
Understory fruit	4360	20
Epiphytes	4585	11
Litter	4141	13
Roots	4034	48

leaves contained least tannin. Buckley concluded that *R. mucronata* was the best source of tannin because the yield of bark per tree was good. *Carapa obovata* contained the highest concentration of tannin, but its bark was thin and the yield per tree was small. Drabble and Nierenstein (1907) reported that older *R. mangle* trees contained more tannin than young trees.

Although mangroves are not used extensively as a source of dyes, when the bark of *R. mangle* is treated with copper or iron salt, brown, olive, rust, and slate-colored dyes are obtained (Fanshaws 1950, Morton 1965). According to Morton, a boiled concentrate of the bark may be used for staining wood in floors and furniture, dyes for textiles may be obtained from the roots, and a red dye from the shoots may be used for coloring leather.

Energy Relationships

Except for a few reports on yield of wood (see "silviculture" section of this report), little is known about production of organic matter in mangrove swamps. The first detailed study of photosynthesis, respiration, biomass, and export of organic matter was made by Golley et al. (1962) in a Puerto Rican *R. mangle* forest in May. Gross photosynthesis was 8.23 g $C/m^2/day$; total respiration was 9.16 g $C/m^2/day$. The greatest rates of photosynthesis (7.33 g $C/m^2/day$O and respiration (4.31 g $C/m^2/day$) occurred in the upper canopy of leaves. Shaded leaves accounted for gross photo synthesis of only 0.40 g $C/m^2/day$ and respiration of 0.48 g $C/m^2/day$. Seedling photosynthesis was 0.12 g $C/m^2/day$ and respiration was 0.36 g $C/m^2/day$. At the soil surface, respiration by prop roots was 2.03 g $C/m^2/day$. At and below the soil surface respiration was 1.64 g $C/m^2/day$. Gross photosynthesis and respiration above ground was related to dry leaf biomass (1017 gm/m^2), leaf area (4.4 m^2/m^2), and chlorophyll *a* content of the leaves (1.19 g/m^2). The trees were approximately 8 m tall, and the factors measured attained their greatest values at between 4 and 6 m height. Unfortunately, the subterranean algal flora was not studied. This might have been important as Marathe (1965) showed 12 algal species in the soil of mangals near Bombay. Another source of primary production was algae on the roots and mud. Dawson (1954) described many attached algal species from roots and mud in Vietnam. Golley et al. stated the *R. mangle* community was more fertile than most marine and terrestrial communities. It was not, however, as efficient as the montane rain forest or coral reefs of Puerto Rico in conversion of sunlight into organic matter under similar light regimes.

Miller (1972), using a model, calculated gross photosynthesis, net photosynthesis, and respiration of *R. mangle* in Florida (Table 8). He contrasted these data with those of Golley et al. (1962). Using Miller's model, the estimates of Golley et al. corresponded to 9.4 g organic matter/m^2/day for gross photosynthesis, 3.4g organic matter/m^2/day for net photosynthesis, and 5.9 g organic matter/m^2/day for respiration. Miller ascribed differences between his data and Golley's to different leaf areas.

Miller's model predicted that maximum photosynthesis occurrs at a leaf area index of approximately 2.5 if no acclimation to shade within the canopy

Table 8. Gross primary production (P_g), net primary production (P_n), and respiration (R) of R. mangle leaves in Florida (after Miller 1972). Data are expressed as grams organic matter/m²/day.

Height (m)	Sunny			Cloudy		
	P_g	P_n	R	P_g	P_n	R
June						
1.75-2.00	0.08	0.03	0.05	0.07	0.03	0.04
1.50-1.75	0.31	0.12	0.19	0.29	0.12	0.17
1.25-1.50	2.17	0.93	1.24	2.02	0.92	1.09
1.00-1.25	30.8	1.48	1.61	2.59	1.13	1.46
0.75-1.00	3.39	1.64	1.75	2.48	0.84	1.65
0.50-0.75	2.16	0.87	1.33	1.61	0.35	1.26
0.25-0.50	0.97	0.31	0.66	0.73	0.10	0.63
0.00-0.25	0.67	0.17	0.51	0.51	0.02	0.49
Total	12.83	5.55	7.34	10.30	3.51	6.79
January						
1.75-2.00	0.09	0.05	0.03	0.09	0.05	0.03
1.50-1.75	0.34	0.21	0.13	0.34	0.21	0.12
1.25-1.50	2.14	1.32	0.82	2.08	1.28	0.80
1.00-1.25	2.43	1.33	1.10	2.52	1.45	1.07
0.75-1.00	2.15	0.91	1.24	2.41	1.19	1.21
0.50-0.75	1.31	0.36	0.95	1.55	0.62	0.93
0.25-0.50	0.57	0.10	0.47	0.70	0.23	0.46
0.00-0.25	0.39	0.03	0.37	0.49	0.13	0.36
Total	9.42	4.31	5.10	10.18	5.16	4.98

is present and predicted that production decreased with increase in leaf area index and leaf width. Also, the environmental variables with the greatest influence on primary production were air temperature and humidity. Increase in solar radiation up to a point, increased primary production as did increasing amounts of diffuse energy. Infrared variation decreased production.

Gill and Tomlinson (1971) reviewed phenological phenomena associated with growth of *R. mangle* in Florida. Although the general progression of development appeared to be mediated endogenously, climatic factors were a strong governing influence. Environmental control of growth was through effects on development of the apical bud. Vegetative branches, inflorescences, and axillary buds are developed within the apical bud of *R. mangle.* The rates of leaf expansion and fall were highest in the summer when temperature and radiation were maximal. Throughout the year, leaf fall was closely correlated with leaf expansion so there was a fairly constant number of leaves on a shoot. Flower buds appeared in greatest abundance between May and July, open flowers between June and September, and fruit between September and March. The hypocotyl appeared in March and greatest fall of propagules was between June and October.

In some swamps, phytoplankton in the water contribute appreciably to synthesis of organic matter. This was the subject of extensive studies by Teixeira and Kutner (1962), Teixeira et al. (1965, 1967, 1969), Watanable and Kutner (1965), Tundusi and Tundusi (1968), and Tundusi and Teixeira (1964, 1968) in Brazil. Tundusi (1969) summarized the work in a Brazilian mangal. Gross primary production of surface water ranged between 2.10 and 91.3 mg C/m^3/hr. Respiration values were between 1.0 and 21.3 mg C/m^3/hr. Nannophytoplankton (size range 5-65 u) accounted for 61.8% of the total carbon uptake. Diatoms were the numerically dominant unicells, a phenomenon also reported by Mattox (1949) in Puerto Rico, Walsh (1967) in Hawaii, and Bacon (1971) in Trinidad.

An important finding in the work of Golley et al. (1962) was that tidal export of particulate matter was 1.1 g C/m^2/day. Heald and Odum (1971) reported production, consumption, and export of organic detritus in a *R. mangle* stand in southern Florida. Heald and Odum pointed out that many commercially important finfish and shellfish live in the mangrove environment and that vascular plant detritus is the primary source of food for many estuarine organisms.

Heald (1971) estimated that production of mangrove debris averaged 2.4 g C/m^2/day, oven dry weight. This was equivalent to almost nine tons/ha/yr. Annually, plants other than mangrove accounted for less than 15% of the total organic debris. Rate of degradation of mangrove detritus was related to conditions of the environment. Breakdown was most rapid in brackish water. The amphipods *Melita nitida* Smith and *Corophium lacustre* and the crab *Rithropanopeus harrisii* Gould were important consumers of detritus in brackish water.

The actively photosynthesizing leaf of *R. mangle* was reported by Heald to contain 6.1% protein, 1.2% fat, 67.8% carbohydrate, 15.7% crude fiber, and 9.2% ash. During abscission, protein and carbohydrate contents were 3.1 and 59.6%, respectively. After falling into brackish water, the carbohydrate content of leaf detritus fell to approximately 36%, but protein content rose to approximately 22%. Heald speculated that increase in protein was due to growth of bacteria and fungi on the detritus particles, and stated that the food value of detritus, which was in the water for one year, was more nutritious than that in the water for one or two months.

Odum (1971) studied the use of mangrove detritus as food by many animal species. He concluded that vascular plant detritus which originated from *R. mangle* leaves was the main source of food for the aquatic animal community. As the detritus decomposed into finer particles, it became covered with bacteria, fungi, and protozoans. The caloric content rose from 4.742 Kcal/g in ash-free fresh leaves to 5.302 Kcal/g in ash-free remains of leaves which were submerged for two months.

There were at least four pathways by which mangrove leaves were utilized by heterotrophs: (1) dissolved organic substances from the leaves → microorganisms → higher consumers, (2) dissolved organic substances → sorption on sediment and aged detritus particles → higher consumers, (3) particulate leaf material → higher consumers, and (4) particulate leaf material → bacteria, fungi, and protozoans → higher consumers. Odum believed the last pathway to be the most important. He speculated that microorganisms convert compunds such as cellulose and lignin into digestible protein utilized by invertebrates and fishes.

SILVICULTURE

Mangrove is one of the most important sources of timber, fuel, posts, poles, railroad ties, and tannin in the tropics. It also has resins which are used as plywood adhesives, and the bark, leaves, shoots, and roots contain dyes. Chatterjee (1958) gave the following uses for mangrove in India: *Heritiera* (boat building, planking, fuel), *Amoora cuculata* (wooden pipes for hookahs and wooden toys), *Aegiatilis rotundiflora* (extraction of high-grade salt after burning), *Avicennia* (fuel wood for brick burning), *Exocoecaria agallocha* (match boxes), *Xylocarpus granatum* (pencils), and *Salicornia brachiata* (source of sodium carbonate).

Because of their many uses, silviculture of mangroves has been practiced for many years in southeastern Asia. Banerji (1958) reported that *R. mucronata* and *B. gymnorhiza* were grown successfully on a plantation of 685 acres in the Andaman Islands between 1898 and 1908. Banerji stated that *B. gymnorhiza* was an excellent source of poles for transmission lines and that *Rhizophora* produced 30 cords of fuel wood per acre, whereas *Bruguiera* produced 11 cords. The annual yield of firewood was estimated at 130,000 tons.

In the Andamans, clearfelling and planting of *B. gymnorhiza* was the best silviculture method. Because this species grows slowly, a rotation to 100 years was recommended for an exploitable breast-height girth of 27 inches. On some plantations, *Bruguiera* attains a height of 30-35 feet (9.2-11.7 m) and a girth of 9-12 inches (23-30 cm) in 15 years. The tree provides one of the strongest timbers in India, and has a durability life of 10 years after treatment with creosote.

In Thailand, Walker (1937, 1939) reported that *R. conjugata* and *R. mucronata* were the mangrove of choice for poles and firewood and have been used on a large scale for planting. There was an abundance of seedlings at all times of the year. Seedlings of *R. mucronata* were less susceptible to attack by crabs, and their long radicle was an advantage in competition with growth of *A. aureum* and *B. parvifloraa*. Seedlings were deep-planted at six-foot (1.8 m) intervals and the maximum felling girth was eight inches (20 cm). The prescribed time for planting was two years after felling. Unfortunately, Walker did not give production figures for the Thai mangroves. He did point out that *B. gymnorhiza* was used for fuel, *S. griffithii* for fishing stakes, and *C. candolleana* as fuel and tanbark.

Becking et al. (1922) divided the mangrove of southeastern Asia into three classes based upon the diameters of mature trees: Class A, less than 20 cm, *A. corniculatum,, Scyphiphora hydrophyllacease* Gaert., *Ceriops* spp.; Class B, 20-40 cm, *A. marinaa, Lumnitzera racemosa, Bruguiera* spp.; Class C, greater than 40 cm., *A. officinalis, S. alba, Rhizophora* spp., *Bruguiera* spp., *Xylocarpus* spp. The authors showed that production of wood per unit area by trees of larger diameter was greatest. Although *Sonneratia* was one of the more productive genera, Backer and van Steenis (1954) stated that in Malaya its economic usage was small. Small amounts were used for fuel and in boats and houses. The young berries of *Sonneratia* can be consumed by humans and pectin can be extracted from them.

According to Banijbatana (1958), approximately 133,400 ha of mangrove forests were available for silviculture in Thailand. The shelterwood system was judged best, and for young forest with trees of 20 cm girth and under, clearing and thinning was recommended. For forests in which the majority of trees were 20-50 cm, heavier thinning was recommended with seedling felling for control of *C. roxburghiana* and *B. cylindricaca*. The *Rhizophoras* reached 65-70 cm in girth between the ages of 39 to 43 years and a rotation system of 40 years was adopted. Yield was calculated at 50-60 cm per acre.

Approximate rates of growth in girth of several mangrove species in Malaya were given by Durant (1941 (Table 22). The volume of wood per acre of mixed *Rhizophora* forest increased from 1,375 cubic feet at 10 years to 5,600 cubic feet at 50 years. Durant recommended harvesting at 22-23 years when the mean annual volume increment was at its maximum of 147.7. The volume of wood at 22 years was 3,250 cubic feet per acre.

Noakes (1955) stated that the total area of mangrove forest in Malaya was approximately 760 square miles. Of this, 460 square miles were under

Table 22. Growth in girth of several mangrove species in Malaya (Durant 1941).

Age Years	R. conjugata	R. mucronata	B. gymnorhiza	B. hainesii	B. parviflora	B. caryophylloides	C. candolleana
5	2	5	3	2	2	1	2
10	6	10	7	5	5	5	5
15	9	13	9	8	8	7	8
20	12	16	12	11	11	9	10
25	16	18	15	14	13	11	13
30	19	20	18	17	16	13	15
35	21	22	20	19	18	15	18
40	24	24	23	21	20	17	20
45	27	26	25	24	23	18	22
50	29	28	28	27	25	20	24
55	32	30	30	30	27	22	-
60	35	32	32	33	29	24	-
65	37	34	35	36	-	26	-
70	39	36	37	40	-	27	-
75	41	-	39	45	-	29	-
80	42	-	41	-	-	30	-

sustained yield management. *Rhizophora mucronata* and *R. conjugata* were the most important Malayan species, covering two-thirds or more of the total area. Fuel was the main product of the mangrove forest, but poles for houses and fish traps were also important.

According to Noakes, fruiting of *Rhizophora* occurred at the age of four years, was annual and highly prolific. Full stocking by water-borne seedlings occurred rapidly after clearcutting. As in Thailand, *Rhizophora* grew slowly. The annual growth increment of boles was slightly greater than one inch in the early stages of growth to just over one-half inch above 12 inches diameter. The trees grew to a maximum height of 70-120 feet. Normal felling size was 50-60 feet in height and 1.5-2.5 feet in girth. Felling size was achieved in 20-30 years. The mean annual volume increment of mixed *Rhizophora* forest culminated at approximately 25 years, allowing for a three-year regeneration period. The yield was approximately 3,106 cubic feet of wood per acre.

Wadsworth (1959) reported silviculture of *Laguncularia racemosa* in Puerto Rico. Undisturbed stands 22 years old attained an average diameter at breast height of 5.0 inches and 2,680 cubic feet of wood per acre. Natural regeneration by water-borne seedlings occurred within two years after clearfelling. Holdridge (1940) recommended a cutting cycle of five years and a rotation cycle of 25 years for exploitation of mangroves in Puerto Rico.

Golley et al. (1962) reported that in a *R. mangle* forest in Puerto Rico, annual production of wood was 0.84 g/m^2/day (0.42 g C/m^2/day). This was much less than that reported by Noakes in Malaya (14 g C/m^2/day).

It should be pointed out here that in some areas, silviculture of mangrove is practiced in the belief that the wood is resitant to marine boring organisms. Southwell and Boltman (1971) tested resistance to marine borers by *R. mangle*, *R. brevistyla*, *A. marina*, *C. erectus*, and *Laguncularia racemosa*. Only *C. erectus* showed natural resistance to teredo, pholad, and limnorid borers. The *Rhizophora* species were almost completely destroyed after immersion in Pacific Ocean water for 14 months.

HERBICIDES

Herbicides have been used for almost 20 years for control of mangrove. In Africa (Sierra Leone), Ivens (1957) reported that application of the auxin-type herbicides 2, 4-dichlorophenoxyacetic acid (2, 4-D), 2, 4, 5-trichlorophenoxyacetic acid (2, 4, 5-T), and 2-methyl, 4-chlorophenoxy-acetic acid (MCPA) were effective in eradication of both *R. racemosa* and *A. nitida* when applied to the bases of trunks at concentrations of 4-20% in diesel oil. Frilled *R. racemosa* were killed by the butyl ester of 2, 4-D at concentrations between 0.5 and 1.0%, whereas 4% was required to kill unfrilled trees. *Avicennia nitida* was slightly more resistant to 2, 4-D than *R. racemosa*; concentrations between 2 and 4% being required to kill frilled trees. 2, 4, 5-T and MCPA were not as effective as 2, 4-D. Recovery of trees after treatment with 2, 4, 5-T was reported.

The first signs of herbicidal effect were noted approximately three weeks after application, when the leaves turned yellow. Extensive defoliation occurred by seven months after treatment, at which time many trees of both genera were dead. Seedlings and young trees were more resistant than old trees.

Ivens also reported that 3-(4-chlorophenyl)-1, 1-dimethylurea (CMU) killed all trees when applied to the pneumatophores of *A. nitida* at the rate of 20 lb/acre. Dalapon (2, 2-dichloropropionic acid) caused complete kill with no regrowth at 40 and 80 lb/acre. There was a small amount of regrowth after application of dalopon at 20 lb/acre.

Truman (1961) reported total kill of treated *A. marina* in Australia by 1% 2, 4-D applied to the basal bark. Only 54% were killed by treatment with 1% 2, 4, 5-T. Spotted gum (*Eucalyptus maculata* Hook.), an upland tree, was only slightly affected by the same treatment. Truman concluded that *A. Marinna* was very susceptible to auxin-type herbicides.

The concept of high susceptibility of mangrove to auxin-type herbicides was extended by Tschirley (1969), Orians and Pfeiffer (1970), and Westing (1971a, b), who stated that mangrove forests in Vietnam were destroyed after a single application of 6.72 kg/ha of the triisopropanolamine salt of 2, 4-D in combination with 0.61 kg/ha of the triisopropanolamine salt of 4-amino-3, 5, 6-trichlorophcolinic acid (picloram). The forests were composed mainly of *S.*

alba, B. parviflora, B. gymnorhiza, A. marina, A. intermedia, R. conjugata, C. candolleana, and *N. fruiticans.* Westing (1971a) reported that treated areas in Vietnam remained uncolonized by mangrove six years after treatment. Westing (1971c) also published a list of references to effects of herbicides in Vietnam. For reviews of herbicidal effects in Vietnam, see Boffey (1971) and Aaronson (1971).

Westing (1971a) pointed out that application of 2, 4-D in combination with picloram both defoliated and killed nearly all trees in the sprayed areas. He also stated that herbicides seemed to prevent recolonization by mangrove, although he observed rapid recolonization of an area cleared by cutting. It is true, however, that a large portion of the denuded mud flats of Vietnam are only occasionally inundated by tidal water and that sufficient numbers of seedlings for regeneration are not carried in. Another possible reason for lack of recolonization may be related to texture of the denuded soil. Natural regeneration of mangrove is greatly retarded when soil becomes too stiff or hardens after exposure to the sun (Banijbatana 1958).

We (Walsh et al., in press) have studied affects of Tordon® 101 on seedlings of *R. mangle* from Florida. Tordon 101 is a mixture of the triisopropanolamine salts of 2, 4-D (39.6%) and picloram (14.3%). Seedlings that had no leaves and one or two pairs of leaves were treated with 1.12, 11.2, and 112.0 kg/ha (1, 10, and 100 lb/acre). These rates were equivalent to 0.44, 4.40, and 44.0 kg/ha 2, 4-D, and 0.16, and 16.0 kg/ha picloram. A combination of 0.44 kg/ha 2, 4-D and 0.16 kg/ha picloram caused stunted growth of seedlings without leaves, but had no permanent effects upon seedlings with one or two pairs of leaves. Higher concentrations caused death of all treated seedlings by 50 days after treatment.

We were never able to quantify tissue residues in seedlings without leaves that had been treated at the lowest concentrations. The limits of quantification were 0.02 ppm (parts per million) 2, 4-D and 0.01 ppm picloram. Even though tissue residues were very low, seedling development was greatly inhibited. In seedlings with leaves, greatest herbicidal residues occurred in the highest leaves and hypocotyl. Table 20 shows distribution of 2, 4-D and picolinic acid in the organs of seedlings treated when two pairs of leaves were present.

At the tissue level, symptoms of herbicide poisoning were desiccation of leaves, plugging of vessel elements, and destruction of root cortex. Root destruction probably impared the ability of seedlings to regulate salt and water balance. For example, concentrations of sodium and potassium in seedlings were directly related to application rate and time (Table 21). No changes were found in concentrations of magnesium, manganese, calcium, iron, or zinc. Strogonov (1964) said that symptoms of salt poisoning in plants include bleaching of chlorophyll accompanied by browning of the leaves. Both symptoms were observed in our treated seedlings.

Table 20. Concentrations of 2,4-D and picolinic acid, in parts per million (± 20%), of wet tissue in organs of R. mangle seedlings treated with Tordon 101 when two pairs of leaves were present. Residues were detected in every analysis of seedlings treated with 1.12 kg/ha but were below quantifiable levels (0.01 for picolinic acid, 0.02 for 2,4-D) (Walsh et al. in press).

Treatment kg/ha	Day	Roots		Hypocotyl		Stem		1st leaves		2nd leaves	
		2,4-D	PA	2,4-D	PA	2,4-D	PA	2,4-D	PA	2,4-D	PA
11.2	30	0.02	0.01	0.10	0.03	0.02	0.01	0.02	0.01	0.13	0.06
	40	0.02	0.01	0.23	0.10	0.23	0.10	0.29	0.10	0.35	0.10
112.0	10	1.23	0.39	1.68	0.49	1.02	0.43	0.63	0.24	0.87	0.41

Table 21. Concentrations of sodium and potassium in the stems of *R. mangle* seedlings treated with Tordon 101 when two pairs of leaves were present; 50 days after treatment (Walsh, unpubl.)

Treatment	ppm Dry Weight	
	Na	K
Control	37,500	4,375
1.12 kg/ha	49,800	4,821
11.2 kg/ha	64,900	5,295
112.0 kg/ha	96,200	6,321

Acknowledgements

I thank Drs. Frank B. Golley and S. S. Sidhu and Mr. Terry A. Hollister for data on elemental composition. Special thanks are given to Mrs. Ann Valmus, Librarian at the Gulf Breeze Laboratory, for aid in obtaining copies of many publications and for her patience in checking the references. The quotation from "The Night Country" was made with permission of Charles Scribner's Sons, Publishers, New York. Thanks are also given to Mrs. Steven Foss for making the illustrations.

® Registered trademark, Dow Chemical Co., Midland, Michigan. Reference to trade names in this publication does not constitute endorsement by the Environmental Protection Agency.

REFERENCES

Aaronson, T. 1971. A tour of Vietnam. Environment 13: 34-43.
Abe, N. 1937. Ecological survey of Iwayama Bay, Palao. Palao Trop. Biol. Stn. Stud. (Tokyo) 1: 217-324.
Abe, N. 1942. Ecological observations on *Melaraphe (Littorinopsis) scabra* (Linnaeus) inhabiting the mangrove tree. Palao Trop. Biol. stn. Stud. (Tokyo) 2: 391-435.
Abel, O. 1926. Amerikafahrt: Eindrucke, Beobachtungen und Studien eines Naturforschers auf einer Reise nach Nordamerika und Westindien. Gustav Fischer, Jena, 462 pp.
Abel, O. 1927. Fossile Mangrovesumpfe. Palaont. Z. 8: 130-140.
Abeywickrama, B. A. 1964. The estuarine vegetation of Ceylon. *in* Les problemes scientifiques des deltas de la zone tropicale humide et leurs implications. Colloque de Dacca. UNESCO, Paris, pp. 207-209.
Abreu, S. F. 1926. Notes on the tanning materials of Brazil. Am. Leather Chem. Assoc. J. 21: 357-358.
Acosta-Solis, M. 1947. Commercial possibilities of the forests of Ecuador — mainly Esmeraldas Province. Trop. Woods Yale Univ. Sch. For. 89: 1-47.
Acosta Solis, M. 1959. Los manglares del Equador. Contrnes Inst. Ecuat. Cienc. Nat. No. 29, 82 pp.
Acuna, R. A. 1953. Mangrove swamps. For. Leaves 6: 35-41.
Adriani, M. J. 1937. Sur la transpiration de quelques halophytes cultivies dans les milieux different en comparison avec celle de quelques nonhalophytes. Proc. K. ned. Akad. Wet. 40: 524-529.
Ahmad, N. 1967. Seasonal changes and availability of phosphorus in swamp-rice soils of north Trinidad. Trop. Agric. Trinidad 44: 21-32.
Alexander, T. R. 1967. Effect of hurricane Betsy on the southeastern Everglades. Q. J. Fla. Acad. Sci. 30: 10-24.
Almodovar, L. R. and R. Biebl. 1962. Osmotic resistance of mangrove algae around La Parguera, Puerto Rico. Rev. Algol. 3: 203-208.
Almodovar, L. R. and F. Pagan. Notes on a mangrove lagoon and mangrove channels at La Parguera, Puerto Rico. Nova Hedwigia *in press*.
Alston, A. H. G. 1925. Revision of the genus *Cassipourea*. Kew Bull. 6: 241-276.
Altrock, A. von. 1897. Uber Mangrove-Nutzung in St. Catharina Brasilien. Tropenpflanzer 1: 263.
Altevogt, R. 1955. Beobachten und Untersuchungen an Indischen Winkerkrabben. Z. Morph. Oekol. Tiere. 43: 501.
Altevogt, R. 1957. Untersuchungen zur Biologie und Physiologie Indischer Winkerkrabben. Z. Morph. Oekol. Tiere 46:1.
Anderson, J. A. R. 1964. The structure and development of the peat swamps of Sarawak and Brunei. Malay. J. Trop. Geogr. 18: 7-16.
Anderson, J. A. R. 1964. Observations on climatic changes in peat swamp forests in Sarawak. Comm. For. Rev. 43: 145-158.

Anon. 1904. Mangrove barks, and leather tanned with these barks, from Pemba and Zanzibar. Bull. Imper. Instit., London, Sept., pp. 163-166.

Anon. 1951. Three hundred homes blackened by mangrove root gas. Miami Herald, Nov. 14.

Anon. 1957. Establishment of mangroves. Rep. For. Admin., Malaya, No. 34.

Anon. 1958. Poisoning of *Exoecaria agallocha*. Rep. For. Dept., Fiji, No. 7.

Anon. 1961. Mangrove gas kills two men in Fort Lauderdale ditch. Miami News, July 28.

Anon. 1965. New fruit-fly found in mangroves. Aust. Nat. Hist. 15: 60.

Arbelaez, E. P. 1956. Plantas utiles de Colombia. Libreria Colombiana, Bogota, pp. 642-643.

Arenes, J. 1954. Rhizophoraceae Madagascarienses nova. Not. Syst. 15: 1-4.

Areschoug, F. W. C. 1902a. Untersuchungen uber den Blattbau der Mangrovepflanzen. Bibl. Bot. 56: 1-90.

Areschoug, F. W. C. 1902b. Om bladbyggnaden hos Mangrove-vaxterna. Bot. Notiser, 1902, pp. 129-140.

Areschoug, F. W. C. 1904. Zur Frage der Salzausscheidung der Mangrovepflanzen und anderer mit ihnen zusammen wachsender Strandpflanzen. Flora, Marburg 93: 155-160.

Argo, V. N. 1953. Root growth claims soil from the sea: mangrove (*Rhizophora* and *Avicennia*) spreads by unique adaptation. Nat. Hist. 72: 52-55.

Armstrong, P. W. 1954. Shorelines and coasts of the Gulf of Mexico. U. S. Dept. Int. Fish. Bull. 89: 39-65.

Arnott, G. W. 1838. On the Rhizophorae. Ann. Nat. Hist. 1: 359-374.

Artz, T. 1936. Die Kutikula einiger Afrikanischer Mangrove Pflanzen. Ber. Dtsch. bot. Ges. 54: 247-260.

Ascherson, P. 1903. Der nordlichste Fundort der Mangrove in Agypten. Bot. Z. 61, Abt. S, pp. 235-238.

Asenjo, C. F. and J. A. Goyco. 1942. El aceite del almendro tropical. Puerto Rico Dept. Agr. Com. Bol. Mens. 1: 5-7.

Ashby, W. C. and N. C. Beadle. 1957. Studies in Halophytes, III. Salinity factors in the growth of Australian saltbushes. Ecology 38: 344-352.

Ashton, P. S. 1969. Speciation among tropical forest trees: some deductions in the light of recent evidence. Biol. J. Linn. Soc. London 1: 155-196.

Asprey, G. F. 1959. Vegetation in the Caribbean area. Caribb. Q. 5: 245-263.

Asprey, G. F. and R. G. Robbins. 1953. The vegetation of Jamaica. Ecol. Monogr. 23: 359-412.

Atkinson, M. R., G. P. Findlay, A. B. Hope, M. G. Pitman, H. D. W. Saddler, and K. R. West. 1967. Salt regulation in the mangroves *Rhizophora mucronata* Lam. and *Aegialitis annulata* R. Br. Aust. J. Biol. Sci. 589-599.

Aubert, G. 1954. Les sols hydromorphes d'Afrique Occidental Francaise. Trans. 5th Int. Congr. Soil Sci., pp. 447-450.

Aubreville, A. 1964. Problems de la mangrove d'hier et d'aujourd'hui. Addisonia 4: 19-23.

Ayers, A. D. 1951. Seed germination as affected by soil moisture and salinity. Agron. J. 44: 82-84.

Axelrod, D. I. 1952. A theory of angiosperm evolution. Evolution 6: 29-60.

Backer, G. A. and C. G. G. J. van Steenis. 1951. Sonneratiaceae. Fl. Males. 1: 280-289.

Bacon, P. R. 1967. Life in the estuaries of the Caroni Swamp. J. Biol. Univ. West Indies, Trinidad 2: 10-13.

Bacon, P. R. 1970. The ecology of Caroni Swamp, Trinidad. Spec. Publ. Central Statistical Off., Trinidad, 68 pp.

Bacon, P. R. 1971c. Plankton studies in a Caribbean estuarine environment. Caribb. J. Sci. ½: 81-89.

Bacon, P. R. 1971b. The maintenance of a resident population of *Balanus eburneus* (Gould) in relation to salinity fluctuations in a Trinidad mangrove swamp. J. Exp. Mar. Biol. Ecol. 6: 189-198.

Bacon, P. R. 1971C. Studies on the biology and cultivation of the mangrove oyster in Trinidad with notes on other shellfish resources. Trop. Sci. 12: 265-278.

Baillaud, E. 1912. La situation et la produktion des matieres tannates tripicals. J. d'Agric. Trop. 12: 105-107.

Baillon, H. 1875. Combretacees. Hist. Pl. 6: 260-283.

Baillon, H. 1875. Rhizophoracees. Hist. Pl. 6: 284-304.

Baker, R. T. 1915. The australian grey mangrove. J. R. Soc. N.S.W. 49: 257.

Bakhuizen van der Brink, R. C. 1921. Revisio generis Avicenniae. Bull. Jard. bot. Buitenz., Ser. 3, 3: 199-226.

Baltzer, F. 1969. Les formations vegetales associees au delta de la Dumbea (Nouvelle Caledonie). Cah. ORSTOM, Ser. Geol. 1, 1: 59-84.

Banerji, J. 1958a. The mangrove forests of the Andamans. Trop. Silvic. 20: 319-324.

Banerji, J. 1958b. The mangrove forests of the Andamans. World For. Congr. 3: 425-430.

Banijbatana, D. 1957. Mangrove forest in Thailand. Proc. 9th Pac. Sci. Congr., Bangkok, pp. 22-34.

Barbour, M. G. 1970. Is any angiosperm an obligate halophyte? Am. Midl. Nat. 84: 105-120.

Barbour, W. R. 1942. Forest types of tropical America. Carrib. For. 3: 137-150.

Barnard, R. C. and G. G. K. Setten. 1947. Investigation scheme for growth and increment studies in Jahore mangrove forest. Malay. For. 16: 40.

Barroweliff, M. 1912. Chemical notes on some malayan economic products. Malay. Agric. J. 1: 178.

Barry, J. P., L. C. Kiet, and V. V. Cuong. 1961. La vegetation des plages vasosablonneuses de la presquile de Cam Ranli. Ann. Fac. Sci. 1961, pp. 129-140.

Bascope, F., A. L. Bernardi, R. N. Jorgenson, K. Huek, H. Lamprecht, and P.

Martinez. 1959. Descripciones de arboles forestales No. 5. Los manglares en America. Inst. for. lat. Invest. Capacit. Merida 5: 1-53.

Baxter, P. H. 1968. Vegetation notes on Fraser Island. Queensland Nat. 19: 11-20.

Baylis, G. T. S. 1940. Leaf anatomy of the New Zealand mangrove. Trans. R. Soc. New Zealand 70: 164-170.

Beadle, N. C. W. 1952. Studies in halophytes. I. The germination and establishment of the seedlings of five species of *Atriplex* in Australia. Ecology 33: 49-62.

Beard, J. S. 1944. The natural vegetation of the island of Tabago, British West Indies. Ecol. Monogr. 14: 135-163.

Beard, J. S. 1944. Climax vegetation in tropical America. Ecology 25: 127-158.

Beard, J. S. 1946. The Natural Vegitation of Trinidad. Clarendon Press, Oxford, 152 pp.

Beard, J. S. 1955. The classification of tropical American vegetation types. Ecology 36: 89-100.

Beard, J. S. 1967a. An inland occurrence of mangrove. West Austr. Nat. 10: 112-115.

Beard, J. S. 1967b. Some vegetation types of tropical Australia in relation to those of Africa and America. J. Ecol. 55: 271-290.

Becking, J. H., L. G. den Berger, and H. W. Meindersma. 1922. Vloed-of mangrovebosschen in Ned.-Indie. Tectona. 15: 561-611.

Behrens Motta, A. 1960. Estudios preliminares para la produccion de extracto curtiente de corteza de mangle (*Rhizophora mangle*). La Escuela de Farmacia 20: 28-32.

Belcher, C. and G. P. Smooker. 1934. Birds of the colony of Trinidad and Tobago. Ibis 4: 572-595.

Benecke, W. and A. Arnold. 1931. Kulturversuche mit Keimlingen von Mangrovepflanzen. Planta (Berl.) 14: 471-481.

Bennett, H. and R. D. Coveney. 1959. Mangrove bark from Sarawak. Trop. Sci. 1: 116-130.

Benson, A. A. and M. R. Atkinson. 1967. Choline sulphate and phosphate in salt secreting plants. Fed. Proc. 26: 394.

Bentham, G. 1859. Synopsis of Legnotideae, a tribe of Rhizophoraceae. Proc. Linnean Soc. Bot. 3: 65-80.

Bentham, G. and J. D. Hooker. 1865a. Rhizophorae. Gen. Pl. 1: 677-683.

Bentham, G. and J. D. Hooker. 1865b. Combretaceae. Gen. Pl. 1: 683-690.

Bergman, H. F. 1954. Oxygen deficiency as a cause of disease in plants. Bot. Rev. 25: 417-485.

Bernstein, L. 1964. Salt tolerance of plants. U.S.D.A. Bull. 283. Govt. Print. Off., Washington, D. C.

Bernstein, L. and H. E. Hayward. 1958. Physiology of salt tolerance. Ann. Rev. Plant Physiol. 19: 25-46.

Berry, A. J. 1963. Faunal zonation in mangrove swamps. Bull. Singapore

Natl. Mus. 32: 90-98.
Berry, A. J. 1968. Fluctuations in the reproductive condition of *Cassidula auris felis,* a malayan mangrove ellobiid snail. (Pulmonata, Gastropoda). J. Zool. 154: 377-390.
Berry, A. J., S. C. Loong, and H. H. Thum. 1967. Genital systems of *Pythia, Cassidula* and *Auricula* (Ellobiidae, Pulmonata) from malayan mangrove swamps. Malacol. Soc. London Proc. 375: 325.
Berry, E. W. 1913. A fossil flower from the Eocene. Proc. U. S. Natl. Mus. 45: 261-263.
Berry, E. W. 1924. The Middle and Upper Eocene floras of southeastern North America. U. S. Geol. Surv. Prof. Pap. 92: 85-87 and 189-190.
Berry, E. W. 1930. Revision of the Lower Eocene Wilcox Flora of the southeastern states. U. S. Geol. Surv. Prof. Pap. 156: 1-196.
Berry E. W. 1936. Miocene plants from Colombia, South America. Bull. Torrey Bot. Club 63: 53-66.
Beschel, R. E. and P. J. Webber. 1962. Gradient analysis in swamp forests. Nature 194: 207-209.
Bews, J. W. 1916. The vegetation of Natal. Ann. Natal Mus. 2: 253-331.
Bews, J. W. 1920. The plant ecology of the coast belt of Natal. Ann. Natal Mus. 4: 368-469.
Bharucha, F. R. and B. S. Navalkar. 1942. Studies in the ecology of mangroves. 3. The chloride-content of sea-water, soil-solution and the leaf cell-sap of mangroves. J. Univ. Bombay 10: 97-106.
Bharucha, F. R. and V. S. Shirke. 1947. A study of the important katabolic changes in the seedlings of *Avicennia officinalis* Linn. J. Univ. Bombay 15: 1-14.
Biebl, R. 1962. Protoplasmatisch-okologische Untersuchungen an Mangrovealgen von Puerto Rico. Protoplasma 55: 572-606.
Biebl, R. 1964. Temperaturresistenz Tropischer Pflanzen auf Puerto Rico. Protoplasma 59: 133-156.
Biebl, R. and H. Kinzel. 1965. Blattbau und Salzhaushalt von *Laguncularia racemosa* (L) Gaertn. f. und anderer Mangrovenbaume anf Puerto Rico. Ost. bot. Z. 112: 56-93.
Bigarella, J. J. 1946. Contribuicao do estudo da planicie littornea do Estado do Parana. Agric. de Biol. Tecnol. 1: 75-111.
Birch, W. R. 1963. Observations on the littoral and coral vegetation of the Kenya coast. J. Ecol. 51: 603-615.
Biswas, K. 1927. Flora of the saltmarshes. J. Dept. Sci. Univ. Calcutta 8: 1-47.
Biswas, K. 1934. A comparative study of Indian species of *Avicennia*. Notes Bot. Gard. Edinburgh 18: 159-166.
Blake, S. T. 1940. The vegetation of Goat Island and Bird Island in Mareton Bay. Old. Nat. 11: 94-101.
Blake, S. T. 1968. The plants and plant communities of Fraser, Moreton, and Stradbroke Islands. Old Nat. 19: 23-30.

Blatter, E. 1905. The mangrove of Bombay presidency and its biology. J. Bombay Nat. Hist. Soc. 16: 644-656.

Blatter, E. 1908. On the flora of Cutch. I. J. Bombay Nat. Hist. Soc. 18: 756-777.

Blatter, E. 1909. On the flora of Cutch. II. Bombay Nat. Hist. Soc. 19: 157-176.

Blomquist, H. L. and L. R. Almodovar. 1961. The occurrence of *Gelidiella tenuissima* Feldm. et Hamel in Puerto Rico. Nova Hedwiga 111: 67-69.

Blum, G. 1941. Uber osmotische Untersuchungen in der Mangrove. Ber. Schweiz. bot. Ges. 51: 401-420.

Boaler, S. B. 1959. *Conocarpus lancifolius* Engler in Somaliland Protectorate. Empire For. Rev. 38: 371-379.

Boer, N. P. de, Th. van der Hammen, and T. A. Wijmstra. 1965. A palynological study on the age of some borehole smaples from the Amazonas delta area, N. W. Brazil. Geol. Mijnb. 44: 254-258.

Boergesen, F. 1909. Notes on the shore vegetation of the Danish West Indian Islands. Bot. Tidsskrift. 29: 201-259.

Boergesen, F. and O. Poulsen. 1900. La vegetation des Antilles danoises. Rev. gen. de Bot. 12: 99.

Boffey, P. M. 1971. Herbicides in Veitnam: AAAS study finds widespread devastation. Science 171: 43-47.

Bole, P. V. and F. R. Bharucha. 1954. Osmotic relations of the leaves of *Avicennia alba* B1. J. Univ. Bombay 22: 50-54.

Bond, G. 1956. A feature of the root nodules of *Casuarina*. Nature 177: 192.

Booberg, G. 1933. Die malayische Strandflora-ein Revision der Schimperschen Artenliste. Bot. Jb. 66: 1-38.

Bonfils, P. and J. Faure. 1961. Etude des sols due Bao Bolon. Agron. Trop. (Maracay) 16: 127-147.

Boone, R. S., M. Chudnoff, and E. Goytia. 1969. Chemical control of Bostrichidae during air drying of fence posts. U. S. For. Ser. Res. Pap. IFT, No. 8: 1-8.

Boughey, A. S. 1957. Ecological studies of tropical coast-lines. I. The Gold Coast, West Africa. J. Ecol. 45: 665-687.

Boughey, A. S. 1963. Dating of tropical coastal regression (based on roots of mangrove (*Rhizophora mucronata*). Nature 200: 600.

Bouillene, R. 1930. Un voyage botanique dans le Bas-Amazone. In Une Mission Biologique Belge au Bresil 2: 13-185. Imprimerie Medicale et Scientifique, Bruxelles.

Bouillene, R. and F. Went. 1933. Reserches experimentales sur la neoformation des racines les plantules et les boutures de plantes superecures. Ann. Jard. Bot. Buitz. 43: 25.

Bournot, K. 1913. Gewinnung von Lapachol aus dem Kernholz von *Avicennia tomentosa*. Archiv der Pharm. 1913: 351-356.

Bower, C. A. and C. H. Wadleight. 1948. Growth and cationic accumulation by four species of plants as influenced by various levels of exchangeable

sodium. Soil Sci. Am. Proc. 13: 218-223.
Bowman, H. H. M. 1915. Report on botanical work at the Tortugas Laboratory for the season of 1915. Carnegie Inst. Wash. Yearb. No. 14, p. 200.
Bowman, H. H. M. 1916a. Report on botanical investigation at the Tortugas Laboratory, season 1916. Carnegie Inst. Wash. Yearb. No. 15, pp. 188-192.
Bowman, H. H. M. 1916b. Physiological studies on *Rhizophora*. Proc. Natl. Acad. U. S. 2: 685-688.
Bowman, H. H. M. 1917. Ecology and physiology of the Red Mangrove. Proc. Am. Philos. Soc. 56: 589-672.
Bowman, H. H. M. 1918. Botanical ecology of the Dry Tortugas. Carnegie Inst. Wash., Pub. No. 252, Pap. Dept. Mar. Biol., vol. 12, paper V, pp. 109-138.
Bowman, H. H. M. 1921. Histological variations in *Rhizophora mangle*. Pap. Michigan Acad. Sci. 22: 129-134.
Brandis, D. 1898. Combretaceae. Nat. Pflanzenfam. III. 7: 106-130.
Brass, L. J. 1938. Botanical results of the Archibold Expeditions. IX. Notes on the vegetaion of the Fly and Wassi Kussa Rivers, British New Guinea. J. Arnold Arb. 19: 175-190.
Breen, C. M. and B. J. Hill. 1969. A mass mortality of mangroves in the Kosi estuary. Trans. R. Soc. S. Africa 38: 285-303.
Breen, C. M. and I. D. Jones. 1969. Observations on the anatomy of foliar nodes of young *Bruguiera gymnorrhiza*. J. S. Afr. Bot. 35: 211-218.
Brelie, G. v. d. and M. Teichmuller. 1953. Beitrage zur Geologie El Salvadors. III. Mikroskopische Beobachtungen an Mangrove — Sedimenten aus El Salvador. Neues Jb. Geol. Paleontol., Mh., No. 6, pp. 244-251.
Brenner, W. 1902. Ueber die Luftwurzeln von *Avicennia tomentosa*. Ber. Dtsch. bot. Ges. 20: 175-189.
Briolle, C. E. 1969. Le cocotier sur les terres de mangrove au Cambodge: amenagement de plantations familiales. Oleagineaux 24: 545-549.
Broekhuysen, G. J. and H. Taylor. 1959. The ecology of South African Estuaries. Part viii: Kosi Bay system. Ann. S. Afr. Mus. 44: 279-296.
Bronnimann, P. and L. Zaninetti. 1965. Notes sur *Lituola salsa* (Cushman et Bronnimann 1948), un Foraminifere da la mangrove de l'ile de la Trinitie, W. I. Arch. Sci. 18: 608-615.
Brown, D. S. 1971. Ecology of Gastropoda in a south African mangrove swamp. Proc. Malacol. Soc. London. 39: 263-279.
Brown, J. M. A., H. A. Outred, and C. F. Hill. 1969. Respiratory metabolism in mangrove seedlings. Plant Physiol. 44: 287-294.
Brown, R. 1814. *In* Matthew Flinders' A Voyage to Terra Australis, Vol. II, Appendix No. III. G. and W. Nicol, London.
Brown, W. H. and A. F. Fischer. 1918. Philippine mangrove swamps. Philippines Dept. Agric. Nat. Resources Bull. No. 17, pp. 1-132.
Brown, W. H. and A. F. Fischer. 1920. Philippine mangrove swamps.

Philippines Bur. For. Bull. No. 22, pp. 9-125.
Browne, F. G. 1955. Forest trees of Sarawak and Brunei and their products. Sarawak Gov. Printer, Kuching.
Brownell, P. F. and J. G. Wood. 1957. Sodium, an essential micro-nutrient for *Atriplex vesicaria* Heward. Nature. 179: 635-636.
Brunnich, J. C. and F. Smith. 1911. Some Queensland mangrove barks and other tanning materials. Qld Agric. 27: 86.
Buchanan, J. 1874. On the flowering plants and ferns of Chatham Islands. Trans. N. Z. Inst. 7: 333-341.
Buckley, T. A. 1929. Mangrove bark as a tanning material. Malay. For. Rec. 7, 40 pp.
Bunning, E. 1944. Botanische Beobachtungen in Sumatra. Flora 137: 334-344.
Burkill, I. H. 1935. A dictionary of the Economic products of the Malay Peninsula. 2 vols., 2402 pp. Publ. by Governments of the Straits Settlements and the Federated Malay States, London.
Burkholder, P. R. and L. M. Burkholder. 1958. Studies on B. vitamins in relation to productivity of the Bahia Fosforescente, Puerto Rico. Bull. Mar. Sci. Gulf Caribb. 8: 201-223.
Burkholder, P. R. and L. R. Almodovar. Species composition and productivity of algal mangrove communities in Puerto Rico. *In press Q. J. Fla. Acad. Sci.*
Burtt, Davy, J. 1938. The classification of woody vegetation types. Imperial For. Inst. Pap. No. 13, Oxford.
Calderon, S. E. and P. C. Stanley. 1941. Flora Salvadorena, 2nd. Ed., El Salvador.
Candolle, A. P. de. 1828a. Combretaceae, Prodr. Syst. Nat. 3: 9-24.
Candolle, A. P. de 1828b. Rhizophoraceae, Prodr. Syst. Nat. 3: 31-34.
Carey, G. 1934. Further investigations on the embryology of viviparous seeds. Proc. Linn. Soc. N. S. W. 59: 392-410.
Carey, G. and L. Fraser. 1932. The embryology and seedling development of *Aegiceras majus* Gaertn. Proc. Linn. Soc. N. S. W. 57: 341-360.
Carlquist, S. 1967. The biota of long distance dispersal. V. Plant dispersal to Pacific islands Bull. Torrey Bot. Club 94: 129-162.
Carter, G. S. 1930. The fauna of the swamps of Paraguayan Chaco in relation to its environment. I. Physicochemical nature of the environment. J. Linn. Soc. London, Zool. 37: 206-258.
Carter, J. 1959. Mangrove succession and coastal change in South-West Malaya. Trans. Inst. Br. Geogr. 26: 79-88.
Case, G. O. 1938. The use of vegetation for coast protection. Agric. J. Br. Guiana 9: 4-11.
Catala, R. L. A. 1957. Report on the Gilbert Islands: some aspects of human ecology. Atoll Res. Bull. 59: Oct. 31.
Cawkell, E. M. 1964. The utilization of mangroves by African birds. Ibis 106: 251-253.

Chandler, M. E. J. 1951. Notes on the occurrence of mangroves in the London clay. Proc. Geol. Assoc. 62: 271-272.

Champion, H. G. 1936. A preliminary survey of the forest types of india and Burma. Indian Forest Rev. Bot. 1, 265 pp.

Chang, Hung-ta, Chao-Chang Chang, and Paisun Wang. 1957. The mangrove vegetation of the Liochow Peninsula. Sunyatsenia Nat. Sci. Ed. 1957: 122-143 (in Chinese).

Chapman, V. J. 1936. The halophyte problem in the light of recent investigations. Q. Rev. Biol. 11: 209-220.

Chapman, V. J. 1940. The botany of the Jamaica shoreline. Geogr. J. 96: 312-323.

Chapman, V. J. 1942. A new perspective in halophytes. Q. Rev. Biol. 17: 291-373.

Chapman, V. J. 1944a. 1939 Cambridge University Expedition to Jamaica. I. A study of the botanical processes concerned in the development of the Jamaican shore-line. J. Linn. Soc. London Bot. 52: 407-447.

Chapman, V. J. 1944b. 1939 Cambridge University Expedition to Jamaica. II. A study of the environment of *Avicennia nitida* Jacq. in Jamica. J. Linn. London Soc. Bot. 52: 448-486.

Chapman, V. J. 1944c. 1939 Cambridge University Expedition to Jamica. III. The morphology of *Avicennia nitida* Jacq. and the function of its pneumatophores. J. Linn. Soc. London Bot. 52: 487-533.

Chapman, V. J. 1962a. Respiration studies of mangrove seedlings. I. Material and some preliminary experiments. Bull. Mar. Sci. Gulf Caribb. 12: 137-167.

Chapman, V. J. 1962b. Respiration studies of mangrove seedlings. II. Respiration in air. Bull. Mar. Sci. Gulf Caribb. 12: 245-263.

Chapman, V. J. 1966a. Some factors involved in mangrove establishment. *In* Les problemes scientifiques des deltas de la zone tropicale humide et leurs implications. Colloque de Dacca. UNESCO, Paris, pp. 219-225.

Chapman, V. J. 1966b. Vegetation and salinity. *In* Salinity and Aridity, H. Boyko (ed.), Junk Publishers, The Hague, Netherlands, pp. 23-42.

Chapman, V. J. 1969. Lagoons and mangrove vegetation. *in* Lagunas Costeras, un Simposio, Mem. Simp. Intern. Lagunas Costeras. UNAM-UNESCO, Nov. 28-30, 1967. Mexico, D. F., pp. 505-514.

Chapman, V. J. 1970. Mangrove phytosociology. Trop. Ecol. 11: 1-19.

Chapman, V. J. and J. W. Ronaldson. 1958. The mangrove and salt-grass flats of the Auckland Isthmus. N. Z. Dept. Sci. Ind. Res., Bull. 125, 79 pp.

Chapman, V. J. and C. B. Trevarthen. 1953. General schemes of classification in relation to marine coastal zonation. J. Ecol. 41: 198-204.

Chatelain, G. 1932. L'exploitation forestiere a la Guayane Francaise. Actes e Comp. Rendus de l'Association Colonies-Sciences, Paris 8: 217-222.

Chattaway, M. M. 1932. The wood of the Sterculiaceae. 1. Specialization of the vertical wood parenchyma within the sub-family Sterculieae. New

Phytol. 31: 119-132.
Chattaway, M. M. 1938. The wood anatomy of the family Sterculiaceae. Phil. Trans. R. Soc. B. 228: 313-365.
Chatterjee, D. 1958. Symposium on mangrove vegetation. Sci. Cult. 23: 329-335.
Chaudhri, I. I. 1967. The vegetation of Karachi. Vegetatio 10: 229-246.
Chayapongse, C. 1955. Yield of mangrove forest. Vanasarn For. J. 2, April 1955.
Cherrie, G. K. 1891. Notes on Costa Rican birds. Proc. U. S. Natl. Mus. 14: 517-537.
Chevalier, A. 1924. Exploitation des Paletuviers a tanin et leur valeur d'apres quelques travaux recents. Rev. Bot. Appl. Agric. Colon 4: 320.
Chevalier, A. 1929. Possibilite de cultiver le Bananier nain sur les terrains a Paletuviers ou Bresil et en Afrique occidentale. Rev. Bot. Appl. Agric. Colon 9: 334-337.
Chevalier, A. 1931. Graines d'Avicennia
comme aliment de famine. Rev. Bot. Appl. Agric. Colon 11: 1000.
Chidester, G. H. and E. R. Schafer. 1959. Pulping of Latin American woods. Products Lab. Rep. No. 2012, Madison, Wisconsin, 11 pp.
Chiovenda, E. 1929. Flora Somala. Rome, 436 pp.
Chipp, C. G. 1921. A list of the fungi of the Malay Peninsula. Bull. Straights Settlements 11: 311-418.
Chipp, T. F. 1931. The vegetation of northern tropical Africa. Scot. Geog. Mag. 47: 193-214.
Chowdhury, K. A. 1934. The so-called terminal parenchyma cells of the wood of *Terminalia tomentosa* W. and A. Nature 133: 215.
Christian, C. S. and G. A. Stewart. 1953. General report on a survey of the Katherine-Darwin region, 1946. CSIRO Austr. Land Res. Ser. 1.
Christophersen, E. 1935. Flowering plants of Samoa. B. P. Bishop Museum Bull. No. 128, Honolulu, Hawaii.
Ciferri, R. 1939. La associazioni de litorale marino della Somalia meridionale. Riv. Biol. Colon No. 2.
Clarke, L. D. and N. J. Hannon. 1967. The mangrove swamp and salt marsh communities of the Sydney district. I. Vegetation, soils and climate. J. Ecol. 55: 753-771.
Clarke, L. D. and N. J. Hannon. 1969. The mangrove swamp and salt marsh communities of the Sydney district. II. The holocoenotic complex with particular reference to physiography. J. Ecol. 57: 213-234.
Clarke, L. D. and N. J. Hannon. 1970. The mangrove swamp and salt marsh communities of the Sydney district. III. Plant growth in relation to salinity and waterlogging. J. Ecol. 58: 351-369.
Clarke, L. D. and N. J. Hannon. 1971. The mangrove swamp and salt marsh communities of the Sydney district. IV. The significance of species interaction. J. Ecol. 59: 535-553.
Coaldrake, J. E. 1961. The ecosystem of the coastal lowlands (Wallum) of

southern Queesland. CSIRO Austr. Bull. No. 28, 138 pp.
Cockayne, L. 1958. The vegetation of New Zealand. Hafner Pub. Co., New York, 456 ppb.
Coelho, P. A. 1965-1966. Os crustaceos decapodos de alguns manguezais pernambucanos. Trab. Inst. Oceanogr., Univ. Fed. Pernambuco, Recief (7/8): 71-89.
Cogger, H. G. 1959. Australain goannas. Aust. Mus. Mag. 13: 71-75.
Cohen, E. 1939. The marine angiosperms of Inhaca Island. S. Afr. J. Sci. 36: 246.
Collins, M. I. 1921. On the mangrove and salt marsh vegetation near Sydney, N. S. W., with special reference to Cabbage Creek, Port Hacking. Proc. Linn. Soc. N. S. W. 46: 376-392.
Colman, J. 1940. Zoology of the Jamaican shoreline. Geogr. J. 96: 323-327.
Compere, P. 1963. The correct name of the Afro-American black mangrove (*Avicennia germinans* L., Verbenaceae). Taxonomy 13: 150-152.
Connor, D. J. 1969. Growth of grey mangrove (*Avicennia marina*) in nutrient culture, Biotropica 1: 36-40.
Cook, E. P. and R. C. Collins. 1903. Economic plants of Puerto Rico. Cont. U. S. Natl. Herb. 8: 2. 229-230.
Cook, M. T. 1907. The embryology of *Rhizophora mangle*. Bull. Torrey Bot. Club 34: 271-277.
Cook, M. T. 1908. The hypertrophied fruit of *Bucida buceras*. Bull. Torrey Bot. Club 35: 305-306.
Cooper, R. E. and S. A. Pasha. 1935a. Osmotic and suction pressure of some species of mangrove vegetation. J. Indian Bot. Soc. 14: 109-120.
Cooper, R. E. and S. A. Pasha. 1935b. Osmotic pressure and the H-ion concentration of sea weeds in relation to those of sea-water. J. Indian Bot. Soc. 14: 3.
Corner, E. J. H. 1940. Wayside trees of Malaya, Vol. 1. Government Printer, Singapore, 772 pp.
Cotton, B. C. 1943. Australian Gastropoda of the families Hydrobiidae, Assimineidae, and Acmaeidae. Trans. R. Soc. Austr. 66: 124-129.
Cotton, B. C. 1950. An old mangrove mud-flat exposed by wave scouring at Glenelg, South Australia. Trans. R. Soc. S. Austr. 73: 59-61.
Craighead, F. C. 1964. Land, mangroves and hurricanes. Fairchild Trop. Gard. Bull. 19: 5-32.
Craighead, F. C. 1968. The role of the alligator in shaping plant communities and maintaining wildlife in southern Everglades. Florida Nat. 41: 2-7.
Craighead, F. C. 1969. Vegetation and recent sedimentation in Everglades National Park. Fla. Nat. 42: 157-166.
Craighead, F. C. 1971. The Trees of South Florida. Vol. 1. The Natural Environments and Their Succession. Univ. Miami Press, Coral Gables, 212 pp.
Craighead, F. C. and V. C. Gilbert. 1962. The effects of hurricane Donna on

the vegetation of southern Florida. Q. J. Fla. Acad. Sci. 25: 1-28.

Crane, J. 1941. Eastern Pacific expeditions of the New York Zoological Society. XXVI. Crabs of the genus *Uca* from the west coast of Central America. Zoologia, N. Y. 26: 145-208.

Crane, J. 1947. Intertidal brachygnathous crabs from the west coast of tropical America with special reference to ecology. Zoologia, N. Y. 31: 69-95.

Creager, D. B. 1962. A new Cercospora of *Rhizophora mangle*. Mycologia 54: 536-539.

Crossland, C. 1903. Note on the dispersal of mangrove seedlings Ann. Bot. Fenn. 17: 267-270.

Cruz, A. A. de la and J. F. Banaag. 1967. The ecology of a small mangrove patch in Matabung Kay Beach, Batangas Province. Univ. Philipp. Nat. Appl. Sci. Bull. 20: 486-494.

Cuatrecasas, J. 1947. Vistazo a la vegetacion natural del calima. Acad. Colombiana de Cient. Exact. Fis. y Nat. Rer. 7: 306-312.

Cuatrecasas, J. 1952. Mangroves of the Pacific coast of South America. Abstract of a series of lectures presented before the Department of Botany Seminar, Northwestern University, by the botanical staff of the Chicago Natural History Museum (cited in West, 1956).

Cuatrecasas, J. 1958a. Introduction al estudia de los manglares. Boln. Soc. Bot. Mex. 23: 84-98.

Cuatrecasas, J. 1958b. Aspectos de la vegetacion natural de Colombia. Rev. Acad. Colombiana Cienc. Exactas, Fis., Nat. 10: 221-264.

Curtis, A. H. 1888. How the mangrove forms islands. Gard. For. 1: 100.

Daiber, F. C. 1960. Mangroves: The tidal marshes of the tropics. Univ. Delaware Estuarine Bull. 5: 10-15.

Daiber, F. C. 1963. Tidal creeks and fish eggs. Univ. Delaware Estuarine Bull. 7: 6-14.

Dale, I. R. 1938. Kenya mangroves. Z. Weltforstwirtsch. 5: 413-421.

Danhof, G. N. 1946. Rotation and management of mangrove in Riouw-Singga Archipelago. Tectona 36: 59-72.

Dabsereau, P. 1947. Zonation et succesion sur la restinga de Rio de Janeiro. I. La holosere. Rev. Can. Biol. 6: 448-477.

Dansereau, P. 1957. Biogeography, an Ecological Perspective. Ronald Press. New York, 394 pp.

Darteville, E. 1949. Les mangroves d'Afrique equatorial. Atti. Acad. Ligure. Sci. Lett. 6: 1-48.

Darteville, E. 1950. Les mangroves de Congo et les autres mangroves d'Afrique occidentale. Bull. Seances Inst. R. Coll. Belge 21: 946-971.

Davies, J. B. 1967. The distribution of sand flies (*Culicoides* supp.) breeding in a tidal mangrove swamp in Jamaica and the effects of tides on the emergence of *C. furens* Poey and *C. barbosai* (Wirth and Blanton). West Indies Med. J. 16: 39-48.

Davies, J. B. 1969. Effect of felling mangroves on emergence of *Culicoides*

spp. in Jamaica. Mosq. News 29: 566-571.
Davis, C. C. 1966. Notes on the ecology and reproduction of *Trichocorixa reticulata* in a Jamaican salt-water pool. Ecology 47: 850-852.
Davis, C. C. and R. H. Williams. 1950. Brackish water plankton of mangrove areas in southern Florida. Ecology 31: 519-531.
Davis, J. H., Jr. 1938. Mangroves, makers of land. Nat. Mag. 31: 551-553.
Davis, J. H. 1939a. Vivipary and dispersal of mangrove seeds. J. Tennessee Acad. Sci. 15: 415.
Davis, J. H. 1939b. The role of mangrove vegetation in land building in southern Florida. Am. Phil. Soc. Yearb., 1938, pp. 162-164.
Davis, J. H. 1940a. Peat deposits of Florida. Fla. Geol. Sur. Bull. 30: 1-247.
Davis, J. H. 1940b. The ecology and geologic role of mangroves in Florida. Pap. from the Tortugas Lab. Vol. 32. Carnegie Inst. Wash. Pub. No. 517, pp. 303-412.
Davis, J. H. 1942. The ecology of the vegetation and topography of the sand keys of Florida. Pap. from the Tortugas Lab. Vol. 33. Carnegie Inst. Wash. Pub. No. 524, pp. 113-195.
Davis, J. H. 1943. The natural features of southern Florida. Florida Dept. Conserv. Geol. Bull. No. 25. 311 pp.
Davis, J. H. 1945. Jamaican shoreline ecology. Ecology 26: 312.
Davy, J. B. 1938. The classification of tropical woody vegetation-types. Imperial For. Inst., Univ. Oxford, Inst. Pap. No. 13. 85 pp.
Dawson, E. Y. 1954. Marine plants in the vicinity of the Institut Oceanographique de Nha Trang, Viet Nam. Pac. Sci. 8: 372-481.
Day, J. H. 1951. The ecology of South African estuaries, Part I. General considerations. Trans. R. Soc. S. Afr. 33: 53-91.
Day, J. H., N. A. H. Millard, and A. D. Harrison. 1952. The ecology of South African estuaries. Part VII. The biology of Durban Bay. Ann. Natal Mus. 13: 259.
Day, J. H., N. A. H. Millard, and G. J. Broekhuysen. 1953. The ecology of South African estuaries. Part IV. The St. Lucia system. Trans. R. Soc. South Africa 34: 129.
Day, J. H. and J. F. C. Morgans. 1956. The ecology of South African estuaries. Part VIII. The biology of Durban Bay. Ann. Natal Mus. 13: 259.
Decaisne, J. 1835. Observations sur quelque nouveau generes et especes des plantes de l'Arabie-Hereuse. Rhizophoraceae. Ann. des Sci. Nat., Bot., Paris, pp. 75-77.
Degener, O. 1946. Flora Hawaiiensis, Books 1-4. The Patton Col., Honolulu.
Deignan, H. G. 1961. Type specimens of birds in the United States National Museum. U. S. Natl. Mus. Bull. 221, 718 pp.
Delevoy, G. 1945. Les mangroves africaines. Bull. Soc. Cient. For. Belg. 52: 84-89.
Delf, E. M. 1912. Transpiration in succulent plants. Ann. Bot. 26: 409-412.
Dent, J. M. 1947. Some soil problems in empoldered rice lands in Sierra Leone. Empire J. Exp. Agric. 15: 206-212.

Derijard, R. 1965. Contribution a l'etude du peuplement des sedments sable-vaseux, et vaseux intertidaux, compactes ou fixes par la vegetation de la region de Tulear (Madagascar). Rec. Trav. Stat. Mar. Endoume. Fasc. h. s. suppl. No. 43in-8, 94 pp.

Detwiler, S. B. 1948. Some basic information regarding tannin and tannin-yielding crops, prepared for use by tannin-research workers. U. S. Off. For. Agric. Washington, D. C. 112 pp.

D'Hoore, J. 1963. Soil Map of Africa (5th revision). C.C.T.A./C.S.A., Joint Project No. 11, Leopoldville.

Dickerman, R. W. and G. Gavino T. 1970. Studies of a nesting colony of green herons at San Blas, Mexico. Living Bird 8: 95-111.

Dickerman, R. W. and C. Juarez L. 1971. Nesting studies of the boat-billed heron *Cochlearius cochlearius* at San Blas, Nayarit, Mexico. Ardea 59: 1-16.

Dickey, D. R. and A. J. van Rossem. 1938. The birds of El Salvador. Field Mus. Nat. Hist. Publ. Zool. Ser. 23.

Diels, L. 1915. Vegetationstypen vom untersten Kongo. Vegetations-bilder 12, VIII, 16 pp.

Ding Hou. 1957. A conspectus of the genus *Bruguiera* (Rhizophoraceae). Nova Guinea (n.s.) 8: 163-171.

Ding Hou. 1958. Rhizophoraceae. F. Males. I. 5: 429-493.

Ding Hou. 1960. A review of the genus *Rhizophora* with special reference to the Pacific species. Blumea 10: 625-634.

Ding Hou. 1965. Studies in the Flora of Thailand. Rhizophoraceae. Dansk. Bot. Arkiv. 23: 187-190.

Dodd, J. R. and C. T. Siemers. 1971. Effect of Late Pleistocene karst topography on Holocene sedimentation and biota, lower Florida Keys. Geol. Soc. Am. Bull. 82: 211-218.

Doderlein, L. 1881. Botanische Mitteilungen aus Japan. Bot. Centrabl. 8: 27-31.

Dolianiti, E. 1955. Frutos de Nipa no Palaeocene de Pernambuco. Div. Geol. e Mineral Brasil Bol. 158: 1-36.

Doyne, H. C. 1937. A note on the acidity of mangrove swamp soils. Trop. Agric. Trinidad 14: 236-237.

Doyne, H. C. and R. R. Glanville. 1933. Some swamp rice growing soils Sierra Leone. Trop. Agric. Trinidad 10: 132-138.

Drabble, E. 1908. The bark of the red and the white mangroves. Q. J. Instit. Comm. Res. Tropics 3: 33-39.

Drabble, E. and L. Hilda. 1905. The osmotic strength of cell-sap in plants growing under different conditions. New Phytol. 4: 8.

Drabble, E. and M. Nierenstein. 1907. A note on the West-African mangroves. Q. J. Instit. Comm. Res. Tropics 2.

Drakenstein, H. A. Rheede tot. 1678-1703. Hortus indicus malabaricus. 12 vols. (*Sonneratiaa*, vol. 3; *Aegiceras, Rhizohora*, and *Bruguiera*, vol. 6, *Avicennia*, vol. 4). Van Someren and van Dijk, Amsterdam.

Drar, M. 1933. A note on some plants of Gebel Elba: *Avicennia officinalis*. Hart. Rev. Egypt. Hart. Soc. 21: 7-10.

Dubois, J. 1954. Sur une classification de sols de delta soumis a des influences salines appliquee au bas-Senegal. Proc. 2nd. Int. Afr. Soil Conf. 2: 1119-1124.

Dugros, M. 1937. Le domaine forestier inonde de la Cochine. Bull. Econ. de l'Indochine 40: 283-314.

Durand, J. H. 1964. La mise en valeur des mangroves de la cote N. W. de Madagascar. Bull. AFES Versailles 5: 200-206.

Durant, C. C. L. 1941. The growth of mangrove species in Malaya. Malay. For. 10: 3-15.

Eddy, W. H. and B. Sokoloff. 1953. Mangrove meal as a cattle food ingredient. Southern Bio-Research Lab., Fla. Southern Col., Lakeland, Florida.

Edwall, P. W. 1877. Enseio para una synonimia dos nomes populares das plantas indigenas do estado de Sao Paulo. Sao Paulo, Boletin No. 16.

Eggers, H. von. 1877. Rhizophora mangle L. Vidensk. Meddel. Dansk Naturhist. Foren. Kjobenhaven 28: 177-181.

Eggers, H. von. 1892. Die Manglares in Ecuador. Bot. Centralb. 52: 49-52.

Eggers, H. von. 1894. Das Kustenbebiet von Ecuador. Geog. Gesell. Bremen 17: 265-289.

Egler, F. E. 1948. The dispersal and establishment of red mangrove in Florida. Caribb. For. 9: 299-310.

Egler, F. E. 1952. Southeast saline Everglades vegetation, Florida, and its management. Veg. Acta Geobot. 3: 213-265.

Egler, F. E. 1961. A cartographic guide to selected regional vegetation literature — where plant communities have been described. II. Southeastern United States. Sarracenia 6: 1-87.

Eisley, L. 1971. The Night Country, Charles Scribner's Sons, New York, 240 pp.

Ekman, S. 1953. Zoogeography of the Sea. Sidgewick and Jackson Ltd., London, 417 pp.

Emery, K. O. and R. E. Stephenson. 1957. Estuaries and lagoons. *in* Treatise on Marine Ecology and Paleoecology. Vol. 1, J. S. Hedgepeth (ed.). Geol. Soc. Am. Mem. 67: 673-750.

Engler, A. 1876. Rhizophoraceae in Martius. Flora Brazil. 12: 426.

Engler, A. 1921. Die Pflanzenwelt Afrikas. Erster Band, I Heft; Driter Band, II Heft. Leipzig.

Engler, A. and L. Diels. 1899. Monographieen Afrikanisher Pflanzenfamilien und - gattungen. III. Combretaceae-*Combretum*. 116 pp., Leipzig.

Engler, A. and L. Diels. 1900. Monographieen Afrikanisher Pflanzenfamilien und - gattungen. IV. Combretaceae excl. *Combretum*. 44 pp. Leipzig.

Enns, T. 1965. Tracer studies of water and sodium transport in mangroves (*Laguncularia racemosa*). Havalradets Dkr. 48: 161-163.

Epstein, E., W. E. Schmid, and D. W. Rains. 1963. Significance and technique of short-term experiments on solute absorption by plant tissue. Plant Cell Physiol 4: 79-84.

Erichson, R. 1921. Die Mangrove-vegetation ihre Verbreitung und ihre Bedeuting fur Schwemmland-bildungen. Ungerdrukte Diss. (Maschinenschrift), Halle.

Erichson, R. 1925. Die Mangrove-Vegetation. Z. f. Naturfr. 16: 190-199.

Ernould, M. 1921. Recherches anatomiques et physiologiques sur les racines respiratoires. Mem. Acad. R. de Bel. 6: 1-52.

Exell, A. W. 1931. The genera of Combretaceae. J. Bot. 69: 113-128.

Exell, A. W. 1954. Combretaceae. Fl. Males 1: 533-589.

Exell, A. W. 1958. Combretaceae. Ann. Mo. Bot. Gar. 45: 143-164.

Exell, A. W. and C. A. Stace. 1966. Revision of the Combretaceae. Bol. Soc. Broteriana 40: 5-25.

Faber, F. C. von. 1913. Uber Transpiration und osmotischen Druch bei den Mangroven. Ber. Dtsch. bot Ges. 31: 277-286.

Faber, F. C. von. 1923. Zur Physiologie der Mangroven. Ber. Dtsch. bot. Ges. 41: 227-234.

Fairbridge, R. W. and C. Treichert. 1947. The rampart system at Low Isles. Rep. Great Barrier Reef Comm. 6: 1016.

Fanshawe, D. 1948. Forest products of British Guiana. Part I. Principal timbers. For. Bull. 1 NS, For. Dept. Brit. Guiana, 39 pp.

Fanshaw, D. 1950. Forest products of British Guiana. Part II. Minor forest products. For. Bull. 2 NS, For. Dept. Brit. Guiana, pp. 16-18, 49.

Fanshaw, D. 1952. The Vegetation of British Guiana, A Preliminary Review. Pap. For. Inst., Oxford, 96 pp.

Feldman, J. and R. Lami. 1936. Sur la vegetation de la mangrove a la Guadelope. C. R. Acad. Sci. (Paris) 203: 883-885.

Feliciano, C. 1962. Notes on the biology and economic importance of the land crab *Cardiosoma guanhumi* Latreille of Puerto Rico. Contrib. Inst. Mar. Biol. Univ. Puerto Rico, Mayaguez, 1-29.

Fenner, C. and J. B. Cleland. 1932. The geography and botany of the Adelaide coast. S. Aust. Nat. 14: 45-48, 55-56, 109-120, 128-133.

Fernandos, D. A. 1934. Ueber Mangroven Cultuven. Tectona 27: 299-303.

Ferns, G. W. 1955a. Tidal forests in southeast Kalimantan. Indian J. For. 4: 3-5.

Ferns, G. W. 1955b. Hutan pajan sekitar Kalimantantenggara (Tidal forests of S. E. Borneo). Rimba Indonesia Penerb. pop. 4: 90-98.

Ffrench, R. P. 1966. The utilization of mangroves by birds in Trinidad. Ibis 108: 423-424.

Ffrench, R. P. 1970. The scarlet Ibis in Surinam and Trinidad. Living Bird 9: 147-165.

Field, G. D. 1968. Utilization of mangroves by birds on the Freetown peninsula, Sierra Leone. Ibis 110: 354-357.

Finucane, J. H. 1965. Threadfish in Tampa Bay, Florida. Q. J. Fla. Acad. Sci. 28: 267-270.

Fischer, P. H. 1940. Notes sur les peuplements littoraux d'Australia. III. Sur le faune de la Mangrove australienne. Mem. Soc. Biogeogr. 7: 315-329.

Fisher, C. E. C. 1927. Contributions to the flora of Burma, 2. Kew Bull. 1927: 81-94.

Flamm, B. R. and J. H. Cravens. 1971. Effects of war damage on the forest resources of South Vietnam. J. For. 69: 784-790.

Fly, L. B. 1952. Preliminary pollen analysis of the Miami, Florida, area. J. Allergy 23: 48-57.

Forattini, O. P. 1957. *Culicoides* daaregiao neotropical (Diptera, Ceratopogonidae). Arch. Hig. Sao Paulo 11, 526 pp.

Forattini, O. P., E. X. Rabello, and D. Pattoli. 1958. *Culicoides* de regiao neotropical. 2. Observacoes sobre biologia em conicoes naturais. Arch. Hig. Sao Paulo 12: 1-52.

Fosberg, F. R. 1947. Micronesian mangroves. J. N. Y. Bot. Gard. 48: 128-138.

Fosberg, F. R. 1953. Vegetation of central Pacific atolls. Atoll Res. Bull. 23, Sept. 30.

Fosberg, F. R. 1960. The vegetation of Micronesia: 1. General description of the vegetation of Guam. Bull. Am. Mus. Nat. Hist. 119: 1-75.

Fosberg, F. R. 1961. Vegetation-free zone on dry mangrove coasts. U. S. Geol. Soc. Prof. Pap. No. 424D, pp. 216-218.

Fosberg, F. R. 1964. Vegetation as a geological agent in tropical deltas. *In* Les problems scientifiques des deltas de la zone tropicale humide et leurs implications. Colloque de Dacca. UNESCO, Paris, pp. 227-233.

Fourmanoir, P. 1953. Notes sur la faune de la mangrove dans la region de Majunga: crabs, crevettes, poissons. Nat. Malgache 5: 87-92.

Foxworthy, F. W. 1910. Distribution and utilization of the mangrove-swamp of Malaya. Ann. Jar. Bot. de Buitz., Suppl. 3, Part I, pp. 319-344.

Foxworthy, F. W. and D. M. Matthews. 1916. Mangrove and nipa swamps of British North Borneo. For. Bull. British N. Borneo 1: 1-67.

Freise, F. W. 1932. Brasilianische Gerbstoffpflanzen und ihre Ausbeutung. Tropenpflanzen 35: 70-74.

Freise, F. W. 1935. A importancia de conversazao dos mangues como viveiros de peixes. Anais do 1º Congr. Nac. de Presca, Rio de Janeiro, 1934. pp. 315-319.

Freise, F. W. 1938. Untersuchungen am Schlick der Mangrovekuste Brasiliens. Chemie der Erde 2: 333-355.

Freyberg, B. von. 1930. Zerstorung und Sedimentation and der Mangrovekuste Brasiliens. Leopoldina 6: 69-118.

Fryckberg, M. 1945. Mangrove bark. Agric. Am. 5: 199.

Fuchs, H. P. 1970. Ecological and palynological notes on *Pelliciera rhizophoraee.* Acta Bot. Neerl. 19: 884-894.

Ganapati, P. N. and M. V. L. Rao. 1959. Incidence of marine borers in the mangroves of the Godavari estuary. Curr. Sci. 28: 332.

Gardner, C. A. 1923. Botanical notes from the Kimberly division of Western

Australia. W. Aust. For. Dept., Bull. No. 32, Government Printer, Perth, Australia.

Gardner, C. A. 1942. The vegetation of Western Australia with special reference to the climate and soils. J. Proc. R. Soc. West Austral. 28: xi-lxxxvii.

Gehrmann, K. 1911. Zur Blutenbiologie der Rhizophoraceae. Ber. Dtsch. bot. Ges. 29: 308-318.

Genkel, P. A. 1963. K. ecologii restanii mangrov. *In* Fizioligiya drevesnykh rastenii. Akad. Nauk SSR, Moscow. Referat. Zh. Biol. No. 7V151, pp. 223-232.

Genkel, P. A. and I. S. Fan. 1958. On the physiological significance of viviparity in mangrove plants. (In Chinese: Russian summary). Acta Bot. Sin. 7: 51-71.

Gerlach, S. A. 1958. Die Mangroveregion tropischer Kusten als Lebensraum. Z. Morph. Okol. Tiere 46: 636-731.

Gerlach, S. A. 1963. Okologische Bedeutung der Kuste als Grenzraum zwischen Land und Meer. Naturw. Rdsch. 16: 219-227.

Germeraad, J. H., C. A. Hopping, and J. Muller. 1965. Palynology of Tertiary sediments from tropical areas. Rev. Palaeobotan. Palynol. 6: 139-348.

Gessner, F. 1955. Die Mangroven. Hydrobotanik, Band I. Veb Deutscher Verlag der Wissenchaften, Berlin, 517 pp., pp. 249-262.

Gessner, F. 1967. Untersuchungen an der Mangrove in Ost-Venezuela. Int. Rev. Ges. Hydrobiol. 52: 769-781.

Giglioli, M. E. C. and I. Thornton. 1965a. The mangrove swamps of Keneba, Lower Gambia River basin. I. Descriptive notes on the climate, the mangrove swamps and the physical composition of their soils. J. appl. Ecol. 2; 81-103.

Giglioli, M. E. C. and I. Thornton. 1965b. The mangrove swamps of Keneba, Lower Gambia River basin. II. Sulphur and pH in the profiles of swamp soils. J. appl. Ecol. 2: 257-269.

Giglioli, M. E. C. and D. F. King. 1966. The mangrove swamps of Keneba, Lower Gambia River basin. III. Seasonal variations in the chloride and water content of swamp soils, with observations on the water level and chloride concentration of free soil water under a barren mud flat during the dry season. J. appl. Ecol. 3: 1-19.

Gill, A. M. 1969. Tidal trees: orient and occident. Bull. Fairchild Trop. Gard. July 1969, pp. 7-10.

Gill, A. M. 1970. The mangrove fringe of the Eastern Pacific. Bull. Fairchild Trop. Gard. July 1970, pp. 7-11.

Gill, A. M. Endogenous control of growth-ring development in *Avicennia*. For. Sci. *in press*.

Gill, A. M. and P. B. Tomlinson. 1969. Studies on the growth of red mangrove (*Rhizophora mangle* L.). I. Habit and general morphology. Biotropica 1: 1-9.

Gill, A. M. and P. B. Tomlinson. 1971. Studies on the growth of red mangrove (*Rhizophora mangle* L.). II. Growth and differentiation of aerial roots. Biotropica 3: 63-77.

Gill, A. M. and P. B. Tomlinson. 1971. Studies on the growth of red mangrove. (*Rhizophora mangle* L.). III. Phenology of the shoot. Biotropica, 3: 109-124.

Gilmour, A. J. 1965. The implication of industrial development on the ecology of a marine estuary. Victoria Fish. Wildl. Dept. Fish Contrib. 20: 1-12.

Gilroy, A. B. and L. J. Chwatt. 1945. Mosquito-control by swamp drainage in the coastal belt of Nigeria. Ann. Trop. Med. Parasitol. 39: 19-40.

Glanville, R. R. and H. C. Doyne. 1933. Some swamp soils of Sierra Leone. Trop. Agric. 10: 132.

Glassman, S. R. 1952. Flora of Ponape. B. P. Bishop Mus. Bull. No. 209, Honolulu, Hawaii.

Gleason, H. A. and M. T. Cook. 1927. Plant ecology of Puerto Rico: Scientific survey of Puerto Rico and the Virgin Islands. Proc. N. Y. Acad. Sci. 7: 1-96.

Gledhill, D. 1963. The ecology of the Aberdeen Creek mangrove swamp. J. Ecol. 51: 693-703.

Glynn, P. W. 1964. Common marine invertebrate animals of the shallow waters of Puerto Rico. Historia Natural de Puerto Rico. Univ. Puerto Rico, Mayaguez, pp. 12-20.

Glynn, P. W., L. R. Almodovar, and J. C. Gonzalez. 1964. Effects of Hurricane Edith on marine life in La Parguera, Puerto Rico. Caribb. J. Sci. 4: 335-345.

Goebel, K. von. 1886. Ueber die Rhizophoren-vegetation. Sitz. ber. d. Naturf. Ges. Rostock. December 1886.

Goebel, K. von. 1889. Ueber die Luftwurzeln von *Sonneratia*. Ber. Dtsch. bot. Ges. 5: 249-255.

Gokhale, A. V. 1962. Control of mangroves. Poona Agric. Col. Mag. 53: 30.

Gola, G. 1905. Studi sui rapporti de la distribuzione della piante e la constituzione fisico-chimica del suolo. Ann. di Bot. 3: 455-512.

Golley, F. B. 1969. Caloric value of wet tropical forest vegetation. Ecology 50: 517-519.

Golley, F., H. T. Odum, and R. F. Wilson. 1962. The structure and metabolism of a Puerto Rican red mangrove forest in May. Ecology 43: 9-19.

Gomez-Pompa, A. 1966. Estudios botanicos en la region de Misantla, Veracruz, Publ. Inst. Mex. Rec. Nat. Renov. Mexico, 173 pp.

Gonzales, J. G. and T. E. Bowman. 1965. Planktonic copepods from Bahia Fosforescente, Puerto Rico, and adjacent waters. Proc. U. S. Natl. Mus. 117: 241-304.

Gonzalez Ortega and J. Candelon. 1934. *Rhizophora mangle* L. Bol. Pro-Cultura Regional, Mazatlan, Mexico 1: 14-16.

Gooch, W. L. 1944. Survey of tannin resources in Mexico. J. For. 43: 56-58.

Good, R. 1953. The Geography of Flowering Plants. 2nd Ed. Longmans, Green and Co., London. xiv, 452 pp.

Gooding, E. G. B., A. R. Loveless, and G. R. Proctor. 1965. Flora of Barbados. H.M.S.O., London, 486 pp.

Goossens, H. J. 1936. Tambakherverkaveling in het ressort Tegal de Binnenvisserij. Landbouw 12: 191-200.

Gosh, D. 1967. Mangrove: A potential Indian tanning material. Tanner (Bombay) 21. 379-383.

Graham, B. M. 1929. Notes on the mangrove swamp of Kenya. J. E. Afr. and Uganda Nat. Hist. Soc. 29: 157-164.

Graham, C. L. 1934. Flora of the Kratabo region, British Guiana. Ann. Carnegie Mus. 22: 204.

Graham, J. 1929. Mangroves of East Africa. J. East Afr. Nat. Hist. Soc. 29: 157.

Graham, S. A. 1964. The genera of Rhizophoraceae and Combretaceae in the southeastern United States. J. Arnold Arbor. 45: 285-301.

Grant, D. K. S. 1938. Mangrove woods of Tanganyika territory, their silviculture and dependent industries. Tanganyika Notes and Rec. 1938 (April), pp. 5-15.

Grass, T. H. 1904. Forstatatistik fur die Waldungen des Rufiyideltas, angefangen im Jahre 1902. Ber. uber Land-u. Forstwirtsch (Deutsch Ostafrika) 2: 165-186.

Gray, J. 1960. Temperate pollen genera in the Eocene (Clairborne) flora, Alabana. Science 132: 808-810.

Greenway, H. 1962. Plant responses to saline substrates. III. The offsets of nutrient concentration on the growth and ion uptake of *Hordeum vulgare* during a sodium stress. Aust. J. Biol. Sci. 16: 616-628.

Greenway, H. and A. Rogers. 1963. Growth and ion uptake of *Agrophyron elongatum* on saline substrates, as compared with a salt-tolerant variety of *Hordeum vulgare*. Plant and Soil 18: 21-30.

Gregory, D. P. 1958. Rhixophoraceae. Ann. Miss. Bot. Gard. 45: 136-140.

Grewe, F. 1941. Afrikanische Mangrovelandschaften, Verbreitung und wirtschaftsgeographische Bedeutung. Wiss. Veroff. D. Mus. F. Landerk. N. F. 9: 103-177.

Griffith, A. L. 1950. Working scheme for the mangroves of the Zanzibar Protectorate. Gov. Printer, Zanzibar, 42 pp.

Groom, P. and S. E. Wilson. 1925. On the pneumatophores of paludal species of *Amoora*, *Goaurapa* and *Heritiera*. Ann. Bot. 39: 9-24.

Guillaumin, A. 1964a. Rhizophoraceae. Scientific Results of the French and Swiss Botanical Mission to New Caldonia (1950-1952). III. Mem. Mus. Nat. Hist. Natur. Nouv. Ser., Ser. B. Bot. 15: 76-77.

Guillaumin, A. 1964b. Revision des Rhizophoracees. Not. Syst. 3: 55.

Guppy, H. B. 1906. Observations of a Naturalist in the Pacific Between 1896 and 1899. Vol. II, Plant Dispersal. Macmillan and Co., Ltd., London. 627 pp.

Guppy, H. B. 1917. Plants, Seeds and Currents in the West Indies and Azores. Macmillian and Co., Ltd., London. 531 pp.

Gurke, A. N. 1895/1897. Notizen uber die Verwertung der Mangroverinde als Gerbmaterial. Not. Konigl. Bot. Gard. Mus. Berlin, 1895/1897, No. 1.

Haan, J. H. De. 1931. Het een en auder over de Tjilatjapsche vloedbosschen. Tectona 24: 39-76.

Haberlandt, G. 1895. Uber die Ernahrung der Keimlange und die Bedeutung des Endosperms bei viviparen Mangrovepflanzen. Ann. Jar. Botan. Buiten. 12: 105-114.

Haberlandt, G. 1910. Botanische Tropenreise; Indo-Malayische Vegetationsbilder und Reiseskizzen. Wilhelm Engelman, Leipzig, 296 pp.

Haden-Guest, S. J. K. Wright, and E. M. Teclaff. 1956. A World Geography of Forest Resources. The Ronald Press Co., New York, 736 pp.

Hagen, B. 1890. Die Pflanzen-und Thierwalt von Deli auf der Ostkuste Sumatra. Tidj. van het Kon. Nederl. Aardijks. Gen. 7: 1-240.

Hager, L. C. 1959. Anatomical study of *Avicennia marina* (Forsk.) Vierh. (Formerly called *A. officinalis*). M. S. Botany Dept., Univ. Witwatersrand, Johannesburg.

Hamilton, A. A. 1919. An ecological study of the salt marsh vegetation of the Port Jackson District. Proc. Linn. Soc., N.S.W. 44: 463-513.

Hammen, T. van der. 1962. A palynological study on the Quaternary of British Guiana. Leidse Geol. Meded. 29: 125-180.

Hammen, T. van der and T. A. Wijmstra. 1964. A palynological study on the Tertiary and Upper Cretaceous of British Buiana. Leidse Geol. Meded. 30: 183-241.

Hammer, L. 1961. Sobre la ecologia de aqua de lus mangles. Bol. Inst. Oceanogr. 1: 249-261.

Hammer, H. E. 1929. The chemical composition of Florida Everglades peat soils, with special reference to their inorganic constituents. Soil Sci. 28: 1-14.

Hammond, W. 1971. A case for mangroves. Littoral Drift. 1: 2-5.

Harms, J. W. 1941. Beobachtungen in der Mangrove Sumatras, insbesondere uber Lebensweise und Entwicklung der Solzlandturbellarie *Stylochus* sp. und der *Oncis stuxbergi.* Z. wiss. Zool. 154: 389-408.

Harms, J. W. and O. Dragendorff. 1933. Die Realisation von Genen und die consecutive Adaption. 3. Mitt. Osmotische Untersuchungen an Phycosoma lurco Sel. und DeMan aus den Mangrove-Vorl andern der Sumatra Inseln. Z. wiss. Zool. 143: 263-322.

Harper, R. M. 1917. Geography of central Florida. Fla. St. Geol. Surv., 18th Ann. Rep., pp. 27-276.

Harper, R. M. 1927. Natural resources of southern Florida. Fla. Geo. Surv. Ann. Rep. 18: 27-192.

Harrer, F. 1939. Die Mangroven des Rufiji-Deltas in Deutsch-Ostafrika. Kolonialforstl. Mitt. 2: 160.

Harris, J. A. and J. V. Lawrence. 1917. The osmotic concentration of the sap

of the leaves of mangrove trees. Biol. Bull. 32: 202-211.
Harrison, S. G. 1961. Notes on the vegetation of the coastal region of British Guiana. J. R. Agric. Comm. Soc. B. G. 40: 93-104.
Harrisson, T. 1965. Some quantitative effects of vertebrates on the Borneo flora. in Symp. Ecolo. Res. Humid Trop. Veg. UNESCO Sci. Coop. Of. Southeast Asia, Kuching, Sarawak, pp. 164-169.
Harshberger, J. W. 1908. The comparative leaf structure of the sand dune plants of Bermuda. Proc. Am. Phil. Soc. 47: 97-110.
Harshberger, J. W. 1914. The vegetation of south Florida, south of 27°30' North, exclusive of the Florida Keys. Wagner Free Inst. Sci. Philadelphia Trans. 7: 49-189.
Hart, M. G. R. 1959. Sulphur oxidation in tidal mangrove soils of Sierra Leone. Plant and Soil 11: 215-236.
Hart, M. G. R. 1962. Observations on the source of acid in empoldered mangrove soils. I. Formation of elemental sulphur. Plant and Soil 17: 87-98.
Hart, M. G. R. 1963. Observations on the source of acid in empoldered mangrove soils. II. Oxidation of soil polysulphides. Plant and Soil 19: 106-114.
Hartmann, G. 1956a. Zur Kenntnis des Mangrove-Estero-Gebietes von El Salvador und seiner Ostracoden-Fauna I. Meeresforsch. 12: 219-248.
Hartmann, G. 1956b. Zur Kenntnis des Mangrove-Estero-Gebietes von El Salvador und seiner Ostracoden-Fauna. II. Meeresforsch. 13: 134-159.
Hartman-Schroder, G. 1959. Zur Oekologie der Polychaete des Mangrove-Estero-Gebietes von El Salvador. Beitr. z. neotrop. Fauna 1: 69-183.
Hartnoll, R. G. 1965. Notes on the marine grapsoid crabs of Jamaica. Proc. Linn. Soc. London 176: 113-147.
Hatheway, W. H. 1953. The land vegetation of Arno Atoll, Marshall Islands. Attoll Res. Bull. 16, April 30.
Hanheway, W. H. 1955. The natural vegetation of Canton Island, an equatorial Pacific atoll. Res. Bull. 43, Aug. 15.
Haug, T. A. 1909. Die Mangrove Deutsch-Ostafrikas. Naturw. Z. Forst-u. Landwirt. 7: 413-425.
Haverschmidt, F. 1962. Notes on some Surinam breeding birds. II. Ardea 50: 173-179.
Haverschmidt, T. 1965. The utilization of mangroves by South American birds. Ibis 107: 540-542.
Haverschmidt, F. 1969. Notes on the Boat-billed Heron in Surinam. Auk 86: 130-131.
Hayata, B. 1912. Rhizophoraceae. Icones Plant. Formosarum 2: 15-16.
Heald, E. 1971. The production of organic detritus in a south Florida estuary. Univ. Miami, Sea Grant Tech. Bull. No. 6, 110 pp.
Heald, E. 1970. The contribution of mangrove swamps to Florida Fisheries. Proc. Gulf Caribb. Fish. Inst., 22nd Ann. Session, pp. 130-135.

Heiden, H. 1893. Anatomische Characteristik der Combretaceen. Bot. Centralbl. 55: 353-391.
Heiden, H. 1893. Anatomische Characteristik der Combretaceen. Bot. Centralbl. 56: 1-12, 65-75, 129-136, 163-170, 193-200, 225-230, 355-360, 385-391.
Hellmayer, C. E. and B. Conover. 1948. Catalogue of birds of the Americas. Field Mus. Nat. Hist. Publ. Zool. Ser. 8, Part 1, No. 2.
Hemsley, W. B. 1894. The flora of the Tonga or Friendly Islands. J. Linn. Soc. London Bot. 30: 176.
Henckel, P. A. 1963. On the ecology of the mangrove vegetation. Mitteil. d. forist.-soziol. Srbeitsgem., N.F. 10: 201-206.
Henkel, J. S., S. St. C. Ballenden, and A. W. Bayer. 1936. Account of the plant ecology of the Dukuduku forest reserve and adjoining areas of the Zululand coast belt. Ann. Natal Mus. 8: 95-123.
Herbert, D. A. 1932. The relationships of the Queensland Flora. Proc. R. Soc. Queensland 44: 2-22.
Herklots, G. A. C. 1961. The Birds of Trinidad and Tobago. Collins Press, London, 287 pp.
Hesse, P. R. 1961a. Some differences between the soils of *Rhizophora* and *Avicennia* mangrove swamps in Sierra Leone. Plant and Soil 14: 335-346.
Hesse, P. R. 1961b. The decomposition of organic matter is a mangrove swamp soil. Plant and soil 14: 249-263.
Hesse, P. R. 1962. Phosphorus fixation in mangrove swamp muds. Nature 193: 295-296.
Hesse, P. R. 1963. Phosphorus in mangrove swamp mud with particular reference to aluminum toxicity. Plant and Soil 19: 205-218.
Heylighers, P. C. 1965. Lands of the Port Moresby-Kairuka area, Papua, New Guinea. VIII. Vegetation and ecology of the Port Moresby-Kairuka area. C.S.I.R.O. Austr. Land. Res. Ser. 14: 146-173.
Hicks, G. H. 1894. Nourishment of the embryo and the importance of the endosperm in viviparous mangrove plants. Bot. Gaz. 19: 327-330.
Hill, B. J. 1966. A contribution to the ecology of the Umalalazi estuary. Zool. Afr. 2: 1-24.
Hill, T. G. 1908. Observations on the osmotic properties of the root ahirs of certain salt marsh plants. New Phytol. 7: 133-142.
Hillier, J. M. 1907. Anti-opium plants. Kew Bull. 1907: 198-199.
Hillier, J. M. 1908. The malayan anti-opium plant. Kew Bull. 1908: 235-236.
Hillis, W. E. 1956. The production of mangrove extract in the delta region of Papua. Empire For. Rev. 35: 220-236.
Hirakawa, Y. 1933. Synoekologische Studien uber die Strandflora der Insel Formosa. I. Rep. Taihoku Bot. Gar. 3: 1-47.
Hirakawa, Y. 1936. Mangrove vegetation in Takao. J. Taiwan Mus. Ass. 4: 1-5 (in Japanese).
Ho, Ching 1957. Ecology of mangrove plants. Biol. News 8: 1-5.
Ho, R. 1962. Physical geography of the Indo-Australian tropics. *in* Impact of

Man on Humid Tropics Vegetation. Goroka, New Guinea, pp. 19-35.
Ho, T. 1899. Rhizophorae in Japan. Ann. Bot. 456-468.
Hodge, W. H. 1956. The trees that walk to the sea. Nat. Mag. 49: 456.
Hoffmeister, J. E. and H. G. Multer. 1965. Fossil mangrove reefs of Key Biscayne, Florida. Geol. Assoc. Am. Bull. 76: 845-852.
Hofmeister, W. 1859. Neuere Beobachtungen uber Embryobildung bei Phanerogamen. Pring. Jahrb. 1: 82.
Holdridge, L. R. 1940. Some notes on the mangrove swamps of Puerto Rico. Caribb. For. 1: 19-29.
Holdridge, L. R. 1942. Trees of Puerto Rico, USDA, For. Serv., Trop. For. Experiment. Stat. Occas. Pap. No. 1.
Holdridge, L. R., L. V. Teesdale, J. E. Meyer, E. Z. Little, E. F. Horn, and J. Marrero. 1947. The forests of Central Equador. U. S. For. Ser. Bull. 1947, 134 pp.
Honel, Fr. von. 1882. Beitrage zur Pflanzenanatomie und Physiologie V. Zur Anatomie der Combretaceen. Bot. Z. 40: 177-182.
Hooker, W. J. 1830. *Terminalia catappa* L. Bot. Mag. 57.
Horn, C. L. 1945. Plant resources of Puerto Rico. *in* Plants and Plant Science in Latin America, F. Verdoorn (ed.), Chronica Botanica Co., Waltham, Mass., pp. 83-84.
Horn, E. F. 1946. Brazilian tanning materials. Trop. Woods 88: 33.
Horn, E. F. 1948. Durability of brazilian corssties. Trop. Woods 93: 30-35.
Hosakawa, T. 1957. Outline of the mangrove and strand forests of the micronesian islands. Mem. Fac. Sci. Kyushu Uni. Ser. E, 2: 101-118.
How Koon-Chew and Ho Chun-Nien. 1953. Rhizophoraceae in the Chinese Flora. Acta Phytotaxon. Sin., Vol. 2.
Howard, R. A. 1950. Vegetation of the Bimini Group, Bahamas. Ecol. Monogr. 20: 317-349.
Howe, M. A. 1911. A little-known mangrove of Panama. J. New York Bot. Gard. 12: 61-72.
Howes, F. N. 1953. Vegetable tanning materials. Butterworth's Sci. Publ., London, pp. 77-79.
Howmiller, R. and A. Weiner. 1968. A limnological study of a mangrove lagoon in the Galapagos. Ecology 49: 1184-1186.
Huberman, M. A. 1957. Mangrove silviculture. Trop. Silvic. 13: 188-195.
Huek, K. 1969. Los mangalares de America. Description de arboles forstales, No. 5. Bol Inst. For. Lat. Amer., Merida, Venezuela.
Huggard, E. R. 1960. Logging road construction. Proc. 5th World For. Congr. 3: 1921-1926.
Hummel, K. 1930. Tierfahrtenbilder vom Tropenstrand. Natur und Mus. 60.
Hunter, J. B. 1970. A survey of the oyster population of the Freetown estuary, Sierra Leone, with notes on the ecology, cultivation, and possible utilization of mangrove oysters. Trop. Sci. 11: 276-285.
Ito, T. 1899a. Rhizophoraceae in Japan. Ann. Bot. 13: 465-466.
Ito, T. 1899b. Mangroves growing in Japan. Nature 60: 79.

Ivens, G. W. 1957. Arboricides for killing mangroves. W. Afr. Rice Res. Stat. Per. Sci. Rep. No. 8, 6 pp.
Jackson, C. R. 1952. Some topographic and edaphic factors affecting distribution in a tidal marsh. Q. J. Fla. Acad. Sci. 15: 137-146.
Jackson, G. 1964. Notes on west African vegetation: I. Mangrove vegetation at Ikorodu, western Nigeria. West Afr. Sci. Assoc. J. 9: 98-111.
Jackson, M. L. 1958. Soil Chemical Analysis. Prentice-Hall, Englewood Cliffs, N. J. 498 pp.
Jacques-Felix, H. 1957. Les*Rhizophora* de la mangrove atlantique d'Afrique. J. Agric. Trop. Bot. Appl. 4: 343-347.
Jakobi, H. 1955. O genero Enhydrosoma no manguezal da costa S. Paulo-Parana (Harpacticoidea-Crustacea). Dusenia 6: 89-96.
Janssonius, H. H. 1950. Vessels in the wood of Javan mangrove trees. Blumea 6: 465-469.
Jardine, F. 1928. The topography of the Townsville littoral. Rep. Great Barrier Reef Comm. 2: 70-87.
Jenik, J. 1967. Root adaptations in west African trees. J. Linn. Soc. London Bot. 60: 136-140.
Jenik, J. and G. W. Lawson. 1968. Zonation of microclimate and vegetation on a tropical shore in Ghana. Oilos 19: 198-205.
Jennings, D. H. 1968. Halophytes, succulence, and socium-a unified theory. New Phytol. 67: 899-911.
Jennings, J. N. and C. F. Bird. 1967. . Regional geomorphological characteristics of some Australian estuaries. *in* Estuaries, G. H. Lauff (ed.), AAAS, Washington, pp. 121-128.
Johnson, I. M. 1949. The botany of San Jose Island (Gulf of Panama). Sargentia 8: 1-306.
Johow, F. 1884. Vegetationsbilder aus West-Indien und Venezuela. I. Die Mangrove-Sumpfe. Kosmos, Z. ges. Entwicklungs. 14: 415-426.
Joly, A. B. and N. Yamaguishi-Tomita. 1967. *Dawsoniella bostrychiae:* a new parasite of mangrove algae. Sellowia 19: 63-70.
Jong, B. de. 1934. Ueber Mangrove Cultiven. Tectona 27: 258-298.
Jonker, F. P. 1942. Rhizophoraceae in Pulle. Fl. Suriname 3: 36.
Jonker, F. P. 1959. The genus *Rhizophora* in Suriname. Acta Bot. Neerl. 8: 58-60.
Jonker, H. A. J. 1933. De vloedbosschen van den Riouw lingga Archipel. Tectona 26: 717-741.
Jordan, H. D. 1954. The development of rice research in Sierra Leone. Trop. Agri. Trinidad 31: 27-32.
Jordan, H. D. 1959. The utilization of saline mangrove soils for rice growing. Trans. 3rd int. Afr. Soils Conf., Dalaba, pp. 327-331.
Jordan, H. D. 1964. The relation of vegetation and soil to the development of mangrove swamps for rice growing in Sierra Leone. J. appl. Ecol. 1: 209-212.
Jost, L. 1887. Kenntnis der Athmungsorgane der Pflanzen. Bot. Z. 45:

601-606, 617-628, 633-642.

Kalk, M. 1959. The zoogeographical composition of the intertidal fauna at Inhaca Island, Mozambique. S. Afr. J. Sci. 55: 178.

Kalk, M. 1959. The fauna of the shores of northern Moxambique. Rev. de Biol., Lisbon 2: 1.

Kanehira, R. 1938. On the genus *Sonneratia* in Japan. J. Jap. Bot. 14: 421-424.

Karny, H. 1928a. Signs of life in mangrove vegetation of Java. Palaeobiologica 1: 475-580.

Karny, H. 1928b. Lebensspuren in der mangroveformation Javas (Ein Beitrag zur Losung des Flyschproblems-Aus einum Briefan O. Abel). Paleobiologica 1: 475-480.

Karsten, G. 1891. Ueber die Mangrove-Vegetation im Malayschen Archipel. Eine Morphologische-biologische Studie. Bibl. Bot. 22: 1-71.

Karsten, G. 1905. Die Mangrovevegetation. Vegetationsbilder 2: 7-12.

Karyone, T. 1927. The mangrove of the South Sea Islands. J. Jap. Bot. 116-120.

Kassas, M. and M. A. Zahran. 1967. On the ecology of the Red Sea littoral salt marsh, Egypt. Ecol. Monogor. 37: 297-316.

Kato, K. 1966a. Geochemical studies on the mangrove region of Cananeia, Brazil. I. Tidal variations of water properties. Bol. Inst. Oceanogr. 15: 13-20.

Kato, K. 1966b. Geochemical studies on the mangrove region of Cannaneia, Brazil. II. Physico-chemical observations on the reduction states. Bol. Inst. Oceanogr. 15: 21-24.

Kawano, Y., H. Matsumoto, and R. Hamilton. 1961. Plant products of economic importance in Hawaii. II. Tannins. Tech. Prog. Rep. No. 130, Univ. Hawaii, Agric. Exper. Stn., Honolulu. pp. 1-14.

Keast, A. 1959. Australian birds, their zoogeography and adaptations to an arid climate. *in* Biogeogrphy and Ecology in Australia, A. Deast, R. L. Cracker, and C. S. Christian (eds.), pp. 89-114.

Keay, R. W. J. 1953. *Rhizophora* in west Africa. Kew Bull. 1: 121-127.

Kehar, N. D. and S. S. Negi. 1953. Mangrove leaves as cattle feed. Sci. Cult. 18: 382-382.

Kehar, N. D. and S. S. Negir. 1954. Mangrove leaves as cattle feed. Sci. Cult. 19: 556-557.

Keller, B. 1925. Halophyten und Zerophyten-studien. J. Ecol. 13: 224-261.

Kellett, J. 1969. Looking back on the Florida freeze of '62. Principle 13: 23-35.

Keng, H. 1952. Mangrove in Taiwan. Taiwan Agric. For. Month. 6: 38-39.

Kerr, A. 1928. Kaw Tao, its physical features and vegetation. J. Siam Soc. Nat. Hist. Suppl. 7: 137-149.

Khan, S. A. 1956. Mangrove forests of Andhra State. Proc. Silvic. Conf., Dehra Dun 9: 182-186.

Khan, S. A. 1961. Regeneration of *Avicennia officinalis* in the coastal forests of West Pakistan. Pak. J. For. 11: 43-45.
Kiener, A. 1966. Contributions a l'etude ecologique et biologique des eaux saumatres malagaches. Les poissons euryhalins et leur role dans le developpement des peches. Vie Milieu 16: 1013-1149.
Keinholz, R. 1928. Environmental factors of Philippine beaches with particular reference to the beach at Puerto Galera, Mindoro. Philipp. J. Sci. 36: 199-213.
Kint, A. 1934. De Luchtfoto en de topografische terreingsteldheid in de Mangrove. De Trop. Natuur 23: 173-189.
Kipp-Goller, T. 1940. Uber Bau and Entwicklung der viviparen Mangrovekeimlinge. Z. f. Bot. 35: 1-40.
Koerner, T. 1904. Encorces tannantes de mangliers. J. Agric. Trop. 4: 113.
Kohlmeyer, J. 1966. Neue Meirespilze an Mangroven. Ber. Dtsch. bot. Ges. 79: 27-37.
Kohlmeyer, J. 1968. A new *Trematosphaeria* from roots of *Rhizophorra racemosa*. Mycopathol. Mycol. Appl. 34: 1-5.
Kohlmeyer, J. 1969. Ecological notes on fungi in mangrove forests. Trans. Brit. Mycol. Soc. 53: 237-250.
Kohlmeyer, J. 1968. Marine fungi from the tropics. Mycologia 40: 252-270.
Kohlmeyer, J. 1969. Marine fungi of Hawaii including the new genus *Heliascus*. Can. J. Bot. 47: 1469-1487.
Kohlmeyer, J. and E. Kohlmeyer. 1965. New marine fungi from mangroves and trees along eroding shorelines. Nova Hedwiga 9: 89-104.
Kohlmeyer, J. and E. Kohlmeyer. 1971. Marine fungi from tropical America and Africa. Mycologia 63: 831-863.
Kolwitz, R. 1933. Zur Okologie der Pflanzenwelt Brasiliens. Ber. Dtsch. bot. Ges. 51: 396-406.
Koriba, K. 1958. On the periodicity of tree growth in the tropics. Gard. Bull. Singapore 17: 11-81.
Korner, A. 1900. Bericht uber weiter Untersuchungen von Mangrove-Extrakten. Jahresbericht der Deutschen Gerberschule Freiberg in Sachsen. Ferb. Freiberg, 1900.
Kostermans, A. J. G. H. 1959. Monograph of the genus *Heritera* Aitron (Stercul.). Reinwardtia 4: 465-583.
Kostermans, A. J. G. H. 1961. A monograph of the genus *Brownlowia* Roxb. (Tiliaceae). Communs For. Res. Inst. Bogor, No. 73, 62 pp.
Krejci-Graf, K. 1935. Beobachtungen am Tropenstrand I – IV. Senckenbergiana 17: 1-242.
Krejci-Graf, K. 1937. Uber Fahrten und Bauten tropischer Krabben. Geol. Meere und Binnengewasser 1: 39-46.
Kudo, Y. 1932. The mangrove of Formosa. Bot. Mag. Tokyo 46: 147-156.
Kuenzler, E. J. 1969. Mangrove swamp systems. *in* Coastal Ecological Systems of the United States. H. T. Odum, B. J. Copeland and E. A. McMahon (eds.), Inst. Mar. Sci. Univ. North Carolina, Vol. 1, pp. 353-383.

Kumar, L. S. S. and W. V. Joshi. 1942. False polyembryony in viviparous *Rhizophora mucronata* Lam. Curr. Sci. 11: 242.

Kunkel, G. 1965. Der Standort: Kompetenzfactor in der Stelzwurzelhildung. Biol. Centrablatt 84: 641-651.

Kunkel, G. 1966. Uber die Struktur und Sukzession der Mangrove Liberias und deren Randformationen. Ber. Schweiz. bot. Ges. 75: 20-40.

Kylin, A. and R. Gee. 1970. Adenosine triphosphatase activities in leaves of the mangrove *Avicennia nitida* Jacq. Pl. Physiol. 45: 169-172.

Laessle, A. M. and C. H. Wharton. 1959. Northern extensions in recorded ranges of plants on Seahorse and associated keys, Levy County, Florida. Q. J. Fla. Acad. Sci. 22: 105-113.

Lafond, J. 1957. Apercu sur la sedimentologie de l'estuaire de la Betsiboka (Madagascar). Rev. Inst. Fr. Ret. 12: 425-431.

Lagerwerff, J. V. 1958. Comparable effects of absorbed and dissolved cations on plant growth. Soil Sci. 86: 63-69.

Lakhanpal, R. W. 1952. *Nipa sahnii*, a palm fruit in the Tertiary of Assam. Paleobotanist 1: 289-294.

Lamb, F. B. 1959a. Prospects for forest land management in Panama. Trop. Woods 110: 16-28.

Lamb, F. B. 1959b. The coastal swamp forests of Navico, Colombia. Caribb. For. 20:79-89.

Lambert, F. and P. Polk. 1971. Observations sur l'ostreculture a Cuba. Hydrobiologia 38: 9-14.

Lamberti, A. 1969. Contribuicao ao Conhecimento de Ecologic das Plantas do Manguez al de Itanhaem. Bol. No. 317, Bot. No. 23, Fac. Fil., Cien. e Let. Univ. Sao Paulo, Brazil, 217 pp.

Landon, F. H. 1933. Planting in mangrove forests. Malay. For. 2: 131-133.

Landon, F. H. 1942. Mangrove volume tables. Malay. For. 11: 117-123.

Langenheim, J. H., B. L. Hackner, and A. Bartlett. 1967. Mangrove pollen at the depositional site of Oligo-Miocene amber from Chiapas, Mexico. Bot. Mus. Leafl. Harv. Univ. 21: 289-324.

LaRue, C. D. and T. J. Musik. 1951a. Mangrove vivipary. Rep. Fed. Exp. Stn. Puerto Rico, No. 13.

LaRue, C. D. and T. J. Musik. 1951b. Does the mangrove really plant its seedlings? Science 114: 661-662.

LaRue, C. D. and T. Jm Musik. 1954. Growth, regeneration, and prevocious rooting in *Rhizophora mangle*. Pap. Mich. Acad. Sci. Arts Lett. 39: 9-29.

Lawrence, D. B. 1949. Self-erecting habit of seedling red mangroves (*Rhizophora mangle* L.). Am. J. Bot. 36: 426-427.

Lawson, G. W. 1966. The littoral ecology of West Africa. Ocean et Mar. biol. Ann. Rev. 4: 405-448.

Le Cointe, P. 1947. Arvores e Plantas Uteis. Amazonia Brasiliera III. 2nd ed. Companhia Editora Nacional, Sao Paulo, Brazil, p. 283.

Lee, B. K. H. and G. E. Baker. 1972. An ecological study of the soil microfungi in a Hawaiian mangrove swamp. Pac. Sci. 26: 1-10.

Lee, B. K. H. and G. E. Baker. 1972. Environment and the distribution of microfungi in a Hawaiian mangrove swamp. Pac. Sci. 26: 11-19.

Leechman, A. 1908. The genus *Rhizophora* in British Guiana. Kew Bull. 1908: 4-8.

Leeuwen, W. D. von. 1911. Uber die Ursache der widerholten Verzweigung der Stutzwurzeln von *Rhizophora*. Ber. Dtsch. bot. Ges. 29: 476-478.

Lendrum, N. G. 1961. Problems of coastal development of Florida. Shore and Beach 29: 27-32.

Leopold, A. S. 1950. Vegetation zones of Mexico. Ecology 31: 507-518.

Lever, R. J. A. W. 1952. A malayan leaf tying caterpiller from mangroves periodically submerged in sea water. Entomol. Mon. Mag. 88: 95-98.

Lewin, R. A. 1970. Toxin secretion and tail autotomy by irritated *Oxynoe panamensis* (Opistobranchiata; Sacoglossa). Pac. Sci. 24: 356-358.

Lewis, J. 1956. Rhizophoraceae. Fl. Trop. E. Afr. Oct., 20 pp.

Lewis, O. A. M. and G. Naidoo. 1970. Tidal influence on the apparent transpirational rhythms of the white mangrove. S. Afr. J. Sci. 66: 268-270.

Li, Shang-Kuei. 1948. Distribution of mangroves in Taiwan. Taiwan For. Exp. Stn. Rep. 41-48: 337-339.

Liebau, O. 1914. Beitrage zur Anatomie und Morphologie der Mangrove-Pflanzen, insbesondere ihres Wurzelsystems. Beitr. Biol. Pfl. 12: 181-213.

Lindeman, J. C. 1953. The vegetation of Suriname. Vol 1, Part 1. The vegetation of the coastal region of Suriname. Paramaibo I - IV: 1-135.

Linke, O. 1939. Die Biota des Jadebusenwattes. Helgol. wiss. Meeres. 1: 201-348.

Little, E. L., Jr. 1945. Miscellaneous notes on nomenclaure of United States trees. Am. Midl. Nat. 33: 495-513.

Little, E. L., Jr. 1953. Check List of Native and Naturalized Trees of the United States (Including Alaska). Agric. Handbook 41, U.S.D.A. For. Serv., Washington, D. C. p. 365.

Little, E. L., Jr. 1961. *Avicennia nitida*. phytologia 8: 49-57.

Little, E. L., Jr. and F. H. Wadsworth. 1964. Common Trees of Puerto Rico and the Virgin Islands. Agric. Handb. 249, U.S.D.A. For. Serv., Washington, D. C.

Liu, T. 1955. Notes on mangroves in Taiwan. Taiwan For. 1: 1-4.

Liu, T. 1956. Mangrove forests. For. Bull. (Natl. Taiwan Univ.) 2: 1-25.

Loder, J. W. and G. B. Russell. 1969. Tumor inhibitory plants. The alkaloids of *Bruguiera sexangula* and *Bruguiera exaristata*. Aust. J. Chem. 22: 1271-1276.

Long, R. W. and O. Lakela. 1971. A Flora of Tropical Florida. Univ. Miami Press, Coral Gables, Florida, 962 pp.

Lotschert, W. 1955. Die Mangrove von El Salvador. Unsch. wiss. Tech. 55: 47-50.

Lotschert, W. 1960. Die Mangrove von El Salvador. Ber. D. Senckenberg Naturforsch. Ges. 90: 213-224.

Lotschert, W. 1968. Speichern die keimlinge von Mangrovepflanzen Salz? Umsch. wiss. Tech. 68: 20-21.

Lotschert, W. and F. Liemann. 1967. Die Salzspeicherung im Keimling von *Rhizophora mangle* L. wahrend der Entwicklung auf den Mitterpflanze. Planta 77: 136-156.

Loveless, A. R. 1960. The vegetation of Antigua. J. Ecol. 48: 495-527.

Loveless, C. M. 1959. A study of the vegetation of the Florida Everglades. Ecology 40: 1-9.

Lowe-McConnell, R. H. 1967. Notes on the nesting of the Boat-bill, *Cochlearius cochlearius*. Ibis 109: 179.

Luederwaldt, H. 1919. Os manguesaes de Dantos. Rev. Mus. Paul. 11: 309-408.

Luytjes, A. 1923. De vloedbosschen van Atjeh. Tectonia 16: 575-601.

Macbride, J. F. 1960. Flora of Peru. Field Mus. Nat. Hist. Bot. Ser. 13: 72.

MacCaughey, V. 1917. The mangrove in the Hawaiian Islands. Hawaiian For. Agr. 14: 361-366.

MacCaughey, V. 1918. The strand flora of the Hawaiian archipelago. Bull. Torrey Bot. Club 45: 259-277.

MacKay, J. B. and V. J. Chapman. 19544. Some notes on *Suaeda australis* Moq. var. *nova zelandica* var. nov. and *Mesembryanthemum australe* Sol. ex Forst. Trans. R. Soc. N. Z. 82: 41-47.

Macluskie, H. 1952. The reclamation of mangrove swamp areas for rice cultiviation. World Crops 2: 129-132.

Macnae, W. 1956. Aspects of life on muddy shores in South Africa. S. Afr. J. Sci. 53: 40-43.

Macnae, W. 1957. The ecology of the plants and animals in the intertidal regions of the Zwartkops estuary, near Port Elizabeth, South Africa. J. Ecol. 45: 113-131 and 361-387.

Macnae, W. 1962. The fauna and flora of the eastern coasts of southern Africa in relation to ocean currents. S. Afr. J. Sci. 58: 208-212.

Macnae, W. 1963. Mangrove swamps in South Africa. J. Ecol. 51: 1-25.

Macnae, W. 1966. Mangroves in eastern and southern Australia. Aust. J. Bot. 14: 67-104.

Macnae, W. 1967. Zonation within mangroves associated with estuaries in north Queesland. *in* Estuaries, G. H. Lauff (ed.), AAAS Publ. 83: 432-441.

Macnae, W. 1968a. A general account of the fauna and flora of mangrove swamps and forests in the Indo-West-Pacific region. Adv. Mar. Biol. 6: 73-270.

Macnae, W. 1968b. Mangroves and their fauna. Aust. nNat. Hist. 16: 17-21.

Macnae, W. and M. Kalk. 1962a. The mangrove swamps of Inhaca Island. J. Ecol. 50: 19-34.

Macnae, W. and M. Kalk. 1962b. The fauna and flora of sand flats at Inhaca Island, Mozambique. J. Anim. Ecol. 31: 93-128.

Mahabale, T. S. and J. V. Deshpande. 1959. The genus *Sonneratia* and its fossil allies. Palaeobotanist 6: 51-64.

Malaviya, M. 1963. On the distribution, structure and ontogeny of stone cells in *Avicennia officinalis* L. Proc. Indian Acad. Sci., Sect. B 58: 45-50.

Malcolm, C. V. 1964. Effects of salt, temperature, and seed scarification on germination of two varieties of *Arthrocnemum halocnemoides*. Jour. Proc. R. Soc. W. Aust. 47: 72-74.

Many authors. 1959. Los Manglares en America. Inst. For. Latinoamer., Merida, Venezuela, 52 pp.

Marathe, K. V. 1965. A study of the subterranean algal flora of some mangrove swamps. J. Indian Soc. Soil Sci. 13: 81084.

Marco, H. F. 1935. Systematic anatomy of the woods of the Rhizophoraceae. Trop. Woods 44: 1-26.

Marcus, B. 1941. Over den slibbodem van zoutwatervijvers. Landbouw 17: 891-910.

Margolef, F. 1962. Comunidades naturales. Inst. Biol. Mar. Univ. Puerto Rico, Spec. Pub., 499 pp.

Marshall, R. C. 1939. The Silviculture of the Trees of Trinidad and Tobago, British West Indies. Oxford Univ. Press, London, 247 pp.

Martinez, M. 1959a. Plantas Medicinales de Mexico, 4th ed., Ediciones Botas, Mexico, pp. 212-215.

Martinez, M. 1959b. Plantas Utiles de la Flora Mexicana. Revised edition, Ediciones Botas, Mexico, pp. 397-401.

Martyn, E. B. 1934. A note on the foreshore vegetation in the neighborhood of Georgetown, British Guiana, with special reference to *Spartina braziliensis*. J. Ecol. 22: 292-298.

Masamune, G. 1930. Contributions to be phytogeography of Japan. I. Trans. Nat. Hist. Soc. Formosa 20: 299-304 (in Japanese).

Mattox, N. T. 1949. Studies on the biology of the edible oyster, *Ostrea rhizophorae* Guilding, in Puerto Rico. Ecol. Monogr. 19: 339-356.

Mauritzon, J. 1939. Contributions to the embryogy of the orders Rosales and Myratales. Lunds Univ. Arsskr. II. Sec. 2, 35: 1-120.

Mayr, E. 1944. Wallace's Line in the light of recent zoogeographical studies. Q. Rev. Biol. 19: 1-14.

MaAdam, J. B. 1952. Forestry in New Guinea. Sci. Soc. Ann. Rep. Proc., New Guinea, pp. 41-61.

McArthur, W. M. 1957. The plant ecology of Garden Island in relation to the neighboring islands and the adjacent mainland. Proc. R. Soc. W. Aust. 40: 46-64.

McGill, J. T. 1958. Map of coastal landforms of the world. Geog. Rev. 48: 402-405.

McMillan, C. 1971. Environmental factors affecting seedling establishment of the black mangrove on the central Texas Coast. Ecology 52: 927-930.

Mead, J. P. 1912. The mangrove forests of the west coast of the Federated Malay States. Govt. Print. Office, Kuala Lumpur.

Meer Mohr, J. C. 1929. Abnormal fruits in *Rhizophora conjugataa*. Rec. Trans. Bot. Neerl. 26: 15-18.

Meindersma, H. W., J. H. Becking, and L. G. den Berger. 1922. Vloed-of mangrovebouschen in Ned-Indie. Tectona 15: 561-611.

Mell, C. D. 1919. The mangroves of tropical America. Sci. Am. Suppl. 2292: 388-389.

Merrill, A. C. 1905. A review of the identification of the species described in Blanco's Flora de Filipinas. Dept. Int., Bur. Govt. Labs., Manila, Bull. No. 27.

Metcalf, C. R. 1931. The breathing roots of *Sonneratia* and *Bruguieraa*, a review of recent work by Troll and Dragendorff. Kew Bull. 11: 465-467.

Millard, N. A. H. and A. D. Harrison. 1953. The ecology of South African estuaries. Part V. Richards Bay. Trans. R. Soc. S. Afr. 34: 157-171.

Miller, A. H., H. Friedmann, L. Griscom, and R. T. Moore. 1957. Distributional check-list of the birds of Mexico. Part 2. Pacific Coast Avifauna, No. 33, 436 pp.

Miller, O. B. 1926. Mangroves at Sordivana Bay, N. Zululand. S. Afr. J. Nat. Hist. 6: 54-55.

Miller, P. C. 1972. Bioclimate, leaf temperature, and primary production in red mangrove canopies in south Florida. Ecology 53: 22-45.

Miller, R. C. 1957. Marine wood-boring molluscs of the Philippines. Pac. Sci. Congr. 3A: 1575-1577.

Millspaugh, C. F. 1907. Flora of the sand keys of Florida. Field Mus. Publ. No. 118. pp. 191-243.

Miquel, F. A. G. 1855-1859. Flora van Nederlandsch Indie. Flora Indiae Batavae, vols. 1 and 2. Leipzig.

Miranda, F. 1958. Estuadios acerea de la vegetacion. *in* Los Recursos Naturales del Surests y su Aprovechamiento, E. Beltran (ed.), Publ. Inst. Mex. Rec. Nat. Renov. 2: 215-271.

Miranda, M. 1952. La vegetacion de Chiapas. Part II. Dept. de Prensa y Turismo, Tuxtla Gutierrez, Chiapas, Mexico, pp. 119-120.

Miyata, I., N. Odani, and Y. Ono. 1963. An analysis of Jo-method of dispersion of the population of *Bruguiera conjugata* (L.) Merrill. Rep. Comm. For. Sci. Res., Kyushu Univ., No. 1, pp. 43-48.

Mogg, A. O. D. 1963. A preliminary investigation of the significance of salinity in the zonation of species in salt-marsh and mangrove swamp associations. S. Afr. J. Sci. 59: 81-86.

Mokievskii, O. B. 1969. Biogeotsenotischeskaya sistema litorali. Okeanologiya 9: 211-222.

Moldenke, H. N. 1958. Hybridity in the Verbenaceae. Am. Midl. Nat. 59: 333-370.

Moldenke, H. N. 1960a. Materials toward a monograph of the genus *Avicennia*. I. Phytologia 7: 123-168.

Moldenke, H. N. 1960b. Materials toward a monograph of the genus *Avicennis*. II. Phytologia 7: 179-232.

Moldenke, H. N. 1960c. Materials toward a monograph of the genus *Avicennia*. III. Phytologia 7: 259-292.

Moldenke, H. N. 1963. Avicenniaceae, Symphoremaceae and Verbenaceae. Studies on the flora of Thailand, 1963. Dansk Bot. Ark. 23: 83-92.
Moldenke, H. N. 1967a. Additional notes on the genus *Avicennia* I. Phytologia 14: 301-320.
Moldenke, H. N. 1967b. Additonal notes on the genus *Avicennia* II. Phytologia 14: 326-336.
Moldenke, H. N. 1968a. Additional notes on the genus *Avicennia*. III. Phytologia 15: 71-72.
Moldenke, H. N. 1968b. Additonal notes on the genus *Avicennia*. IV. Phytologia 15: 470-478.
Monod, T. and R. Schnell. 1952. Les groupements et les unites geobotaniques de la region Guineenne (Mangroves). Mem. De l'Instit. Franc. d'Afrique Noire 18: 197-218.
Montfort, C. and W. Brandrup. 1927. Physiologische und Pflanzen geographische See Salzwirkungen. II. Okologische Studien uber Keimung und erste Entwicklung bei Halophyten. Jahrb. Wiss. Bot. 66: 902-946.
Moore, H. B., L. T. Davies, T. H. Frazer, R. H. Gore, and N. R. Lopez. 1968. Some biomass figures from a tidal flat in Biscayne Bay, Florida. Bull. Mar. Sci. 18: 261-269.
Moreira Filho, H. and M. B. Kutner. 1962. Contribuicao para o conhecimento das diatomaceas do manguesae de Alexandra. Bol. Univ. Parana Bot. 4: 1-24.
Morton, J. F. 1965. Can the red mangrove provide food, feed and fertilizer? Econ. Bot. 19: 113-123.
Moul, E. T. 1957. Preliminary report on the flora of Onotoa Atoll. Atoll Res. Bull. 57, Sept. 15.
Moss, C. E. 1954. The species of *Arthrocnemum* and *Salicornia* in southern Africa. J. S. Afr. Bot. 20: 1-22.
Muir, J. 1937. The seed drift of South Africa, and some influences of ocean currents on the strand vegetation. Mem. Bot. Sur. S. Afr. 16: 7-14.
Mullan, D. P. 1931. On the occurrence of glandular hairs (slat glands) on the leaves of some Indian halophytes. J. Indian Bot. Soc. 10: 184-189.
Mullan, D. P. 1931. Observations on the water storing devices in the leaves of some Indian halophytes. J. Indian Bot. Soc. 10: 126-133 and 184-191.
Mullan, D. P. 1932. Observations on the biology and physiological anatomy of some Indian halophytes. I. J. Indian Bot. Soc. 11: 103-118 and 285-302.
Mullan, D. P. 1933a. Observations on the biology and physiological anatomy of some Indian halophytes. II. J. Indian Bot. Soc. 12: 165-182.
Mullan, D. P. 1933b. Observations on the biology and physiological anatomy of some Indian halophytes. III. J. Indian Bot. Soc. 12: 235-256.
Muller, J. 1961. A palynological contribution to the history of the mangrove vegetation. Pac. Sci. Congr., 1960, pp. 146-147.
Muller, J. 1964. A palynological contribution to the history of the mangrove vegetation in Borneo. *in* "Ancient Pacific Floras", Univ. Hawaii Press,

Honolulu, pp. 33-42.

Muller, J. 1968. Palynology of the Pedowan and Plateau sandstone formations (Cretaceous-Eocene) in Savanak, Malaysia. Micropalaentology 14: 1-37.

Muller, J. and S. Y. Hou-Liu. 1966. Hybrids and chromosomes in the genus *Sonneratia* (Sonneratiaceae). Blumea 14: 337-343.

Muller, J. and G. G. J. van Stennis. 1968. The genus *Sonneratia* in Australia. North Queensland Not. 35: 6-8.

Murphy, R. C. 1938. The littoral of the Pacific coasts of Colombia and Equador. Geogr. Rev. 29: 1-33.

Myers, J. G. 1933. Notes on the vegetation of the Venezuelan Llanos. J. Ecol. 21: 335-349.

Myers, J. G. 1935. Zonation along river courses. J. Ecol. 23: 356-360.

Nakanishi, S. 1964. An epiphytic community on the mangrove tree, *Kandelia kandel* (Rhizophoraceae). Hikobia 4: 124.

Nash, G. V. 1908. The story of mangrove. Torreya 8: 73-78.

Navalkar, B. S. 1940. Studies in the ecology of mangroves. I. Determination of the osmotic pressure of *Avicennia alba* Blume. J. Univ. Bombay 8: 1-19.

Navalkar, B. S. 1942. Studies in the ecology of mangroves. II. Physical factors of the mangrove soil. J. Univ. Bombay 9: 97-106.

Navalkar, B. S. 1948. Studies in the ecology of mangroves III. J. Univ. Bombay 16: 35-45.

Navalkar, B. S. 1952. Succession of the mangrove vegetation of Bombay and Salsette Island. J. Bombay Nat. Hist. Soc. 50: 157-161.

Navalkar, B. S. 1953. The analytical aspects of some of the marshy vegetation of Bombay and Salsette islands. J. Bombay Nat. Hist. Soc. 51: 636-652.

Navalkar, B. S. 1956. Geographical distribution of halophytic plants of Bombay and Salsette islands. J. Bombay Nat. Hist. Soc. 53: 335-343.

Navalkar, B. S. 1959. Studies in the ecology of mangroves. VII. Humus content of mangrove soils of Bombay and Salsette islands. J. Univ. Bombay 28: 6-10.

Navalkar, B. S. and F. R. Bharucha. 1948. Studies in the ecology of mangroves. IV. The hydrogen ion concentration of the seawater, soil solution, and the leaf cell sap of mangroves. J. Univ. Bombay 17: 35-45.

Navalkar, B. S. and F. R. Bharucha. 1949. Studies in the ecology of mangroves. V. Chemical factors of mangrove soils. J. Univ. Bombay 17: 17-35.

Navalkar, B. S. and F. R. Bharucha. 1950. Studies in the ecology of mangroves. VI. Exchangeable bases of mangrove soils. J. Univ. Bombay 18: 7-16.

Newell, N., J. Imbrie, E. G. Purdy, and D. Thurber. 1959. Organism communities and bottom facies, Great Bahama Bank. Bull. Am. Mus. Nat. Hist. 117: 181-228.

Nicolic, M. and S. A. Melendez. 1968. El ostion del mangle, *Crassostrea rhizophorae* Guilding 1928. Experimentos indicales en el cultivo. Nota

Invest Cent. Pesqueras Buta 7: 1-30.
Nisbet, I. C. T. 1968. The utilization of mangroves by malayan birds. Ibis 110: 348-352.
Noakes, D. S. P. 1950. The mangrove charcoal industry. Malay. For. 24: 201-203.
Noakes, D. S. P. 1951. Notes on the silviculture of the mangrove forests of Matang, Perak. Malay. For. 14: 183-196.
Noakes, D. S. P. 1955. Methods of increasing growth and obtaining natural regeneration of the mangrove tupe in Malaya. Malay. For. 18: 23-30.
Noakes, D. S. P. 1957. Mangrove. Trop. Silvic. 2: 309-318.
Noakes, D. S. P. 1958. Mangrove. Wld. For. Congr. 3: 415-424.
Noamesi, G. K. 1959. A revision of the Zylocarpeae (Meliaceae). Diss. Abstr. 19: 1531.
Noel, A. R. A. 1959. The vegetation of the fresh water swamps of Inhaca Island. S. Afr. J. Bot. 25: 189-201.
Nudaliar, C. R. and H. Sunandar. 1952. Distribution of *Rhizophora mucronata* Lam. in the "backwater" of the west coast and its economic importance. Madras Agric. J. 610-645.
Odani, N. 1964. An analysis of the mangrove communities of Iriomoti-jima, Yaeyama Group, the Tyukyus. Rep. Comm. For. Sci. Res., Kyushu Univ., No. 2, June.
Odum, E. P. 1971. Fundamentals of Ecology. 3rd ed., W. B. Saunders Co., Philadelphia, 574 pp.
Odum, H. T. 1971. Environment, Power and Society. John Wiley and Sons, Inc., New York, 331 pp.
Odum, W. E. 1971. Pathways of energy flow in a south Florida estuary. Univ. Miami, Sea Grant Tech. Bull. No. 7, 162 pp.
Ogura, Y. 1942. Mangroves in eastern Asia. Bot. and Zool. 10: 145-149 (in Japanese).
Ogura, Y. 1958. Some problems on the mangrove. Bull. Yokohama Municip. Univ. Soc. 10: 545-578.
Okuda, T., A. J. Garcia, and J. B. Alvarez. 1965. Variacion estacional de los elementos nutritivos en el agua de la Laguna Y el Rio Unare. Bol. Inst. Oceanogr. Univ. Oriente 4: 123-135.
Orians, G. H. And E. W. Pfeiffer. 1970. Ecological effects of the war in Vietnam. Science 168: 544-554.
Orr, A. P. and F. W. Newhouse. 1933. Physical and chemical conditions in mangrove swamps. Br. Mus. Nat. Hist. Sci. Rep., Great Barrier Reef Exped. 1928-1929. 2: 102-110.
Ostenfeld, C. H. 1918. Stray notes from tropical west Australia. Dan. Bot. Ark. 2: 1-29.
Outred, H. A. 1966. The effects of low oxygen tensions upon the intensity and pattern of respiration in certain mangrove seedlings. M. Sc. Thesis, Auckland Univ., Auckland, N. Z.

Oyama, K. 1950. Studies of fossil biocoenosis. No. 1. Biocoenological studies on the mangrove swamps, with descriptions of new species from Yatuo Group. Rep. Geol. Surv. Japan 132: 1-14.

Padmanabhan, D. 1960. The embryology of *Avicennia officinalis* L. I. Floral Morphology and gametophytes. Proc Indian Acad. Sci. Sect. B, 52: 131-145.

Padmanabhan, D. 1962a. The embryology of *Avicennis officinalis* L. III. The embryo. J. Madras Univ. 32B: 1-19.

Padmanabhan, D. 1962b. The embryology of *Avicennia officinalis* L. The seedling. Proc. Indian Acad. Sci. 56: 114-122.

Paijmans, K. 1967. Lands of the Sofia-Pourgani area, Territory of Papua and New Guinea. VI. Vegetation of the Sofia-Pourgani area. CSIRO Aust. Land Res. Ser. 17: 142-167.

Paijmans, K. 1969. Lands of the Kerema-Vailala area, Papua-New Guinea. VIII. Vegetation and ecology of the Kerema-Vailala area. CSIRO Aust. Land Res. Ser. 23: 95-116.

Pal, N. 1951. On the embryology of Terminalia catappa. Sci. Cult. 17: 178-179.

Pammel, L. H. and C. N. King. 1948. Gerination studies of some trees and shrubs. Proc. Iowa Acad. Sci. 35: 184-197.

Pannier, P. 1959. El efecto de distintas concentrationes sotenas sobre el desarrolo de *Rhizophora mangle* L. Acta cient. Venez. 10: 68-78.

Pannier, P. 1962. Estudio fisiologico sobre la viviparia de *Rhizophora mangle* L. Acta cient. Venez. 13: 184-197.

Panshin, A. J. 1932. An anatomical study of the woods of the Philippine mangrove swamps. Philipp. J. Sci. 48: 143-208.

Panzer, W. 1935. Sandkrabben-Spuren an der Kuste von Neu-Guinea. Nat. Volk 65: 36-46.

Parham, J. W. 1964. Plants of the Fiju Islands. Govt. Press, Suva, Fiji.

Parkes, K. C. and R. W. Dickerman. 1967. A new subspecies of mangrove warbler (*Dendroica petechia*) from Mexico. Ann. Carnegie Mus. 39: 85-90.

Passler, J. 1912. Die untersuchgensergebnisse der Deutsch-Ostafrika eingesandtern Mangroverinden. Pflanzer Jg. 8: 65.

Patil, R. P. 1961. A note on the vegetation at Sazina Khali in the Sunderbans. Indian For. 87: 481-483.

Patil, R. P. 1964. Cultivation of mangrove seedlings in pots at Allahabad U.P. Sci. Cult. 30: 43-44.

Patil, R. P. and S. S. Sidhu. 1961. Seedling anatomy of *Bruguiera parviflora* W. and A. Proc. 48th Indian Sci. Congr. 3: 280.

Patton, R. T. 1942. Ecological studies in Victoria. VI. Salt marsh. Proc. R. Soc. Vict. 54: 131-144.

Paulsen, O. 1918. Chenopdiaceae from western Australia. Dan. Bot. Ark. 2: 56-66.

Payet, M., P. Pene, M. Sankale, G. Delorme, and H. Delatoure. 1962. Le micro-climat de la presque ile du Cap-Bert: son determinisme

physiologique. Bull. Soc. med. Afr. noire Lang. fr. 13: 40-58.
Paynter, R. A., Jr. 1955. The ornithogeography of the Yucatan Peninsula. Peabody Mus. Nat. Hist., Yale Univ., Bull. No. 9, 374 pp.
Pearman, R. W. 1955. Tanning materials of the British Commonwealth. Colonial Plant and Animal Products 5: 96.
Pearman, R. W. 1957. Mangrove bark — its value as a tanning material. Leather Trades Rev. 125: 315-316.
Pearse, A. S. 1932. Animals in brackish water pools at Dry Tortugas. Pap. Tortugas Lab. 28: 125-142.
Pellegrin, F. 1952. Les Rhizophoracees de l'Afrique Equatiriale Francaise. Mus. Nat. Hist. Nat. Notulae Syst. 14: 292-300.
Penfound, W. T. 1952. Southern swamps and marshes. Bot. Rev. 18: 413-446.
Penfound, W. T. and E. S. Hathaway. 1938. Plant communities in the marshlands of southeastern Louisiana. Ecol. Monogr. 8: 1-56.
Perrier de la Bathie, H. 1953. La mangrove (paletuviers) de Madagascar: biologie et utilite. Rev. Int. Bot. Appl. Agric. Trop. 33: 373-374 and 581-582.
Petit-Thouars, A. du. 1813. Notice sur le manglier. J. Bot. 2: 27-41.
Phillips, J. 1959. Agriculture and ecology in Africa. Flora 131: 1-40.
Phillips, O. P. 1903. How the mangrove tree adds new land to Florida. J. Geogr. 2: 10-21.
Pidgeon, I. 1940. The ecology of the central coastal area of New South Wales. III. Types of primary succession. Proc. Linn. Soc. N.S.W. 65: 221-249.
Pickening, C. 1879. Chronological History of Plants. Little, Brown and Co., Boston, 1222 pp.
Piexoto, A. R. 1959. Tropical-almound *Terminalia;* Fruit, oil, and tanning. Bol. Agric. Minas Gerais Dept. Prod. Veg. 8: 69-71.
Pijl, L. van der. 1969. Evolutionary action of tropical animals on the reproduction of plants. Biol. J. Linn. Soc. London 1: 85-96.
Pitot, A. 1958. Rhizophores et racines chez *Rhizophora* sp. Bull. Inst. Franc. Afr. Noire 20: 1103-1138.
Pittier, H. 1898. Primitiae florae costaricensis. Anal. Inst. fis.-geogr. Nacl. 8: 1-26.
Pittier, H. 1908. Ensayo Sobre las Plantas Usuales de Costa Rica. H. L. and J. B. McQueen, Washington, D. C., p. 112.
Pittier, H. 1926. Manual de las Plantas Usuales de Venezuela. Pub. by the author, Caracas, p. 286.
Pollard, C. L. 1902. Plant agencies in the formation of the Florida keys. Pl. World 5: 8-10.
Pope, W. T. 1929. Manual of Wayside Plants of Hawaii. Advertiser Pub. Co., Honolulu, pp. 157-159.
Porsch, O. 1924. Vogel blumenstudien. I. Jb. wiss. Bot. 63: 553-706.
Post, E. 1963a. Systematische und pflanzen-Geographische Notizen zur *Bostrichia-Caloglassa* Assoziation. Rev. Algel. 9: 1-84.

Post, E. 1963b. Zur Verbreitung und Okologie der *Bostrychia-Caloglossa* Assoziation. Int. Rev. ges. Hydrobiol. Hydrogr. 48: 47-152.

Prain, D. 1903. Flora of the Sunderbans. Rec. Bot. Surv. India 2: 231.

Prakash, U. 1960. A Survey of the Deccan Intertrappean beds flora of India. J. Palaentol. 34: 1027-1040.

Price, J. L. 1955. A survey of the freshwater fishes of the island of Trinidad. J. Agric. Soc. Trinidad and Tobago, Pap. No. 863: 1-28.

Price, W. A. 1954. Shorelines and coasts of the Gulf of Mexico. U. S. Fish. Bull. 89: 39-65.

Pustelnik, W. 1953. Analisis de la Corteza de Mangle Colorado (*Rhizophora mangle* L.) del oriente de Venezuela y su importancia en la elaboracion de extractos curientes solidos. Agron. Trop. 3: 107-116 (also in Rev. Soc. Venez. 5: 19-31).

Putnam, R. C. and A. V. Bowles. 1955. The chemistry of vegetable tannins. XI. Mangrove (Borneo cutch). J. Am. Leather Chemists Assoc. 50: 42-46.

Pynaert, L. 1933. La mangrove congolaise. Bull. Agric. Congo Bel. 23: 184-207.

Quesnel, V. C. 1964. Suitability of sulphurous acid for hydrolysis of condensed tannins. Tetrahedron Letters 48: 3699-3702.

Quinones, L. R. and J. F. Puncochar. 1943. Informe preliminar sobre la utilizacion practica de la corteza del mangrove. Caribb. For. 5: 44-47.

Rai, J. N., J. P. Tewari, and K. G. Mukerji. 1969. Mycoflora of mangrove mud. Mycopathol. Mycol. Appl. 38: 17-31.

Rains, D. W. and E. epstein. 1967. Prederential absorption of potassium by leaf tissue of the mangrove, *Avicennia marina:* an aspect of halophytic competence in coping with salt. Aust. J. Biol. Sci. 20: 847-857.

Rajagopalan, V. R. 1949. Cytological studies in Combretaceae I. Proc. Indian Sci. Congr. 36: 137-138.

Rao, A. N. and L. C. Wan. 1969. A new record of folian sclerids in *Scyphiphora hydrophyllacea.* Curr. Sci. 38: 594-595.

Rao, R. S. 1957. Observations on the mangrove vegetation of the Godavri Estuary. Proc. Mangrove Swamp Symp., Ministry of Food and Agriculture, Calcutta, India, pp. 36-44.

Raunkaiaer, C. 1934. The Life Forms of Plants and Statistical Plant geography, Being the Collected Papers of C. Raunkiaer. Oxford Univ. Press, Oxford, 632 pp.

Rawitscher, F. K. 1944. Algumas nocoes sobre a vegetacao do litoral brasileiro. Bol Assoc. dos Geogr. Bras. 5: 13-28.

Reinders-Gouwentak, C. A. 1953. Sonneratiaceae and other mangrove-swamp families, anatomical structure and water relations. Fl. Males. 4: 513-515.

Reintjes, J. W. 1963. The importance of the occurrence of menhaden in the coastal waters and estuaries of peninsular Florida. Proc. Gulf Caribb. Fish. Inst. 16: 108-113.

Rendle, A. B. and S. LeM. Moore. 1921. A systematic account of the plants collected in New Caldonia and the Isle of Pines by Prof. R. H. Compton, 1914. I. Flowering plants (angiosperms). J. Linn. Soc. London 15: 245-418.

Revelle, R. and R. W. Fairbridge. 1957. Carbonates and carbon dioxide. Mem. geol. Soc. Am. 67: 239-295.

Reyes, L. J. 1938. Philippine woods. Dept. Agric. Commerce, Manila, Tech. Bull. No. 7.

Richards, P. W. 1952. The Tropical Rain Forest: an Ecological Study. Cambridge Univ. Press. Cambridge, 299-312.

Ricklefs, R. E. 1971. Foraging behavior of mangrove swallows at Barro Colorado Island. Auk 88: 635-651.

Ridgway, R. 1873. On some new forms of American birds. Am. Nat. 7: 602-619.

Ridgway, R. 1902. The Birds of North and Middle America, Part 2. U. S. Natl. Mus. Bull. No. 50, 834 pp.

Ridley, H. N. 1920. New and rare plants from the Malay Peninsula. J. Fed. Malay. St. Mus. 10: 128-156.

Ridley, H. N. 1922. The flora of the Malay Peninsula, Vol. 1, pp. 692-703, L. Reeve and Co., London.

Ridley, H. N. 1930. The Dispersal of Plants Throughout the World. L. Reeve and Co., London, 744 pp.

Rimbach, A. 1932. The forests of Equador. Trop. Woods. 31: 1-9.

Robas, A. K. 1970. South Florida's mangrove-bordered estuaries: their role in sport and commercial fish production. Univ. Miami Sea Grant Inf. Bull. No. 4, 28 pp.

Robertson, R. 1959. The mollusk fauna of Bahamian mangroves. Am. Macacol. Union Ann. Rep. Bull. 1959: 22-23.

Robertson, W. B. 1962. Fire and vegetation in the Everglades. Proc. 1st Ann. Tall Timbers Fire Ecol. Conf., pp. 67-80.

Robyns, W. 1971. Over viviparie en biotecnose bij planten. Meded. Kon. Vlaamse. Acad. Wetensch. Lett. Schone Kusten Belgie 33: 3-14.

Rodriguez, G. 1959. The marine communities of Margarita Island, Venezuela. Bull. Mar. Sci. Gulf Caribb. 9: 237-280.

Rodriguez, G. 1963. The intertidal estuarine communities of Lake Maracaibo, Venezuela. Bull. Mar. Sci. Gulf Caribb. 13: 197-218.

Rodriguez, G. 1964. Physical parameters of the Maracaibo estuary and their implications. Proc. Gulf. Caribb. Fish. Inst., 17th Ann. Sess., pp. 42-50.

Rogers, J. S. 1951. Native sources of tanning materials. U.S.D.A. Yearb., Wash., D. C., pp. 709-715.

Roig y Mesa, J. T. 1945. Plantas Medicinales, Aromaticas o Venenosas de Cuba. Cultural, S. A., Habana, pp. 447-449.

Roonwall, M. L. 1954. The marine borer, *Bactronophorus thoracites* (Gould) Mollusca, Eulamellibranchiata, Teredinidae) as a pest of living trees in the mangrove forests of the Sundurhaus, Bengal, India. Proc. Zool. Soc. Bengal 7: 91-105.

Roos, P. J. 1964. The distribution of reef corals in Curacao. Natuurwetensch. Studiekring Suriname Ned. Antillen Uitganen, 1-51.

Rosevear, D. R. 1947. Mangrove swamps. Farm For. 8: 23-30.

Russell, A. 1943. *Conocarpus erecta* (buttonwood, Zaragoza-mangrove), a new domestic source of tannin. Chemurg. Dig. 2: 27-29.

Roth, I. 1965. Histogenese der Lentizellenam Hypocotyl von *Rhizophora mangle* L. Ostero. Bot. Z. 112: 640-653.

Ruhland, W. 1915. Untersuchungen uber die Hautdrusender Plumbaginaceen. J. Wiss. Bot. 55: 409-498.

Russell, A. 1942. Natural tanning materials of the southeastern United States. J. Am. Leather Chemists Assoc. 37: 340-356.

Putzler, K. 1969. The mangrove community, aspects of its structure, faunistics and ecology. *in* Lagunas Costeros, Un Simposio, Mem. Simp. Intern. Lagunas Costeras, UNAM-UNESCO, pp. 515-536.

Rutzler, K. 1970. Oil pollution: damage observed in tropical communities along the Atlantic seabord of Panama. Bioscience 20: 222-224.

Rzedowski, J. and R. McVaugh. 1966. La vegetacion de Nuera Galicia. Con. Univ. Mich. Herb. 9: 1-133.

Safford, W. E. 1905. The useful plants of the island of Guam. Contrib. U. S. Natl. Herb., Vol. II.

Sagrario, C. D. del. 1957. Recovery of tannin from mangrove bark and its conversion into a colorless tannin extract. Chem. Abstr. 53: 4792D.

Salvosa, F. M. 1936. *Rhixophora.* Nat. Appl. Sci. Bull., Univ. Philippines 5: 179-237.

Sawman-Wanakit, L. 1936. Forests in the south. Vanasarn For. J., Thailand, Sp. Ed., No. 3, December.

Sanchez, R. and M. Elana. 1963. Datos relativos a los mangalares de Mexico. Ann. Esc. Nac. Cienc. Biol. Mex. 12: 61-72.

Sandison, E. E. 1966. The effect of salinity fluctuations on the life cycle of *Balanus pallidus stutsburi* Darwin in Lagos Harbour, Nigeria. J. Anim. Ecol. 35: 363-378.

Sandison, E. E. and M. B. Hill. 1966. The distribution of *Balanus pallidus stutsburi* Darwin, *Gryphaea gasar* (Adanson) Dautzenberg, *Mercierella enigmatica* Fauvel and *Hydroides uncinata* Philippi in relation to salinity in Lagos Harbour and adjacent creeks. J. Anim. Ecol. 35: 235-250.

Santos, T. A. and I. B. Siapno. 1968. Cutting operations of mangrove forest in Zamboanga peninsula (Mindanao). Philipp. Lumberman 14: 14-20.

Sargent, C. S. 1893. The mangrove tree. Gard. For. 6: 97-98.

Sargent, C. S. 1903. Combretaceae. Silvic. N. Am. 5: 19-29.

Sargent, C. S. 1922. Manual of Trees of North America (Exclusive of Mexico). 2nd ed., Houghton Mifflin Co., Boston, pp. 763-764.

Sargent, C. S. 1962. Effects of recent tropical cyclones on the coastal vegetation of Mauritius. J. Ecol. 50: 275-290.

Sasaki, S. 1912. Plants of the Kojurin Forest, Formosa. Trans. Nat. Hist. Soc. Formosa 2: 45-50.

Sasaki, S. 1923. *Kandelia rheedii* is disappearing from the Taiwanian flora. Trans. Nat. Hist. Soc. Formosa 13: 37-39.

Satayanaraynan, Y. 1959. Ecological studies of the Elephante Island. Inst. Franc. de Pondicherry, Trav. Sec. Scien. Tech. 1: 99-114.

Sauer, J. D. 1961. Coastal plant geography of Mauritius. Coast. Stud. Ser., Louisiana St. Univ. 5: 1-153.

Sauer, J. D. 1967. Geographic reconnaissance of seashore vegetation along the mexican gulf coast. Coastal St. Res. Inst. La. St. Univ., Tech. Rep. No. 56.

Saunders, J. B. 1958. Recent foraminifera of mangrove swamps and river estuaries and their fossil counterparts in Trinidad. Micropaleontology 4: 79-92.

Saver, J. D. 1965. Geographic reconnaissance of western Australian seashore vegetation. Aust. J. Bot. 13: 39-69.

Savory, H. J. 1953. A note on the ecology of *Rhizophora* in Nigeria. Kew Bull. 1: 127-128.

Scheffen, W. 1937. Strandbeobachtungen im Malayischen Archipel. Geol. Meere Binnengew. 1.

Scheffer, T. C. and C. G. Duncan. 1947. The decay resistance of certain Central American and Ecuadorian woods. Trop. Woods. 92: 1-24.

Schenck, H. 1889. Ueber die Luftwurzeln von *Avicennia tomentosa* und *Laguncularia racemosa*. Flora 72: 83-88.

Schimper, A. F. W. 1891. Die Indo-Malayische Strandflora. Bot. Mit. Trop. 3: 1-204.

Schimper, A. F. W. 1892-1893. Rhizophoraceae. Nat. Pflanzenfam. 7: 42-56.

Schimper, A. F. W. 1903. Plant Geography on a Physiological Basis. Oxford Univ. Press. Oxford, 839 pp.

Schimper, A. F. W. and F. C. von Faber. 1935. Pflanzengeographie auf Physiologischer Grundlage. 2 vols., 3rd Ed., Jena.

Schmidt, J. 1902. Rhizophoraceae. Bot. Tidsskr. 24: 249-554.

Schmidt, J. 1903. Skuddene has den gamle verdens mangrovetraer. Nordiske Forlag Econst. Bojeuen, 113 pp.

Schmidt, J. 1904. Zur Frage der Salzausscheidung der Mangrovepflanzen. Flora 93: 260-261.

Schmidt, J. 1905. Bidrag til Kundskab om Skuddene hos den Gamle Verdens Mangrove-traer. Bot. Tidsskr. 26: 1-113.

Schmidt, J. 1906. Die Kustenvegetation von Koh Chang. Vegetationsbilder 3: 78-91.

Schneider, E. E. 1916. Commercial woods of the Philippines: their preparation and uses. Philipp. Bur. Forest., Bull. No. 14, pp. 179-182.

Schneider, J. 1921. Uber Mangrove auf den Philippinen. Gerger 47: 53-54.

Schnepper, W. C. R. 1933. Vloedbosch culturen. Tectona 26: 907-919.

Schnetter, M. 1969. Observaciones ecologicas en la isla de Salamanca (Departmento del Magdalene) Colombia. Caldasia 10: 299-315.

Scholander, P. F. 1965. From the frozen forest to the tropical mangroves. *in* Science in Alaska: Proc. 15th Alaskan Sci. Conf., College, Alaska, 31 Aug. -

4 Sept. 1964, AAAS, Washington, D. C.

Scholander, P. F. 1968. How mangroves desalinate seawater. Physiol. Plantarum 21: 258-268.

Scholander, P. F., E. D. Bradstreet, H. T. Hammel, and E. A. Hemmingsen. 1966. Sap concentrations in halophytes and some other plants. Plant Physiol. 41: 529-532.

Scholander, P. F., L. van Dam, and S. I. Scholander. 1955. Gas exchange in the roots of mangroves. Am. J. Bot. 42: 92-98.

Scholander, P. F., H. T. Hammel, E. A. Hemmingsen, and E. D. Bradstreet. 1964. Hydrostatic pressure and osmotic potential in leaves of mangroves and some other plants. Proc. U. S. Natl. Acad. Sci. 52: 119-125.

Scholander, P. F., H. T. Hammel, E. Hemmingsen, and W. Carey. 1962. Salt balance in mangroves. Plant Physiol. 37: 722-729.

Scholander, P. F., H. T. Hammel, E. D. Bradstreet, and E. A. Hemmingsen. 1965a. Sap pressure in vascular plants. Science 148: 339-346.

Scholander, P. F., H. T. Hammel, E. D. Bradstreet, and E. A. Hemmingsen. 1965b. Sap pressure in vascular plants (Addendum). Science 148: 1488.

Scholander, P. F. and M. de Oliveira Perez. 1968. Sap tension in flooded trees and bushes of the Amazon. Plant Physiol. 42: 1870-1873.

Scholl, D. W. 1963. Sedimentation in modern coastal swamps, southwestern Florida. Bull. Am. Assoc. Pet. Geol. 47: 1581-1603.

Scholl, D. W. 1964. Recent sedimentary record in mangrove swamps and rise in sea level over the southwestern coast of Florida. I. Mar. Geol. 1: 344-366.

Scholl, D. W. 1965a. High interstitial water cholorinity in estuarine Mangrove swamps. Florida. Nature 207: 284-285.

Scholl, D. W. 1965b. Recent sedimentary record in mangrove swamps and rise in sea level over the southwestern part of Florida. II. Mar. Geol. 2: 343-364.

Scholl, D. W. Modern coastal swamp statigraphy and the ideal cyclotherm. *in* Environments of Coal Deposition, E. C. Dapples and M. B. Hopkins (eds.), Geol. Soc. Amer. Spec. Pap., *in press*.

Scholl, D. W. and M. Stuijer. 1967. Recent submergence of southern Florida: a comparison with adjacent coasts and other eustatic data. Geol. Soc. Am. Bull. 78: 437-454.

Schone, H. 1963. Menotaktische Orientierung nach polarisiertem und unpolarisierten Light bei der Mangrovekrabbe *Goniopsis*. Z. Vgl. Physiol. 46: 496-514.

Schone, H. and H. Schone. 1963. Balz und Verhaltensweisen der Mangrovekrabbe *Goniopsis cruentata* Latr. und das Winkverhalten der eulitoralen Brachyuren. Z. Tierpsychol. 20: 641-656.

Schratz, E. 1934. Beitrage zur Biologie der Halophyten. I. Zur Keimungphysiologie. Jahrb. f. wiss. Bot. 80: 112-142.

Schuster, W. H. 1962. Das marine Littoral als Lebensraum Terrestrischer Kleinarthropoden. Int. ges. rev. Ges. Hydrobiol. 47: 359-412.

Schuster, W. H. 1952. Fish culture in brackish water ponds of Java. Indo-Pacific Fish. Counc. Spec. Publ., No. 1, pp. 1-143.

Seifriz, W. 1943. The plant life of Cuba. Ecol. Monogr. 13: 375.426.

Sen Gupta, J. C. 1935. Die Osmotischen Verhaltnisse bei einigen Pflanzen in Bengal (Indian). Ber. Dtsch. bot. Ges. 53: 783-795.

Sen Gupta, J. C. 1938a. Ueber die osmotische Werte und den Chloranteil in Pflanzen einiger Salzgebietes Bengal. Ber. Dtsch. bot. Ges. 56: 474-485.

Sen Gupta, J. C. 1938b. Der histochemische Nachweis vor Natrium bei einigen Pflanzen von Bengal. Ber. Dtsch. bot. Ges. 56: 486-494.

Sen Gupta, J. C. 1942. Investigation of the annual variation of the water content, osmotic value and chloride fractions of some plants in Port Canning, near calcutta. 50th Ann. Vol R. Bot. Gard. Calcutta, 1942.

Seshadri, T. R. and B. Venkataramani. 1959. Leucocyanin from mangroves. Indian J. Sci. Res. 18B: 261-262.

Setchell, W. L. 1926. Phytogeographical notes on Tahiti. Univ. Cal. Publ. Bot. 12: 241-290.

Setten, G. G. K. 1947. A note on the growth and yield of Beruus (*Bruguiera cylindrica* (L.) (Bl.). Malay. For. 16: 74.

Setten, G. G. K. 1955. A note on the growth and yield of Berus (*B. cylindrica* (L.) Bl. syn. *B. caryophylloides* Bl.) in Selangor. Res. Pamphl. For. Res. Instit. Fed. Malaya, No. 16, 12 pp.

Shantz, H. L. and C. F. Marbut. 1923. The vegetation and soils of Africa. Am. Geogr. Soc. Res. Ser. 13: 1-242.

Shah, J. J. and K. P. Sundarraj. 1965. Stipular sclerids in *Rhizophora mucronata* L. Curr. Sci. 34: 155-156.

Shelford, V. E. 1963. The Ecology of North America. Univ. Ill. Press, Urbana, 610 pp.

Shier, D. E. 1969. Vermetid reefs and coastal development in the Ten Thousand Islands, southwest Florida. Geol. Soc. Bull. 80: 485-508.

Shieve, F. 1937. The vegetation of the cape region of Baja California. Madrona 4: 105-113.

Shimada, Y. 1924. Investigations on *Kandelia rheedii* in Komoko, Taiwan. Trans. Nat. Hist. Soc. Formosa 14: 25-29 (in Japanese).

Sidhu, S. S. 1960. Studies on mangroves. Natl. Acad. Sci. India, No. 1960, pp. 111-112.

Sidhu, S. S. 1961. Vegetation of East Godavari. Proc. 48th Indian Sci. Congr. 3: 358-359.

Sidhu, S. S. 1961. Chromosomal studies on some mangrove species. Proc. 48th Indian Sci. Congr. 3: 302-304.

Sidhu, S. S. 1962. Chromosomal studies of some mangrove species. Indian For. 88: 585-592.

Sidhu, S. S. 1963a. Studies on the mangroves of India. I. East Godavari region. Indian For. 89: 337-351.

Sidhu, S. S. 1963b. Studies on mangroves. *in*Symposium on Ecological Problems in the Tropics. Proc. Natl. Acad. Scr. India, Sect. B. 33: 129-136.

Sidhu, S. S. 1968. Further studies on the cytology of mangrove species in India. Caryologia 21: 353-357.

Simpson, C. T. 1932. Florida Wild Life. Macmillan, New York, pp. 9-11, 118.

Simpson, J. G. and R. C. Griffiths. 1967. The fisheries resources of Venezuela and their exploitation. Serie Recursos y Expl. Presqueros 1: 171-206.

Silveira, F. 1937. Mangrove. Rodriguesia 3: 131-154.

Simberloff, D. S. and E. O. Wilson. 1969. Experimental zoogeography of islands: the colonization of empty islands. Ecology 50: 278-296.

Singh. T. C. N. and A. T. Natarajan. 1953-1954. Morphology of the pollen grains of the constituents of the mangrove vegetation (abstr.) Proc. Indian Sci. Congr. 40: 99-100.

Slooten, D. V. van. 1937. Die Verbreitung von *Lumnitzera* und anderen Mangrovegewachsen. Blumea Suppl. 1: 162-175.

Slooten, H. J. van der. 1960. Resina de Fenol-Formaldehido para contrachapeado obtenida del tanio de *Rhizophora mangle*. Inst. For. Latino Amer. de Invest. Capacit., Merida, Venezuela, Bol. 6: 34-39.

Slud, P. 1964. The birds of Costa Rica, distribution and ecology. Bull. Am. Mus. Nat. Hist. 128.

Small, J. K. 1914. Exploration in the Everglades and on the Florida keys. J. N. Y. Bot. Gard. 15: 69-79.

Small, J. K. 1923. Mangroves, J. N. Y. Bot. Gard. 24: 211.

Small, J. K. 1924. Mangroves. J. N. Y. Bot. Gard. 25: 73-74.

Small, J. K. 1933. Manual of the Southeastern Flora. Univ. North Carolina Press, Chapel Hill, pp. 938-939.

Smith, A. C. and I. M. Johnson. 1945. A phytogeographic sketch of Latin America. *in* Plants and Plant Science in Latin America, F. Verdoorn, (ed.), Chronica Botanica Co. Waltham, Massachusetts.

Smith, F. G. W., R. H. Williams, and C. C. Davis. 1950. An ecological survey of the sub-tropical inshore waters adjacent to Miami. Ecology 31: 119-146.

Smith, J. 1882. A Dictionary of Popular Names of the Plants Which Furnish the Natural and Acquired Wants of Man in All Matters of Domestic and General Economy. Macmillan, London, pp. 263-264.

Sodiro, R. P. L. 1901. El mangle rojo. Estudios de Medicina, Quito ¾: 3-18.

Sokoloff, B., J. B. Redd, and R. Dutscher. 1950. Nutritive value of mangrove leaves (*Rhizophora mangle* L.). Q. J. Fla. Acad. Sci. 12: 191-194.

Solereder, H. 1885. Zur Anatomie und Systematik der Combretaceen. Bot. Centralbl. 23: 161-166.

Solereder, H. 1908. Systematic Anatomy of the Dicotyledons. Clarendon Press, Oxford, 423 pp.

Southwell, C. R. and J. D. Bultman. 1971. Marine borer resistances of untreated woods over long periods of immersion in tropical waters. Biotropica 3: 81-107.

Spackman, W., C. P. Dolsen, and W. Riegal. 1966. Phytogenic organic sediments and sedimentary environments in the Everglades-mangrove complex. Part I. Evidence of a transgressing sea and its effect on environments of the Shark River area of southwest Florida. Paleontrophica 117: 135-152.

Spackman, W., D. W. Scholl, and W. H. Taft. 1964. Field Guide Book to Environments of Coal Formation in Southern Florida. Geol. Soc. Am., 67 pp.

Specht, R. L. and C. P. Mountford. 1958. Records of the American-Australian Scientific Expedition to Arnhem Land. 3. Botany and Plant Ecology. Melbourne Uni. Press, Melbourne.

Speck, N. H. 1960. Vegetation of the North Kimberly area. CSIRO Aust. Land Res. Ser. 4: 41-63.

Speck, N. H. and M. Lazarides. 1964. Vegetation and pastures of the West Kimberly area. CSIRO Aust. Land Res. Ser. 9: 140-174.

Spooner, G. M. and H. B. Moore. 1940. the ecology of the Tamar estuary. 6. An account of the macroflora of the intertidal muds. J. mar. biol. Assoc. U.K. 24: 283-291.

Sprague, T. A. and L. A. Boodle. 1909. Kokoti (*Anopyxis ealaensis* Sprague). Bull. Misc. Inf., R. Bot. Gard., Kew, London, pp. 309-312.

Stace, C. A. 1961. Cuticular characters as an aid to the taxonomy of the South-West African species of *Combretum*. Mitt. Bot. Staatssam. Munchen 4: 9-17.

Stace, C. A. 1965. The significance of the leaf epidermis in the taxonomy of the Combretaceae. J. Linn Soc. London Bot. 59: 229-252.

Stace, C. A. 1966. The use of epidermal characters in phylogenetic considerations. New Phytol. 65: 304-318.

Stamp, L. D. 1925. The aerial survey of the Irrawaddy delta forests. J. Ecol. 13: 262-276.

Standley, P. C. 1924. Trees and shrubs of Mexico. Cont. No. 23., U. S. Natl. Herb., Washington, D. C.

Standley, P. C. 1928. Flora of the Panama Canal Zone. Contrib. No. 27, U. S. Natl. Herb., Washington, D. C. p. 283.

Standley, P. C. 1930. Flora of Yucatan. Field Mus. Nat. Hist. Bot. Ser. III, No. 3, p. 371.

Standley, P. C. 1937. Flora of Costa Rica, Part II. Field Mus. Nat. Hist., Chicago, pp. 766-767.

Standley, P. C. and S. J. Record. 1936. The forests and flora of British Honduras. Field Mus. Nat. Hist. Pub. No. 350, Bot. Ser. XII, pp. 275-276.

Stearn, W. T. 1958. A key to the West Indian mangroves. Kew Bull. 13: 33-37.

Steenis, C. G. G. J. van. 1935. Maleische vegettieschetsen. Tijdschr. K. ned. aardrijsk. Genoot. 52: 363-398.

Steenis, C. G. G. J. van. 1941. Kustaanwas en mangrove. Natuurwet. Tijdschr. Ned. Ind. 101: 82-85.

Steenis, C. G. G. J. van. 1954a. Vegetatie en Flora. New Guinea 2: 218-275.

Steenis, C. G. G. J. van. 1954b. Plumbaginaceae. Fl. Males. 1: 107-112.

Steenis, C. G. G. J. van. 1958a. Rhizophoraceae. Flora Malesiana 5: 431-493.

Steenis, C. G. G. J. van. 1958b. Tropical shore formations. Proc. Symp. Humid Trop. Veg., Tijiawi (Indonesia), pp. 215-217.

Steenis, C. G. G. J. van. 1962a. The distribution of mangrove plant genera and its significance foe paleogeography. Kon. Neder. Akad. van Wetensch. 65: 164-169.

Steenis, C. G. G. J. van. 1962b. The land bridge theory in botany. Blumea 2: 235-372.

Steenis, C. G. G. J. van. 1963. Pacific Plant Areas. 1. Monogr. Natl. Inst. Sci. Technol. Philipp., Manila, Monogr. No. 8, 279 pp.

Steenis, C. G. G. J. van. 1965. Concise plant geography of java. *in* Flora Java, C. G. G. J. van Steenis (ed.), Vol. 2, pp. 1-72.

Steenis, C. G. G. J. van. 1968. Do *Sonneratia caseolaris* and *S. ovata* occur in Queensland or the Northern Territory? N. Queensland Nat. 35: 3-6.

Steenis, C. G. G. J. van and M. M. J. van Balgoog. 1966. Pacific Plant Areas. 2. Blumea, Suppl. 5, 312 pp.

Steers, J. A. 1929. The Queensland coast and the Great Barrier Reef. Geogr. J. 74: 232-257.

Steers, J. A. 1937. The coral islands and associated features of the Great Barrier Reef. Geogr. J. 89: 1-28 and 119-146.

Steers, J. A. 1940. Sand cays and mangroves in Jamaica. Geogr. J. 96: 305-323.

Steers, J. A., V. J. Chapman, J. Colman, and J. A. Lofthouse. 1940. Sand cays and mangroves in Jamaica. Geogr. J. 96: 305-328.

Stehle, H. 1945. Forest types of the Caribbean Islands, Part 1. Caribb. For. Suppl. to Vol. 6, pp. 1-16.

Stephens, W. M. 1962. Trees that make land. Sea Front. 8: 219-230.

Stephens, W. M. 1963. Mangroves: trees that make land. Smithson. Inst. Rep. 1962, pp. 491-496.

Setphens, W. M. 1969. The mangrove. Oceans 2: 51-55.

Stephenson, T. A. 1939. The constitution of the intertidal fauna and flora of South Africa. Part 1. J. Linn. Soc. London 40: 487-536.

Stephenson, T. A. and A. Stephenson. 1950. Life between tide marks in North America. I. The Florida Keys. J. Ecol. 38: 354-402.

Stephenson, T. A., A. Stephenson, G. Tandy, and M. A. Spender. 1931. The structure and ecology of Low Isles and other reefs. Br. Mus. Nat. Hist. Sci. Rep., Great Barrier Reef Exped., 1928-1929. 3: 17-112.

Stephenson, W., R. Endean, and I. Bennett. 1958. An ecological survey of the marine fauna of Low Isles, Queensland. Aust. J Mar. Freshwater Res. 9: 261-318.

Stern, W. L. and G. K. Voight. 1959. Effect of salt concentration on growth of red mangrove in culture. Bot. Gaz. 121: 36-39.

Stewart, A. 1916. Some observations concerning the botanical conditions of the Galapagos Islands. Wis. Acad. Sci. Trans. 18: 272-340.

Steup, F. K. M. 1946. Boschbeheer in de vloedbosschen van riouw. Tectona 36: 289-298.

Stocker, O. 1924. Beitrage zum Halophytenproblem. Okologische untersuchungen der Strandund Dunenpflanzen des Dars (Vorpommern). Z. Bot. 16: 289-330.

Stocker, O. 1925. Beitrage zum Halophytenproblem. Okologische untersuchungen der Strandund Dunenpflanzen des Dars (Vorpommern). Z. Bot. 17: 1-24.

Stocker, O. 1935. Transpiration und Wasserhaushalt in verschieden Klima Zonen. III. Jb. wiss. Bot. 81: 464-471.

Stocking, C. R. 1956. Osmotic pressure and osmotic value. Handb. Pflphysiol. 2: 63-64.

Stoddart, D. R. 1962. Three Caribbean atolls: Turneffe Islands, Lighthouse Reef, and Glover's, British Honduras. Tech. Rep. Coastal Stud. Inst. La. State Univ. 11, 151 pp.

Stokes, A. P. D. and R. C. Roberts. 1934. Saline peat profiles of Puerto Rico. J. Wash. Acad. Sci. 24: 175-182.

Story, R. 1969. Vegetation of the Adelaide-Alligator area. CSIRO Aust. Land REs. Ser. 25: 114-130.

Straatmans, W. 1954. Reclamation of tidal mud flats in Tonga. Tech. Pap. South Pac. Comm., No. 53, 18 pp.

Strenzke, K. 1958. *Axelsonia tubifera* n. sp., ein neuer arthropleoner Collembole mit Geschlechtsdimorphismus aus der brasilianischen Mangrove. Acta. Zool. Cracov. 2: 607-619.

Strunk, L. 1906. Chemische untersuchungen von mangrovenrinden. Tropenpflanzer 10: 116.

Studholme, W. P. and W. R. Philipson. 1966. A comparison of the cambium in two woods with included phloem: *Heimerliodendron brunonianum* and *Avicennia vesinifera.* N. Z. J. Bot. 4: 332-341.

Sundararaj, D. D. 1954. Mangrove (*Avicennia officinalis* L.) leaves as cattle feed. I. Sci. Cult. 19: 339.

Sukhamananda, M. 1951. Sutan mangrove Forest. Vanasarn For. J. 3: 1-16.

Suter, H. H. 1960. A General and Economic Geology of Trinidad and Tobago. H.M.S.O., London, 145 pp.

Svenson, H. K. 1945. A brief review of the Galapagos flora. *in* Plants and plant Science in Latin America. F. Verdoorn (ed.), Cronica Botanica Co., Waltham, Massachusetts.

Svenson, H. K. 1946. Vegetation of the coast of Equador and Peru and its relation to the Galapagos Islands. Am. J. Bot. 33: 394-426.

Swain, T. 1965. The tannins. *in* Plant Biochemistry, J. Bonner and J. E. Varner (eds.), Academic Press, New York, 1054 pp.

Swart, H. J. 1958. An investigation of the mycoflora in the soil of some

mangrove swamps. Acta Bot. Neerl. 7: 741-768.
Swart, H. J. 1963. Further investigations of the mycoflora in the soil of some mangrove swamps. Acta Bot. Neerl. 12: 98-111.
Symington, C. F. 1940. Three malayan species of *Bruguiera*. Malay. For. 9: 131-139.
Tabb, D. C. and R. B. Manning. 1961. A checklist of the flora and fauna of northern Florida Bay and adjacent brackish waters of the Florida mainland collected during the period July 1957 through September 1960. Bull. Mar. Sci. Gulf Caribb. 11: 552-649.
Tabb, D. C., D. L. Bubrow, and R. B. Manning. 1962. The ecology of northern Florida Bay and adjacent estuaries. Fla. State Board Conserv. Tech. Ser. Publ. No. 39.
Tadeo, C. B. 1962. Plants with edible fruits in the Philippines. For. Leaves 14: 57.
Tanner, W. F. 1960. Florida coastal classification. Trans. Gulf Coast Assoc. Geol. Soc. 10: 259-266.
Tansley, A. G. and F. E. Fritsch. 1905. The flora of the Ceylon littoral. New phytol. 4: 27-36.
Taylor, B. W. 1959. The classification of lowland swamp communities in northeastern Papua. Ecology 40: 703-711.
Taylor, J. S. 1931. A note on the flora of mangrove. Entomol. Rec. J. Var., N. S. 43: 41-42.
Taylor, W. R. 1925. The marine flora of the Dry Tortugas. Rev. Algol. 2: 113-135.
Taylor, W. R. 1928. The marine algae of Florida, with special reference to the Dry Tortugas. Carnegie Inst. Wash., Publ. No. 379; Pap. from the Tortugas Lab. No. 25, 219 pp.
Taylor, W. R. 1954. Sketch of the character of marine algae vegetation of the shores of the Gulf of Mexico. *in* The Gulf of Mexico. It's Origin, waters, and marine life. P.S. Galtsoff, Coordinator, U.S.D.I., Fish. Bull. 55: 177-192.
Teas, H. J. and F. Montgomery. 1968. Ecology of red mangrove seedling establishment. Bull. Assoc. Southeast. Biol. 15: 56-57.
Teijsmann, J. E. 1957. Dagverhaal eener Botanische Reis over de Westkust van Sumatra. Nat. Tijd. Ned. Ind. 14: 249- 376.
Teixerira, C. and M. B. Kutner. 1961. Contribuicao para o conhecimento das diatomaceas da regiao de Cananeia. Bol. Inst. Oceanogr. Sao Paulo 11: 41-74.
Teixeira, C. and M. B. Kutner. 1963. Plankton studies in a mangrove environment. I. First assessment of standing stock and principal ecological factors. Bol. Inst. Oceanogr., Sao Paulo 12: 101-124.
Teixeira, C., M. B. Kutner, and F. M. S. Torgo. 1965. O efeito da respiracao bacteriana no estudo da producao primaria. Rev. bras. Biol. 25: 287-294.
Teixeira, C., J. Tundisi, and M. B. Kutner. 1965. Plankton studies in a mangrove environment. II. The standing stock and some ecological factors. Biol. Inst. Oceanogr., Sao Paulo, Publ. 14: 13-41.

Teixeira, C., J. Tundisi and J. Santoro Ycaza. 1967. Plankton studies in a mangrove environment. IV. Size fractionation of the phytoplankton. Biol. Inst. Oceanogr. Sao Paulo 16: 39-42.

Teixeira, C., J. Tundisi, and J. Santoro Ycaza. 1969. Plankton studies in a mangrove environment. VI. Primary production, zooplankton standing-stock and some environmental factors. Int. rev. Ges. Hydrobiol. 54: 289-301.

Teng, K. Forests in the sea at Hainan Island. Chin. For. 2: 30-31.

Thom, B. G. 1967. Mangrove ecology and deltaic geomorphology: Tobasco, Mexico. J. Ecol. 301-343.

Thome, S. and A. F. Moller. 1901. Mangrove. Tropenpflanzen 5: 339.

Thompson, J. M. 1954. The genera of oysters and the Australian species. Aust. J. Mar. Freshwater Res. 5: 132-168.

Thorne, R. F. 1954. Flowering plants of the waters and shores of the Gulf of Mexico. U. S. Fish Wildl. Serv., Fish. Bull. 89: 193-202.

Thornton, I. and M. E. C. Giglioli. 1965. The mangrove swamps of Keneba, Lower Gambia River Basin. II. Sulphur and pH in the profiles of swamp soils. J. appl. Ecol. 2: 257-269.

Tieghem, Ph. van. 1898. Avicenniacees et Symporemacees. Place de ces deux noubelles familles dans la classification. J. Bot., Paris 12: 345-358.

Tinley, K. L. 1958. A preliminary report on the ecology of the Kosi lake system. Unpublished report of the Natal Parks, Fish and Game Preservation Board, Natal, 90 pp.

Tischler, G. 1910. Untersuchungen an Mangrove-und Orchideen-Wurzeln mit specieller Bezichung auf die Statolithen des Geotropismus. Ann. Jard. Bot. Buit., Suppl. 3, pp. 131-186.

Tobler, F. 1914. Die Mangrove der Insel Ulenge (Deutsch-Ostafrika). Bot. Jahrb. 50: 398-411.

Todd, R. and P. Bronnimann. 1957. Recent foraminifera and Thecamoebina from the eastern Gulf of Paria. Cushman Found. Foram. Res. Spec. Publ. No. 3.

Tokioka, T. 1942. Ascidians found on the mangrove trees in Iwayama Bay, Palao. Palao Trop. Biol. Stn. Stud. (Tokyo) 2: 497-506.

Talken, H. R. 1967. The species of *Arthrocnemum* and *Salicornia* (Chenopodiaceae) in southern Africa. Bothalia 9: 255-307.

Tomlinson, T. E. 1957. Relationship between mangrove vegetation, soil texture and reaction of surface soil after empoldering saline swamps in Sierra Leone. Trop. Agric. Trinidad 34: 41-50.

Tomlinson, T. E. 1957. Changes in a sulphide-containing mangrove soil on drying and their effect upon the suitability of the soil for the growth of rice. Empire J. Exp. Agric. 25: 108-118.

Tralau, H. 1964. The genus *Nypa* van Wurmb. K. Svensk, Vetensk. Akad. Handl., Ser. 5, 10: 5-29.

Treub, M. 1883. Notes sur l'embryon, le sac embryonaire et l'ovule de *Avicennia officinalis*. Ann. Jard. Bot. Buit. 3: 79-87.

Triana, I. and I. E. Planchon. 1862. Prodromus Florae Novo- Grandensis ou enumeration des plantes de la Nouvelle-Grenade, avec descriptions des especes nouvelles. Ann. Soc. Nat. Quart. Ser., Bot. 17: 1-382.

Trimble, H. 1892. Mangrove tannin. Contr. Bot. Lab. Univ. Penn. 1: 50-55.

Troll, W. 1930. Ueber die sogenannten Atemwurzeln der Mangroven. Ber. Stsch. bot. Ges. 48: 81-99.

Troll, W. 1931. Botanische Mitteillungen aus den Tropen. III. *Dictyotposis paopagalifera* W. Troll, eine neue Brackwasseralge ostindischer Mangrovegebiete. Flora, N. F. 25: 474-502.

Troll, W. 1932. Uber die sogenannten Atemwurzelin der Mangroven. S. Br. Ges. Morph. Physiol. 40: 46-48.

Troll, W. 1933a. Uber *Acrostichum aurem* und *Acrostichum* speciosum und nestene Formen des letzeren. Flora, N. F. 28' 301-328.

Troll, W. 1933b. *Camptostemon schultzii* Mart. und *C. philippensis* Becc. als neue Vertreter der australasiatischen Mangrove-vegetation. Flora, N. F. 28: 348-352.

Troll, W. 1942. Die Mangrovegewache. *in* "vergleichende Morphologie der Hoheran Pflanzen" Erster Band, Dritter Teil, Berlin-Zehlendorf.

Troll, W. and O. Dragendorff. 1931. Uber die Luftwurzeln von *Sonneratia* Linn. und ihre biologische Bedeutung. Planta 13: 311-473.

Truman, R. 1961. The eradication of mangroves. Aust. J. Sci. 24: 198-199.

Trumbull, S. 1954. Mangrove wood, leaves, may bar expansion of Everglades Park. Miami Herald, Apr. 14.

Tschirley, F. H. 1969. Defoliation in Vietnam. Science 163: 779-786.

Tsopa, E. 1939. La vegetation des halophytes du Nord de la Roumanie en connexion avec celle du reste du pays S.I.G.M.A. 70: 1-22.

Tundisi, J. 1969. Plankton studies in a mangrove environment — its biology and primary production. *in* Lagunas Costeras, un Simposio. Mem. Simp. Intern. Lagunas Costeras. UNAM — UNESCO, Nov. 28-30, 1967, Mexico, D. F., pp. 485-494.

Tundisi, J. and C. Teixeira. 1965. Plankton studies in a mangrove environment. *in* Simposio sobre a Oceanografic do Atlantico Sul Occidental, Rio de Janiero, Brazil, 14-18 Setembro. 1964. Ann. Acad. Brasil, Cienc. 37 (Suppl.): 192-193.

Tundisi, J. and C. Teixeira. 1968. Plankton studies in a mangrove environment. VII. Size fractionation of the phytoplankton: some studies on methods. Bol. Inst. Oceanogr. 17; 89-94.

Tundisi, J. and T. M. Tundisi. 1968. Plankton studies in a mangrove environment. V. Salinity tolerances of some planktonic crustaceans. Bol. Inst. Oceanogr., Sao Paulo 17: 57-65.

Turrill, W. B. and E. Milne-Redhead. 1956. Flora of tropical East Africa: Rhizophoraceae. Crown Agents, London, 311 pp.

Uhvits, R. 1946. Effect of osmotic pressure on water absorption and germination of alfalfa seed. J. Bot. 33: 278-285.

Ule, E. 1901. Die vegetation von Cabo Frio an der Kuste von Brasilien. Bot. Jb. 28: 511-528.
Ulken, A. 1970. Phycomyceten aus der Mangrove bei Cananeia (Sao Paulo, Brazilien). Veroff. Inst. Meeresforsch. Bremerhaven 12: 313-319.
Unwin, A. H. 1920. West african forests and forestry. E. P. Dutton and Co., New York, 416 pp.
Uphof, J. C. T. 1935. Die holzigen Halophyten-vereinen des Kustengebietes Floridas. Mitt. Dut. Dendrol. Ges. 47: 39-53.
Uphof, J. C. T. 1937. Die nordliche Verbreitung der Mangroven im atlantischen Gebiet. Arch. Hydrobiol. 31: 141-144.
Vahrmeijer, J. 1966. Notes on the vegetation of north Zululand. Afr. Wildl. 20: 151-161.
Vann, J. H. 1959a. The physical geography of the lower coastal plain of the Guiana coast. Geogr. Branch, O.N.R., Washington, D.C., Project NR. 388-028, Tech. Rep. 1, 91 pp.
Vann, J. H. 1959b. Landform-vegetation relationships in the Atrato delta. Ann. Assoc. Am. Geogr. 49: 345-360.
Vanwijk, C. L. 1951. Soil survey of the tidal swamps of South Borneo in connection with agricultural practices. Tectona 41: 75-110.
Vaughan, R. E. and P. O. Wiehe. 1937. Studies on the vegetation of Mauritius. I. J. Ecol. 25: 301-311.
Vaughan, T. W. 1909. The geologic work of mangroves in southern Florida. Smithson. Misc. Collect. 52: 461-464.
Vazquez Bota, J. 1963. Classification de las melsas forestales de campeche. Bol. Tech. Inst. Macl. For. Mex., Vol 10.
Venkateswarlu, J. 1935. A contribution to the embryology of Sonneratiaceae. Proc. Indian Acad. Sci. 5: 23-29.
Venkateswarlu. J. 1944. The estuarial flora of the Godavari. J. Bombay Nat. Hist. Soc. 44: 431-435.
Benkateswarlu J. and R. S. P. Rao. 1964. The wood anatomy and taxonomic position of Sonneratiaceae. Curr. Sci. 33: 6-9.
Versteegh, F. 1952. Problems of silviculture and management of mangrove forests in Indonesia. Doc. FAO Asia-Pac. For. Comm. 77, 5 pp.
Verwey, J. 1930. Einiges uber die Biologie Ost-Indisher Mangrovekrabben. Terubia 12: 167-261.
Vibul Vanakij, L. 1937. Mangrove forest. Vanasarn For. J., No. 6.
Vieillard, E. 1862. Plantes utiles de la Navelle-Caledonie. Ann. Sci. Nat. Paris, ser. IV, vol. 16.
Visser, S. A. 1963. The occurrence of microorganisms in tropical swamps. Biochem. J. 89: 83-84.
Vu Van Cuong Humbert. 1964a. Flore et vegetation de la mangrove de la region de Saigon-Cap Saint Jacques, Sud Viet-Nam. These Doct., 3rd Cycle, spec. Sci. Biol., Paris, 199 pp.
Vu Van Cuong Humbert 1964b. Nouveautes pour la Flore du Camboge, du Laos, et de Vietnam (Rhizophoraceae, Sonneratiaceae, Myrtacae).

Addisonia 4: 343-347.
Wadsworth, F. H. 1959. Growth and regeneration of white mangrove in Puerto Rico. Caribb. For. 20: 59-71.
Wadsworth, F. H. and G. H. Englerth. 1959. Effects of the 1956 hurricane on forests in Puerto Rico. Caribb. For. 20: 38-57.
Waisel, Y. 1958. Germination behavior of some halophytes. Bull. Res. Coun. Israel (Botany) 60: 187-189.
Walker, F. S. 1937. The management and exploitation of the Klang mangrove forests. Malay. For. 6: 71-78.
Walker, F. S. 1938. Regeneration of Klang mangroves. Malay. For. 7: 71-76.
Walker-Arnott, G. A. 1938. On the Rhizophoraceae. Ann. Mag. Nat. Hist. 1: 359-374.
Walsh, G. E. 1967. An ecological study of a Hawaiian mangrove swamp. in Estuaries, G. H. Lauff (ed.), AAAS Publ. 83, Washington, pp. 420-431.
Walter, H. 1936. Die Okologischen Verhaltnisse in der Nabel Wuste, Namib (Deutsch Sudwestafrika). Ber. Dtsch. bot. Ges. 54: 17-32.
Walter, H. 1936a. Der wasser-und Salzgehalt der Ost-Afrikanischen Mangroven. Ber. Dtsch. bot. Ges. 54: 33-49.
Walter, H. 1936b. Uber den Wasserhaushalt der Mangrovpflanzen. Ber. Schweiz. bot. Ges. 46: 217-228.
Walter, H. and M. Steiner. 1936. Die Okologie der Ost-Afrikanischen Mangroven. Z. Bot. 30: 65-193.
Warburg, O. 1893. *Rhizophora mangle*, tropische Fragmente. Bot. Jb. 40: 517-548.
Ward, J. M. 1967. Studies in ecology on a shell barrier beach. I. Vegetatio 14: 241.
Waring, G. A. 1926. Geology of the Islands of Trinidad and Tobago, B.W.I. John Hopkins Press, Baltimore, 180 pp.
Warming, E. 1877. Om *Rhizophora mangle* L. Bot. Notes 1877: 14-21.
Warming, E. 1883. Tropische Fragmente. II. *Rhizophora mangle* L. Bot. Jb. 4: 519-548.
Warming, E. and M. Vahl. 1925. Oecology of Plants. Oxford Univ. Press, London, 422 pp.
Warner, G. F. 1967. The life history of the mangrove tree crab, *Aratus pisoni* L. Zool. 153: 321-335.
Warner, G. F. 1969. The occurrence and distribution of crabs in a Jamaican mangrove swamp. J. Anim. Ecol. 38: 379-389.
Watanabe, K. and M. Kutner. 1965. Bacteriological analysis of mangrove waters at Cananeia. in Simposio Sobre a Oceanografia do Atlantico Sul Occidental, Rio de Janeiro, Brazil, 14-18 Septembro, 1964. Ann. Acad. Brazil Cienc. 37 (Suppl.): 3.
Watanabe, K. and M. Kutner. 1965. Plankton studies in a mangrove environment. III. Bacterial analysis of waters in Cananeia. Bol. Inst. Oceanogr. 14: 43-51.
Waterman, A. M. 1946. The effect of water soluble extracts from the

heartwood of tropical American woods on the growth of two wood-decay fungi. Trop. Woods. 88:3.

Watson, J. G. 1928. Mangrove forests of the Malay Peninsula. Malay. For. Rec. 6: 1-275.

Watson, J. G. 1934. Stick thinnings. Malay. For. 3: 12-17.

Webb, K. L. 1966. NaCl effects on growth and transpiration in *Salicornia bigelovii* salt marsh halophyte. Plant and Soil 24: 261-268.

Webb, L. J. 1966. The identification and conservation of habitat types in the wet tropical lowlands of North Queensland. Proc. R. Soc. Queensland. 78: 59-86.

Webber, J. H. 1898. Strandflora of Florida. Science (N.S.) 8: 658-671.

Weberling, F. 1960. Weitere Untersuchungen uber das Vorkommen rudimentarer Stipeln bei den Myrtales (Combretaceae, Melastomataceae). Flora 149: 189-205.

Webster, L. 1967. Distribution of mangroves in South Australia. A.A. (hon.) thesis, Geogr. Dept., Univ. Adelaide, Adelaide, Australia.

Wei, S. 1947. The relation of typhoons to plant distribution in Taiwan. Bull. Taiwan For. Res. Inst. 6: 1-14.

Weiss, M. H. 1966a. Apercu preliminaire sur les rapports entre le milieu et la distribution des paletuviers le long d'une mangrove littorale proche de Tulear. Rec. Trav. Stn. Mar. Endoume, Marseille, Suppl. 5: 165-173.

Weiss, M. H. 1966b. Apercu preliminaire sur une mangrove naturelle a l'interieur des terres au Sud de Tulear. Rec. Trav. Stn. Mar. Endoume, Marseille, Suppl. 5: 175-178.

Weiss, M. H. and A. Kiener. 1971. Observations relatives a la nature chimique des eaux de la region de Tulear. Diversite, variations, relations avec les zonations biocenotiques. Tethys (Suppl.) 1: 215-236.

Welch, B. L. 1963. From coral reef to tropical island via *Thalassia* and mangrove. Va. J. Sci. 14: 213-214.

Wenzel, H. 1925. Die Mangroverinden als Gerbmaterialen. Anatomische Untersuchungen der gerbstoffreichsten Mangroverinden. Bot. Arch. 59-96.

West, E. and L. E. Arnold. 1956. Native Trees of Florida. 2nd ed., Univ. Florida Press, Gainesville, p. 162.

West, O. 1945. Distribution of mangroves in the eastern Cape Provence. S. Afr. J. Sci. 41: 238-242.

West, R. C. 1954. The mangrove swamp of the Pacific littoral of Colombia. Coast. Geogr. Conf., Feb. 1954, OWR-NRC, Washington, D. C., pp. 45-50.

West, R. C. 1956. Mangrove swamps of the Pacific coast of Colombia. Ann. Assoc. Am. Geogr. 46: 98-121.

West, R. C. 1957. The Pacific lowlands of Colombia, a negroid area of the American tropics. La. St. Univ. Press, Baton Rouge, 278 pp.

Westermaier, M. 1900. Zur Kenntnis der Pneumatophoren. Bot. Unters. im Anaschluss an eine Tropenreise, vol. 1.

Westing, A. H. 1971a. Ecological effects of military defoliation on the forests of South Vietnam. Bioscience 21: 893-898.

Westing, A. H. 1971b. Forestry and the war in South Vietnam. J. For. 69: 777-784.

Westing, A. H. 1971c. Herbicides as weapons in South Vietnam, a bibliography. Bioscience 21: 1225-1227.

Wetmore, A. 1965. The birds of the Republic of Panama. I. Tinamidae (Tinamous) to the Rhynchopidae (Skimmers). Smithson. Misc. Coll., No. 150.

Weyl, R. 1953a. Lithogenetische Studien in den Mangroven der Pazifik-Kuste. Neues Jb. Beol. Palaontol., Mh, No. 5, pp. 202-218.

Weyl, R. 1953b. In den Mangroven El Salvadors. Nat. Volk 83: 26-30.

Wharton, W. J. L. 1883. Mangrove as a destructive agent. Nature 29: 76-77.

Wherry, E. T. 1920. Plant distribution around salt marshes. Ecology 1: 42-48.

White, C. T. 1926. A variety of *Ceriops tagal* C. B. Rob. (=C. *candolleana* W. and A.). J. Bot. London 64: 220-221.

Whiteleather, R. T. and H. H. Brown. 1945. An Experimental Fishery Survey in Trinidad, Tobago and British Guiana. U. S. Govt. Printing Office, Washington, D. C. 130 pp.

Whitford, H. 1911. The forests of the Philippines. Bur. For. Bull. 10 (1): 1-94.

Whitford, H. 1911. The forests of the Philippines. Bur. For. Bull. 10 (2): 1-113.

Whitlo, C. K. 1947. Animal life in mangroves. W. Aust. Nat. 1: 53-56.

Whittaker, R. H. 1953. A consideration of climax theory: the climax as a population and pattern. Ecol. Monogr. 23: 41-78.

Wiehe, P. O. 1935. A quantitative study on the influence of tide upon populations of *Salicornia europaea*. J. Ecol. 23: 323-333.

Wiele, N. 1882. Om Stammens og Bladets Byning hos *Avicennia nitida* L. Bot. Tidsskr. 13: 156.

Wieler, A. 1898. Die function der Pneumathoden und das Aerenchym. Jb. wiss. Bot. 32: 503-524.

Wiens, H. J. 1956. The geography of Kapingamarangi Atoll in the eastern Carolines. Atoll Res. Bull. 48, June 1956.

Wiens, H. J. 1962. Atoll Environment and Ecology. Yale Univ. Press, New Haven, 532 pp.

Wijk, C. L. van. 1953. Soil survey of the tidal swamps of South Borneo in connection with the agricultural possibilities. Contrib. Cent. Agr. Res. Stat., Bogor, Indonesia, No. 123.

Williams, R. R. 1911. The economic possibilities of the mangrove swamps of the Philippines. Philipp. J. Sci. A. 6: 45-61.

Willis, J. H. 1944. Excursion to Seaholm salt marsh flora and mangroves. Viet. Nat. 61: 40-41.

Wilson, E. O. and D. S. Simberloff. 1969. Experimental zoo-geography of islands: defaunation and monitoring techniques. Ecology 50: 267-278.

Wilson, O. 1924. Latin America as a source of tanning materials. Chem. Metall. engin. 30: 303-305, 344-346, 398-399.

Wilson, W. P. 1889. The production of aerating organs on the roots of swamp and other plants. Proc. Acad. Nat. Sci. Philadelphia, pp. 67-69.

Winkler, H. 1931. Einige Bemerkungen uber Mangrove-Pflanzen und den *Amorphophallus titanum* im Hamburger Botanischer Garten. Ber. Dtsch. bot. Ges. 49: 87-102.

Womersley, H. B. S. and S. J. Edmonds. 1952. Marine coastal zonation in southern Australia in relation to a general scheme of classification. J. Ecol. 40: 89-90.

Womersley, H. B. S. and S. J. Edmonds. 1958. A general account of the intertidal ecology of south Australian coasts. Aust. J. Mar. Freshwater Res. 9: 217-260.

Wyatt-Smith, J. 1953a. The malayan species of *Bruguiera.* Malay. For. 16: 156-161.

Wyatt-Smith, J. 1953b. The malayan species of *Sonneratia.* Malay. For. 16: 213-216.

Wyatt-Smith, J. 1954a. The malayan species of *Avicennia.* Malay. For. 17: 21-25.

Wyatt-Smith, J. 1954b. Mangrove flora replaced by fresh water forest. Malay. For. 17: 25-26.

Wyatt-Smith, J. 1960. Field key to the trees of mangrove forests in Malaya. Malay. For. 23: 126-132.

Wyatt-Smith, J. and A. J. Vincent. 1962. Progressive development in the management of tropical lowland evergreen rain forest and mangrove forest in Malaya. Malay. For. 25: 199-223.

Yamamoto, Y. 1940. Enumeration of the mangrove plants, including herbs, climers, epiphytes, and the lower forms of plant life from the Dutch East Indies. Trop. Agric. 12: 157.

Yamanaka, T. 1957. The forest vegetation of the Amani islands with special reference to the *Shiia sieboldii* forest. Rep. U.S.A. Biol. Stat. Kochi Univ. 4, 14 pp.

Yamashiro, M. 1961. Ecological study on *Kandelia candel* (L.) Druce, with special reference to the structure and falling of seedlings. Hikobia 2: 209-214.

Young, C. M. 1930. A Year on the Great Barrier Reef. Putnam, New York, 246 pp.

Yonge, C. M. 1953. Aspects of life in muddy shores. *in* Essays in Marine Biology, S. M. Marshall and A. P. Orr (eds.), Oliver and Boyd, Edinburgh, pp. 29-49.

Zieman, J. C. Origin of circular beds of *Thallasia* (Spermatophyta: Hydrocharitaceae) in south Biscayne Bay, Florida and their relationship to mangrove hammocks. Bul. Inst. Mar. Sci. *in press.*

Zilch, A. 1954. Moluscos de los manglares de El Salvador. Com. Inst. Trop. Invest. Cient. (El Salvador) 3: 77-87.

Zinderen Bakker, E. M. van. 1965. On swamp vegetation and structure of swamps in south and east Africa. Bot. Jb. Syst. Pflanzengeogr. 84: 215-231.

Zullo, V. A. 1969. Additional records of *Tetrabalanus polygenus* Cornwall, 1941 (Cirripedia, Thoracica). Occas. Pap. Calif. Acad. Sci. 74: 1-8.

BEACH AND SALT MARSH VEGETATION

OF

THE NORTH AMERICAN PACIFIC COAST

Keith B. Macdonald
Department of Geological Sciences,
University of California, Santa Barbara,
Santa Barbara, California 93106

and

Michael G. Barbour
Botany Department,
University of California, Davis,
Davis, California 95616

Introduction

The objective of this paper is to survey beach and salt marsh vegetation along the Pacific Coast of North America between Point Barrow, Alaska (71°N) and Cabo San Lucas, at the southern tip of Baja California (23° N; Fig. 1).

In order to best meet the symposium objectives, the terms beach and salt marsh are narrowly defined. *Beach* is that strip of land from just above Mean Low Water (MLW) to just beyond the reach of storm waves, or, if there is a foredune, to the top of the foredune. The term "strand" has often been applied to this strip of land, but it is clear from a number of floras that the conception of strand may vary widely, from the very edge of shore only, to the beach plus adjacent inland dunes. To avoid ambiguity, we have chosen the more intuitive term, beach. The survey of beach vegetation, then, will not include species of dunes and ocean-facing cliffs unless those species are also characteristic of beach vegetation. The exclusion of dune species is certainly justified in a symposium devoted to halophytes, for levels of both soil- and air-borne salinity drop appreciably behind the beach (Barbour et al. 1972; Boyce 1954; Martin and Clements 1939; Oosting and Billings 1942). Plants of dunes may be xerophytes, but they are not halophytes (Barbour 1970b;

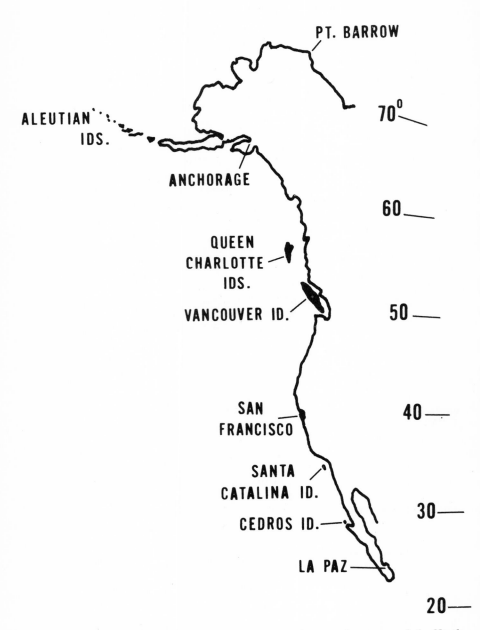

Figure 1. Area of coverage of beach and salt marsh vegetation survey of the North American Pacific coast.

Martin and Clements 1939; Purer 1934, 1936). Kearney (1904) has even raised the question of whether beach plants are halophytes, but the data accumulated since then (see review by Barbour 1970d) indicates that levels of salt spray are high, even if soil salinity levels are not.

The exclusion of cliff species, although necessary, is regrettable. On windy, rocky, open coasts, plants of ocean-facing cliffs receive considerable amounts of salt spray, enough spray to cause some authors to call them "quite salt tolerant" (Calder and Taylor 1968) or "strongly halophytic" (Peck 1961). However, the number of species found in this habitat is quite high, and slight changes in aspect or in access to fresh water permits normally inland species to occur almost side-by-side with more strictly halophytic species. Given the information presented in the average flora, it is quite difficult without personal observation to decide which species are halophytes. At Bodega Head, California, for example, ocean-facing cliffs support truly restricted species such as *Artemisia pycnocephala, Eriophyllum staechadifolium, Armeria maritima* var. *californica, Jaumea carnosa,* and *Plantago maritima*, but if a seep runs down the bank, then typically grassland species such as *Anagallis arvensis, Polypogon monspeliensis*, and *Sonchus asper* follow (Barbour 1970a). Fortunately, a number of halophytic cliff species also occur on shingle beaches (such as *Plantago maritima* spp. *juncoides*) or in salt marshes (such as *Spergularia* species), and so are included in this review. Some future survey of this diverse cliff community would be valuable.

Marine angiosperms of the lower intertidal zone, species of *Phyllospadix* and *Zostera,* are also excluded from this review. Their omission reflects the lack of data on them in the literature; field identification of species in these genera is difficult. As Calder and Taylor (1968) refreshingly admit: "We have experienced difficulty in determining whether sterile vegetative specimens are *Phyllospadix* or *Zostera* (marina)...and to identify *P. scouleri* and *P. torreyi.*" Possibly, what little we "know" about these three Pacific coast species will need considerable revision once herbarium material can be unravelled.

Coastal salt marshes are defined here as occurring on shorelines sheltered from excessive wave action - typically around the margins of lagoons, bays, and estuaries or behind barrier spits and islands; they represent vegetated portions of the upper intertidal zone and are usually carpeted with a rather dense cover of phanerogams, occasionally interspersed with macro-algae. Irrespective of the local tidal range, Pacific Coast salt marsh plants generally become established at about Mean Lower High Water (MLHW - the average height of the lower of two unequal high tides that occur each day under the Pacific Coast's "mixed tide" regime; Doty 1946). While the lower (outer or seaward) limit of a coastal salt marsh is never in doubt, the upper (inner or landward) margin is much harder to recognize. It is probably best defined to coincide with Extreme High Water (EHW). At many sites this level is marked by a "debris line" of stranded algae, logs, or tidal trash and by a rather abrupt change of slope. Other salt marshes, however, extend well inland on gently rising slopes that reach well above the tides. Their vegetation undergoes changes to be sure, but no abrupt upper salt marsh boundary is evident. This problem is often further compounded by human disturbance - establishment of footpaths, roads, etc. - which frequently begins within this transition zone.

Restricting our treatment of coastal salt marshes to areas lying below EHW still includes marshes developed under a wide variety of salinity regimes. Not only does the frequency and duration of tidal submergence change markedly with elevation across each marsh, but the dynamic interaction between salt-water flooding fresh-water run-off is modified by pronounced latitudinal variations in precipitation, temperature, and evaporation (Doty 1946; Macdonald 1969). Mason (1969), in California, distinguishes between "...salt-water marshes, seasonally salt- and fresh-water marshes, and fresh-water marshes subject to tidal flooding...," but for most Pacific Coast sites, the presently available environmental data are insufficient to establish these distinctions.

As with the beaches, the vegetation of the more saline coastal marshes is dominated by a rather small number of halophytes - often less than 10 and only rarely as many as 20 species per site. As salinities decline, as local fresh-water sources appear, or as human disturbance increases - maritime, upland, or weedy species appear and the diversity of the marsh flora rapidly increases.

The objective of this review is description of vegetation rather than flora; thus we have limited our survey to those species of vascular plants which most characterize beach and salt marsh vegetation. Exhaustive lists of "occasional" species have been omitted. Nomenclature, unless stated otherwise, follows that in three floras, depending on species distribution: Hulten (1968) for the northern portion of the coast, Munz and Keck (1963) for the central portion, and Shreve and Wiggins (1964) for the southern most portion.* Hitchcock, Cronquist, Ownbey, and Thompson's excellent five volume flora of the Pacific Northwest (1969) was also particularly useful because of its full treatment of nomenclatural synonyms.

Despite differences in the literature coverage (which favors salt marshes), our treatments of beach (Barbour) and salt marsh vegetation (Macdonald) are generally similar. We have summarized available descriptions of the two vegetation types, starting at Point Barrow, Alaska, and working south. Emphasis has been placed on regional changes in species composition, descriptions of vertical zonation, and impressions of local vegetational variability. Broad perspectives, within which latitudinal vegetation changes and differences between beach and salt marsh vegetation of the Pacific and Atlantic Coasts can be considered, are provided by reviewing climatic and geological data.

---

* **For those species whose nomenclature varied even among these three floras, that cited in the northern-most flora was accepted.**

Climate

The Pacific Coast of North America sweeps in a broad diagonal across 48° of latitude and 55° of longitude and extends from Arctic tundra to the subtropical southern fringe of the Sonoran Desert. Climatic variations are considerable, yet because of the maritime influence and the fact that the warm North Pacific Current bathes the Aleutians, southern Alaska, and British Columbia, and the cold California Current bathes the southern coastline, the change in climate is not as severe as experienced inland over shorter distances.

Both the general nature of climatic trends along the Pacific Coast and their marked differences from comparable trends on the Atlantic Coast are illustrated by the data presented in Figs. 2 and 3. (Each plot is based on data from at least 25, approximately equally spaced, strictly coastal weather stations.) Total annual precipitation data (Fig. 2) indicate a relatively steady increase in rainfall on the Atlantic seaboard, from a low of 25 cm (10 in.) on the northern shores of Baffin Island to a high of 150 cm (60 in.) at Miami, Florida. The Arctic Slope of Alaska is drier than Baffin Island but rainfall increases dramatically along the mountainous coast of the Gulf of Alaska (33 cm/129 in. at Yakutat Bay). It remains much higher on the Pacific Coast than at comparable Atlantic Coast latitudes until Eureka, California, from which point it declines equally dramatically to totals of only a few centimeters in the deserts of Baja California (5 cm/2 in. at Guerrero Negro). Mean monthly temperature data (Jan. and July; Fig. 3) indicate that sites along the Pacific Coast are generally both warmer and more equable than sites at similar latitudes on the Atlantic Coast.

Additional climatic parameters for selected Pacific Coast stations are shown in Table 1. Note that the climate changes from a severely cold polar tundra type (ET in Koppen's system) with low annual precipitation peaking in summer to a uniformly cool, increasingly wet mesothermal type (C) with cloudy and foggy weather and the start of a pronounced winter wet/summer dry pattern to a warmer, drier (but still foggy) semi-arid or arid type (BS or BW) with a complete absence of frost. Further notes describing local conditions more completely are summarized below.

Climate on the beach of the Alaskan Arctic Slope, according to Wiggins and Thomas (1962), is severe. Ice action, blast from strong winds, and inundation by cold water can decimate stretches of beach of all plants. Temperatures can drop to -57°C and rarely rise above 21°C. Freezing temperatures can occur on any summer day, and the growing period is often much shorter than 90 days (the frost-free season at Point Barrow, as seen in Table 1, averages only 17 days). Most of the precipitation falls in summer, but mean relative humidity is 60-93% during the entire year. Mean wind velocity is 12 mph.

Climate on the Aluetian Islands, according to data for Amchitka (Armstrong 1971) and Unalaska (Hulten 1937), is much more moderate. Pack

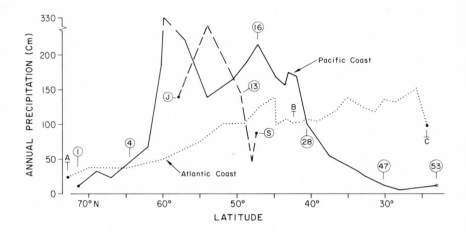

Figure 2. Latitudinal change in mean annual total precipitation (cm) along the Pacific and Atlantic Coasts of North America. (Dashed line for inner Pacific Coast, Juneau to Seattle. Circled numbers refer to selected localities from Fig. 4. A = Baffin Bay, B = Boston, C = Key West.)

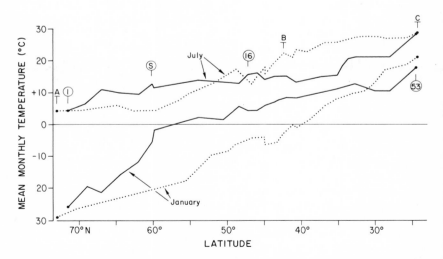

Figure 3. Latitudinal change in mean monthly temperatures (January, July; °C) along the Pacific and Atlantic Coasts of North America. (Solid lines = outer Pacific Coast, dotted lines = Atlantic Coast, S = Seward, Alaska; see Fig. 2 for additional details.)

Table 1. Climatic data for selected stations along the Pacific Coast. Data include °N latitude, precipitation (cm), number of frost-free days per year, and temperature means for the year, for the coldest month, and for the warmest month (°C). The last column is the climatic classification according to Koppen's scheme. [Sources: Arias (1942), Calder and Taylor (1968), Climatological data (1968), Eber/ et al. (1968), Hambidge (1941), Koeppe (1931), Trewartha (1954), and Wiggins and Thomas (1962).]

Station Name	Lat.	Ppt.	Frost-free	Air Temperature			Water Temperature		K
				Yr.	Cold Mo.	Warm Mo.	Cold Mo.	Warm Mo.	
Point Barrow, Alaska	71°20'	10	17	-12	-22	4	0	8	ET
Anchorage, Alaska	61°20'	36	110	1	-12	14	4	13	Ccf
Sandspit, Queen Charlotte Ids.	53°15'	125	207	6	3	14	7	15	Cbf
Vancouver, B.C.	49°17'	147	231	9	2	17	8	15	Cbf
Newport, Oreg.	44°35'	155	248	10	7	14	10	15	Cbs
Eureka, Calif.	40°45'	97	328	11	8	14	11	13	Cbs
San Francisco, Calif.	37°45'	50	356	12	10	15	11	13	Cbs
Morro Bay, Ca.	35°20'	43	320	14	11	18	13	16	BSs
Santa Catalina Id., Calif.	33°22'	33	365	16	12	19	14	17	BSs
Ensenada, Baja California	31°40'	24	365	17	13	20	14	19	BSs
La Paz, Baja California	24°00'	11	365	23	18	29	20	28	BWh

ice, for example, does not extend to the Aleutians in winter. Frost can still occur at practically any time of the year, but the record low at Amchitka is only -9°C. (As the record high is only 20°C, one can see that annual variation in temperature is small.) Annual precipitation at Amchitka is 83 cm but it increases eastward, and summers become drier eastward. Calm, clear days are rare (2-5% of the days each month are clear); 50% of summer days have fog. Mean wind speed is 20 mph.

The Queen Charlotte Islands at sea level are only a few degrees warmer than the Aleutians, but they experience more rain and a more pronounced summer dry period (Calder and Taylor 1968). In January, the coldest month, 45% of the days at Sandspit experience temperatures below freezing, but the months of June through September are frost-free. Apart from rather dry, sunny conditions in summer caused by the North Pacific High, overcast days are typical. Average annual number of hours of bright sunshine at Sandspit is only 32% of the maximum possible - a value similar to that for sites along much of the British Columbia coast. Average wind speed is 10 mph at Sandspit, but it ranges up to 21 mph at more exposed locations such as Cape St. James. Relative humidity continues to be quite high, 82-98% during the year. Snowfall accounts for only 6% of total precipitation.

The coast of Oregon (Peck 1961) shows declining amounts of rain and rising summer temperatures from north to south. Annual precipitation averages 200 cm in the north, 150 cm in the south, and maximum summer temperatures rise from above 23° to 32°C. Fog incidence also increases southward, often prevailing for weeks in the south. The average number of cloudy, foggy, or rainy days is close to 280 each year.

The climate of Bodega Head, California, just north of San Francisco, continues to be cool and wet in winter, but virtually no rain falls in summer (Barbour et al. 1972). Only 10% of the mean 77 cm rain falls in the months May-September. Fogs are frequent, up to 80% of summer days and 10% of winter days experiencing fog, but they usually dissipate by 10 a.m. Hours of sunshine are 61% of maximum, and daily radiation ranges from an average of 0.15 kcal/cm^2 in January to 0.5 kcal/cm^2 in June. Average wind speed is 9 mph.

Santa Catalina Island, according to Thorne (1967), has an oceanic, Mediterranean climate, with warm, dry summers and mild, moist winters with much fog. Effective rainfall is limited to the period late September to late April, and the amount can fluctuate greatly from year to year (49 cm in 1965, 25 cm in 1966). Frosts are absent, but high temperatures are still moderate enough for the annual range of temperature between extremes to be only 31°C (the range at Amchitka was 29°C).

The climate of the northwest coast of Baja California is much like that of San Diego (Nelson 1922). Yearly minima seldom reach freezing, and maxima average 38°C. From Cedros Island south, however, rainfall drops and temperatures rise. The rainfall pattern also shifts southward, from summer dry to winter dry. Copious dews are common all along the Pacific Coast. Fogs are common in all seasons, but are most frequent in summer. However, as is true for much of southern California, the fog generally clears by late morning. Hours of sunshine are 70% of maximum.

Geology

The Pacific Coast of North America, backed by both the Coast Ranges and the Cordilleran Mountain Complex, is a tectonically active region subject to continuing volcanism, faulting, and regional uplift. For much of its length, the coast is renowned for its spectacular cliffs and rocky shores. From Puget Sound north Pleistocene glacial action has further increased coastal relief, producing an intricate network of fjords and coastal archipelagos. The only sizable coastal plain is that developed on the Alaskan Arctic Slope (Black 1969). Long, uninterrupted stretches of broad ocean beach are rare; smaller beaches isolated by rocky headlands are more typical. Salt marshes are similarly restricted, and they usually occur as relatively narrow fringes around the few protected bays and lagoons scattered along the coast or in river estuaries. Single salt-water marshes of more than a few thousand acres are extremely rare.

Evidence from old maps and aerial photographs confirms a recent origin (100 years old) for many Pacific Coast marshes. Cores penetrating marsh sediments at several sites have shown them to be a thin veneer (20-180 cm thick) overlying sandy bay-floor sediments. It is not necessary to invoke land-subsidence or sea level changes to account for these deposits for they have not accumulated at depths greater than the physiological tolerances of the marsh plants and present tidal ranges would allow (Macdonald 1969).

The Atlantic Coast north of New York has also been heavily glaciated; however, the rocks of the Canadian Shield and Appalachian Mountains that outcrop in this region are of relatively subdued relief. In this northern area, rocky coasts are again common and extensive beaches and salt marshes rare. South of New York an increasingly broad, flat coastal plain is developed that persists south to Florida and around the Gulf of Mexico. Steady subsidence throughout this region, together with the rise of sea level that followed the last glaciation, have combined to produce many long barrier spits and islands. Extensive, continuous salt marshes covering thousands of acres have developed within the protected lagoons formed between these barrier islands and the mainland coast (Clark 1967, Stanley 1969). The sediments of these marshes, unlike those of the Pacific Coast, frequently include thick sequences of peat that extend well below present sea level and reflect an extensive history of marsh development. Redfield (1967) at Barnstable Estuary, for example, records peat sequences up to 9 m thick that date back 4,000 years.

Beach Vegetation

In his book on coastal vegetation, Chapman (1964) wrote that the strand "...has not been subjected to serious study...and very little is known about the requirements of the individual species, or even about the nature of the environment." Synecological and autecological studies of beach vegetation are quite rare in the literature. Consequently, most of the information summarized in this section comes from floras. Cooper's fine survey of Pacific Coast strand and dune flora (1936) served as an initial, central reference. Information about Alaskan and Baja Californian floras, however, was quite limited at that time, and therefore his survey was incomplete. His approach was not ecological, but his notes on distribution do permit one to distinguish beach species from dune species.

More recent information for the north portion of the coast can be obtained from Wiggins and Thomas' flora of the Alaskan Arctic Slope (1962), Hulten's flora of the Aleutian Islands (1937), and his flora of Alaska which includes areas south to Vancouver Island (1968), and Calder and Taylor's flora of the Queen Charlotte Islands (1968). Of the four, Calder and Taylor's flora contains the most detailed community descriptions and habitat notes, the Alaskan flora contains the least (but it has the best information on overall ranges of species distribution). References to relative abundance, coupled with photographs of habitats in the texts, led to some conclusions as to which species were most characteristic of each zone along the beach (i.e., open beach nearest shore, driftwood zone, foredune, transition zone, intertidal). A recent synecological paper for Amchitka Island (Amundsen and Clebsch 1971), though short, was very helpful.

Along the central portion of the coast, we utilized Jones' survey of the Olympic Peninsula (1936), which devoted all of half a page to beach and dunes, Peck's Oregon flora (1961), Kumler's study of succession on Oregon dunes (1969), Munz and Peck's California flora (1968), Barbour's extensive floristic and ecological data for Bodega Head, California (Barbour 1970a-d, 1972a, 1972b; Barbour and Rodman 1970; Barbour et al. 1972), Howell's Marin County flora (1949), Hoover's San Luis Obispo County flora (1970), an ecological survey of Morro Bay State Park by Williams and Potter (unpublished manuscript), Smith's brief flora of the Santa Barbara region (1952), and the autecological studies of Martin and Clements at Santa Barbara (1939). The overview of Pacific Coast vegetation by Knapp (1965) was so brief as to be only marginally useful.

For the southern part of California and all of Baja California, we used Boughey's checklist of Orange County plants (1968), Thorne's flora of Santa Catalina Island (1967), the lay-oriented booklet on prominent plants of coastal San Diego County by Higgins (1956), and Shreve and Wiggins' classic work on the Sororan desert (1964). Information on Baja California was especially difficult to find, and likely sources such as Nelson's major reference work on the natural resources of Baja California (1922) and the reports of many miscellaneous botanizing expeditions (Brandegee 1889, Johnson 1958, and Orcutt 1885, for example) were very disappointing in their refusal to talk about beach plants.

A number of other floras were examined, but these did not add new material beyond that found in the above references so they will not be cited here. We are pleased, however, to acknowledge the voluminous personal communications from Dr. Ira Wiggins on the beach vegetation of Baja California. His help was critically important to the accurate completion of our coast survey.

Our review indicates that there are some 57 species which characterize the beaches of the Pacific Coast. These have been arranged in Table 2 by their northern range limit. The range limits given are probably accurate within 30' for most species; but given the state of floristic information along some portions of the coast, the limits of some species may be in error by 10-20' or even more. This much error may be particularly true for *Ammophila arenaria* and *Mesembryanthemum edule*. Given the chance of error, then, we do not necessarily think it is significant that there are 40-50° gaps in northern limits:

that is, a group of species appears to suddenly begin at 60°N, at 54°N, and at 50°N, and none begins in the intervals 65–60, 58–54, or 54–50. It could be that these gaps correlate with major climatic changes, but at this stage we would not like to make that correlation.

Many of the species have very broad ranges, for example, *Elymus arenarius* ssp. *mollis* occurs over a 34°30' latitude stretch of coastline and extends west out onto the Aleutian Islands. *Honckenya peploides, Mertensia maritima, Plantago maritima, Atriplex patula, Cakile edentula,* and *Galium aparine* all range for 27° or more of latitude. *Cakile edentual,* however, may now be absent south of 39°N, and this would reduce its range to 21° (Barbour and Rodman 1970). Those which range south of 23°N appear to have truncated distribution ranges because we have followed their distribution only to the tip of Baja California (at 23°N); consequently, even though it appears from Table 2 that northern species range over wider distances than southern species, this may not be the case, in fact.

None of the 57 species were restricted to an insular area (the Aleutians, the Queen Charlottes, Vancouver Island, or the Channel Islands as typified by Santa Catalina Island), and few were restricted to less than 10° of latitude on the mainland (excluding those which were truncated by the tip of Baja).

A final comment on Table 2: the distribution ranges given are for the species, not for the habitat. That is, over portions of the range the species may be absent or may not actually be present on the beach per se.

The contents of Table 2 are basically floristic in nature. To give an impression of the vegetation and zonation of communities back from mean low tide, one has to go back to the original references and piece together as much as possible. The material summarized below attempts to reconstruct an impression of beach vegetation

In the north, the most explicit vegetation description is provided for Amchitka in the Aleutians by Amundsen and Clebsch (1971):

> "A grass stand is the dominant beach community above high-tide line on those beaches frequently flooded by storms. The aspect dominance is *Elymus arenarius*, but *Festuca rubra* and *Poa eminens* are also common...A rather diffuse community of decumbent succulent herbs occurs generally between the grass community and the mean-high-tide mark. These halophytes, often storm-washed, do not provide very dense cover. The principal species are *Senecio pseudo-arnica, Mertensia maritima, Honckenya peploides,* and *Lathyrus maritimus.*"

Table 2. Characteristic beach species along the Pacific Coast. Data include north and south distribution limits (°N latitude), latitudinal range, and additional presence in the Aleutian, Queen Charlotte, and Channel Islands (the latter includes Santa Catalina, Santa Barbara, Santa Cruz, San Clemente, San Nicolas, Anacapa, Santa Rosa, and San Miguel in any combination). Presence is noted with an asterisk (*).

Species	Limits			Also occurs on		
	North	South	Range	Aleut.	Qn.Ch.	Channel
Cochlearia officinalis L.	71°	48°	23°	*	*	
Elymus arenarius L. ssp. Mollis (Trin.) Hult.	71°	36°30'	34°30'	*	*	
Honckenya peploides (L.) Ehrh.	71°	43°30'	27°30'	*	*	
Mertensia maritima (L.) S.F. Gray	71°	54°	27°	*	*	
Puccinellia langeana (Berl.) Sorens.	71°	53°	18°	*		
Stellaria humifusa Rottb.	71°	48°	23°		*	
Poa eminens Presl.	69°	47°	22°	*		
Senecio pseudo-arnica Less.	69°	50°	19°	*	*	
Ligusticum scoticum L. ssp. hultenii (Fern.) Calder and Taylor	67°	49°	18°	*	*	
Plantago maritima L. ssp. juncoides (Lam.) Hult. (range includes var. californica)	65°	34°	31°		*	*
Atriplex patula L. (range includes var. hastata)	60°	33°	27°		*	*
Cakile edentula (Bigel.) Hook. ssp. californica (Heller) Hult.	60°	32°30'	27°30'		*	*
Carex macrocephala Willd.	60°	43°30'	16°30'		*	
Festuca rubra L. ssp. acuta (Krecz. an Bobr.) Hult.	60°	43°30'	16°30'	*	*	
Fragaria chiloensis (L.) Duchesne ssp. pacifica Staudt.	60°	35°	25°	*	*	
Galium aparine L.	60°	33°	27°	*	*	*
Glehnia littoralis F. Schm. ssp. leiocarpa (Math.) Hult.	60°	39°	21°		*	
Lathyrus maritimus L.	60°	41°	19°	*	*	
Puccinellia nutkaensis (Presl.) Fern. and Weath.	60°	49°	11°		*	

Hulten (1937) more or less agrees with this account for the entire Aluetians, but is not as careful in distinguishing between zones:

"The *Elymus* association found on the not very frequent sandy beaches is about the same as in other places on the Northern Pacific. *Elymus arenarius, Senecio pseudo-arnica, Lathyrus maritimus, Honckenya peploides,* and *Mertensia maritima* are primary species."

He adds that *Ligusticum hultenii* is also characteristic, but less abundant.

Table 2, Cont.

Species				
Puccinellia pumila (Vas.) Hitchc.	60°	48°	12°	*
Sagina crassicaulis S. Wats.	60°	36°30'	23°20'	* *
Vicia gigantea Hook.	60°	35°	25°	*
Zostera marina L.	60°	32°30'	27°30'	* *
Phyllospadix scouleri Hook.	58°	33°30'	24°30'	* *
Agrostis pallens Trin.	54°	37°30'	16°30'	*
Ammophila arenaria (L.) Link.	54°	37°	17°	*
Cakile maritima Scop.	54°	28°	26°	* *
Convolvulus soldanella L.	54°	32°30'	21°30'	*
Lathyrus littoralis (Nutt. ex T. & G.) Endl.	54°	36°30'	17°30'	*
Abronia latifolia Esch.	54°	34°30'	19°30'	* *
Ambrosia chamissonis (Less.) Greene ssp. bipinnatisecta (Less.) Wiggins and Stockw.	50°	30°	20°	*
Tanacetum camphoratum Less.	49°	38°30'	10°30'	
Tanacetum douglasii DC.	49°	40°	9°	
Ambrosia chamissonis (Less.) Greene	48°30'	34°30'	14°	*
Polygonum paronychia C. & S.	48°30'	36°30'	12°	
Calystegia soldanella L.	47°	32°	15°	*
Phyllospadix torreyi S. Wats.	45°	23°	22°	*
Juncus phaeocephalus Engelm.	44°	34°	10°	
Oenothera cheiranthifolia Hornem. & Spreng. (range includes var. suffruticosa)	43°30'	33°30'	10°	*
Mesembryanthemum chilense Mol.	42°30'	32°30'	10°	*
Heliotropium curassavicum L. (mainly var. oculatum)	42°	23°	19°	*
Tetragonia expansa Murr.	42°	32°	10°	
Atriplex leucophylla (Moq.) D. Dietr.	41°	32°30'	8°30'	*
Mesembryanthemum edule L.	39°	32°	7°	
Abronia umbellata Lam.	38°30'	32°30'	6°	*
Mesembryanthemum crystallinum L.	36°30'	23°	13°30'	*
Abronia maritima Nutt. ex Wats.	35°	23°	12°	*
Mesembryanthemum nodiflorum L.	34°	23°	11°	*
Iva hayesiana Gray	33°	28°	5°	
Mammilaria dioica K. Bdg.	33°	23°	10°	
Lotus nuttallianus Greene	32°30'	30°	2°30'	
Atriplex julacea S. Wats.	32°	28°	5°	
Rhizophora mangle L.	25°30'	23°	2°30'	
Avicennia germinans (L.) L.	24°30	23°	1°30'	
Ipomoea pes-caprae (L.) R. Br.	24°30'	23°	1°30'	
Laguncularia racemosa (L.) Gaertn.	24°30'	23°	1°30'	
Sporobolus virginicus (L.) Kunth.	24°30'	23°	1	

From floristic lists, the grass zone and the herb zone must then give way to the intertidal zone with *Puccinellia langeana* in the upper intertidal and *Zostera marina* in the lower intertidal.

Further north, on the Arctic Slope, Wiggins and Thomas (1962) summarize the most conspicuous species on narrow, active beaches as including *Mertensia maritima, Elymus arenarius, Honckenya peploides,* and *Lathyrus maritimus,* with the latter being less abundant. On wider, active beaches, and more common southwest of Barrow, three more are common: *Stellaria humifusa, Cochlearia officinalis* ssp. *arctica* and ssp. *oblongifolia.* The authors point out that "...the tension zone between the tundra and the strand is comparatively narrow." The species which they list in the transition zone to typical tundra appear to be most representative of inland habitats, so they are not listed here.

In the Queen Charlotte Islands, Calder and Taylor (1968) present detailed species lists for a number of beach communities. They divide beaches first into shingle and sand beaches, and each of these major divisions is subdivided.

On shingle beaches influenced by wave action, the common species are *Atriplex patula, Cakile edentula, Mertensia maritima,* and *Honckenya peploides* in the tide-washed coarse gravel zone closest to shore. Behind this is a driftwood zone with *Elymus arenarius, Lathyrus maritimus, Vicia gigantea, Fragaria chiloensis, Ligusticum scoticum,* and *Galium aparine.* The authors add that *Lolium perenne* has been planted on some spits for erosion control, and it has since spread to dominate some upper beaches.

On shingle beaches influenced only by tidal action, a lower zone just above *Fucus* is dominated by *Puccinellia pumila, Cochlearia officinalis, Atriplex patula, Sagina crassicaulis,* and *Plantago maritima.* Just above this is a zone occasionally covered by high tides with several inland species and *Festuca rubra.* A third zone above this is dominated by essentially inland species.

On crescent sand beaches with extensive surf action, the number of species is few, density is low, and they first occur in the driftwood zone: *Cakile edentula, Lathyrus maritimus, Mertensia maritima, Ambrosia chamissonis, Senecio pseudo-arnica, Honckenya peploides, Carex macrocephala,* and *Elymus arenarius.* Behind this is a transition zone to forest, dominated essentially by inland species. On continuous sand beaches, *Glehnia littoralis* additional occurs in the driftwood zone, and *Ambrosia chamissonis, Fragaria chiloensis, Lathyrus littoralis, Abronia latifolia,* and *Convolvulus soldanella* come in behind.

Puccinellia pumila is common in the upper intertidal, and *Zostera marina* and *Phyllospadix scouleri* occur in the lower intertidal. Calder and Taylor refreshingly admit the difficulty in identifying vegetative material of *Zostera* and *Phyllospadix*: "We have experienced difficulty in determining whether sterile vegetative specimens are *Phyllospadix* or *Zostera*...and to identify *P. scouleri* and *P. torreyi*." Possibly the range limits of these three species need considerable refinement, once herbarium material can be unravelled.

Moving south to the Olympic Peninsula, Jones (1936) included the following as being characteristic of early dune stabilization (hence pioneers, close to the shore): *Elymus arenarius, Ammophila arenaria,* and *Abronia umbellata.* Of 18 other species he lists under the category "sand dune and high sea beach," the following are probably found on the beach: *Carex macrocephala, Abronia latifolia, Cakile edentula, Fragaria chiloensis, Lathyrus littoralis, Glehnia littoralis, Convolvulus soldanella, Ambrosia chamissonis,* (including ssp. *bipinnatisecta*), and *Tanacetum douglasii.*

Along the north shore of Oregon, Peck (1961) included the following as being among the "most abundant and characteristic" of the beach and young dune habitats: *Agrostis pallens, Elymus arenarius, Carex macrocephala, Juncus phaeocephalus, Honckenya peploides, Cakile edentula, Fragaria chiloensis, Lathyrus maritimus, L. littoralis, Glehnia littoralis, Convolvulus soldanella, Ambrosia chamissonis,* and *Tanacetum camphoratum.* As one passes south of the Coquille River (43°N), Peck says the vegetation becomes "...decidedly Californian in character, with an aspect quite different from that of the northern coast." Unfortunately, Peck does not describe exactly how the aspect of the vegetation changes. Beaches of Oregon and northern California that we have visited (Newport to Fort Bragg)—especially those fully exposed to the open ocean—have very depauperate plant cover restricted to a few scattered clumps of *Cakile maritima* or *C. edentula* in the driftwood zone. Kumler (1969) listed pioneer plants along the coast of central Oregon as including *Glehnia littoralis, Carex macrocephala,* and *Polygonum paronychia* in most abundance, *Convolvulus soldanella, Abronia latifolia,* and *Ambrosia chamissonis* in less abundance. He noted that ground cover behind the foredune was nine times that on the beach.

The beach vegetation at Bodega Bay (Barbour 1969, 1970c; Barbour et al 1972) is very depauperate, including only *Cakile maritima, Ambrosia chamissonis, Mesembryanthemum chilense,* and *Ammophila arenaria.* Of the four, *Cakile* occurs closest to the high tide line, and *Ammophila* only becomes abundant on the foredune and behind. A few other species listed in Table 2 occur at Bodega Bay, but only behind the foredune: *Abronia latifolia, Fragaria chiloensis, Mesembryanthemum edule, Oenothera cheiranthifolia,* and *Ambrosia chamissonis* ssp. *bipinnatisecta.* Probably the most abundant plant on the beach is *Cakile,* but actual transects show that on the average there is only one clump of *Cakile* every 60 m of beach.

Just south of Bodega Bay, in Marin County, Howell (1949) states that both *Cakile* species grow on the outermost beaches, and that *Abronia latifolia, Glehnia littoralis, Abronia chamissonis,* "and many other plants" grow on higher, more stable beaches. Of the 46 other species he lists, the following appear to be beach species: *Atriplex leucophylla, Abronia umbellata, Mesembryanthemum chilense, Lathyrus littoralis, Convolvulus soldanella, Tanacetum camphoratum, Elymus arenarius,* and *Ammophila arenaria.*

Jumping south to San Luis Obispo County, Hoover's recent flora (1970) does not contribute much information on vegetation. He describes a "beach-dune" community as comprising 10 common species. Those which appear to be beach species are: *Mesembryanthemum chilense, Atriplex leucophylla, Abronia umbellata, A. maritima, A. latifolia, Oenothera cheiranthifolia,* and *Ambrosia chamissonis* ssp. *bipinnatisecta.* And, above all, is *Cakile maritima,* which "...has become probably the most universally present member of this community." Having all three *Abronia* species present permitted Hoover to look for evidence of hybridization, and he concluded hybrids do form in all possible crosses, but especially between *A. umbellata* and *A. latifolia.*

Williams and Potter's as yet unpublished work on the dunes at Morro Bay gives a much better picture of beach zonation. They conclude that the primary invaders, found closest to shore, are *Cakile maritima* and *Abronia maritima.* The hillocks which they form are then invaded by *Ambrosia chamissonis* ssp. *bipinnatisecta* and *Mesembryanthemum chilense.* Larger hillocks further inland then come to be dominated by the dune scrub community (*Lupinus chamissonis, Eschscholtzia califronica, Croton californicus,* and others).

According to Smith (1952), the prominent components of upper beach vegetation include *Atriplex leucophylla, Abronia maritima, Ambrosia chamissonis,* ssp. *bipinnatisecta, Mesembryanthemum edule,* and *Ammophila arenaria.* We have collected *Cakile maritima* there, but apparently at the time Smith made his observation, it was only "locally common" and *C. edentula* was "scattered." For their autecological studies, Martin and Clements (1939) described the beaches around Santa Barbara this way:

> "The beach proper is bordered by a well-developed line of foredunes from 10 to 15 feed in height, back of which lies a small valley. The major dominant of the foredunes is *Abronia maritima,* with *Convolvulus soldanella, Franseria bipinnatifida,* and *Oenothera spiralis* less abundant, and *Atriplex leucophylla* sparse."

Using the nomenclature adopted for this review, *Franseria=Ambrosia chamissonis* ssp. *bipinnatisecta,* and *Oenothera=O. cheiranthifolia.*

On Santa Catalina Island, Thorne (1967) lumps beach species into the habitat category "dunes." His community list includes many species, but the following appear to be beach species: *Mesembryanthemum crystallinum, M. nodiflorum, M. chilense, Franseria chamissonis, Cakile maritima, Atriplex leucophylla, Abronia maritima, A. umbellata,* and *Calystegia soldanella.*

For coastal San Diego County, Higgins (1956) made it clear that she was only discussing "...plants which are conspicuous components of the flora of the salt marshes, coastal sand dunes, and ocean bluffs...no effort has been made to be exhaustive." The characteristic plants, however, are exactly what we are after in this survey. She included the following beach species: *Abronia umbellata, A. maritima* (noted as much less common than the first species), *Mesembryanthemum crystallinum, M. chilense, M. edule, Tetragonia expansa, Lotus nuttallianus, Mammillaria dioica, Convolvulus soldanella, Heliotropium curassavicum,* and *Oenothera cheiranthifolia.*

Perhaps the best vegetation account of the beach for southern California was written by Purer (1936) for the region between El Segundo and Silver Strand (Los Angeles and San Diego Counties). In that paper, she discussed the autecology of seven beach species. Those species, and some of her comments, appear below:

"*Abronia maritima*...grows upon the first or foredunes, the embryonic dunes which rise from the uppermost part of the strand...It is associated with few plants on the strand. Occasionally it may be found growing with *Franseria* (=*Ambrosia*) and *Atriplex*."

"*Convolvulus soldanella*...is found growing on the strand and in open places in the dunes. It is not usually associated with other plants, but is found in flat areas between hummocks overgrown by *Abronia maritima* and *Franseria*..."

"*Atriplex leucophylla*...generally forms small hummocks in the open areas of the strand. It grows rather close to the ocean, often associated with *Abronia maritima* and sometimes with *Franseria*."

"*Mesembryanthemum chilense*...is a sand-binder and dune-former making its appearance usually after such a true pioneer as *Abronia maritima*...it may be classed as a secondary pioneer."

"*Ambrosia chamissonis* ssp. *bipinnatisecta*...is a sand-binder and stabilizer rather than a dune-former..."

"*Oenothera cheiranthifolia*...is found on the upper portions of the strand beyond the storm-swept areas, and in stabilized locations farther from the ocean...it may be classified as a pioneer..."

The seventh species, *Abronia umbellata*, Purer felt needed a bit more protection than *A. maritima*, and depending on the aspect could be a beach species. We have seen beaches near Santa Barbara with small hillocks dominated by *Abronia* species, and often *A. maritima* is on the windward side, *A. umbellata* is on the leeward.

The only information we have on beach vegetation in Baja California comes from correspondence with Dr. Ira Wiggins. Despite the publication of Shreve and Wiggins' Sonora desert flora, which covers all of Baja California, floristic and ecological data on Baja beach vegetation is essentially nil. The accounts of vegetation types by Shreve and Wiggins (1964) covers only desert communities, and passes over the beach without a word. Even the floristic treatment of the beach is incomplete. Dr. Wiggins' communications indicated that *Ambrosia chamissonis* and *Abronia umbellata* are common beach plants in Baja California, but the book does not list these species. According to Table 2, a number of beach species reach their southern limit at the United States border ($32°30'N$). Given the limited floristic information we have for Baja California, however, those limits should be viewed with suspicion, and it is probable that many extend for some distance down the coast into Baja.

Characteristic beach plants along the west coast of Baja appear to include: *Abronia umbellata*, *A. maritima*, *Ambrosia chamissonis* (including ssp. *bipinnatisecta*), *Mesembryanthemum chilense* (at least south to Rosario), *M.*

crystallinum, M. nodiflorum, Atriplex julacea (only in the northern half), *Iva hayesiana* (in the north), and *Heliotropium curassavicum*. From Magdalena Bay south, a few other species become common: *Sporobolus virginicus, Ipomoea pes-caprae,* and the mangroves *Rhizophora mangle, Laguncularia racemosa,* and *Avicennia germinans*.

Salt Marsh Vegetation

The literature describing Pacific Coast salt marshes, although still sparse, has expanded considerably in the twelve years since Chapman (1960) published his world-wide synthesis of salt marsh vegetation and ecology. Geographic coverage remains erratic with most papers describing marshes near California's major population centers. In less traveled areas such as Alaska, British Columbia, and Baja California, data have been gleaned from the same floras as mentioned previously. Many contributions are limited to qualitative descriptions or brief floral lists; more recently, however, the work of Vogl (1966), Mudie (1969, 1970), and Thorsted (1972) has emphasized quantitative data describing species frequencies, vegetation structure, and compositional changes with elevation. Barbour et al. (1972), Mudie (1970), and Jefferson (in progress) have studied the soil-salinity regimes of several marshes in Oregon and California and the first two of these authors, along with Scholander et al. (1965) have conducted experimental studies on the salt tolerance and physiology of a variety of Pacific Coast salt marsh plants. Extensive personal communications from Carol Jefferson, Peta Mudie, and Ira Wiggins have added substantially to our data on the marshes of Oregon and Baja California, respectively, and these we gratefully acknowledge.

Because of more complete literature coverage, a general summary of salt marsh floristics was produced by a quite different technique from that used for the beach vegetation. Instead of listing distributional ranges, the actual presence or absence of a particular species at a given locality was noted. All of the species recorded from "salt marsh habitats" (generally undefined) at over 50 localities along the Pacific Coast were listed. (The distribution of these localities, together with the references describing them, is shown in Fig. 4.) The resulting master list contained over 140 species. Many of these were "occasionals" recorded from single localities (the taxonomist almost inevitably stresses unusual or rare species at the expense of common forms); other uncommon forms were clearly not salt marsh species but reflected local fresh-water influences or the presence of coastal dunes. These species were eliminated and the resulting distributional lists are presented in Tables 3 and 4. Table 3 includes 41 species considered here as characteristic of salt marsh vegetation north of the Canadian Border. Table 4 includes 51 of the more frequently encountered salt marsh species from Washington, Oregon, California, and Baja California. A total of 78 different species are tabulated and 14 of these occur both north and south of the Canadian Border. The

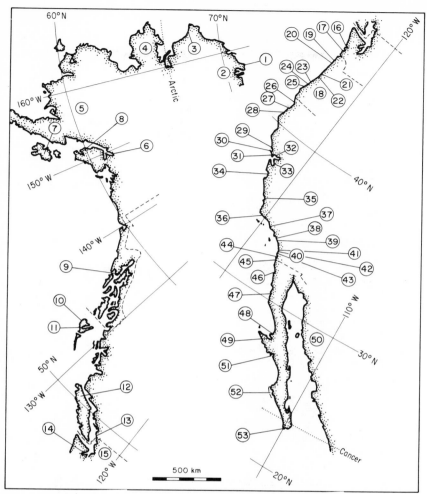

Figure 4. The North American Pacific Coast. The circled numbers refer to the following salt marsh localities mentioned in the text: 1 - Point Barrow, 2 - Arctic slope, 3 - Cape Thompson, 4 - Seward Peninsula, 5 - SW Alaska, 6 - Goose Bay and Knik Arm, Anchorage, 7 - Kodiak Is., 8 - Homer, 9 - Glacier Bay, 10 - Queen Charlotte Is., 11 - Massett Inlet, 12 - Vancouver Is., 13 - Fraser R. Delta*, 14 - Hood Canal, 15 - Whatcom County, 16 - Grays Harbor*, 17 - Willapa Bay, 18 - Oregon Coast, 19 - Nehalem Bay, 20 - Tillamook Bay, 21 - Yaquina Bay, 22 - Alsea Estuary, 23 - Umpqua Estuary, 24 - Coos Bay*, 25 - Coquille Estuary, 26 - Lake Talawa, 27 - Big Lagoon, 28 - Humboldt Bay*, 29 - Bodega Bay, 30 - Tomales Bay*, Drake's Estero, 31 - Bolinas Bay, 32 - San Pablo Bay, Suisun Bay, 33 - San Francisco Bay, 34 - Elkhorn Slough*, 35 - Morro Bay, 36 - Goleta, Carpinteria, 37 - Mugu Lagoon*, 38 - Anaheim Bay, Bolsa Chica Bay, 39 - Newport Bay, 40 - San Diego County, 41 - Santa Margarita R., 42 - Las Penasquitas Lagoon, 43 - Mission Bay*, 44 - S. San Diego Bay, 45 - Tijuana Slough, 46 - Estero de punta Banda, 47 - Bahia de San Quintin*, 48 - Laguna Guerrero Negro*, 49 - Ojo de Liebre, 50 - Sonoran Coast Lagoons, 51 - Laguna San Ignacio, 52 - Bahia de la Magdalena, 53 - Cabo San Lucas. (* indicates sites included in Macdonald 1967, 1969).

Species	\multicolumn{12}{c}{Localities}
	1	2	3	4	5	6	6A	8	9	10	11	13
Carex ursina Dew.	●	●										
Carex subspathacea Wormsk.	●	●	●									
Puccinellia phryganodes (Trin.)S.&M.	●	●	●			●	●	●				
Stellaria humifusa Rottb.	●	●	●	●						●	●	
Salix ovifolia Trautv.	●	●										
Carex glareosa Wahl.	●	●				●						
Carex ramenskii Kom.		●				●		●				
Calamagrostis deschampsioides Trin.	●	●			●			●				
Triglochin maritima L.		●			●	●	●	●	●	●	●	●
Dupontia fischeri R. Br.			●									
Potentilla egedii Wormsk.												
var. groenlandica (Tratt.)Pol.			●									
Chrysanthemum arcticum L.		●				●	●					
Puccinellia langeana (Berl.)Sorens.		●							●			
Puccinella pumila (Vas.)Hitchc.											●	
Deschampsia caespitosa (L.)Beauv.		●								●		
Puccinellia borealis Swallen				●								
Potentilla egedii Wormsk.												
ssp. grandis (T.&.G.)Hult.					●		●	●			●	●
Carex lygnbyei Hornem.					●	●		●	●		●	●
Castilleja hyetophila Pennell					●							
Juncus arcticus Willd.					●							
Plantago maritima L.												
ssp. juncoides (Lam.)Hult.					●	●		●	●			
Triglochin palustris L.						●	●					
Poa eminens Presl.						●						
Puccinellia glabra Swallen						●						
Puccinellia trifolia Swallen						●	●					
Spergularia canadensis (Pers.)G. Don						●						
Suaeda maritima (L.)Dum.						●						
Glaux maritima L.						●			●			
Salicornia europaea L.						●		●				
Atriplex patula L.												
(inc. ssp. hastata, obtusa)*						●						●
Scirpus maritimus L.							●					●
Hordeum brachyantherum Nevski.							●			●	●	●
Agrostis exarata Trin.										●	●	
Festuca rubra L.										●		
Grindelia integrifolia D.C.										●		
Plantago macrocarpa Cham. & Schl.										●	●	
Trifolium wormskjoldii Lehm.										●		○
Lilaeopsis occidentalis C.&R.										●		
Scirpus cernuus Vahl.										●		
Juncus lesueurii Bol.												●

TABLE 3: Pacific Coast salt marsh floras: Species composition at 12 localities between Point Barrow and the Fraser River Delta. (Localities as in Fig. 4. Total species per locality indicated below; see text for additional species recorded from single localities. Asterisk indicates introduced species.)

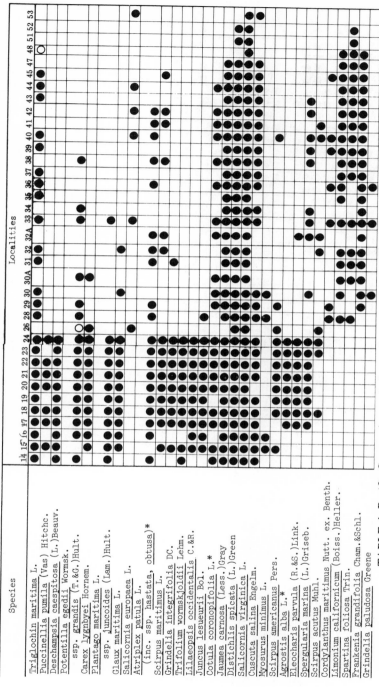

TABLE 4: Pacific Coast salt marsh floras: Species composition at 35 localities between the Olympic Peninsula, Washington, and Cabo San Lucas, Baja California. (Localities as in Fig. 4. Total species per locality indicated below; majority of additional species recorded once or twice only. Asterisk indicates introduced species; open circle (º) indicates species identification doubtful.)

TABLE 4 continued

Lasthenia glabrata Lindl.
Atriplex semibaccata R.Br.*
Scirpus californicus (C.A. May)Steud.
Sesuvium verrucosum Raf.
Salicornia bigelovii Torr.
Suaeda californica Wats.
Salicornia subterminalis Parish.
Cressa truxillensis HBK.
Monanthochloe littoralis Engelm.
Batis maritima L.
Heliotropium curassavicum L.
Juncus acutus
  var. sphaerocarpus Engelm.
Mesembryanthemum chilense Mol.
Mesembryanthemum crystallinum L.*
Mesembryanthemum nodiflorum L.*
Amblyopappus pusillus H.&A.
Atriplex watsonii A. Nels.
Hordeum leporinum Link.*
Frankenia palmeri S. Wats.
Allenrolfea occidentalis (S.Wats.)Kuntze
Maytenus phyllanthoides Benth.
Laguncularia racemosa (L.)Gaertn.
Rhizophora mangle L.
Avicennia germinans (L.)L.

reader is again cautioned to note that geographic coverage is generally better to the south of the Canadian Border, than to the north.

Comparisions of observed patterns of species occurrence with those predicted from published geographic range records, suggest that most of the tabulated species fall into one of three distributional types. These include: (1) species that occur with great regularity at almost every salt marsh locality examined within their known latitudinal range (*P. Phryganodes* and the various species of *Carex* from Alaska, for example, as well as: *T. maritima, D. spicata, S. virginica, L. californicum, F. grandifolia, S. california, M. littoralis,* and *S. subterminalis,* all occurring further south), (2) species that are almost equally predictable in their occurrence, yet are conspicuously absent at a few specific localities (e.g., *S. foliosa, S. bigelovii, B. maritima*—interestingly all "low marsh" species) and (3) species presently recorded from only a few salt marsh sites widely scattered through their known ranges (*P. maritima* ssp. *juncoides, G. maritima, S. europaea,* and *G. integrifolia* are good examples here, along with opportunistic, introduced forms such as *A. patula* ssp. and *C. coronopifolia*). The first two categories include most of the "typical" salt marsh plants of the Pacific Coast; the third includes a diverse group of "occasional" species whose presence at a particular locality probably reflects both the historical development of the site and locally important environmental variables (i.e., local differences in soil salinities, substrate types, degree of protection, nature of maritime and upland vegetation, degree of disturbance, etc.).

As expected on a coastline that spans 48° of latitude between Arctic tundra and sub-tropical desert, there are very significant latitudinal changes in the species composition of the salt marsh vegetation. When the geographic ranges of the more frequently encountered species are plotted and ordered by their northern limits (cf. Table 2), new species are seen to appear, and previously recorded ones drop out, at irregular intervals throughout the length of the Pacific Coast. Under these circumstances, the recognition and separation of phytogeographical provinces is perhaps a highly subjective exercise of dubious value. It is true, however, that at certain latitudes and along relatively short stretches of the coast, compositional changes are more marked and small clusters of new species appear more or less simultaneously. (Interestingly, this feature appears to be more noticeable for northern species limits than southern ones.)

The salt marshes of the Alaskan Arctic Slope contain few species and are dominated by *Puccinnellia phryganodes* (the usual primary colonist), several species of *Carex* and *Calamagrostis deschampsioides*. *Triglochin maritima* also reaches this far north and it reappears regularly at least as far south as the Mexican Border. *Carex lygnbyei, Potentilla egedii* var. *grandis,* and *Plantago maritima* ssp. *juncoides* all appear along the west Alaskan coast and are represented by scattered occurrences well south into California. *Deschampsia caespitosa,* which co-occurs with the last three species, is consistently represented in the higher marshes of Washington and Oregon and has been recorded south to Point Conception. Many of the characteristic Alaskan taxa (*Puccinellia, Carex* and *Calamagrostis*) drop out along the northern shores of the Gulf of Alaska and *Glaux maritima, Atriplex patula* ssp. (introduced), and *Salicornia europaea* appear.

Between Anchorage and Vancouver, field data are sparse and salt marsh floras only poorly known. *Salicornia virginica*, a dominant species along much of the United States and Baja coast, is first recorded from Chichagof Island in the Alexander Archipelago (Hulten 1968). In northern Washington this important species is joined by several others (*Juncus lesuerii, Scirpus americanus, S. robustus, Cotula coronopifolia*—another introduced species, *Distichlis spicata*, and *Jaumea carnosa*. Most of these forms extend to the Mexican Border. *S. virginica* is also joined by *Cuscuta salina* and both species extend beyond the tip of Baja California.

The next major group of floristic changes occur between Cape Blanco, Oregon, and San Francisco Bay. Several wide ranging species that first appeared in central or south Alaska now begin to drop out and the new forms that appeared in Washington are first joined by *Spartina foliosa* (an important primary colonist), *Frankenia grandifolia* and *Limonium californicum*, and a little to the south by *Atriplex semibaccata* (an introduction) and *Sueada californica*. South of Point Conception almost all of the species that first occurred north of the Canadian Border have also disappeared and new species including *Batis maritima, Cressa truxillensis, Salicornia bigelovii, S. subterminalis*, and *Monanthochloe littoralis* begin to occur in the salt marshes.

The salt marsh floras reach their maximum diversity between Los Angeles and Tijuana. This is deceptive, however, for much of this diversity reflects human disturbances that have modified salt marsh environments and permitted normally maritime or upland species to invade the littoral zone. Table 4 indicates that nine such species, including three introductions, occur rather consistently in these southern California marshes.

The final sequence of floristic changes occurs south of the Vizcaino Desert in Central Baja California. Here several species important in California begin to drop out (*F. grandifolia, L. californicum, T. maritima*). The first stunted growth of mangroves appears at Laguna San Ignacio and they play an increasingly important role at more southerly salt marsh localities.

Tables 3 and 4 summarize the basic floristics of Pacific Coast salt marsh vegetation; more detailed descriptions of its general appearance and characteristic vertical species zonation are presented below. The marsh vegetation is described region by region from Alaska, south; when possible, the qualitative descriptions have been supplemented by quantitative distribution data (Tables 5-10). Species names included in the several brief quotes cited below have, when necessary, been changed to conform to the nomenclature adopted for this review.

Table 5. Lower Coquille Estuary, Oregon: Average percentage estimated cover for species in each zone. Ranges for each group of stands given in parentheses (adapted from Johannessen, 1961).

Species	"New Marsh"		"Old Marsh"		
	Outer Edge	Low	Low	Int.	High
Spergularia marina	25(0-50)				
Salicornia virginica	45(40-50)	45(30-90)	42(30-60)	23(10-40)	25(0-40)
Triglochin maritima	30(0-60)	20(5-50)	20(10-30)	6(0-18)	
Scirpus americanus		14(0-80)	7(0-30)		
Carex lyngbyei		2(0-15)	1(0-5)	17(0-50)	
Distichlis spicata		11(0-40)	11(0-25)	13(0-20)	17(0-30)
Jaumea carnosa		5(0-35)	17(0-30)	7(5-10)	11(0-30)
Juncus effusus				2(0-5)	
Agrostis alba				20(0-35)	
Deschampsia caespitosa		<1(0-1)		9(5-12)	41(0-100)
Others		<1(0-5)			5(0-20)
Number of stands	2	7	4	3	4

Table 6. Bodega Bay, California: Average ground cover (%) for all plant species at different distances from mean high tide line in the salt marsh, June 1970 (from Barbour et al., 1972).

Species	Meters From Mean High Tide										
	0	5	10	15	20	25	30	35	40	45	50
Distichlis spicata	63	26	16	7	28	32	22	27	16	31	27
Salicornia virginica	21	70	80	83	65	47	36	4	2	12	20
Scirpus americanus			3	4	4	11	12	7	5	5	6
Triglochin maritima						1		2	3		
Jaumea carnosa						22	60	74	45	31	
Scirpus koilolepis										7	6
Cordylanthus maritimus											7
Plantago lanceolata											1
Bare ground	16	4	1	6	3	7	8				2

Table 7. South San Francisco Bay: Lower and upper limits of some common salt marsh plants expressed as percentages of the local tidal range. (MLLW to EHW = 3.35m, MHW = 2.38m, MHHW = 2.56m; data from Hinde, 1954.)

Species	Lower Limit	Upper Limit
Spartina foliosa	50	76
Salicornia virginica	58	94+
Jaumea carnosa	65	77
Distichlis spicata	65	93
Frankenia grandifolia	69	94+

Table 8. Upper Newport Bay, California: Average percentage cover for the common species in each zone. Each value is an average based on 240 quadrats (25 cm x 25 cm). The range for six separate stands is given in parentheses. Additional species present indicated by asterisks (from Vogl, 1966).

Species	Littoral			Maritime
	Lower	Middle	Upper	
Spartina foliosa	38(4-87)	1(0-2)	*	
Batis maritima	4(0-23)	15(3-42)	1(1-6)	
Salicornia virginica	4(1-10)	23(6-67)	40(2-65)	11(0-29)
Suaeda californica	*	*	2(1-7)	19(9-25)
Frankenia grandifolia		3(0-11)	2(2-12)	*
Distichlis spicata		2(0-10)	5(0-29)	
Triglochin maritima		11(0-42)	1(0-8)	*
Limonium californicum		1(0-4)	1(0-10)	
Monanthochloe littoralis			15(0-66)	
Cuscata salina		*	2(0-5)	
Juncus acutus			2(0-14)	
Scirpus californicus			14(0-90)	
Salicornia subterminalis				24(0-54)
Mesembryanthemum crystallinum				13(0-26)
Encelia californica				4(0-14)

Table 9. South San Diego Bay: Average percentage cover for the common species of the "low marsh" (MLHW-EHW) and "high marsh" (∿EHW). Ranges for three separate low marshes and two high marshes are given in parentheses (from Mudie, 1970).

Species	Low Marsh*	High Marsh
Salicornia pacifica	47 (36-64)	14 (8-19)
Batis maritima	24 (10-38)	2 (0-3)
Salicornia bigelovii	12 (<1-20)	
Frankenia grandifolia	6 (1-11)	3 (0-6)
Spartina foliosa	5 (0-10)	
Distichlis spicata	2 (0-5)	
Suaeda californica	2 (0-5)	
Jaumea carnosa	1 (0-3)	
Limonium californicum	1 (<1-2)	
Monanthochloe littoralis	<1 (0-<1)	20 (11-28)
Triglochin maritima	<1 (0-<1)	
Salicornia subterminalis		39 (31-46)
Atriplex semibaccata		1 (0-1)
A. watsonii		<1 (0-1)
Cressa truxillensis		1 (0-2)
Miscellaneous annuals		7 (<1-14)
Bare ground		14 (7-21)

* Includes "middle marsh-middle littoral," of several authors.

Table 10. Bahia de San Quintin, Baja California: Average percentage cover for the common species in each zone. Each value is a rounded average based on 112 quadrats (25 cm x 25 cm). The range for seven separate stands is given in parentheses (from Thorsted, 1972).

Species	Littoral		
	Lower	Middle	Upper
Spartina foliosa	47 (42-53)	1 (0-4)	
Salicornia virginica	20 (6-43)	33 (27-61)	23 (5-58)
Batis maritima	6 (4-8)	17 (3-35)	1 (0-6)
Frankenia grandifolia	2 (0-5)	10 (2-17)	14 (0-35)
Jaumea carnosa	1 (0-8)	1 (0-4)	4 (0-24)
Suaeda californica	1 (0-3)	7 (3-11)	8 (0-22)
Triglochin concinna		5 (0-25)	
Limonium californicum		2 (0-6)	6 (0-17)
Salicornia bigelovii		7 (0-41)	
Monanthochloe littoralis		1 (0-8)	46 (1-81)
Salicornia subterminalis			23 (0-45)
Atriplex watsonii			2 (0-7)
Bare ground	3 (0-11)	14 (3-30)	9 (0-24)

Alaska (Salt marsh acreage unknown)

Polunin (in Chapman 1960) describes the northernmost salt marshes of the Pacific Coast, near Point Barrow on the shores of the Arctic Ocean. The primary invader is *Puccinellia phryganodes* which forms "mats" low on the shore; higher up *Carex ursina* and *C. subspathacea* are abundant, and *Stellaria humifusa* is present. Polunin notes that these marshes are less luxuriant than others developed under less rigorous environmental conditions in southwest Alaska. Spetzman (1959) confirms the importance of *P. phryganodes* and lists its habitat as, "...silty tidal-zone flats along the Arctic Ocean (elevation 0–3 m above sea level)."

Wiggins and Thomas (1962) do not detail salt marsh habitats in their study of the Arctic Slope (68–71°N) but species notations in the flora confirm their presence. The following species are important: *Triglochin maritima* (coastal marshes and along streams), *P. phryganodes* (margins brackish lagoons, mudflats, seashores), *C. subspathacea* (tidal flats and tundra periodically covered with seawater), *C. ursina* (sandy upper beaches and along lagoons and estuaries), *C. maritima* (sandy shores of estuaries and lagoons), *C. glareosa* ssp. *glareosa* (brackish marshes and just above sea level along beaches and ocean bluffs), and *C. g.* var. *amphigena* (margins brackish lagoons). All but the first of these species occur widely between Point Barrow and

Greenland, and most are not restricted to salt marsh habitats.

In a broader context, Hulten's (1968) comprehensive "Flora of Alaska" lists only 11 species from "coastal salt marsh" or "tidal marsh" habitats and three more from "brackish marshes." Of these *Carex glareosa* (brackish), *C. ramenskii, C. subspathacea, P. phryganodes,* and *Salix ovalifolia* are either largely or entirely restricted to the Arctic Slope and N.W. Coast, while *Astragalus polaris, Carex lyngbyei, Castilleja hyetophila, Juncus arcticus* ssp. *sitchensis, Plantago maritima* ssp. *juncoides, Puccinellia geniculata,* and *T. maritima* (brackish) are found mostly in the south and southwest. *Calamagrostis deschampsioides* (brackish) is widely distributed throughout coastal Alaska and the final species, *Puccinellia grandis,* is presently known only from Norton Sound.

At Cape Thompson, N.W. Alaska (68°N), Johnson et al. described a "saline meadow" community developed, "...on wet alluvium near the mouth of Ogotoruk Creek. During on-shore storms this meadow is flooded by the Chukchi Sea, which deposits driftwood in large quantities. Most of the area is covered by an almost continuous sedge-grass mat, including *Eriophorum angustifolium, Carex glareosa* (var. *amphigena*)*, *Deschampsia caespitosa, Dupontia fishcheri* (ssp. *psilosantha*)*, *Calamagrostis deschampsioides*, Arctagrostis latifolia* (var. *arundinacea*), *Puccinellia phryganodes*, P. langeana,* and *Arctophila fulva*. The presence of such maritime species as *Chrysanthemum arcticum* (ssp., polaris), *Potentilla egedii* (var. *groenlandica*)*, and *Matricaria ambigua* reveals the influence of the sea on the composition of this vegetation type."

The authors equate this community with the "Saline sedge marshes herb type," described by Hanson (1953) and the "Lagoon and salt marsh subject to floods" of Porsild (1951). Examination of distributional notes in the annotated flora suggests that only those species marked with asterisks are both common in, and relatively restricted to, these saline meadow habitats. *P. phryganodes* again appears to be the primary colonist on muddy shores. These notes further suggest that *Stellaria humifusa* and *Pedicularis parviflora* are common saline meadow species and that additional forms such as *Salix ovalifolia, Primula borealis,* and *Gentiana tenella* occur within the zone of storm tides or at the edge of the driftwood zone.

Further south, Hanson (1953), summarizing vegetation types of the Seward Peninsula and north to the Noatak River (64–67°N), points out "...that saline marshes occur extensively on estuaries, borders of lagoons, and other areas near the sea." He cites the salt tolerant sedges *Carex lyngbyei* and *C. subspathacea* as important vegetational components and lists other characteristic species on saline estuaries, often early invaders on mudflats, as *Puccinellia borealis, Stellaria humifusa,* and *Potentilla egedii* ssp. *grandis*. He adds that the total number of species is few in saline marshes but rises rapidly (30 species cited) in fresh-water marshes.

Hanson's earlier work (1951) provides our first photographs and adequate description of the zonal succession of these Alaskan salt marshes. Describing the shores of Goose Bay estuary, northwest of Anchorage, Hanson outlines a sequence of eight zones extending from bare saline mudflats to brackish or fresh-water meadow communities that are invaded first by willows and alders, and later by Alaska birch and spruce trees.

The most common primary invader, subject to frequent tidal submersion, is *Puccinellia phryganodes*, whose widely branched, creeping stems grow rapidly and root readily to form rather open mats on the slippery mud. *Salicornia europaea* and *Suaeda maritima* are present and scattered tufts of *Puccinellia trifolia* appear a little higher up. *Glaux maritima* and *Spergularia canadensis* may occur in this lowest zone, along with a few scattered individuals of species more characteristic of the next zone. The second zone develops about 60 cm above ordinary high tides; it is very wide in places and much of the ground is bare. The most abundant species is *Plantago maritima*, which co-occurs in the lower parts with *P. trifolia* and *P. glabra* and higher up with *Triglochin maritima*. *S. europaea* is present in low, poorly drained spots and *Elymus arenarius* occurs on many scattered mounds. Other species include *Triglochin palustris, Ranunculus cymbalaria, Potentilla egedii* ssp. *grandis, Atriplex patula, Poa eminens,* and *Chrysanthemum arcticum*. (An example of this community from Middle Bay estuary, Kodiak Island, yielded a rather sparse cover of 10 species: *P. maritima, Stellaria humifusa, C. lyngbyei,* and *P. pacifica* made up 92% of the cover.)

The third zone, slightly higher than the second, was developed on firm, compact, fairly well-drained silt. *T. maritima* and *P. egedii* were the characteristic species, with *T. palustris, Carex glareosa, Poa eminens* and on wetter spots *Carex ramenskii*, also commonly present. The remaining zones became progressively wetter and probably less brackish. *C. ramenskii* was the dominant in the lowest of these, making up as much as 84% of the dense cover; *T. maritima* and *P. egedii* were also present.

A gently sloping silt beach just east of Anchorage (Knik Arm) yielded a simpler succession. The first invaders—*P. phryganodes, P. triflora,* and *T. maritima*—were subject to frequent overflow by high tides and formed a zone up to 3 m wide. All three species persisted at higher elevations but *T. maritima* became dominant and was joined by *T. palustris*. *Scirpus pacificus* and *Carex lyngbyei* appeared at the inner edge of this zone and dense growth of the latter dominated higher levels. *Calamagrostis canadensis, Chrysanthemum arcticum,* and *Potentilla egedii* were scattered among the *C. lyngbyei* in the driftwood zone.

Hanson also describes a generally similar zonation developed on the silty shores of a small bay near Homer ($60°N$). *P. phryganodes* was again the first species to invade the mud and was soon joined by scattered individuals of *T. maritima* and *Plantago maritima*. The most abundant species in the next higher zone was *Carex ramenskii* which usually co-occurred with scattered patches of *C. mackenziei* and tufts of *T. maritima*. In places where the latter became dominant, *P. maritima* and *P. phryganodes* were again present. The third zone extended back to the driftwood line and was developed on saturated peaty soils subject to brackish or fresh-water influences. *Carex lyngbyei* was dominant here and scattered individuals of *Calamagrostis deschampsioides* and *Potentilla palustris* were present. *Calamagrostis* became dominant at the storm-tide strand line which was backed by willow thickets and *Equisetum*.

Cooper's (1931) account of the vegetation of Glacier Bay(58°N) includes a brief description of the succession from the bare mudflats bordering Icy Strait to "beach meadows" backed by willow-alder thickets and spruce forest. After listing 12 species characteristic of the beach meadows, a common vegetation type along the shores of southeastern Alaska, he states that: "...As we pass toward the Strait, the meadow plants one by one disappear, *Elymus* (*E. arenarius*) remaining longest. *Plantago maritima* L. appears and becomes dominant, with it growing *Hordeum brachyantherum* Nevski and *Triglochin maritima* L. Where the plant carpet becomes discontinuous *Glaux maritima* L. is found scattered thickly and last of all the grass, *Puccinellia langeana* ...makes a broken carpet upon the otherwise bare mud flat."

British Columbia (salt marsh acreage unknown)

In the Queen Charlotte Islands (52–54°N; Calder and Taylor, 1968)) salt marsh communities, dominated by grasses and sedges, represent the only extensive lowland meadows. Salinity levels vary considerably but are generally higher in marshes subject to daily tidal flooding than in those that are either flooded only at extreme tides or are inundated by both fresh and salt water. Calder and Taylor distinguish between high and low salinity salt marshes and provide floral lists for each. Their description of high salinity marshes is quoted below:

"These marshes are fronted by shingle beaches or mud flats. The lower part of the marshes is often reticulately dissected by drainage courses which become active during periods of high tides. ...The marsh surrounding these drainage courses is considerably raised and represents a vegetative terrace only inundated during extreme tides or storms. The following species predominate in the frontal marsh zone bordering the shoreline: *Deschampsia caespitosa, Hordeum brachyantherum, Festuca rubra, Agrostis exarata, Carex lyngbyei, Plantago macrocarpa, Stellaria humifusa, Triglochin maritima* and *Trifolium wormskjoldii*. Between this densely vegetated portion of the marsh and the adjacent closed forest, additional species such as *Apargidium boreale, Carex pluriflora, Galium trifidum* and *Calamagrostis nutkaensis* occur. Two showy composites, *Aster subspicatus* and *Grindelia integrifolia*, are occasionally found in these marshes. These high salinity marshes characteristically have few species and are remarkably uniform throughout the Islands."

Low salinity marshes are best developed on river deltas bordering Masset Inlet where tidal waters that partially flood the marsh meadows are constantly diluted by fresh-water run-off. *Triglochin maritima, Puccinellia pumila, Lilaeopsis occidentalis,* and *Scirpus cernuus* occur along the margins of the muddy drainage courses, and grasses and sedges predominate on the vegetated terraces between them. More species are present here than on the higher salinity, open coast marshes. Interestingly, Calder and Taylor make no mention of *Salicornia* in their floral lists, although Heusser (1960) lists *S. europaea* among characteristic species of the "strand vegetation" around Sandspit and Rose Point.

Descriptions of the Fraser River Delta marshes (49°N) have not been found in the literature. A brief visit by Macdonald (May 1964) suggests that while fresh-water influences predominate, some more saline marshes are developed along the borders of the Strait of Georgia at the western ends of Sea Island and Smokey Tom Island. The succession at these sites begins with rather dense, pure stands of *Scirpus robustus* that extend out onto the bare, rather sandy, tide flats. *Triglochin maritima* and *Carex lyngbyei* appear next in the succession and are soon joined by *Juncus lesuerii, Potentilla egedii,* and *Trifolium* sp. to form a very uniform vegetation type that covers most of the higher marsh. Scattered individuals of *Atriplex patula* were also present.

Washington (12,000 acres*)

Muenscher (1951), cataloging the flora of Whatcom County (49°N), northern Washington, cites 14 salt marsh localities along the southeastern shores of the Strait of Georgia and lists 27 of their "...more common or interesting plants." He describes the salt marshes as occurring behind gravelly or sandy beaches, or along tidal streams, and being covered with "...a rather uniform vegetation frequently forming a dense meadow." He makes no mention of any characteristic species zonation patterns.

Some of the species included by Muenscher are introduced weeds, others probably occurred in fresh-water habitats; however, all are listed below, together with other salt marsh species listed in the main body of the flora. They have been arranged in alphabetical order and those described as "abundant," "common," or "frequent" in salt marshes are marked with asteriks.

Carex Lyngbyei	**Plantago elongata*
C. obnupta	*P. maritima*
Chenopodium rubrum	*Polypogon monspeliensis*
Cotula coronopifolia	*Potentilla anserina*
Cuscuta salina	*Puccinellia distans*
Deschampsia caespitosa	*P. pumila*
*Distichlis spicata	*Ranunculus cymbalaria*
Eleocharis palustris	*Rumex maritimus*

* 48 km^2; minimum estimate, includes Grays Harbor (1964) and Willapa Bay (1970) only.

Glaux maritima
*Grindelia integrifolia
*Hordeum jubatum
Jaumea carnosa
Juncus gerardi
J. lesuerii
Lilaeopsis occidentalis
Myosurus minimus
Orthocarpus pusillus

*R. occidentalis
*Salicornia virginica
Scirpus americanus
*S. robustus
Sidalcea hendersonii
Spergularia canadensis
Suaeda maritima
Trifolium wormskjoldii
*Triglochin maritima

Salt marshes dominated by *Salicornia* near Roche Harbor, San Juan Island, are mentioned in Shelford et al. (1935), but no additional details are given. Further south, Jones' (1936) botanical survey of the Olympic Peninsula includes a brief description of salt marshes developed at the mouths of the Dosewallips and Duckabush Rivers on the western shores of Hood Canal (48°N). *Distichlis spicata* is listed the common dominant species with *Salicornia virginica* as chief co-dominant. A list of "...the principal species..." of the salt marsh association includes the following additions: *Triglochin maritima, Deschampsia caespitosa, Rumex maritimus, Atriplex patula, Cuscuta salina, Orthocarpus castillejoides, Plantago maritima, Jaumea carnosa,* and *Cotula coronopifolia*. Somewhat less saline "seashore meadows" higher in the succession yielded *Potentilla egedii, D. spicata, Stellaria calycantha, Ranunculus cymbalaria, Trifolium wormskjoldii,* and two well-established introduced species: *Myosotis scorpioides* and *Hypochaeris radicata*.

Apart from a brief account of some salt marshes at Westport (47°N; Macdonald 1967,1969), the extensive marshlands of Grays Harbor and Willapa Bay (National Estuary Study 1970), on the outer coast of Washington, remain virtually undescribed.

The tideflats at Westport consist mostly of coarse, well-sorted sand; extensive salt marshes are developed on a gentle slope, contiguous with the adjacent flats, and extending inland for over a thousand meters. To the north the bayward margin of the marsh is protected by a dune-like ridge (1 m high) covered with *Deschampsia caespitosa*. The tideflats outside this ridge were being colonized by pure stands of *Distichlis spicata*. Three large tidal channels deeply dissected the southern portion of the marsh and here the marsh borders were sometimes truncated by irregular, low (.5 m) cliffs. Here the primary colonist was usually *Salicornia virginica*, although the succession sometimes began with scattered clumps of *Triglochin maritima*.

Visually, the salt marsh was readily divisible into two zones: an outer, lower zone covered with rather short (10—20 cm) vegetation and clearly dominated by *S. virginica*, and a higher, inner zone in which tall (1 m), closely-spaced tussocks of *D. caespitosa* accounted for up to 70% of the cover. Forty-two random samples (circles, 50 cm diam.) taken in the lower zone contained the following species, in order of decreasing frequency: *S. virginica* (present, 98% of samples), *Jaumea carnosa* (55%), *T. maritima*

(55%), *D. spicata* (46%), *Plantago maritima* (24%) *and* Glaux maritima (17%). No obvious patterns of changing species abundances with elevation were apparent; almost pure stands, covering tens of square meters, of each of the three most frequently encountered species were noted. Several macro-algae, including fucoids and diminutive forms similar to those described by Chapman (1960), were present throughout the zone and were abundant near the outer edges of the marsh. Eight additional random samples were taken in the higher zone. *D. caespitosa* accounted for a majority of the cover in every sample. *J. carnosa* was present between the grass tussocks in every sample and *S. virginica* co-occurred in seven of them. *D. spicata, Atriplex patula, P. maritima, Lasthenia* sp., and *Cuscuta salina* each occurred scattered throughout the higher marsh. Further inland several additional species appeared, including: *Carex lyngbei, Myosurus minimus, Potentilla egedii,* and *Stellaria calycantha*.

Additional unpublished field observations (Macdonald, May 1964) allow further comments on these Grays Harbor marshes. Marshland at Oyhut, North Bay, was readily divisible into an outer strip (15 m wide) of low lying marsh, highly dissected by pools and channels, and subject to daily tidal submergence, and a much broader, level expanse of meadow-like high marsh that was only flooded by spring tides. Primary colonists on the highly variable (mud, silt through sand, gravel) tideflats were either *Triglochin maritima* or *Salicornia virginica. Distichlis spicata* appeared, and became abundant on slightly higher ground. *Lilaeopsis occidentalis* was present and several varieties of fucoid algae were found living at the margins of the marsh pools and channels. Deep tidal channels that drained the high marsh were bordered with *S. virginica*, but the remainder of the high marsh consisted of a rather uniform mixture of *Juncus lesueurii, Carex lyngbyei, Deschampsia caespitosa,* and *Potentilla egedii*.

Several species from these two associations (*S. virginica, D. spicata, C. lyngbyei, D. caespitosa,* and *P. egedii,* together with *Glaux maritima*) co-occurred at Markham, some 6 miles inside Grays Harbor. The rather uniform vegetation that resulted covered almost level ground that appeared to be undergoing erosion and was separated from the adjacent tideflats by a cliff nearly a meter high.

Oregon (7,600 acres*)

Oregon's salt marshes are much better known than those of Washington. House (1914) provided an early account of the Coos Bay marshes and Peck (1961) includes a partial species list for the marshes north of Cape Blanco in his account of the plant areas of Oregon. More complete, illustrated descriptions of the salt marshes are provided by Johannessen (1961, 1964), who studied shoreline and vegetation changes in six Oregon estuaries:

* 31 km^2; preliminary minimum estimate including larger marshes only, could be doubled through addition of many marshes less than 3 acres (Jefferson 1972).

Nehalem Bay (46°N), Tillamook Bay, Alsea Estuary, Umpqua Estuary, Coos Bay, and Coquille Estuary (43°N).

Through comparative studies of old maps and aerial photographs, Johannessen shows that Oregon's salt marshes have expanded dramatically over the last 100 years. He attributes this expansion to increased rates of soil erosion and tideflat deposition, as European colonists displaced Oregon's Indians and introduced new agricultural and lumbering practices. Supportive evidence includes the occurrence of sawn logs and layers of bark chips buried as much as a meter below present marsh surfaces. Sharp breaks of slope, up to a meter high, often separate broad level surfaces occupied by older "high marsh" from gently sloping mudflats and younger "low marsh," and may represent pre-colonial equilibrium shorelines.

Johannessen (1964) provides an excellent account of the early stages of succession in the tidal marshes of Nehalem Bay. In the lowest habitats, near deep river channels, circular clumps of *Triglochin maritima* up to 2 m in diameter develop. On higher, broader expanses of the mudflats, similar circular clumps of either *Scirpus robustus* (75%) or *Carex lyngbyei* (25%) appear. On ground a few centimeters higher the vegetation becomes continuous and *C. lyngbyei* is the dominant. Further inland, 20–30 cm above the marsh margin, *Deschampsia caespitosa* becomes the dominant and frequently co-occurs with *Scirpus balticus*. Johannessen points out that these nearly circular "outliers" of vegetation are always indicative of rapidly expanding marshes. As more sediments accumulate and other species establish themselves, the patches may maintain their roughly circular form in otherwise homogenous associations. (Perhaps this observation explains the occurrence of large subcircular patches of *Salicornia, Distichlis,* and *Jaumea* in the low marsh at Westport, Grays Harbor.)

Tillamook Bay is the largest of Oregon's estuaries and drains four river basins. Tidal marshes on the deltas of these rivers have expanded rapidly (1867–1939: 4 m per yr.; 1939–1961: 3 m per yr.) in response to unusually high rates of sediment influx following extensive forest fires. Primary invaders, forming circular colonies on sandy mudflats, commonly include *Carex lyngbyei* and *Scirpus americanus; S. robustus* occurs on muddier sediments. Rapid increases in tideflat elevation, due to sediment influx, have allowed *Deschampsia caespitosa*, typical of mature high marshes, to colonize otherwise bare sand flats immediately adjacent to the primary invaders.

Johannessen's brief description of Alsea Estuary again includes *Triglochin maritima, Salicornia virginica,* and *Scirpus americanus* as primary invaders of the tideflats, the first being most characteristic of exposed situations and the second more typical of protected backwaters. *S. americanus* becomes dominant (60%) at intermediate elevations where it co-occurs with *S. virginica* (15%), *Distichlis spicata* (10%), *T. maritima* (5%), and *Deschampsia caespitosa* (10%). Either the latter or *Carex lyngbyei* are prevalent at higher levels.

Many of the marshes of the Umpqua Estuary are truncated by vertical banks 60–120 cm high. Where gentle slopes are present, *Scirpus americanus* is the usual primary colonist. *Distichlis spicata* appears 30 cm above the lowest plants and is still common at elevations of 60 cm when *S. americanus* disappears and *Agrostis alba* becomes the dominant. Higher still, *A. alba* and *Juncus balticus* each account for 40% of the cover, the remainder of which consists of *Potentilla egedii, Elymus triticoides, Carex lyngbyei,* Hordeum *brachyantherum, Deschampsia caespitosa,* and *Trifolium pratense* (probably introduced by grazing cattle). Considerable local variation is apparent for Johannessen describes other low-lying stands, less than 900 m distant, dominated by *Salicornia virginica* (55%) and *Jaumea carnosa* (35%). Additional species included *Scirpus cernuus, S. americanus,* and *Plantago maritima* in order of decreasing frequency. Forty-five to 50 cm above the lowest marsh plants, *Agrostis alba* became dominant (60%); *D. spicata* (20%), *J. carnosa* (10%), and other species, including *D. caespitosa*, completed the cover.

At Pony Slough, Coos Bay, Johannessen cites *Salicornia virginica* as the most abundant species; *Distichlis spicata, Carex lyngbyei, Triglochin maritima,* and *Deschampsia caespitosa* are also present in decreasing order of abundance. An accompanying data table (Johannessen 1961: Table 7.2) yields a slightly different picture, however, the average estimated species composition (as percentage, based on three stands) being: *Scirpus americanus* (40%), *Salicornia virginica* (28%), *Carex lyngbyei* (17%), *Triglochin maritima* (10%), *Deschampsia caespitosa* (5%), with a trace of *Agrostis alba*.

In the Lower Coquille Estuary the lower marsh included *Salicornia virginica, Scirpus americanus, Triglochin maritima,* and *Distichlis spicata* as the most widespread species; *Jaumea carnosa* was sometimes present but it also occurred at higher levels. *S. virginica* remained common at intermediate elevations where it was joined by *Carex lyngbyei* and *Agrostis alba* as co-dominants; *Deschampsia caespitosa* was the most abundant species of the highest marsh. In tabulating the average species composition of 20 stands of vegetation from the Coquille Estuary, Johannessen provides the first quantitative data describing plant succession in Pacific Coast salt marshes (Table 5).

The brief descriptions outlined above suggest that the salt marshes of Oregon form a relatively uniform group; in summarizing his finding, however, Johannessen (1961) provides additional insights into minor latitudinal variations. *Triglochin maritima* is the most distinctive primary colonist, occurring in exposed sites at the lowest elevation of any marsh plant in all six estuaries. *Salicornia virginica* is an important colonist of the lower, saltier flats in the south, but it is less important in the north where it more commonly stabilizes low banks. *Scirpus americanus* is also common at low elevations, particularly in the five southern estuaries, but at less exposed sites; *S. robustus* is a primary colonist on sand at Nehalem Bay and in mud at Tillamook. *Carex lyngbyei* prefers less saline habitats than either species of *Scirpus* and is commonly associated with their upper distributional margins; it may extend well out on tideflats along the banks of fresh-water streams. *Distichlis spicata* is rarely a primary colonist but it rapidly invades *S. virginica* stands developing at low elevations; like the latter it is very restricted at Nehalem and Tillamook Bays but is common at the more southerly sites. In the southern estuaries, *Jaumea carnosa* is an important subdominant at both intermediate and high elevations. Both the diversity and local variation among marsh floras increase at higher levels. *Deschampsia caespitosa* is the most characteristic and widespread high marsh species, and it commonly co-occurs with such species as *Potentilla egedii, Agrostis alba, Hordeum brachyantherum, Juncus balticus, J. effusus,* and many others.

Carol Jefferson (Oregon State University) is currently engaged in the most comprehensive study of salt marsh succession, and its relationship to such factors as marsh age, physiography, elevation, and salinity, yet attempted for Oregon. She has found that 95% of Oregon's salt marshes fall readily into one of three categories—"low marshes" (15%), dominated by *Triglochin maritima* and *Salicornia virginica*; "high marshes" (60%), that look like meadows, are dominated by *Deschampsia caespitosa*, and drop down a low cliff (60–90 cm) to barren tideflats; and "delta marshes" (20%) developed on river deltas and dominated by *Carex lyngbyei*.

Collecting between the Columbia River and Eureka, California, she has thus far documented a marsh flora of 57 vascular plants and 13 macro-algae. A gradual shift in species composition from north to south reflects the greater percentage of "high marshes" found in northern Oregon and some elevational species composition changes may reflect different successions into fresh-water swamp, coniferous forest, or sand dunes. *Carex lyngbyei, Scirpus acutus,* and *Juncus lesueurii*, for example, appear related to fresh-water influence, while transition to forest or dunes may include various of the following species: *Agrostis alba, Grindelia integrefolia, Potentilla egedii, Rumex* sp., *Plantago maritima,* and *Atriplex patula.*

Preliminary studies of tidal fluctuations and estuarine salinities have also produced interesting tentative conclusions. Maximum tidal heights within the Oregon estuaries are reached in winter when rainfall and run-off are heavy and salinities, particularly of surface waters, are relatively low (e.g., Yaquina Bay: Nov.–Feb., 2-15ppt; June–Sept., 12-36 ppt). Maximum flooding of the high marshes thus typically involves low salinity water and occurs during a period of little or no plant growth. During plant germination and sprouting (mid-March to late April), both tide levels and salinities remain low.

Conductivity determinations taken during low tide on subsurface soil-water samples from a typical, cliffed "high marsh" (Yaquina Bay, Feb.–May 1972 only) indicate a steady decline of average salinity from 21.6 ppt at the outer edge, to 15.5 ppt at the inner fringe of the marsh (personal communication—Carol Jefferson 1972).

California (56,500 acres*)

The salt marshes of northern California are only poorly known. At Lake Talawa (42°N), and Big Lagoon further south, coastal lagoons have been cut off from the ocean by barrier spits that are sometimes breached by storm tides. The sand spits are extensively colonized by *Salicornia* sp. and *Distichlis spicata*, and the lagoons are bordered with brackish marshes composed of *Juncus balticus, J. effusus, Carex lyngbyei, C. obnupta, Scirpus acutus, S. americanus,* and *Potentilla egedii* (Hehnke 1969).

Salt marshes developed at Humboldt Bay (41°N) have been briefly described by Macdonald (1967). Many of these marshes appear to have been adversely affected by the local lumber industry; some show signs of excessive pollution while others are extensively covered with bark chips and drifted logs. Some marshes have low cliff-like margins, while others reminiscent of Oregon's marshes are terraced, perhaps reflecting changing rates of sediment influx. At Samoa, *Spartina foliosa* appears for the first time as primary colonist of the outer mudflats. *Salicornia virginica* is also present on the low flats and is a primary invader in sheltered situations. At this site (May 1964), *Distichlis spicata* and *Jaumea carnosa* dominated the higher marsh terrace. *S. virginica, Triglochin maritima, Myosurus minimus, Limonium californicum,* and *Cuscuta salina* were also present in the high marsh, and *Juncus lesueurii* and *Potentilla egedii* appeared along its inner margins.

Near Arcata, further within the bay, firm mudflats were being colonized by circular patches of *S. virginica* up to a meter or more in diameter. The low marsh, apparently formed through coalescence of such stands, was composed almost entirely of this species, although scattered clumps of fucoid algae were also present. *Spartina foliosa* appeared at intermediate elevations. Instead of occurring in its usual form—a rather dense uniform meadow-like cover—it grew in compact, closely spaced tussocks (70—100 cm tall) occupying irregular areas of the middle marsh and completely surrounded by *Salicornia*. Occasional open spaces between the tussocks were colonized by patches of diminutive brown algae, straggly stems of *S. virginica, Jaumea carnosa,* and sometimes *Triglochin maritima*. The highest parts of the marsh were dominated by *Distichlis spicata* with *S. virginica* as co-dominant; a scattering of short specimens of *S. foliosa* were present and additional species included: *Atriplex patula, Cordylanthus maritimus, Cuscuta salina, Jaumea carnosa, Limonium californicum,* and *T. maritima*.

The California Department of Fish and Game has documented additional areas of marsh land at the mouths of several northern California rivers; however, their vegetation, probably strongly influenced by fresh-water run-off, still awaits study (John Speth, personal communication).

The great majority of California's present salt marsh acreage is developed

* 229 km^2 (Speth 1970) this is a generous maximum estimate; more than 55% of this acreage is in San Francisco Bay.

around San Francisco Bay and several smaller bays just to the north—Bodega, Tomales, Drake's, and Bolinas. Despite a wealth of general descriptions (e.g., Ferris 1970; Filice 1954; Howell 1970; Thomas 1961; Monlina and Rathbun 1967; Pestrong 1972; Williams and Monroe 1967) only Hinde (1954) and Barbour et al. (1972) have published quantitative data describing species distributions across these marshes.

Barbour et al. document zonation within a small fringing marsh developed on saline, sandy flats at the edge of Bodega Harbor ($38^{O}N$). Fifteen species of vascular plants, principally low, rhizomatous perennials, were present on the marsh that began abruptly at mean high tide level and rose gently inland to the highest high tide line (a vertical rise of some 39 cm in a horizontal distance of 50—60 m). Elevational changes in average cover for all plant species were measured in replicate 0.25 m^2 quadrats placed at 5 m intervals from mean high tide. These data (Table 6) suggest that different groups of species reach peak abundance at different distances from mean high tide. Additional species, absent from the quadrats, occurred along the inland fringe of the marsh, rarely submerged by tides; these included *Atriplex patula, Cakile maritima, Cotula coronopifolia, Cuscuta salina, C. subinclusa, Frankenia grandifolia, Oenanthe sarmentosa*, and *Potentilla egedii*.

Barbour et al. (1972) also measured ground water salinities beneath the marsh. Maximum salinities declined landward (June 1970: 5 m from outer edge marsh, 10—23 ppt; 45 m from outer edge, 2—10 ppt), rising and falling with the tide, but showing some lag effects. A ten-fold seasonal drop in salinity was recorded under the marsh center between October (11 ppt) and April (1 ppt) that conincided with winter rains; by July the ground water salinity had returned to its previous high.

A much larger marsh bordering Walker Creek near the entrance of Tomales Bay (Macdonald 1967) contained several of the same species as the Bodega marsh, but in different proportions. Where the marsh sloped gently to the adjacent mudflats (on south and east sides), scattered circular patches of *Salicornia virginica* suggested active colonization. Pure stands of this species also formed a dense, bushy cover over most of the lower parts of the marsh. Several additional species appeared at higher elevations (usually 30 cm above lowest *Salicornia*) to form a diverse assemblage in which *S. virginica* was usually still present, but *Triglochin maritima* was the dominant species and *Limonium californicum* was a common co-dominant. *Distichlis spicata* was widespread in this assemblage but usually accounted for less cover than any of the preceeding species. Additional species present included *Jaumea carnosa, Frankenia grandifolia, Myosurus minimus*, and *Spergularia marina* in decreasing order of frequency. The most distinctive characteristic of this diverse *Triglochin-Limonium* assemblage was that none of the plants present was more than 10 cm tall (Apr.—May 1965). No obvious explanation of this unusual stunting was apparent; possibly it resulted from poor soil conditions or the strong cold on-shore winds that frequently blow across this rather exposed site. On the highest parts of the marsh, bordering Walker Creek, scattered clumps of *Grindelia paludosa* were added to the assemblage and here *J. carnosa* was a common co-dominant.

A generally similar species distribution was developed at Millerton, a smaller, more protected site well within the bay. *D. spicata* was more abundant, and *L. californicum* and *T. maritima* were less common in the higher marsh than at Walker Creek. Distinctive additions to the flora included *Cotula coronopifolia* which was quite common, *Cuscuta salina*, *Glaux maritima* and *Spergularia marina*, all of which occurred in high marsh disturbed by cattle grazing.

Spartina foliosa, conspicuously absent from Tomales Bay, is listed by Ferris (1970) among the characteristic salt marsh species of Drake's Estero, Point Reyes Peninsula. Other common species here again include: *S. virginica*, *D. spicata*, *F. grandifolia*, and *L. californicum*. *C. coronopifolia*, *Potentilla egedii*, and *Lasthenia glabrata* may be present in the higher marsh or on the surrounding flats; *Carex lyngbyei* is present but rare. Similar marshes are developed in Bolinas Bay (Molina and Rathbun 1967) where three well-defined floral zones, indicated by *S. foliosa*, *S. virginica*, and *D. spicata*, succeed one another across the marsh and are sharply separated from supralittoral stabilized sand dunes.

Despite their considerable size and obvious ecologic importance, salt marshes of the San Francisco Bay complex are still poorly known. Floral lists provided by several authors imply variations in marsh vegetation around the Bay that probably reflect the wide range of environmental conditions represented. North and east of the Golden Gate, both the tidal range and efficiency of tidal flushing decline; salinities drop steadily towards the Sacramento-San Joaquin Delta (San Pablo Point 13—29 ppt, Martinez 1—21 ppt, Delta mouth 0—6 ppt), and sedimentation rates are high. The tidal range and flushing rates increase south of the Golden Gate, little run-off is received, and salinities generally range between 27—29 ppt; sedimentation rates are lower than in the north Bay area (Kelley 1966; Pestrong 1972). As with the Oregon marshes, man's early activities in the watershed, in this case extensive hydraulic mining for gold in the Sierra Nevada (late 19th century), tremendously increased sedimentary influx rates and led to extensive bay infilling and marsh accretion (Pestrong 1972).

Felice (1954) describes Castro Creek marsh, developed at about Mean Higher High Water (MHHW), near the entrance of San Pablo Bay. Low-lying tidal channels were bordered with *Spartina foliosa*, fresh-water rivulets by *Scirpus robustus*; in both cases these species were replaced at progressively higher elevations by *Salicornia virginica*, *Grindelia humilis*, and *Atriplex patula*. The highest parts of the marsh were covered by a diverse assemblage, dominated by *Distichlis spicata*, but also containing *Jaumea carnosa*, *Triglochin concinna*, *T. maritima*, *Frankenia grandifolia*, and *Limonium californicum*.

Hinde (1954), studying the marshes of south San Francisco Bay (Palo Alto), recognized two major vegetation zones. The outer perimeter of the marshes was occupied by a narrow (40 m wide) low-lying belt of *Spartina foliosa*. Sometimes this stopped abruptly at the edge of an actively retreating cliff (1.3 m high), but elsewhere the *Spartina* was colonizing the upper margins of the soft mudflats that lay below this cliff. Dense stands of *S. foliosa* also extended inland along the many low-lying tidal creeks that dissected the marsh surface. The areas above and between the tidal creeks were generally dominated by *Salicornia virginica*. Extensive pure stands of the latter were common, although other species such as *Cuscuta salina, Distichlis spicata, Frankenia grandifolia, Jaumea carnosa*, and to a lesser extent, *Limonium californicum* and *Triglochin maritima* occurred scattered among the *Salicornia*. In higher areas of the marsh and on peripheral dikes, built in 1936 to partially protect the area from tidal flooding, local pure stands and mixtures of *D. spicata, J. carnosa*, and *F. grandifolia* were developed. Data included in Hinde (1954) and recently extended by Pestrong (1972) permits tabulation of the vertical range of several important salt marsh species from the Palo Alto area, relative to tide levels (Table 7).

Hinde, comparing his results with those of Cooper (1926), also noted temporal changes in the marsh flora. *Plantago maritima* and *Triglochin maritima*, widespread in Cooper's time, were absent; *L. californicum* and *Grindelia paludosa* had considerably declined in abundance and reclamation activities had apparently permitted species, not recorded by Cooper (e.g., *Atriplex semibaccata, Chemopodium ambrosioides, Cotula coronopifolia*, and *Spergularia marina*), to invade the salt marsh fringes.

Lane (1969) describes a generally similar succession on Dumbarton Point marsh on the east side of the Bay; a significant addition was the presence of *Salicornia europaea*, not recorded by Hinde.

In comparing the salt marshes of the outer coast with those of San Francisco and San Pablo Bays, Howell (1970) notes that *Puccinellia lucida* and *Grindelia integrifolia* are present on the coast marshes but not in the Bay, while *Grindelia humilis* and *Salicornia bigelovii* are present in the Bay marshes but not along the coast. *Cordylanthus maritimus* is present in south San Francisco Bay, while at the less saline sites of the north Bay, *C. mollis, Scirpus acutus, Glaux maritima, Rosa californica*, and *Achillea borealis* are all represented. Howell (1970) goes on to provide the following summary of species distributions across the Bay Area marshes: "...*Spartina foliosa* is the conspicuous colonizer on the lowest tidal flats...It gives way to the usual salt marsh flora that includes the following: *Salicornia virginica, Frankenia grandifolia, Limonium californicum, Plantago maritima* and *Jaumea carnosa*. The *S. virginica* is the most widespread plant in the marshes. Associated with it along its lower border is *Triglochin concinna*, while the related *T. maritima* is more nearly restricted to the upper, less saline borders. *Rumex occidentalis*...and *Grindelia humilis* are also more often found in the upper reaches of the marshes..."

Howell also describes a saline or subsaline belt intermediate between the upper edge of the Bay salt marshes and adjacent grasslands. *Distichlis spicata* and *Spergularia* spp. are characteristic of this belt which also includes other salt-tolerant native and introduced species (e.g., *Cotula coronopifolia, Monerma cylindrica, Parapholis incurva, Potentilla egedii).*

The occurrence of *Suaeda californica* in the Bay marshes (Munz and Keck 1959, Thomas 1961) should also be noted. Here, at its northern distributional limit, the species is apparently uncommon for it is not mentioned in any of the floral lists cited above; further south it becomes an important member of the high marsh flora.

Only two extensive areas of salt marsh are developed along the 250 mile stretch of the Pacific Coast between San Francisco and Point Conception. The first of these is at Elkhorn Slough (37°N), the second at Morro Bay (35°N). The Elkhorn marshes mostly consist of monotonous, almost pure stands of *Salicornia virginica; Spartina* is conspicuously absent. Macdonald (1967) notes the following species frequencies among 45 random samples (circles, 50 cm diam.) taken throughout a 25-acre (0.1 km^2) marsh immediately south of Moss Landing Marine Lab: *S. virginica* (100%, present in every sample), *Jaumea carnosa* (27%), *F. grandifolia* (20%), and *D. spicata* (11%). Several additional species—*Cuscuta salina, Potentilla egedii, Spergularia marina,* and *Suaeda californica,* for example—occurred infrequently and were not recorded in the random samples. Multiple regression analysis confirmed that the abundance of *D. spicata* and *J. carnosa* exhibited a significant (p 0.05) positive correlation with elevation but that of *S. virginica* showed a negative correlation with elevation. This general description of the Elkhorn marshes is confirmed by Browning (1972) who indicates that *S. virginica* accounts for over 90% of the vegetative cover and cites the only additional fairly common species as *D. spicata* and *F. grandifolia.*

Hoover (1970) lists the following species from the Morro Bay salt marshes but does not include additional descriptive details: *Atriplex patula, Distichlis spicata, Frankenia grandifolia, Jaumea carnosa, Limonium californicum, Salicornia virginica,* and *Suaeda californica.* Again note the conspicuous absence of *Spartina*; a brief visit by Macdonald confirms the dominance of *S. virginica.*

Salt marshes developed at Goleta and Carpinteria, just south of Point Conception, are again dominated by extensive stands of *S. virginica* and lack *Spartina*; several new species not present on more northerly marshes now appear, however. Early records (Smith 1952) indicate that these sites contained very diverse marsh assemblages that included: *Triglochin concinna, T. maritima, Distichlis spicata, Monanthochloe littoralis, S. virginica, S. subterminalis, Suaeda california, S. depressa* (var. *erecta), Frankenia grandifolia, Limonium californicum, Cuscuta salina, Cordylanthus maritimus, Jaumea carnosa, Lasthenia glabrata,* and *Solidago occidentalis.* Still more species appeared in saline soils at the marsh peripheries or near local sources of fresh-water (e.g., *Atriplex patula, A. rosea, Cressa truxillensis, Cotula coronopifolia, Grindelia paludosa, Carex* sp., *Juncus* spp., *Scirpus* spp., etc.).

More recent studies (unpublished data, Havlik 1970, Spinner 1970) suggest that many of these species are now either rare or absent, probably victims of increasing urban encroachment.

Mugu Lagoon ($34°N$), protected within the U. S. Naval Missile Center, Point Mugu, contains extensive salt marshes that have been partially dredged but otherwise remain close to their natural state. Studies by Macdonald (1967, 1969) and Warme (1971) suggest that four distinct vegetation types, occupying successively higher parts of the marsh, can be distinguished. The lowest of these, dominated by *Spartina foliosa* and including *Salicornia bigelovii* and *S. virginica*, has been virtually eliminated by dredging. The majority of the marsh surface is occupied by "lower" and "upper" vegetation types that are usually rather abruptly separated (often by a beach ridge, low mud cliff, or sharp change of slope) at about Mean High Water (MHW). The species composition of the higher marsh changes with elevation as it grades into broad, barren salt flats separated by sinous strips of marsh vegetation. The "lower" vegetation type, consisting of a very open cover (5-20%) of short plants (10 cm), is developed on frequently submerged, wet ground. *Salicornia bigelovii* and straggly, prostrate specimens of *S. virginica* predominate; *Suaeda californica* (usually found higher up) and *Jaumea carnosa* are common associates and scattered occurrences of *Frankenia grandifolia, Monanthochloe littoralis* (again more common higher up), and *Triglochin maritima* were recorded (Macdonald, March 1966).

The "upper" vegetation type was a complex mosaic of highly variable composition; cover was generally dense (70-100%) and the plants 30-50 cm tall. *Salicornia virginica* was the most widespread species; it formed extensive pure stands and also co-occurred with many other species. At higher elevations and on several sandy beach ridges that cut across the marsh parallel with the present shoreline, dense, subcircular mats of *Distichlis spicata* and *Monanthochloe littoralis* up to 10 m in diameter were developed. *F. grandifolia, S. californica,* and *T. maritima* were all common, while *Jaumea carnosa* and *Limonium californicum*, as well as *Distichlis, Monanthochloe,* and *Cuscuta salina* were also widespread but somewhat less abundant. As the salt flats were approached, *S. virginica* was joined by *Atriplex* spp. and *Salicornia subterminalis. Batis maritima*, represented by scattered individuals at lower elevations, commonly developed in pure stands around the fringes of the salt flats.

The U. S. Navy has also protected the extensive salt marshes of Anaheim Bay (U. S. Naval Weapons Station, Seal Beach), just south of Los Angeles. The marshes contain prime stands of *Spartina foliosa* that pass upward into a *Salicornia*-dominated marsh much like that at Mugu Lagoon. No detailed account of these marshes has yet been published but they are currently under study by faculty and students from California State College, Long Beach. Very similar marshes were previously developed at Bolsa Chica Bay close by. These were removed from tidal circulation in the 1890's and a complete ecological succession has ensued (Macdonald et al. 1971). *Spartina* has all but disappeared, its place taken in areas receiving some seasonal fresh-water supply by extensive stands of *S. virginica* (50% cover), or, where salt water intrusion apparently produces higher ground water salinities, by *S. subterminalis* and *Mesembryanthemum nodiflorum* (25% cover). Henrickson (unpublished data, 1971) has recorded a flora of 168 species from this site of which 70 are introduced forms.

The marshes of Newport Bay, slightly to the south, have been the subject of excellent studies by Stevenson and Emery (1958) and Vogl (1966). The marsh described by Stevenson and Emery extended between 0.61 m and 1.34 m above MLLW and was backed by steep bluffs. Mean Higher High Water (MHHW) was estimated to be 1.49 m and the highest parts of the marsh were submerged for short periods on an average of five times each month; no salt flats were developed. Three major floral communities were described and mapped, each occupying successively higher elevations across the marsh. The lowest (0.61-0.88 m above MLLW) was dominated by *Spartina foliosa*, the primary colonist of the muddy tidal flats. Small scattered specimens of *Salicornia virginica* co-occurred with *Spartina* in the upper 5 cm of its vertical range. The second community was restricted to a narrow vertical range (0.88-1.07 m above MLLW) centered close to Mean High Water (MHW). Low, prostrate plants of *S. virginica* dominated the community, and *S. bigelovii* and *Frankenia grandifolia* were both common. The first and last of these species were also common in the third community (1.07-1.34 m above MLLW) but both grew considerably taller in the higher zone. The most distinctive species in this third community was *Suaeda california*; *S. virginica*, *S. subterminalis*, and *Distichlis spicata* were all abundant and locally dominant. Other species scattered through the community included: *Triglochin maritima, Limonium californicum, Batis maritima, Cordylanthus maritimus,* and *Cuscuta salina*. Several large, shallow depressions in the high marsh that held water for a few days following tidal flooding were carpeted with nearly pure stands of *Monanthochloe littoralis*. Sandy alluvial fans extending from the base of the bluffs onto the high marsh were covered with dense stands of *D. spicata*. These descriptions are strikingly similar to those for the marshes of both Mugu Lagoon and Mission Bay (Macdonald 1967) further south; the species lists are also almost identical.

Vogl (1966), working on marshes further inside Newport Bay, provides quantitative data describing the nature of the vegetation. Six study areas (each 10 acres) were each divided into four zones: lower littoral—completely or partially submerged each day; middle littoral—only inundated by spring tides; upper littoral—only flooded by unusually high tides; and maritime—the ecotone between littoral and upland vegetation which was never flooded. The frequency and cover of every species in each zone of the six study areas were then sampled by laying out 100 random, 25 cm square quadrats. The percentage cover data are reproduced in Table 8. In addition to the species listed, the following occasionals were noted: middle littoral—*Spergularia marina*; upper littoral—*Jaumea carnosa* and *Mesembryanthemum chilense*;

maritime bluffs—*Atriplex watsonii, A. canescens,* and several others. Interestingly, *S. bigelovii* was not recorded. Vogl notes that while each dominant and subdominant species reaches its peak abundance in a particular "zone", each also occurs in other zones subject to very different environmental conditions. He concludes as follows: "Although the zonal approach (to the classification of salt marsh vegetation) was useful and convenient in this study, the zones did not exhibit discrete boundaries. Instead the floristic composition gradually changed along existing environmental gradients and the plant species of the salt marsh followed a vegetation continuum (Curtis and McIntosh, 1951)..."

Continuing south, Purer, (1942) describes the marshes of San Diego County. Her paper remains a classic for not only do her descriptions predate much of San Diego's coastal development, (Envir. Task Force 1970) but she also includes the only anatomical studies of Pacific Coast salt marsh species published to date. She summarizes the general composition of the salt marshes by listing the follwing very abundant or frequent genera:

Lower littoral (daily tidal coverage)—*Batis, Salicornia, Spartina*, Triglochin,* and several algae; Middle littoral (spring tide coverage)—*Atriplex, Cordylanthus, Cuscuta, Distichlis*, Cyperus, Eleocharis, Salicornia*, Scirpus,* and *Suaeda*;* Upper littoral (occasional storm coverage)—*Atriplex*, Cressa, Distichlis*, Frankenia*, Heliotropium, Jaumea, Limonium*, Monanthochloe*, Pluchea, Spergularia,* and *Suaeda* (* indicate principal genera).*

More recently through the work of Bradshaw (1968), Macdonald (1967), and particularly Mudie (1969, 1970), several quantitative studies of these San Diego County marshes have been completed. McIlwee (1970) has also provided a more complete description of the marshes at Tijuana Slough on the Mexican border. Floral lists from several of these studies are included in Table 4.

It is sufficient here to say that the bays and lagoons containing these marshes fall into two groups, those under the continuous influence of ocean tides (Mission Bay, San Diego Bay, Tijuana Slough) and those which for variable periods are cut off from the ocean by barrier beaches (e.g., Santa Margarita River, Los Penasquitos Lagoon). The marshes under continual tidal influence all share species compositions and a floral zonation generally similar to that already described for Newport Bay, further north. As an example, values for the average percentage cover of common species in South San Diego Bay's "low" and "high" marshes (Mudie 1970) are shown in Table 9.

The marshes subject to seasonal tidal influence exhibit quite different salinity regimes that reflect the local fresh-water supply (Carpelan 1969), and their floras are more variable. Although they frequently contain species characteristic of the upper littoral zone (see above), they usually lack species normally found in the lower littoral, and many weedy introduced forms may be present. Los Penasquitos (Bradshaw 1968), for example, contains three floral associations: a *S. virginica* association subject to seasonal, daily tidal flooding; a *D. spicata* association covered only during high spring tides; and a *S. subterminalis* association that may be partly flooded by storm tides but in some places extends well into the maritime zone. Although a flora of 106 species was recorded from this site, several conspicuous marsh species are absent: *A. watsonii, B. maritima, S. bigelovii, S. foliosa,* and *T. maritima.* *Spartina* was previously present (Purer 1942) but it apparently died out during long periods of lagoon closure and high salinities; *Salicornia europaea*, rather rare in Southern California, is present.

Mudie (1970) provides valuable insights into the salinity regimes characteristic of these lower latitude ($32\frac{1}{2}°N$) marshes. The flood tide brings bay waters 35-40 ppt. into the marsh creeks; salinities rise sharply during ebb flow and may reach 55 ppt. Evaporation is high and isolated pools within the marshes are much saltier (57-115 ppt) than the tidal creeks. Soil-paste extracts indicate that "low marsh" (MLHW–EHW) soil salinities vary considerably over short distances (10 m) but generally increase with soil moisture content and decrease with elevation. Soil moisture shows a positive correlation with peat and sand content, however, there is generally 50% decrease in moisture from low marsh to high marsh areas. Salinity values of 20-78 ppt. were recorded in the low marsh, versus 25-31 ppt. under less variable high marsh conditions. The highest values (78-87 ppt.) were recorded beneath high marsh salt flats (the measuring technique probably underestimates in situ values for these dry soils, Mudie 1970). Mudie also found evidence of seasonal salinity fluctuations in the top 15 cm of the marsh sediments and noted that soil salinities dropped off sharply at the littoral salt marsh—maritime sagebrush transition. She concluded that there were few clear-cut relationships between plant distributions and soil salinities but that the combined factors of tidal submergence, soil moisture (which also reflect sediment composition and oxygen tensions), and salinities probably control vegetation patterns.

Baja California (salt marsh acreage unknown)

Coastal salt marshes developed along the Baja Peninsula have, until recently, generally remained undisturbed by man. Expanding salt production facilities and the increasing threat of "resort development," however, suggest that this situation will change. Perhaps because of this threat, scientific interest in Baja's undisturbed salt marshes has recently increased considerably. The few comments included in Shreve and Wiggin's (1964) Sonoran Desert Flora and brief descriptions of San Quintin (Dawson 1962) and Ojo de Liebre (Phleger and Ewing 1962) marshes have more recently been supplemented by descriptive and quantitative studies by Macdonald (1967, 1969, and unpublished data), Neuenschwander (1972), and Thorsted (1972). Earlier this year Peta Mudie led an expedition from Scripps Institute of Oceanography to examine the marshes south of Laguna San Ignacio

(27°N).

The salt marshes at Estero de Punta Banda just south of Ensenada (32°N), although increasingly disturbed by man, still exhibit a lower frequently flooded zone dominated by *Spartina foliosa*, succeeded by a drier *Salicornia* marsh similar to those of San Diego County.

Bahia de San Quintin (30½°N) contains several rather extensive salt marshes. Studies by Dawson (1962), Macdonald (unpublished data) and Thorsted (1972) all indicate that these too are closely similar to the marshes of southern California. Thorsted (1972), using the quantitative methods of Vogl (1966), divided seven separate vegetation stands between lower (daily tidal flooding), middle (spring tide flooding) and upper (occasional storm tide flooding) littoral zones and used 25 cm square quadrats to sample species frequency and cover. Fifteen species were encountered on the marsh, all but one of which (*Mesembryanthemum nodiflorum*) were indigenous—a marked contrast to the many introduced species of the more disturbed American marshes—and only eight of which were common. Average percentage cover data (reproduced in Table 10) indicated that three distinctive vegetation types dominated by *Spartina*, *S. virginica* and *Monanthochloe*, respectively, occupied the lower, middle, and upper littoral. Interstand comparisons revealed that the lower littoral vegetation was the most uniform and that interstand variation increased with elevation as did species diversity. Thorsted confirms Vogl's view that salt marsh species follow a "vegetation continuum" although his descriptions may suggest differing rates of vegetation change along the environmental gradients. His comparisons with Vogl's Newport Bay data (compare Tables 10 and 8) indicate that *Distichlis* is absent, while *M. littoralis* and *S. subterminalis* (characteristic of dry habitats) are more abundant at San Quintin. A greater degree of vegetation stratification (cover ranged 78-126%) was apparent at the southern site and the lower littoral vegetation was considerably more diverse.

In a closely related study, Neuenschwander (1972) found that the salt marsh vegetation extended inland a maximum of 30 m above the high tide debris line (cf. maritime zone of Vogl 1966). The transition flora included 33 species of perennial angiosperms, 12 characteristic of the salt marshes, and 21 normally associated with upland vegetation. *Monanthochloe* (21%) and *S. subterminalis* (20%) formed the majority of the cover in the lower part of the transition, while three upland species—*Frankenia palmeri* (17%), *Lycium brevipes* (3%), and *Atriplex julacea* (3%)—accounted for most of the cover in the upper part (average total plant cover, 74%). It is noteworthy that all three upland species are succulent halophytes. The width of the transition zone was a function of slope which varied from 2% to 45%. Neuenschwander noted that the salt marsh species cover was rather uniform while the upland species exhibited different substrate preferences and variable cover. Species diversity increased with elevation but frequency and cover declined. He concluded that the lower portion of the transition represented a capillary fringe around the Bay, containing sub-irrigated salt marsh species, while the upper part contained upland species controlled by adjacent desert conditions.

Further south the salt marshes of Laguna Guerrero Negro (28°N) have been briefly described by Macdonald (1967). Additional unpublished data indicate a species composition and zonal distribution very like that described by Thorsted (1972) for San Quintin. Three distinct floral assemblages clearly dominated by *Spartina foliosa*, *Batis*, and *S. bigelovii*, and *M. littoralis*, respectively, again succeeded one another with increasing elevation. Fifty stratified random samples (circles, 0.5 m diam.) distributed among the three floral assemblages in proportion to their relative areal extents yielded the following overall species frequencies: *B. maritima* (62%), *S. bigelovii* (60%), *S. virginica* (54%), *S. foliosa* (24%), *S. subterminalis* (14%), *M. littoralis* (14%), *S. california* (8%), *F. grandifolia* (2%), and *Triglochin* sp.(2%). Additional species not included in the samples were: *L. californicum*, an extensive local stand of *Sesuvium* sp., and *F. palmeri*, present on the salt marsh fringes (Apr. 1966).

The marshes lying below 28°N still await description. Wiggins (written communication to Barbour 1972) indicates that several species prominent to the north may not extend much below 28°N. These include *F. grandifolia*, *F. palmeri*, *L. californicum*, and *T. maritima*. Mobberley (1956) indicates that *S. foliosa* extends further south and is present at both Laguna San Ignacio and Bahia de la Magdalena (25°N). This last site also contains mangroves and mangrove-like shrubs: *Rhizophora mangle*, *Laguncularia racemosa*, *Conocarpus erecta*, and *Avicennia germinans*. These apparently co-occur with *Spartina* as in the marshes at Tabasco on the Gulf of Mexico (Thom 1967). Many of the remaining salt marsh species—*Batis*, *C. salina*, *Monanthochloe*, *S. europaea*, *S. virginica*, *S. subterminalis*, *Sesuvium*, and *S. californica*—certainly extend to Cabo San Lucas and some go well beyond (Wiggins, op. cit.).

The following additional comments on the salt marshes of southern Baja California were generously provided by Peta Mudie (personal communication 1972) following her reconnaissance expedition south:

"Recently a taxonomic and ecological survey of the halophytic flora of Baja California commenced with preliminary studies of the salt marsh and mangrove vegetation in the coastal lagoons and estuaries on the Pacific Coast between Bahia Asuncion (27°05'N, 114° 58'W) and Puerto Chale (24° 30'N, 112° 26'W).

"The lagoons in this region are beach barrier formations (similar in physiography to the Guerrero Negro-Manuela lagoon complex) in which fine sandy sediments predominate. Somewhat hypersaline conditions probably prevail in the lagoons during much of the year; however, at the time of this study (mid-April to early May) the salinities in the inner lagoons were approximately those of the open coast (34-35 ppt). Estuaries occur in the drowned lower valleys of Arroyo San Juan and Rio de la Purisima (north of La Bocana); small beach barriers partially close the entrances. Sediments in the middle and upper estuarine reaches are predominantly clayey silts.

Hypersaline conditions prevail in the estuaries in April and fresh-water influence is probably confined to periods of sporadic river flooding (summer?); evidence of occasional massive flooding was seen at the La Purisima estuary. Diurnal water temperatures in both lagoons and estuaries at the time of study ranged from 16-25°C.

"The general features of the lagoon vegetation are a "low marsh" formation of mangroves and *Spartina foliosa* up to about MHHW, superceded by a "high marsh" formation of salt marsh species in which *Batis maritima, Salicornia virginica* (short form), *S. bigelovii,* or *Monanthochloe littoralis* are dominant, with an abrupt transition at EHW to a zone characterized by *Allenrolfea occidentalis, Salicornia subterminalis,* and/or *Tricerma* (= *Maytenus) phyllanthoides*. A similar zonation occurs in the estuaries; however, mangroves (other than a few *Rhizophora* seedlings) are conspicuously absent from the estuarine low marsh.

Dense beds of *Zostera marina* occur in most of the lagoons, but are apparently absent from the estuaries. Shallow open ponds within the salt marshes and mangroves are frequently colonized by *Ruppia maritima*.

"Low marsh formation in the lagoons appears to commence with the stabilization of low sand bars by *Spartina*, followed by the establishment of *Laguncularia* or *Rhizophora* seedlings within the *Spartina* stands. Later the mangroves shade out the *Spartina* but this species frequently persists on aggrading channel banks in front of the mangroves. The composition and physiognomy of the mangroves changes considerably with latitude.

"In the most wind-exposed northern lagoon (Pond Lagoon, west of Punta Abreojos, at 26° 45'N), dense scrubby *Laguncularia racemosa* (0.7–1.0 m high) comprises 90% of the mangrove association with *Rhizophora mangle* (1.0–1.3 m high) occurring only as isolated stands within the *Laguncularia* scrub or on wind-sheltered channel banks. *Rhizophora* increases in height (as does *Laguncularia*) and frequency in Estero del Coyote and in Estero de la Laguna San Ignacio (southeast of San Ignacio Lagoon), and becomes the dominant mangrove species from the northern end of Laguna Santo Domingo southwards. Towards the southern end of their local range (around Puerto Chale at 24° 34'N), both species are 4–7 m high along the channels, becoming scrubbier with increasing elevation. *Avicennia germinans* apparently only joins the mangrove association from the north end of Laguna Santo Domingo southwards. It appears to be a colonizer on active sand banks within the mangrove formation and it may become dominant (as a low growing shrub) in poorly drained low areas around MHHW. A superficial survey of the mangrove formations on the Gulf coast of Baja California suggests that this species is much more prevalent in the warmer Gulf waters.

"The high marsh vegetation in both the lagoons and estuaries is similar in structure and general appearance to that in northern Baja California lagoons. However, the salt marshes are floristically simpler than in the north: *Limonium, Frankenia grandifolia, Triglochin,* and *Jaumea* are absent, and

Distichlis spicata ssp. *stricta* is rare. *Sesuvium portulacastrum* occurs in the high marsh and on sandy levees within the low marsh from Pond Lagoon southwards.

"Preliminary results of soil water salinities (estimated from electrical conductivity) indicate that the salinity in the low marsh remains close to that of seawater and is 0.5 to 0.75 times higher in most areas of the high marsh. The soils in poorly drained pan areas within the salt marsh (characterized by *Batis* and *Salicornia bigelovii*) and on salt flats are typically 2.5–3 times more saline than the low marsh.

"Where the marshes are backed by dunes, the EHW level is demarcated by the presence of *Allenrolfea, Salicornia subterminalis, Frankenia palmeri,* and *Atriplex* spp. In many of the northerly areas, the EHW level is characterized by extensive highly saline mudflats, devoid of vegetation other than scattered *Allenrolfea*. From Arroyo San Juan southwards, low sandy plains occur behind the marshes; these are sparsely covered by a characteristic assemblage of *Salicornia subterminalis, Suaeda* cf. *taxifolia, Tricerma phyllanthoides, Haplopappus venetus* ssp. *furfurascens, Atriplex barclayana,* and *Sporobolus contractus*."

Nichol's (1964) study of lagoons on the Sonoran Coast (29°N) should also be mentioned here, for it appears to be the only study to date that describes the salt marshes developed well within the Gulf of California. Here mangroves, *Salicornia,* and *Batis* dominate the lower littoral zone, while the upper littoral is dominated by *Monanthochloe* and *Allenrolfea,* and also contains *Distichlis, F. palmeri,* and *Suaeda.* Plant cover declines drastically inland as the salt marshes pass up into barren flats whose high salt content precludes plant growth.

Conclusions

Some 57 species of vascular plants appear to be most characteristic of the beaches from Point Barrow, Alaska, to the tip of Baja California, a latitudinal spread of 48° which extends through several major types of climate. Eight species range over 27° of latitude or more.

Beach vegetation is usually scanty at any given location, but the species do occupy several distinct zones. *Zostera marina, Phyllospadix species,* and *Puccinellia* species occur in the intertidal zone, especially on rocky substrates along the central portion of the coast. Inland from the high tide line is a low beach which may be bare of all plants. Behind this is a high beach with sand hummocks created by plant roots or with some stability created by driftwood (driftwood in the north often consists of large tree trunks so the degree of shelter provided can be great); this zone is dominated by a variety of prostrate or half-buried herbs. Further inland is either a grass-dominated zone, a foredune, a bluff, or a gradual transition zone to typically inland vegetation depending on the particular location.

In the north, the low and high beach typically support *Senecio pseudo-arnica, Mertensia maritima, Honckenya peploides, Lathyrus maritimus, Glehnia littoralis,* and *Cakile edentula.* Shingle beaches may support species more typical of rocky bluffs further south rather than of beach per se (such as *Sagina crassicaulis, Plantago maritima,* and *Vicia gigantea*). The grass zone further inland is dominated by *Elymus arenarius.*

In the central portion of the coast a number of new herbaceous species come to dominate the high beach: *Cakile maritima, Convolvulus soldanella, Abronia latifloria, Mesembryanthemum chilense, Oenothera cheiranthifolia,* and *Ambrosia chamissonis.* The low beach is usually unpopulated, and even the high beach may be very depauperate. If a foredune exists behind the beach, it is dominated by the introduced grass, *Ammophila arenaria*; many of the beaches are narrow, however, and abut onto steep bluffs.

In the south the grass zone is absent altogether, and the vegetation progresses back through hillocks of increasing size and stability until a dune scrub community is reached. Important pioneer beach plants include: *Abronia maritima, A. umbellata, Mesembryanthemum crystallinum, Heliotropium curassavicum,* and — on muddy substrate in the extreme south — the mangroves *Rhizophora mangle, Laguncularia racemosa,* and *Avicennia germinans.*

Ecological studies of Pacific Coast beach vegetation are few, and almost all of them have been conducted in California. Kumler (1969) and Williams and Potter (1972) have examined the sequence of plant succession from beach through scrub in central Oregon and south-central California, respectively, but they did no autecological work. (Kumler did try germination experiments on some beach species, but he got very low germination rates and concentrated instead on species of later successional stages.) Martin and Clements (1939) examined growth of and micro-environment of several beach species at Santa Barbara, but the field stations were actually in the dunes rather than on the beach itself. Purer (1934, 1936) did examine beach species on the beach, but her autecological data emphasizes micro-environmental, morphological, and anatomical observations, and her approach was not experimental. Apparently the only experimental work done on a Pacific Coast beach plant is that done by Barbour (1970b, 1970c, and 1972b) on *Cakile maritima* — which, unfortunately, is not even native to the Pacific Coast! Extrapolating information from *Cakile* to other beach plants may be as tenuous as extrapolating from a weed to an endemic dominant: a very tenuous operation.

For a final comment on the beach vegetation, it is striking to us that so many introduced species have spread so far and contributed so significantly to the vegetation: several *Mesembryanthemum* species, two *Cakile* species, and *Ammophila arenaria.* Why were they able to do so well on the Pacific Coast? Were there more unoccupied niches (if such things do really exist) on this coast than on others? Will further introductions be able to do as well, or is the habitat space now filled? In this connection, it may be ominous to recall the statement by Calder and Taylor (1968) that *Lolium perenne* has been introduced to beaches in the Queen Charlotte Islands for the purpose of erosion control. Already at Sandspit it "...is now one of the dominant grasses in this region."

The sources describing Pacific Coast "salt marsh habitats" examined during this review cited over 140 species of vascular plants and a dozen macro-algae as characteristic vegetational components. Habitat descriptions were disturbingly vague and details of tidal range, salinity regime, etc., were usually omitted. Many of the species cited were either maritime or inland forms. Over 35% of the 140 species were noted from single locations and only 35–40 of them occurred at five or more of the localities shown in Fig. 4.

Species range data indicate that the composition of the salt marsh vegetation changes rather gradually with latitude as previously recorded species drop out and new ones appear. Despite this, there are groups of salt marsh plants that rather regularly co-occur over limited stretches of the coast. *Puccinellia phryganodes, Calamagrostis, deschampsioides*, and several species of *Carex* are restricted to the Alaskan marshes. A more diverse group including *Carex lygnbyei, Deschampsia caespitosa, Distichlis spicata, Glaux maritima, Jaumea carnosa, Juncus lesueurii, Salicornia europaea, S. virginica, Plantago maritima* ssp. *juncoides, Potentilla egedii* var. *grandis, Scirpus americus, S. robustus, Triglochin maritima,* and the parasite *Cuscuta salina,* together with introduced forms such as *Atriplex patula* and *Cotula coronopifolia,* characterizes the marshes of British Columbia, Washington, and Oregon. Several of these species persist to central California or the Mexican Border. At Cape Blanco, San Francisco Bay, and Point Conception several new additions to the marsh vegetation appear and members of the preceding, more northerly group drop out. The new additions, listed in order of their northern range limits, include *Spartina foliosa, Limonium californicum, Frankenia grandifolia, Atriplex semibaccata* (another introduced species), *Suaeda californica, Salicornia bigelovii, Cressa truxillensis* (usually a maritime rather than littoral species), *Salicornia subterminalis, Monanthochloe littoralis,* and *Batis maritima.* All of these species persist well down into Baja California and several represent tropical floristic elements that continue down into Central and South America. Several species of mangroves, capable of colonizing considerably lower levels of the intertidal flats than the more northerly salt marsh plants, appear at Laguna San Ingacio and Bahia de la Magdalena and gain in importance southward (*Avicennia germinans, Laguncularia racemosa,* and *Rhizophora mangle*).

In addition to the latitudinal changes in vegetation composition, the species represented at each locality characteristically occupy discrete vertical zones. Quantitative data indicate that while each species may reach its maximum abundance over a limited elevational range, it will also occur in lesser abundances in other elevational zones under different environmental conditions. As Vogl (1966) points out, if each species reaches its peak abundance at a different elevation, then the floristic composition may change gradually across the marsh to produce a vegetation continuum. Alternatively, if several species reach their peak abundance at similar elevations, then much more discrete "zones" of differing vegetation types can result. Although conclusive quantitative data are still lacking it is Macdonald's impression (1967) that the more northerly, less saline marshes of Oregon and Washington tend to follow vegetation continua while those developed under the drier climates of Southern and Baja California tend to exhibit more discrete "zones" of distinctive vegetation in which several species reach their peak abundances together. There is also some evidence to suggest that while the more northerly marshes may be separated into "high" and "low" marsh vegetation types, the more southerly sites may contain a third distinctive vegetation type (usually with *S. bigelovii, S. virginica* (low form), *Batis*, and *F. grandifolia*) that is restricted to intermediate elevations. If these suggestions are confirmed by future quantitative studies, then an increasing partitioning of upper intertidal habitats at lower latitudes is indicated. Macdonald (1967) has further suggested that such differences may reflect the soil-water salinity regimes developed under the near-desert conditions of southern and Baja California which appear to be very different from those developed to the north.

Perhaps as expected, it is the lower-most zone of salt marsh vegetation, subject to regular tidal submergence, that exhibits the greatest degree of similarity between widely separated sites. Only a few suitably adapted species are present and each characteristically occurs over a broad latitudinal range. *P. phryganodes* is the primary invader of intertidal flats in the Alaskan marshes; *T. maritima, S. virginica*, and *Scirpus* sp. are important colonists along Canadian shores as well as in Washington and Oregon. *Spartina foliosa* starts the succession at Humboldt Bay, California, and persists southward to Laguna San Ignacio and Bahia de la Magdalena where mangroves occur early in the succession. Increasing numbers of species appear in the salt marsh vegetation at high elevations and as subaerial conditions begin to predominate over tidal influences, local factors and the historical background of each site increasingly modify the vegetation.

Although the soil-water salinity regimes of some of the marshes in Oregon and California have been studied, much too little is known of these and other factors that control the distribution patterns of salt marsh plants along the Pacific Coast. Hopefully, future workers will also pay more attention to the topology of salt marsh surfaces and their relationships to local tidal regimes. In the more remote areas of Alaska and British Columbia, even Washington, simple descriptive floristic studies are still badly needed and there is certainly plenty of scope for all sorts of quantitative distributional studies! Finally, as Barbour stated for the beach vegetation, the very rewarding areas of experimental culturing studies have barely started.

Literature Cited

Amundsen, C. C., and E. E. C. Clebsch. 1971. Dynamics of the terrestrial ecosystem vegetation of Amchitka Island, Alaska. BioScience 21:619–623.

Arias, A. C. 1942. Mapa de las provincias climatologicas de la republica Mexicana. Secretaria de Agricultura y Fomento, Inst. Geogr., Mexico. 54 pp.

Armstrong, R. H. 1971 Physical climatology of Amchitka Island, Alaska. BioScience 21:607–609.

Barbour, M. G. 1970a. The flora and plant communities of Bodega Head, California. Madroño 20:289–313.

——————. 1970b. Germination and early growth of the strand plant *Cakile maritima*. Torrey Bot. Club 97:13–22.

——————. 1970c. Seedling ecology of *Cakile maritima* along the California coast. Bull. Torrey Bot. Club 97:280–289.

——————. 1970d. Is any angiosperm an obligate halophyte? Amer. Midl. Natur. 84:105–120.

——————. 1972a. Additions and corrections to the flora of Bodega Head, California. Bull. Torrey Bot. Club (in press).

—————— and J. E. Rodman. 1970. Saga of the west coast sea-rockets: *Cakile edentula* ssp. *californica* and *C. maritima*. Rhodora 72:370–386.

Barbour, M. G., R. B. Craig, and F. R. Drysdale. 1972. Bodega Head: coastal ecology. Univ. California Press, Berkeley (in press).

Black, R. F. 1969. Geology, especially geomorphology, of northern Alaska. Arctic 22:283–299.

Boyce, S. G. 1954. The salt spray community. Ecol. Monogr. 24:29–67.

Boughey, A. S. 1968. A checklist of Orange County flowering plants. Museum of Systematic Biology, Univ. California, Irvine. 89 pp.

Bradshaw, J. S. 1968. Report on the biological and ecological relationships in the Los Penasquitos Lagoon and salt marsh area of the Torrey Pines State Reserve. Calif. State Div. Beaches and Parks Contract 4-05094-033. 113 pp.

Brandegee, T. S. 1889. A collection of plants from Baja California. Proc. California Acad. Sci., 2nd series, 2:117–216.

Browning, B. M. 1970. The natural resources of Elkhorn Slough, their present and future use. Calif. State Dept. Fish and Game, Coastland Wetlands Series 4. 105 pp. plus appendices.

Calder, J. A., and R. L. Taylor. 1968. Flora of the Queen Charlotte Islands, part 1. Systematics of the vascular plants. Canada Dept. Agriculture Monogr. 4(1), Ottawa. 659 pp.

Carpelan, L. H. 1969. Physical characteristics of Southern California coastal lagoons. Pages 319–334 *in* Coastal lagoons, a symposium (UNAM–UNESCO), ed. by A. A. Castanares and F. B. Phleger, Instituto de Biologia, Universidad Nacional Autonoma de Mexico, Mexico City. 687 pp.

Chapman, V. J. 1960. Salt marshes and salt deserts of the world. Interscience. New York. 392 pp.

——————. 1964. Coastal vegetation. MacMillan, N. Y. 245 pp.

Clark, J. 1967. Fish and man, conflict in the Atlantic estuaries. American Littoral Soc. Spec. Pub. 5. 78 pp.

Climatological data, United States by sections. 1968. U. S. Dept. Commerce, Washington, D. C.

Cooper, W. S. 1931. A third expedition to Glacier Bay, Alaska. Ecology 12:61–95.

―――――――――. 1936. The strand and dune flora of the Pacific Coast of North America: a geographic study. Pages 141–187 *in* Essays in geobotany, ed. by T. H. Goodspeed, Univ. California Press, Berkeley. 319 pp.

Curtis, J. T., and R. P. McIntosh. 1951. An upland forest continuum in the prairie-forest border region of Wisconsin. Ecology 32:476–496.

Dawson, E. Y. 1962. Benthic marine exploration of Bahia de San Quintin, Baja California, 1960–61: Marine and marsh vegetation. Pacific Natur. 3:275–280.

Doty, M. S. 1946. Critical tide factors that are correlated with the vertical distribution of marine algae and other organisms along the Pacific Coast. Ecology 27:315–328.

Eber, L. E., J. F. T. Saur, and O. E. Sette. 1968. Monthly mean charts of sea surface temperature, North Pacific Ocean, 1949–62. U. S. Dept. Interior, Circ. 258, Washington, D. C. (not paged).

Environmental Task Force. 1970. The coastal lagoons of San Diego County, County of San Diego. 193 pp.

Ferris, R. S. 1970. Flowers of Point Reyes National Seashore. Univ. California Press. 119 pp.

Filice, F. P. 1954. An ecological survey of the Castro Creek area in San Pablo Bay. Wasmann J. Biol. 12:1–24.

Hambridge, G. (ed.) 1941. Climate and man. Yearbook of Agriculture, U. S. Dept. Agriculture, Washington, D. C. 1248 pp.

Hanson, H. C. 1953. Vegetation types in northwestern Alaska and comparisons with communities in other Arctic regions. Ecology 34:111–140.

―――――――――. 1951. Characteristics of some grassland, marsh, and other plant communities in western Alaska. Ecol. Monogr. 21:317–375.

Havlik, N. 1970. A partial flora of the bluffs and adjacent areas of the main campus, University of California, Santa Barbara. Unpublished report. 7 pp.

Hehnke, M. 1969. Bay and Estuary Report: Big Lagoon, Stone Lagoon, Freshwater Lagoon, Lake Earl, Lake Talawa. Calif. State Dept. Fish and Game Rpt.

Heusser, C. J. 1960. Late-Pleistocene environments of North Pacific North America. Amer. Geog. Soc. Spec. Pub. 35. 308 pp.

Higgins, E. B. 1956. Coastal plants of San Diego. San Diego Soc. Nat. Hist., San Diego. 16 pp.

Hinde, H. P. 1954. Vertical distribution of salt marsh phanerogams in relation to tide levels. Ecol. Monogr. 24:209–225.

Hoover, R. F. 1970. The vascular plants of San Luis Obispo County, California. Univ. California Press, Berkeley. 350 pp.

House, H. D. 1914. Vegetation of the Coos Bay Region, Oregon. Muhlenbergia. 9:81–100.

Howell, J. T. 1937. Marin flora. Univ. California Press, Berkeley. 322 pp. (1970. 2nd Edition with supplement. 366 pp.)

Hulten, E. 1937. Flora of the Aleutian Islands. Bokforlags Aktiebolaget Thule, Stockholm. 397 pp.

─────────. 1968. Flora of Alaska and neighboring territories. Stanford Univ. Press, Stanford, California. 1,008 pp.

Johannessen, C. L. 1961. Some recent changes in the Oregon Coast: Shoreline and vegetation changes in the estuaries. Dept. of Geography, Univ. Oregon Final Rep. Project NR 338–062:100–138.

─────────. 1964. Marshes prograding in Oregon: Aerial photographs. Science 146:1575–1578.

Johnson, A. W., L. A. Viereck, R. E. Johnson, and H. Melchior. 1966. Vegetation and flora. Pages 277–354 in Environment of the Cape Thompson region, Alaska ed. by N. J. Wilimovsky and J. N. Wolfe, U. S. Atomic Energy Comm., Division of Technical Information, Washington, D. C. 1,250 pp.

Johnson, B. H. 1958. The botany of the California Academy of Science expedition to Baja California in 1941. Wasmann J. Biol. 16:217–318.

Jones, G. N. 1936. A botanical survey of the Olympic Peninsula, Washington. Univ. Washington Press, Seattle. 286 pp.

Kearney, T. H. 1904. Are plants of sea beaches and dunes true halophytes? Bot. Gas. 37:424–436.

Kelly, D. W. 1966. Ecological studies of the Sacramento-San Joaquin estuary. Calif. State Dept. Fish and Game, Fish Bull. 133. 133 pp.

Knapp, R. 1965. Die vegetation von Nord und Mittelamerika. Gustav Fischer Verlag, Stuttgart, Germany. 373 pp.

Koeppe, C. E. 1931. The Canadian climate. McNight and McNight, Bloomington, Ill. 280 pp.

Kumler, M. L. 1969. Plant succession on the sand dunes of the Oregon Coast. Ecology 50:695–704.

Lane, R. S. 1969. The insect fauna of a coastal salt marsh. Unpublished M. A. Thesis, San Francisco State College, Calif. 78 pp.

Macdonald, K. B. 1967. Quantitative studies of salt marsh mollusc faunas from the North American Pacific Coast. Ph.D. Dissertation. Univ. Calif., San Diego. (No. 67–12907). 316 pp. Univ. Microfilms. Ann Arbor, Mich.

─────────. 1969. Quantitative studies of salt marsh mollusc faunas from the North American Pacific Coast. Ecol. Monogr. 39:33–60.

─────────, J. Henrickson, R. Feldmeth, G. Collier, and R. Dingman. 1971. Changes in the ecology of a Southern California wetland removed from tidal action. Page 144 in Abstract volume, second national coastal and shallow water research conference, ed. by D. S. Gorsline, University Press, Univ. S. Calif. 327 pp.

Martin, E. V. and F. E. Clements. 1939. Adaptation and origin in the plant world. I. Factors and functions in coastal dunes. Carnegie Inst. Washington, Washington, D. C. 107 pp.

Mason, H. L. 1957. A flora of the marshes of California. Univ. California Press, Berkeley. 878 pp.

McIlwee, W. R. 1970. San Diego County, coastal wetlands inventory: Tijuana Slough. Calif. State Dept. of Fish and Game, Preliminary Rpt. 62 pp.

Mobberley, D. G. 1956. Taxonomy and distribution of the genus *Spartina*. Iowa State College J. Sci. 30:471–574.

Molina, A. and A. Rathbun. 1967. The zonation of the conspicuous Phanerogams on Kent Island, Bolinas Lagoon. College of Marin, Kentfield (unpublished report). 58 pp.

Mudie, P. J. 1969. A survey of the coastal wetland vegetation of north San Diego County. Calif. State Resources Agency, Wildlife Management Admin. Rpt. 70–4. 18 pp. plus appendices.

_____. 1970. A survey of the coastal wetland vegetation of San Diego Bay. Calif. Dept. Fish and Game Contract W26. D25–51. 79 pp. plus appendices.

Muenscher, W. C. 1941. The Flora of Whatcom County, State of Washington. Pub. by author, Ithaca, N. Y. 139 pp.

Munz, P. A. 1968. Supplement to a California flora. Univ. California Press, Berkeley. 224 pp.

_____, and D. D. Keck. 1963. A California flora. Univ. California Press, Berkeley. 1,681 pp.

National Estuary Study. 1970. U. S. Dept. Interior, Fish and Wildlife Service. 7 vols.

Nelson, E. W. 1922. Lower California and its natural resources. Nat. Acad. Sci., Washington, D. C. 194 pp. (Reproduced in 1966 by Manessier Publishing Co., Riverside, California)

Neuenschwander, L. F. 1972. A phytosociological study of the transition between salt marsh and terrestrial vegetation of Bahia de San Quintin. Unpublished M.A. Thesis, Los Angeles State College, Los Angeles, Calif. 60 pp.

Nichols, D. R. 1965. Composition and environment of recent transitional sediments on the Sonoran Coast, Mexico. Unpublished Ph.D. Dissertation, Univ. California, Los Angeles. 400 pp.

Oosting, H. J. and W. D. Billings. 1942. Factors affecting vegetational zonation on coastal dunes. Ecology 23–131–142.

Orcutt, C. R. 1885. Flora of southern and lower California. Pamph. California Bot. 12:3–13.

Peck, M. E. 1961. A manual of the higher plants of Oregon. 2nd ed. Oregon State Univ. Press, Corvallis. 936 pp.

Pestrong, R. 1972. San Francisco Bay Tidelands. California Geology 25:27–40.

Phleger, F. B. and G. C. Ewing. 1962. Sedimentology and oceanography of coastal lagoons in Baja California, Mexico. Geol. Soc. Amer. Bull.

73:145–182.

Porsild, A. E. 1951. Plant life in the Arctic. Canadian Geog. J. 42:120–145.

Purer, E. A. 1934. Foliar differences in eight dune and chaparral species. Ecology 15:197–203.

Purer, E. A. 1936. Studies of certain coastal sand dune plants of southern California. Ecol. Monogr. 6:1–87.

_____. 1942. Plant ecology of the coastal salt marshlands of San Diego County, California. Ecol. Monogr. 12:81–111.

Redfield, A. C. 1967. The ontogeny of a salt marsh estuary. Pages 108–114 in Estuaries, ed. by G. H. Lauff, AAAS pub. 83. 757 pp.

Scholander, P. F., H. T. Hammel, E. D. Bradstreet, and E. A. Hemmingsen. 1965. Sap pressure in vascular plants. Science 148:339–346.

Shreve, F. and I. L. Wiggins. 1964. Vegetation and flora of the Sonoran desert, in 2 vols. Stanford Univ. Press, Stanford, California.

Smith, C. F. 1952. A flora of Santa Barbara. Santa Barbara Botanic Garden, Santa Barbara, California. 100 pp.

Speth, J. 1970. California coastal wetlands inventory, 1969–70. Calif. State Dept. Fish and Game, Preliminary Rpt. 6 pp.

Spetzman, L. A. 1969. Vegetation of the Arctic Slope of Alaska. Geol. Survey Prof. Paper 302–B. 53 pp.

Spinner, R. 1970. Observations and the plant distributions of Goleta Slough. Univ. Calif. Santa Barbara. (unpublished preliminary report). 26 pp.

Stanley, D. J. 1969. The new concepts of continental margin sedimentation. Short Course Lecture Notes, Amer. Geol. Inst., Washington, D. C. 256 pp.

Stevenson, R. E. and K. O. Emery. 1958. Marshlands at Newport Bay, California. Occas. Papers Allan Hancock Foundation 20:109.

Thom, B. G. 1967. Mangrove ecology and deltaic geomorphology: Tabasco, Mexico. J. Ecol. 55:301–343.

Thomas, J. H. 1961. Flora of the Santa Cruz Mountains of California. Stanford Univ. Press, Calif. 434 pp.

Thorne, R. F. 1967. A flora of Santa Catalina Island, California. Aliso 6:1–77.

Thorsted, T. H. 1972. The salt marsh vegetation of Bahia de San Quintin, B. C., Mexico. Unpublished M. A. Thesis, Los Angeles State College, Calif. 64 pp.

Trewartha, G. T. 1954. An introduction to climate. 3rd ed. McGraw-Hill, N. Y. 402 pp.

Vogl, R. J. 1966. Salt-marsh vegetation of Upper Newport Bay, California. Ecology 47:80–87.

Warme, J. E. 1971. Paleoecological aspects of a modern coastal lagoon. Univ. California Pubs. Geol. Sci. 87. 131 pp.

Wiggins, I. L. and J. H. Thomas. 1962. A flora of the Alaskan Arctic Slope. Univ. Toronto Press, Toronto, Canada. 425 pp.

Williams, J. C. and H. C. Monroe. 1967. A field guide to the natural history of the San Francisco Peninsula. McCutchan Pub. Corp., Berkeley, Calif. 285

pp.
Williams, W. T. and J. R. Potter. 1972? Aspection and phytoecological interactions in the coastal strand community at Morro Bay State Park, California. Submitted to Ecology.

INLAND HALOPHYTES OF THE UNITED STATES

Irwin A. Ungar

Department of Botany, Ohio University, Athens, Ohio 45701

Abstract

Distinct community zonation is often found in inland saline marshes even though physical gradients of moisture, aeration, and salinity may vary continuously. However, investigations of marshes over broad geographical areas in the United States and Canada indicate that all possible halophytic species combinations occur, making it more difficult to classify plant communitites. The unifying characteristic of these communitites are the dominants since associated species have more limited distributions. One can find similar species or generic combinations in saline soils from Texas north to Saskatchewan and from Kansas west to California, traversing many climatic zones and soils of varied ionic composition, dominated locally by sulfates, chlorides, or carbonates. This uniformity is due to the fact that few species have evolved adaptations such as succulence, salt glands, or the osmotic adjustment capacity necessary for survival.

The development of distinct communities seems to be a response to edaphic factors, however, laboratory experiments indicate that halophytes have broader tolerances to salinity than one would surmise from field distributions. Many of the inland halophytic species, including the highly salt tolerant *Suaeda depressa*, *Sesuvium verrucosum*, *Puccinellia nuttalliana*, and *Hordeum jubatum*, can grow and reproduce under nonsaline conditions. Halophytes are apparently limited to saline extreme environments because they can tolerate low water potentials and not due to a physiological requirement for excess salts. Competitive exclusion may explain their elimination from moderate sites. Even halophytes when near their tolerance limits, respond to increased salinity with a decrease in vigor, density, and cover. Species diversity is lowest in the most saline soils and reaches its maximum in nonsaline soils.

Dynamics in plant communities of saline soils is best thought of in terms of a cyclical invasion and retrogression pattern. These vegetation cycles are closely related to soil-water potentials which are themselves affected by precipitation and evaporation.

Acknowledgements

Research in this review was supported by National Science Foundation research grants GB 1408 and BG 6009. I would like to thank the John C. Baker Fund for a research award which allowed me to complete this paper.

Introduction

Many halophytic species occurring in inland saline soils have a broad distribution in the western half of North America. Closely related species in some of the genera including *Suaeda, Salicornia, Sesuvium, Distichlis, Spartina, Puccinellia,* and *Baccharis* occur in coastal marshes. Axelrod (1950) and Branson et al. (1967) feel that many of the salt desert and grassland genera including *Atriplex, Suaeda, Salicornia, Agropyron, Hordeum,* and *Eurotia* which have relatives in Europe and Asia may be of polar origin. Either due to a lack of careful studies of taxonomic relationships or isolating mechanisms during the evolutionary development of these groups, not many of the inland halophytic members of the above mentioned genera occur in Atlantic, Gulf of Mexico, or Pacific coastal marshes.

The possibility of migration of coastal species into the Great Lakes region in the post glacial period is discussed by Peattie (1922). He hypothesized that these species could become established in the shallow water bordering the lakes or on their margins, but probably not in the colder deep waters. Svenson (1927) reports that the halophytes *Atriplex hastata, Salicornia herbacea, Hordeum jubatum, Ruppia maritima, Triglochin maritima,* and *Scirpus maritimus* are among the species which are coastal and occur in western New York State and along the Great Lakes. All of the above mentioned species can also be found in saline ponds and soils west of the Mississippi River. Schofield (1959) lists *Atriplex patula, Salicornia europaea, Spergularia marina, Glaux maritima* var. *obtusifolia,* and *Plantago maritima* ssp. *juncoides* as widespread maritime species which are sometimes found inland. The salt desert region of Utah and vicinity appears to be a center of dispersal for inland halophyte species since it has the most diversity in number of species within genera. This area also contains genera which do not occur in the more eastern saline soils. Organisms which occur in the salt desert shrub regions include *Eurotia lanata, Allenrolfea occidentalis, Sarcobatus vermiculatus, Atriplex confertifolia,* and other species cited in Flowers (1934); Kearney, Briggs, Shantz, McLane, and Piemeisel (1914); Billings (1945, 1949); Gates, Stoddart, and Cook (1956); Vest (1962); West and Ibrahim (1968); Brown (1971). An extensive bibliography is available on the vegetation of the salt desert shrub vegetation type (West 1968) and for vegetation research in Utah (Christensen 1967). Other genera including *Aster, Ruppia, Potamogeton, Scirpus, Eleocharis,* and *Atriplex* have broader distributions, and inland species in these genera also occur in coastal areas but whether these organisms are the same genetically or if ecotypes exist is not known at present.

The distribution of these inland halophytes appears to be controlled mainly by edaphic conditions. Species appear to be selected out by the highly saline environment in a gradient from the most to least salt tolerance, with climate, topography, soil moisture, and biotic factors playing secondary roles. Though the distribution of some of these halophytic species ranges from areas with desert to forest climates, they are generally narrowly limited to saline environments, indicating either a requirement for excess salts, a tolerance for excess salts, or an inability to compete (sensu lata) with plants in less extreme environments. A few of the most salt tolerant taxa including *Salicornia rubra*, *Suaeda depressa*, and *Sesuvium verrucosum* are at present generally limited to saline environments, while other species such as *Hordeum jubatum*, *Chenopodium rubrum*, *Polygonum aviculare*, *Kochia scoparia*, and *Tamarix pentandra* have more general distributions.

Foshag (1926) has reported that the chief sources of salts in western playas of the desert region are Tertiary saline bearing sediments, rock decay, volcanic emanations, and hot springs. In Kansas and Oklahoma, salt originates in rocks of Permian age (Gould 1900, Hay 1891, Latta 1950). In South Dakota, rocks containing highly saline conditions are of Cretaceous age (McGregor 1964). Salts in these more deeply buried deposits filter into upper strata and often, either by seepage or artesian action, reach the surface. Svenson (1927) mentions salt springs as a source of salt around Onondaga Lake in western New York and in New Brunswick. Other causes for excess salts in soil are related to man's activities and include irrigation, oil well drilling, highway salt, and salt mining.

Although a great deal of data is available on studies of coastal marshes of the United States (Ganong 1903, Harshberger 1911, Johnson and York 1915, Knight 1934, Steiner 1934, Penfound and Hathaway 1938, Taylor 1938, 1939; Miller and Egler 1950; Hinde 1954; Kurz and Wagner 1957; Stevenson and Emery 1958; Adams 1963; Barbour 1970; Barbour and Davis 1970, Good 1965) only a limited amount of research has been carried out in saline communitites of the western half of North America. Floristic and ecological data have been collected, either in state floras or through local community studies, for many inland halophyte species and some communities, but no general outline is available describing the distribution of halophytes in western North America. An attempt will be made here to summarize the available data and to describe the pattern of distribution, if any exists, with some emphasis on what is known concerning species-soil relationships and using plants as indicators of saline conditions.

Several general changes in vegetational composition that commonly occur with reduced salinity are an increase in growth, increase in density, or an increase in species number (diversity). The latter two phenomena may occur together, or one species may increase greatly in density with few, if any, species being added to a communities composition (Ungar 1965a). Generally there is a sharp decrease in species diversity, with even a low increment of soil salinity, and then further drops in diversity until only one or two species are left which are capable of tolerating the salinity extremes (Flowers 1934; Baalman 1965; Dodd and Coupland 1966b; Ungar 1967b, 1968b).

The earliest data available for vegetation of inland saline soils are the floristic and community studies of Hitchcock (1898), Schaffner (1898), Pound and Clements (1900), and Ortenburger and Bird (1931). These studies describe the floristic composition of saline soils which vary from the succulent or graminoid type to that of the desert shrub. Other studies by Hilgard (1914); Kearney et al. (1914); Flowers (1934); Hanson and Whitman (1938); Gates et al. (1956); Bolen (1964); Dodd, Rennie and Coupland (1964); Hunt and Durrell (1966); and Ungar (1965a, 1965b, 1967b, 1968b, 1969) describe some species-soil relationships in saline soils of western North America. Quantitative vegetational data were not collected in these inland saline areas until the studies of Bolen (1964); Baalman (1965); Dodd and Coupland (1966); Hadley and Buccos (1967); Ungar, Hogan, and McClelland (1969); and Ungar (1965a, 1967b, 1968b, 1970a).

The environment of inland halophytes differs from that of the coastal species in that at least one major environmental extreme tidal action is absent. Coastal species must adapt to salinity regimes and daily submergence, while in the evolution of inland halophytic species, salinity must have been an important factor; only occassional submergence occurs in these habitats after heavy rainfalls.

In coastal areas of the eastern and western United States, a perennial *Spartina* spp. community generally occurs bordering the estuary, salt marsh, or salt pond in areas of greatest submergence and exposure to tidal action (Stevenson and Emery 1958, Chapman 1960). Annual succulents occur beyond this zone on higher ground. In inland marshes where tidal action is not a factor, annual species of *Salicornia* and *Suaeda* invade the center of pans which have the highest soil salinities with perennial grasses and sedges forming zones beyond these succulent annuals in areas of lower salinity. Tidal action acts to moderate salinity extremes in coastal marshes even during drought periods, whereas in inland marshes, there appears to be a direct relationship between drought conditions and increased salinity (Ungar 1968b). This leads to great fluctuations in osmotic stress in inland marsh soils over short time intervals which can only be alleviated by increased precipitation.

For some of the drier climates, there are probably greater moisture stress differences in inland saline soils during particular seasons. These greater stress differences can be attributed to the absence of tidal activity and sparse rainfall (Ungar 1967b). However, in some inland spring fed areas studied by Bolen (1964), little variation occurred between his April and August soil salinity samples. The high evaporative power of the air and sporadic rainfall in much of the grassland and desert regions will also affect the degree of salinity stress the species are exposed to by causing fluctuations in soil solution salt concentrations throughout the growing season. Coastal areas may also have considerable soil salinity fluctuation as Kurz and Wagner (1957) have reported for the coastal marshes of Florida and South Carolina. While in coastal environments the chief salt involved is sodium chloride, inland areas have species which tolerate conditions of high sulfate, bicarbonate, and carbonate as well as chlorides. This general tolerance to high osmotic pressures from various ionic combinations has been shown experimentally for germinating seed of *Suaeda depressa, Iva annua,* and *Puccinellia nuttalliana* (Ungar and Capilupo 1969; Ungar, Hogan, and McClelland 1969; Macke and Ungar 1971). In studies of natural distributions, *Suaeda depressa, Distichlis stricta, Hordeum jubatum, Salicornia rubra, Scirpus americanus, Scirpus paludosus,* and others occurred in conditions where the salt content varied from chloride to sulfate or bicarbonate dominance (Flowers 1934; Tolstead 1942; McCarraher 1962; Dodd, Rennie, and Coupland 1964; Ungar, Hogan, and McClelland 1969; Ungar 1970a).

Fireman and Hayward (1952) have reported that salt desert shrub species differ in their sodium accumulation ability. *Atriplex confertifolia* surface soils have low sodium concentrations below them due to this, while *Sarcobatus vermiculatus* accumulates high concentrations of sodium. These high sodium soils under *S. vermiculatus* have poorer aggregating properties and lower moisture retaining capacities than soils under plants not able to accumulate sodium. Differential accumulation of ions by halophytic species may alter the soil conditions in a manner that they prevent other species from growing with them.

Plant Communities of Inland Salt Deserts

Salt desert shrub communities of inland United States have been described in detail by previously mentioned authors and include plant communities dominated by the following: *Artemesia spinescens, Eurotia lanata, Chrysothamnus* sp., *Atriplex confertifolia, Kochia americana, Sarcobatus vermiculatus, Atriplex nuttallii,* and *Atriplex corrugata* (Branson, Miller, and McQueen 1967). These communities are listed as they occur in soils of increasing order of salinity stress with soils ranging from -0.6 bars to -5.0 bars in osmotic stress in the *Artemesia spinescens* and *Atriplex nuttallii* communities. Branson et al. (1967) reports that the mean total soil moisture stress in these salt desert shrub communities ranges from a low of -50 bars in the *Sarcobatus vermiculatus* stands to a high of -95 bars in the *Atriplex nuttallii* communities. These species usually do not occur in high-water-table marshes, which will be comprehensively reviewed in this paper; and, desert shrub species distributions are primarily determined by their drought as well as salt tolerance.

Plant Communities of Inland Salt Marshes
A. Submerged
    1.   *Potamogeton - Chara - Ruppia*

B. Succulents
   2. *Salicornia rubra (Salicornia* sp.)
   3. *Suaeda depressa (Suaeda* sp.)
   4. *Sesuvium verrucosum*
   5. *Allenrolfea occidentalis*
C. Graminoid
   6. *Distichlis stricta*
   7. *Puccinellia nuttalliana*
   8. *Triglochin maritima*
   9. *Sporobolus airoides*
   10. *Hordeum jubatum*
   11. *Scirpus paludosus (Scirpus* sp.)
D. Shrub
   12. *Tamarix pentandra*

It should be made clear that although the chief dominant is reported here for each community, other species may play important roles locally. In any one inland saline site the controlling factor in plant community recognition is the importance of the dominant species. Intermediate communities occur throughout the range and many species combinations can be found. These communities may be characterized by combinations of the dominants listed above, or locally, other salt tolerant species may become important. Flowers (1934) has reported that the pioneer halophyte communities of the Great Salt Lake tend to be more or less constant, but zones in less saline conditions are more variable and a great number of species combinations may persist.

Inland Saline Lake Community (*Potamogeton-Ruppia-Charophyta*)

A. Macroscopic vegetation and its relationship to chemical factors.

Inland saline lakes and ponds have few vascular plants inhabiting them and may be dominated by the macroscopic algae *Chara, Nitella*, and *Tolypella*. Members of these genera were found to be very abundant in some central Kansas ponds described by Jewell (1927) and Ungar (1964, 1965a). The only submerged vascular plant found occurring in these study areas was *Ruppia maritima*. These central Kansas saline ponds ranged in salinity anywhere from 0.25% to 1.26% total salts, with a chloride content equal to about 50% of the total ionic composition. A pH range from 6.7 to 8.7 was found here, with a median value of 7.7. There was a broad fluctuation in salinity concentrations in these areas depending on the depth of water in a given pond. Several ponds studied on the Big Salt Marsh by Ungar (1965a) had ranges from 0.48% to 1.17%, 0.25% to 0.39%, and 0.28% to 1.26% total salts. Droughts cause a drying up of the shallow ponds and an increase in salt concentration. This drying effect, accompanied by increased salinity, occurred during July in this central Kansas study area. Sporadic rainfalls will counteract the drying effect, filling the ponds, at any time during the growing season. Most of the submerged species begin their growth early in the spring, and the algae are mature by early summer. This has survival value in a shallow pond environment which may dry rapidly later in the summer and prevent organisms from completing their life cycles. The submerged species of vascular plants and algae growing in these areas have a broad tolerance to withstand fluctuations in salinity. If this were not true, each rainy period or dry spell would kill many plants not capable of osmotically adjusting to the changes in the osmotic concentration of the pond waters.

Penfound (1953) reports that *Potamogeton pectinatus* occurs in aquatic habitats of the Great Salt Plains in north-central Oklahoma. Further studies in this area by Baalman (1965) indicate that *P. pectinatus*, *Ruppia maritima*, and *Chara* sp. occur at the mouth of saline streams of the Great Salt Plains. Shallow stream locations on the west side of the Great Salt Plains which contain *R. maritima* have a total salt content ranging from 1.02% to 2.25% with a chloride concentration equal to 50% of the total (Ungar 1968b). These waters had a pH ranging from 6.9 to 7.9, with a median of 7.2.

A cursory study of submerged marginal vegetation of Salt Pond in Lincoln, Nebraska, indicates that two of the common hydrophytes are *P. pectinatus* and *R. maritima* (Ungar, Hogan, and McClelland 1969). These waters have a salinity which ranged from 0.31% to 0.50%. There appeared to be less fluctuation in this deeper water lake than in the shallower ponds of central Kansas. The lake waters pH ranged from 6.6 to 8.2, with a median of 7.8. McCarraher et al. (1961) and McCarraher (1962) has found that *P. pectinatus*, *R. maritima*, and *Chara* sp. are the primary submerged aquatic species in saline lakes of the sand hills region of western Nebraska. These lakes have a total dissolved solid content of 0.1% in Alkali Lake to 1.2% in Smithy Lake No. 2. The pH ranges from 9.0 to 10.8 in the western Nebraska alkali lakes. These lakes differ chemically from the eastern Nebraska lakes, which are high in sodium chloride, in that their ionic make-up appears to be dominated by sodium bicarbonate and sodium carbonate. They have low chloride and sulfate contents.

Young (1923) has reported a rise in total salt content from 0.8% to 1.5% at Devils Lake, North Dakota, from 1889 to 1923. This lake is high in sulfates, 54.1% of the total ions. The pH has been reported to be 10.0 (Wilson 1958). The probable cause of rising salinities during this period was the decrease in the level of the lake water. Flowers (1934) reported similar salt concentration fluctuations in the Great Salt Lake. He found a 1.0% increase in salt concentration with each one foot decrease in water level. Great fluctuations in salt content ranging from 2.3% in July to 4.6% in August 1967 have been reported in sulfate dominated lakes by Ungar (1970a). This was also accompanied by an obvious recession of the shoreline and decrease in water level. Rawson and Moore (1944) has reported similar fluctuations in Saskatchewan, Canada. Closed lakes in the arid and semi-arid regions fluctuate in level much more than lakes in more humid areas (Langbein 1961). Shjeflo (1968) has demonstrated that the major water source for prairie potholes in North Dakota was summer precipitation and spring run-off from snow melting. Evapotranspiration accounted for about a 2.1 foot of water loss per growing season, May to October. The saline lakes of

Saskatchewan range from 0.04% to 11.8% in total salt concentration (Rawson and Moore 1944). The lakes were alkaline and had a pH ranging from 7.7 to 9.5. Halophytic submerged species found in fresh water and saline lakes, ranging from 0.02% to 2.0% total salts, include *Ruppia maritima*, *P. pectinatus*, and *Chara* sp. Salts were found to be mostly sulfates of magnesium and sodium with little chlorides. Seasonal salinity fluctuation in Redberry Lake ranged from 1.4% in May to 1.5% in August 1941. Annual increase in salinity with time has also been observed in the Saskatchewan Lakes, and in Redberry Lake the following values were measured: 1926 - 1.2%, 1931 - 1.2%, 1939 - 1.4%, 1940 - 1.3%, and 1941 - 1.5%. A similar relationship existed in seven other lakes reported on, and there appears to be a relationship in which the annual salinity increase shows an inverse relationship to the mean depth of the lake. Deeper lakes had relatively lower percentage increases and Rawson and Moore (1944) used a hyperbolic curve equation $y = \frac{a}{x} - c$ to explain this relationship (y = rate of salinity increase, x = means depth, a = constant of 11, and c = constant of 3). They do feel that due to great seasonal fluctuations in salinity of some lakes, Little Manitou had 10.5% in May and 14.8% in September 1940; the relationship described above is not yet proved. *Chara* sp. was found in up to 0.8%, *R. maritima* in up to 1.4%, and *P. pectinatus* in up to 2.0% total salts. This data indicates the broad range of salinity tolerance of these "halophytic" aquatic species ranging from nonsaline environments to those of extreme salinity.

Ungar (1970a) reports that Stink Lake and Bitter Lake in north-central South Dakota are highly saline sulfate lakes. Total salt concentrations in Stink Lake ranged from 2.3% on July 9, 1967, to 4.6% on July 4, 1968. This lake contains about 50.0% sulfates, chiefly of sodium and magnesium, and a low chloride content. Its pH ranged from 7.0 to 9.2. The only submerged macroscopic plant found on the margins of the lake was *P. pectinatus*.

Shallow saline ponds and river deltas in the Great Salt Lake region of Utah contained submerged hydrophytes *R. maritima* and *P. pectinatus*. Soil at the bottoms of these pools contained 2.01% salt. This area had a pH of 8.4. Great Salt Lake contains no species of submerged vascular plants or macroscopic algae in the Charophyta. This is most likely due to its extremely high salinity concentrations ranging from 13.8% to 27.7% (Flowers 1934, Flowers and Evans 1966). Of the total salts in the lake, sodium chloride makes up about 80.0% of the total, and the water's pH is normally 7.4 (Flowers and Evans 1966). Fluctuations in salinity are thought to be due primarily to rise and fall of the water level. In studies of spring-fed marshes in western Utah, Bolen (1964) found that *R. maritima* and *Chara* sp. were the most important submerged species. He reports that no *P. pectinatus* could be found in his study area. Absence of *P. pectinatus* is explained by the dense growth of *Chara* and *R. maritima* in the pools which allowed little chance for successful colonization by other submersed species. Salinity averaged about 0.3% in this area with submergent soils having a pH ranging from 7.2 to 8.8 and a meidan pH of 8.1. Spring waters had a salinity ranging from 0.2% to 0.3% and a pH value from 7.2 to 7.6. Sodium chloride was found to be the principal salt causing high salinities in this area.

Anderson (1958a, 1958b) studied the saline lakes of Washington and found that their salinities ranged from 1.4% to 16.1%. Soap and Lenore lakes were equally high in carbonate, bicarbonate, sulfate, and chlorides of sodium. Hot Lake had an abundance of magnesium sulfate and varied considerably in salinity concentration (Table 1). After spring run-off in April 1955 it had a low of 4.0% total salts while by late summer, surface waters due to evaporation and mixing had a high of 20.0% total salts (Anderson 1958b). The pH ranged from 9.4-10.0, with a median of 9.9. McKay (1939) found similar changes in salinity ranging from 1.6% in May 1933 to 6.0% in July 1933 during dry and wet years. In spring 1956 with high run-off, surface salinities dropped to 0.04% total salts; median pH was 8.2. The only macroscopic algae occurring in Hot Lake (Epsom Lake) is *Chara*, while the only vascular species is *Ruppia maritima* (St. John and Courtney 1924, McKay 1934, Anderson 1958a). Some concentration stratification exists in these saline ponds with higher concentrations generally found in deeper water (Anderson 1958a, 1958b). At 3 m there was more than 2 X the dissolved solids content than in the surface waters (Table 1).

Table 1. Mineral analyses of Hot Lake waters, August 22, 1955. Ions expressed as % (adapted from Anderson 1958).

	Surface	3m
Conductivity (micromhos)	57,910	60,440
pH	8.2	7.8
Total Dissolved Solids	16.1	39.2
Ca	0.1	0.1
Mg	2.3	5.4
Na	0.7	1.7
K	0.1	0.2
HCO_3	0.3	0.3
CO_3	0.0	0.0
SO_4	10.4	24.4
Cl	0.2	0.2
NO_3	0.0	0.0

Carpelan (1958), in his study of the chemistry of the Salton Sea, California, found great fluctuations in chlorinity. These correlated directly with surface level of the water; as the surface level dropped, chlorinity increased and vice versa. Total salt concentration ranges from 3.1% to 3.4% at present with chlorides making up about 43% of the total salts. The pH ranges from 8.3 to 8.8 during the year. The ionic concentration has increased from 0.4% in 1907 to 3.4% in 1955. The predominant ions are sodium and chloride. This increased salt content resulted from excess of evaporation over inflow (Carpelan 1958).

St. John and Courtney (1924) found *Ruppia maritima* submerged in Hot Lake (Epsom Lake). The chief salt in this lake was $MgSo_4$, and the lake bottom contained a solid layer of nearly pure $MgSO_4.7H_2O$. McKay (1934) reports that magnesium sulfate makes up 84% to 88% of the total salts in this lake. This is either directly toxic to plants or affects them by increasing osmotic pressures and thus reducing the amount of available water.

B. Physiological ecology of submerged aquatics and their limits of tolerance.

Ruppia maritima L. is a widespread species along the Atlantic and Gulf Coast and in alkaline lakes, ponds, and streams, in the western half of North America (Muenscher 1944). McMillan and Moseley (1967) have reported that it can withstand up to 7.4% total salts of concentrated seawater under experimental conditions. In inland saline waters it has been found growing in waters with as low as 0.02% total salts, and in lakes in which the salinity reaches as high as 2.3% total salts (Rawson and Moore 1944, Ungar 1968b, Steward and Kantrud 1969, Sloan 1970).

Bourn (1935) found that *R. maritima* would tolerate in excess of 3.5% seawater, but that seed set was retarded at 2.8%. The occurrence of *R. maritima* in 1951 after the 1950 hurricane in Florida, but not in the previous season, indicates that lowered salinity may favor *R. maritima* and that it was not as salt tolerant as other submerged halophytes (Strawn 1961). The water's salinity level dropped to as low as 0.9% after the hurricane, which is the period Strawn (1961) feels *R. maritima* became established.

Potamogeton pectinatus can be found growing in brackish waters and streams throughout the United States (Muenscher 1944). It has been reported in fresh - and salt-water locations by Rawson and Moore (1944). Data of McCarraher (1962), Baalman (1965), and Ungar (1968b, 1970a) show that *P. pectinatus* can grow in bicarbonate, chloride, or sulfate lakes, indicating no ionic preference. Experimental data of Bourn (1932) on coastal *P. pectinatus* shows that it could tolerate total salt concentrations up to 1.3%, but higher concentrations became lethal after any lengthy exposure.

Seed germination studies with inland sources of seed prove optimum germination occurs in tap water, with each salt increment decreasing the total germination percentage (Table 2; Teeter 1963). These seeds tolerated up to 1.5% sodium chloride. When they were subjected to treatments of high salinity, seeds were capable of recovery and would germinate when placed in tap water; however, recovery was not as high as in the original control. There was a definite delay in the rate of germination with increased salinity. Germination in tap water began in 5 or 6 days while at higher concentrations, 1.2% and 1.5%, there was an 18 day delay.

Growth was measured by Teeter (1963) for plants of 1, 4, and 8 weeks of age for a period of 5 weeks. Decrease in growth with increased salinity was most marked when experiments were begun with young plants. Effects of high salinity is clearly illustrated as plants age (Teeter 1963). Plants grew

Table 2.   Percentage germination of *P. pectinatus* in different NaCl treatments (adapted from Teeter 1963).

NaCl %	Treatment	Recovery	Total
0.0	71.1*	0.5	71.6
0.3	44.4	6.0	50.4
0.6	35.5	7.7	43.2
0.9	20.0	9.0	29.0
1.2	8.8	14.6	23.4
1.5	5.5	19.4	24.9
2.0	0.0	14.6	14.6

* Control germination significantly different at 1.0% level from other samples.

vigorously in nutrient solution with no NaCl increment indicating *P. pectinatus* does not require high salinities for normal growth. These results differ from those of Bourn (1932), who worked with coastal *P. pectinatus*, since he found optimum growth at about 0.7% seawater. Data in Table 3 indicate a general reduction in growth of *P. pectinatus* with added salt increments.

Table 3.   Growth of *Potamogeton pectinatus* after 5 weeks at varied salinities. Results reported in grams dry weight. (adapted from Teeter 1963).

NaCl %	1 week	4 weeks	8 weeks
0.0	3.6	9.6	14.8
0.3	3.0	9.6	10.9
0.6	3.2	8.3	11.3
0.9	2.6	7.2	8.2
1.2	0.4	5.9	10.1

One week old plants died in 1.5% NaCl, 4 weeks old died in 1.8%, and above this salinity plants lasted 5 days and began decomposing. This agrees well with the 1.3% maximum reported by Bourn (1932) for coastal *P. pectinatus*. Eight week old plants tolerated up to 1.8% NaCl while the 12 week old plants tolerated 1.2%. These experiments indicate that the maximum tolerance lies between concentrations of 1.2% to 1.5% NaCl with a few plants surviving in as high as 2.1%. Teeter (1963) states that the overall vigor of the plant, not size or age, appears to be the determining factor in whether or not plants will survive at high salinities.

Zanichellia palustris has been reported to occur as an under-story species in shallow saline water where *Scirpus paludosus* is dominant (Walker and Coupland 1970). Ungar has also found *Z. palustris* in shallow water associated with *Ranunculus cymbalaria* along Bitter Lake in South Dakota marshes, but it was always associated with some fresh-water run-off.

Communities of *Salicornia*, *Sesuvium*, and *Suaeda* (pioneer Halophytes)
Salicornia rubra Community

A. Vegetation

Salicornia rubra Nels. (*Salicornia* spp.) is one of the most salt tolerant halophytic species in the western half of the United States and Canada. In the United States it occurs in all states west of Minnesota, Iowa, Missouri, Arkansas, and Louisiana, except for the south-western states of Oklahoma, Texas, New Mexico, and Arizona (Muenscher 1944). It does not occur in any of the states mentioned as the eastern limit or in any other part of the eastern United States. In Canada it occurs from British Columbia east to Saskatchewan (Hitchcock et al. 1964). The taxonomic status of *S. rubra* can still be considered uncertain; Mason (1957) retains it while Harrington (1954) suggests that it may not be distinct from the coastal species *Salicornia europaea* L. Munz (1965), with no explanation, includes *S. rubra* in the taxon *S. europaea*. Flowers (1934) and Kearney (1914) consider *S. rubra* one of the most salt tolerant species in the Great Salt Lake region of Utah. In saline soils of Saskatchewan, *S. rubra* was found to dominate the most saline vegetated sites (Dodd, Rennie, and Coupland 1964; Dodd and Coupland 1966b). Other studies in the prairie and plains region indicate that *S. rubra* is highly salt tolerant (Ungar 1965a, 1965b, 1970a; Hadley and Buccos 1967; Keith 1958; Rawson and Moore 1940; Billings 1945; Coupland 1950; Weaver and Clements 1938; Weaver 1918; Pound and Clements 1900). In the desert and Pacific coast states many researchers, including Shreve 1942, Aldous and Shantz 1924, Billings 1945, Branson et al. 1967, St. John and Courney 1924, Nelson 1955, Shantz and Zon 1924, have considered *S. rubra* a pioneer halophytic species.

Flowers (1934) reports that *S. rubra* and *S. utahensis* are the first pioneer types of the halosere of the Great Salt Lake. *S. rubra* penetrates furthest out on the saline flats in this region (Kearney 1914). Flowers (1934) found *Salicornia* growing under a very broad salinity range, 0.5% to 6.0%. Optimum growth occurred on soils with between 2.0% and 3.0% total salts. *Salicornia* is generally a member of unistratal communities on salt pans (Flowers 1934, Keith 1958, Ungar 1970a). These stands usually have a depauperate flora due to the highly saline environment. Occurring with *S. rubra* in areas of highest salinity in the western part of its range are *Salicornia utahensis* and *Allenrolfea occidentalis* (Flowers 1934). In the eastern part of its range, *S. rubra* is found growing with *Suaeda depressa* and *Distichlis stricta* in areas of extremely high salinity. In areas of reduced salinity, it occurs with *Atriplex patula* var. *hastata*, *Puccinellia nuttalliana*, *Hordeum jubatum*, *Triglochin maritima*, *Chenopodium rubrum*, *C. salinum*, and *Salsola pestifer* (Dodd and Coupland 1966b; Keith 1958; Ungar 1969, 1970). When occurring with *P. nuttalliana*, *H. jubatum*, or other taller species, a bistratal community develops with *S. rubra* forming the lower layer and *P. nuttalliana* in the South Dakota and Colorado marshes, the upper stratum (Ungar 1970a). It is generally associated with the most moist saline sites in a depression, and decreases in importance on less moist sites (Weaver 1918). In Nebraska, *S. rubra* was a co-invader with *S. depressa* of pans and dry lake bottoms of salinities averaging 1.5%. It was also found in invading pans of higher salinities, but with droughts during the summer, it disappeared leaving *S. depressa* and some persistent stands of *D. stricta* in areas where salinities averaged 4.0% (Ungar, Hogan, and McClelland 1969). This indicates that *S. rubra* is probably more sensitive to drought periods than the latter species.

Vegetative cover in these areas vary from as low as 0.3% in South Dakota (Ungar 1970a) to as high as 25% in Saskatchewan (Dodd and Coupland 1966b). Frequency distributions are presented in Table 4. Due to the fact that *S. rubra* is a pioneer species, it generally is the sole species present and it can be used as an indicator producing values close to 100% frequency (Ungar 1970a). Some of the differences in vegetation composition in Table 4 must be relegated to differences in method of analysis and not real vegetational differences. Species composition in the *Salicornia* community range from 1 in Keith's (1958) study in Alberta to 12 in Dodd and Coupland's (1966) investigations in Saskatchewan. In investigations of Keith (1958) and Ungar (1970a), *Salicornia* occurred in several communities but the *S. rubra* community was restricted to nearly pure stands, whereas in Dodd and Coupland (1966b) some mixed stands were evidently included. However, even in the latter research it always comprised 88% to 100% of the vegetative cover indicating its dominance in these areas.

Salicornia rubra occurs in several communities which are less extreme in salinization than the *S. rubra* community. Dodd and Coupland (1966b) have found it in their *Triglochin*, *Puccinellia*, and *Distichlis* communities. In studies by Ungar (1969, 1970a) it has also been found in *Puccinellia* and

Table 4. Percentage frequency of species in <u>Salicornia rubra</u> communities. (+ = present but not represented in quantitative data.) 1. Alberta, Canada (Keith, 1958); 2. Saskatchewan, Canada (Dodd and Coupland, 1966); 3. Grand Forks Co., North Dakota (Hadley and Buccos, 1967); 4. Lincoln, Nebraska (Ungar, Hogan, and McClelland, 1969); 5. Codington Co., South Dakota (Ungar, 1970a).

Total Salts [a]	1.9%	2.5%	1.7%	1.6%	4.8%
Conductivity mmhos/cm	45.0	39.8	28.0	24.3	32.0
Location	1.	2.	3.	4.	5.
Species					
Salicornia rubra	100	93.6	100	75	100.0
Suaeda depressa	---	10.3	100	77	10.0
Atriplex patula var. hastata	---	2.4	---	14	-----
Distichlis stricta	---	1.1	---	2	+
Scirpus paludosus	---	----	---	3	-----
Puccinellia nuttalliana	---	4.8	---	--	+
Hordeum jubatum	---	0.3	---	--	-----
Triglochin maritima	---	0.2	---	--	-----
Chenopodium rubrum	---	11.4	---	--	+
C. salinum	---	2.8	---	--	-----
Spergularia salina	---	3.0	---	--	-----
Salsola pestifer	---	2.3	---	--	-----

[a] = Calculated on the basis of: 0.064 L mmho/cm = % total salts

Table 5. _Salicornia_ community soils.

	pH median	Saturation %	Soil Moisture %	Conductivity mmhos/cm	Na %	Cl %	SO₄ %	Total Salts%
North Dakota Hadley and Buccos (1967)	---	---	---	28.0	---	---	---	1.8
Canada Dodd, Rennie and Coupland (1964)	8.3	---	---	39.8	---	---	---	2.6
South Dakota Ungar (1970a)	8.4	48.0	35.5	32.0	0.4	0.1	2.0	4.8
Nebraska Ungar (1969)	7.5	63.3	34.3	24.3	0.6	0.7	0.1	1.6
Kansas Ungar (1965a)	8.2	40.0	---	---	---	1.1	---	1.9
Utah Flowers (1934)	8.8	---	---	---	2.1	3.2	---	6.3

Distichlis communities in South Dakota and in *Salicornia-Suaeda, Suaeda,* and Dwarf *Distichlis* communities. Keith (1958) found *S. rubra* in a *Suaeda-Chenopodium* community, while Hadley and Buccos (1967) found it to be limited to a *Salicornia-Suaeda* community. Flowers (1934) feels its lower limit of distribution are soils with salinities of 0.5% total salts. It has been grown in the laboratory on sand in nutrient solution and occurs rarely in nature at lower salinities denoting that though its distribution usually indicates it is an obligate halophyte, physiologically this may not be the case (Ungar, unpublished data).

Salicornia rubra, being one of the most salt tolerant species, is capable of growing in soils that few other species can survive in. It is therefore, often considered one of the pioneer successional species on saline soils (Kearney 1914, Flowers 1934). This interpretation may or may not be correct. If there is an alleviation of high salinities, other species will enter zones occupied by *Salicornia*, however, if salinities are not reduced, *Salicornia* does not appear to be capable of modifying the environment sufficiently to begin any autogenic succession. It, therefore, should be considered a pioneer on extremely saline soils, but not necessarily a dynamic seral stage in succession. Observation in Nebraska indicate that drying of the soils during the summer growing season limits the distribution of *Salicornia* (Ungar, Hogan, and McClelland 1969). An interesting cyclic pattern may develop as in Lincoln, Nebraska, in which shallow lakes dry allowing *Salicornia* invasion, and then the following year there is more rainfall and no *Salicornia* develops on the lake surface. Another type of cycle, which apparently involves the rainfall before the initiation of germination, was observed in South Dakota (Ungar 1970a). In this case, it was found that the normal rainfall, 29.7 cm, between October and March was apparently correlated with the occurrence of *Salicornia* on salt pans. This preseason rainfall was 33.4 cm in 1966-1967 and *S. rubra* occurred, while in 1967-1968 it was 14.1 cm and no *S. rubra* appeared on the flats. In the 1968-1969 period, preseason rainfall totalled 40.6 cm and *S. rubra* appeared on the salt pans again. During 1966-1967 and 1968-1969 precipitation exceded the normal 113% and 142% in each case, respectively, while in 1967-1968 preseason rainfall was only 49% of the normal. The preseason precipitation either provides enough water for germination or reduces salintiy so that seedlings can get established. In 1967-1968, a year of low pre-growing season precipitation, germination was inhibited either due to a lack of soil moisture or rising osmotic pressures in the surface soils. With high evaporation rates during the summer, salinity stress continues to increase and *Salicornia* cannot establish itself on the salt ponds (Ungar 1970a). A similar situation apparently occurred in the salt pans in the Lincoln, Nebraska, marshes in 1968 (Ungar, Hogan, and McClelland 1969). *Salicornia rubra* was also conspicuously missing from pans it previously occurred on and was present in a telescoped zonal pattern on some pan edges, while in other pans it was completely missing. It did appear in large numbers during the 1967 growing season on a dried out shallow lake north of Lincoln. This indicates that pre-growing season precipitation and soil moisture content are both indirectly and directly, respectively, responsible for *Salicornia* distribution in the Lincoln marshes.

B. Soils

The broad distribution of *Salicornia rubra* in the saline soils from the eastern part of the prairie region to the Pacific states has been reported by several researchers (Muenscher 1944, Mason 1957, Rydberg 1932). It apparently does not occur, or only rarely is found, in saline coastal soils, nor has it been found in saline areas of the arid southwest (Muenscher 1944, Mason 1957). It is found in areas of high salinity with varied ionic composition. *Salicornia rubra* has been found in soils in Washington where 99.6% of the total salts were mangesium sulphate (St. John and Courtney 1924), in soils which contain 86.1% to 93.3% sodium chloride in Nebraska (Shirk 1924), and in South Dakota in soils containing over 85% sulfates of calcium, magnesium, and sodium. This indicates that the specific ion content of the soil may not play a significant role in its distribution.

Tolerance to osmotic stress is apparently the most important factor controlling its distribution. With droughts and a drying of the salt pan, *S. rubra* often disappears indicating it is not a drought tolerant species. It has been found growing in soils with as low as 0.6% total salts and also in soils with up to as high as 8.0% total salts in the surface 10 cm of soil (Ungar 1970a; Ungar 1969, unpublished data). Soil moisture is generally higher in the *Salicornia* community than in other highly saline communities. At Stink Lake, South Dakota, *Salicornia* communities averaged 35.5% soil moisture, while *Suaedea*, *Puccinellia*, and Dwarf *D. stricta* communities averaged 25.2% (Ungar 1970a). In Lincoln, Nebraska, *Salicornia* sites averaged 34.3% soil moisture, while Dwarf *D. stricta*, *Suaeda*, and unvegetated areas averaged 25.4%, 28.7%, and 22.9% each respectively. At the Lincoln sites, *Salicornia* was common in areas with thick algal crusts, while in South Dakota, *S. rubra* did not appear to be as restricted to these locations.

Soil pH does not appear to be a limiting factor in *S. rubra* distribution. It has been found growing in nearly pure stands in soils with pH values ranging from 7.0 to as high as 9.0 (Keith 1958; Ungar, Hogan, and McClelland 1969). Soil texture is quite variable in the sites occupied by *Salicornia*. It is usually found in heavier soils ranging from sandy loam to clay in texture. It was never found in areas of sand which were well aerated and had excellent drainage. These soils which dried rapidly are probably beyond the drought resistance of *S. rubra*.

C. Physiology

Seed germination studies by Hogan (1968) indicate that *Salicornia rubra* is capable of germinating in up to 5.0% NaCl concentrations. Germination percentages ranged from a high of 28% in distilled water to a low of 3% in 5% NaCl solutions. The low germination values may be due to the varied maturity of seed on a single plant. This could be determined by the differences in size of the seed. These germination data indicate that *S. rubra* is capable of germinating at higher salinities than most other halophytic species studied by Chapman (1960) and Ungar (1962). Other species of *Salicornia* have germinated in salinities up to 5.0% NaCl (Chapman 1960, Ungar 1967a). The optimum temperature for germination was 25 C, with 15 and 35 C giving reduced values. Placing ungerminated seed at 5.0% NaCl into distilled water brought about germination of 16% of these ungerminated seeds. This indicates that high salt concentrations are not toxic to *S. rubra* seed. Some survival value for *Salicornia* is illustrated by dormancy at high salinities and the retention of germinability when salinities are reduced. Seedling growth data indicate that young plants can tolerate the range of salinity at which the seeds germinated.

Seed germination studies with *Salicornia europaea* from an inland location at Rittman, Ohio, indicated that seed could germinate in up to 10.0% NaCl (Ungar 1971, unpublished data). Germination was reduced from 80.8% in distilled water to 36.8% in 5.0% NaCl solution. Constant temperature of 35 C yielded 89.0% germination in distilled water and germination percentages gradually decreased to 53% at 15 C. In 1.0% NaCl solutions this gradual change did not occur with the temperature extreme having 47.0% and 45.0% germination, while in 5.0% solutions there was still a further decrease to 18.0% and 20.0% at the temperature extremes. There was no strong temperature influence as in distilled water. It seems that temperature is more limiting at low salinities, but when salinity concentrations increase, this becomes the determining factor in *S. europaea* germination.

Data of Halket (1915), Baumeister and Schmidt (1962), and Keller (1925) indicate that other species of *Salicornia* show better growth in terms of height and increased dry weight with the addition of 1.0% to 2.0% NaCl to the growth medium. There studies and experiments by Solovev show that *Salicornia* can grow with no salt increment in soil or in a nutrient hydroponic solution (Strogonov 1964). Webb (1966) has found in his studies with *Salicornia europaea* and *S. bigelovii* that best growth occurred in media with 1.0% NaCl added. In his transplanted seedlings experiments he found poor survival at salinities below 0.5%. This is contrary to the results of Halket (1915), Keller (1925), and Baumeister and Schmidt (1962) who did not report any die off in controls.

Harward and McNulty (1965) have found that there was a general increase in both chloride and sodium at nearly equal concentrations in the sap of *Salicornia rubra*. This was accompanied by a decline in plant water content during seven weeks in the summer of 1964. These results could be explained by a drying out of the soil during this period. It can be concluded from these experiments that sodium and chloride ions were primarily responsible for ionic adjustment in *S. rubra* at these field sites west of Salt Lake City, Utah.

Sesuvium verrucosum **Community**

A. Vegetation

Sesuvium verrucosum Raf. (*Sesuvium sessile* Pers.) is a highly salt tolerant succulent halophyte. Rydberg (1932) describes its range as from Wyoming to Kansas, south to Texas, and west to California. It is a less widespread species than the *Suaeda depressa* and *Salicornia rubra* salt pan communities. No reports in the literature have been found indicating that *S. verrucosum* occurs from northern Kansas to Canada in the prairie region (Morley 1964; Ungar, Hogan, and McClelland 1969; Hadley and Buccos 1967; Dodd and Coupland 1966b). *S. verrucosum* is one of the most salt tolerant species occurring in inland North America, commonly occurring on saline flats in pure stands or sometimes associated with *Suaeda depressa*. It has been reported as a pioneer species in saline soils of Oklahoma by Ortenburger and Bird (1931), Baalman (1965), and Ungar (1968b). In central Kansas it was found occasionally on the Big Salt Marsh (Ungar 1964, 1965a) and behaved like an annual. In Oklahoma and throughout its range it is described as a perennial herb, flowering from April to November (Rydberg 1932).

Ortenburger and Bird (1931) felt *S. verrucosum* plays a part in a vegetational cycle on the Great Salt Plains. Vegetation advanced beyond this stage during rainy periods, but with increased drought the pioneer stage developed again. Similar conclusions were reached by Ungar (1968b, 1969) in a salt flat community in which *S. depressa* and *S. verrucosum* played the pioneer role in a cyclical series in which *Sporobolus airoides* was the next invader. This indicates that zonation in these inland saline areas may be static and not successional. Cycles within zones of vegetation occur, but they do not necessarily lead to the climax prairie vegetation. *Sesuvium verrucosum* appears to be more drought tolerant than some of the grass species, and it is also certainly more salt tolerant than *S. airoides* or the later invading prairie species.

Typically, *S. verrucosum* has been found growing on salt pans in a unistratal community (Baalman 1965; Ungar 1965a, 1968b). It was found growing in small colonies on highly saline pans in Oklahoma (Ungar 1968b). In this community, *S. verrucosum* made up 93.8% of the relative density and 93.9% of the relative cover, occurring in all plots representing this community type. Baalman (1965) reported it as occurring in pure stands, and the only vascular plant species found associated with it by Ungar (1968b) was *Suaeda depressa*. These salt pan areas had a total surface salt content of 2.3%. In areas of reduced salinity, averaging less than 1.3% total salts, it was occasionally found in a dwarf *D. stricta* community. *Sesuvium* was rarely found in multistratal communities, but did occur in the multilayered *Tamarix pentandra* stands. When associated with *T. pentandra*, it was not found directly beneath the trees, but it occurred in bare patches between trees. Its occurrance in other parts of its range is poorly known, but Mason (1957) describes it as occurring in low alkaline flats and flood plains in California. Since *S. verrucosum* appears to be drought and salt tolerant, one would expect to find it in fresh-water marshes and wetter prairies. It does not occur in these areas, probably due to its low habit, shade in tolerance, and slower growth.

B. Soils

In the Great Salt Plains, soils in the 0-10 cm layer averages 2.2% total salts, ranging from 0.2% to 4.5% (Table 6). Fifty percent of the total ions in this area were chloride, with NaCl being the most common salt (Ungar 1968b). Salinity in the subsurface (40-50 cm) soil ranged from 0.27% to 0.51%, averaging 0.39%. Seasonal change in salinity occurs due to a general drying trend in the mid-summer, and sporadic rains will cause great changes in osmotic levels. During 1964, June surface salinities averaged 1.9%, July 3.09%, Aug. 2.5%, and Sept. 1.3%.

Saturation percentages of the soil ranged from 22% to 68% and averaged 44.0%. Soil moisture measurements ranged from 3% to 45%, averaging 17.5%. Soil moisture content varied but averaged close to what would be field capacity. During heavy rain periods, there would be standing water in the areas *Sesuvium verrucosum* grows, in both Kansas and Oklahoma.

Soil pH ranged from 7.0 to 9.7 with a median of 8.1 in this community. Since pH was quite variable, it is probably not one of the more important determiners of plant distribution in this area.

Factors relating to depth of water table, like microtopography, are of primary importance in *S. verrucosum* distribution. It was nearly always found at the lowest microelevations. Two or more factors varied, however, with increased elevation, and it is difficult to determine which were more important. As elevation increased, surface moisture and salinity decreased. With a decrease in salinity there was more competition from prairie species. In low wet areas, which were less saline *Distichlis stricta* and *Scirpus paludosus* offered competition to *S. verrucosum*, and it was not prominent.

C. Physiology

Seed germination studies with *Sesuvium verrucosum* indicate that it has a seed coat induced dormancy. Unscarified seed, germinated in distilled water, had a germination percentage of 23% after twenty days; scarified seeds had germination percentages of 42% in distilled water. Germination was reduced with added salt increments (Table 7).

Preliminary growth studies indicate that *S. verrucosum* can grow normally without addition of excess salts and that it is drought resistant as well as being resistant to high salinities. Plants averaged 15% dry matter and 85% water.

Suaeda depressa Community

A. Vegetation

Suaeda depressa (Pursh) Wats. is a succulent-leaved halophyte, and one of the primary invaders of saline soils from the prairie and plains region to the Pacific Coast. Distributional-ecological data concerning *Suaeda depressa* have been reported by Kearney et al. (1914) and Flowers (1934) in Utah;

Table 6. Soil Characteristics from Great Salt Plains.

	Sesuvium Community	Bare Ground
Texture	sandy loam to clay	sand to sandy loam
Conductivity mmhos/cm	81.5	154.0
Saturation Percentage	44.0%	25.0%
Soil Moisture Percentage	17.5%	17.7%
Total Salts Percentage	2.2%	2.5%
Chlorides Percentage	1.1%	1.5%
pH median	8.1	7.8
Organic Matter Percentage	7.2%	0.4%

Table 7. Seed Germination in Sesuvium verrucosum.

% NaCl	% Germination Scarified Seed	% Germination Unscarified Seed
0.5	27%	16%
1.0	20%	7%
2.0	10%	2%
3.0	3%	0%
5.0	0%	0%

Coupland (1950), Keith (1958), and Rawson and Moore (1944) in Canada; Schaffner (1898), Gates (1940), and Ungar (1962, 1964, 1965a, 1965b, 1966, 1967b) in Kansas; Pound and Clements (1898), Weaver (1918), and Ungar (1965b) in Nebraska; Ortenburger and Bird (1931), Baalman (1965), and Ungar (1965b, 1967b) in Oklahoma; and Hanson and Whitman (1938) in North Dakota. It is widely distributed west of the Mississippi River, but narrowly limited ecologically to wet saline soils. *S. depressa* is, therefore, never a dominant or characteristic species in a state's flora, except locally in those small areas which are governed by saline soils. The genus *Suaeda*, as well as other halophytes *Salicornia*, *Spergularia*, and *Limonium*, are superficially world-wide in distribution, but all are narrowly limited to saline soil conditions.

Three common growth forms are represented in *S. depressa* natural habitat. In highly saline soils it may be dwarfed and possess a prostrate growth form or appear as a weak upright, single-stemmed plant. In locations of lowered salinity it shows much more robust growth and may have a single main shoot or, commonly, it occurs with several main shoots due to branching at the base. Highly saline soils contained plants ranging from 1 to 20 cm tall, while in areas of reduced salinity they commonly ranged from 30 to 70 cm and some individuals were found up to 100 cm. Maples (1968) and Williams and Ungar (1972) have found that salinity and nitrogen concentration both affect its growth form.

Suaeda depressa occurs in several communities in inland salt marshes and salt deserts. It is a pioneer species on barren salt flats and may dominate a unistratal community or may occur with *Distichlis stricta* (Torr.) Rydb., *Salicornia rubra* Nels., *Sesuvium verrucosum* Raf., *Sporobolus airoides* Torr., or *Sporobolus texanus* Vasey as a pioneer invader. *Suaeda depressa* made up over 90.0% of the relative density in all stands in which it was the pioneer invader (Table 8). Its relative frequency was always over 90.0% and it was the characteristic species under conditions of highest salinity. Actual frequency in plots varied and could be as low as 30.0% due to large amount of bare area in this community. Cover values were low, usually averaging less than 1.0%. *Suaeda* also occurs in less saline communities, bordering salt flats with *Polygonum ramosissumum* Michs., *P. aviculare*, *Atriplex patula* var. *hastata* (L.) Gray, *Hordeum jubatum* L., *Iva annua* L., *Distichlis stricta*, *Leptochloa fascicularis*, and *Atriplex argentea*. In multistratal communities, one finds it with *Scirpus paludosus* Nels., *Sporobolus airoides*, *Spartina pectinata* Link., *Scirpus americanus* Pers., *Tamarix pentandra* Pall., or *Baccharis salicina* T. and G. It is never very important in these bi- or multistratal communitites in which the taller species shade it. Good growth takes place in bare patches between the species mentioned above, but it is never found growing with any vigor directly under them, indicating a possible need for high light intensities. Chapman (1947a) reports that *Suaeda maritima* (L.) Dum. also shows poor growth when shaded by other species. Table 8 presents density-frequency relations of *S. depressa* in several communities in Oklahoma. These data show

Table 8. Density-Frequency relations of *Suaeda depressa* in various plant communities of the Great Salt Plains, Alfalfa County, Oklahoma, and the Big Salt Marsh, Stafford County, Kansas. Density is based on total stems transected by 10 three-meter lines in each community. + = rare

	Density No.	Relative Density %	Frequency %
Great Salt Plains Alfalfa County, Oklahoma			
Suaeda depressa	26	100.0	100.0
Sesuvium verucosum	+	+	+
Big Salt Marsh Stafford County, Kansas			
Suaeda depressa	8	100.0	30.0
Sesuvium verucosum	+	+	+
Salicornia rubra	+	+	+

similar relations to those reported by Ungar (1965a, 1967b) in Kansas and Keith (1958) in Canada.

Field observations in Cloud County, Kansas, indicate that the density of *S. depressa* is very strongly dependent upon the amount of rainfall and standing water. With high rainfall many of the hundreds of young seedlings produced in early spring will survive, causing dense growth, 70-120 plants per square foot. The usual dry periods in July and August caused the soils to dry out and most seedlings will die. Under these conditions only one out of a thousand seedlings can survive. Weaver (1918), in his observations of changes in community make-up in saline areas around Lincoln, Nebraska, noted striking mortality of *S. depressa* from June to July 1917. A drought in July, 0.56 inches of rain compared to the normal 4.01 inches, was the probable cause.

Suaeda depressa

Suaeda depressa also plays a role in an interesting cyclic occurrence on the marshes of southern Kansas and northern Oklahoma. The cycle starts with *Sporobolus airoides* invading the salt flats, either on soil accumulated by succulents *Suaeda depressa* or *Sesuvium verucossum* as suggested by Baalman (1965), or as a primary invader on the barren flats. *Sporobolus* does not have as high a salt tolerance as *Suaeda* and when the salinity in the soil rises sharply, the perennial *Sporobolus* will begin dying off and *S. depressa* occupies these sites. It appears as if we have a replacement cycle here with *S. airoides* replacing *S. depressa* and, then with sharp prolonged increases in salinity, *Suaeda* occupies the hummocks produced by *Sporobolus*. If no sharp increase in salinity occurs, *Sporobolus* will accumulate sand and form well-drained hummocks that are low in moisture and salinity, which are eventually invaded by prairie species.

B. Soil

S. depressa occupies soils with a wide variety of surface soil textures, 0-10 cm depth, ranging from sand to clay. Earlier work in Kansas (Ungar 1965a) indicated that it was most common on heavy soils with sandy clay loam to clay soils predominating. The subsoil in Kansas ranged from sand to sandy clay loam, while in Oklahoma, sand is the predominant type. Soil texture does not appear to be limiting, but this species is commonly found in marshes in which the soils tend to be heavier.

Surface soil organic matter is usually low, 3.6% to 10% in pure stands of *Suaeda*. Much of this may be due to blue-green algal crusts which form on these soils since little if any litter is observed on the soil surface. Subsurface samples are very low in organic matter, 0.0% to 6.1%, averaging approximately half of the surface soil organic content (Table 9).

Suaeda depressa grows in soils which have a water content somewhere between field capacity and saturation. This high soil water content is due to the high-water table in these marshes, and often after heavy rainfalls there are 2-4 inches of standing water on the soil surface. Some wilting occurs if this surface water remains for long periods. During extended dry periods, the soil surface will crack and dry, but even at these times it is rare to find soils much below field capacity. In the Great Salt Plains, Alfalfa County, Oklahoma, the saturation percentage of the surface soils averaged 53% and the actual soil moisture on a dry weight basis averaged 25% which is about field capacity. The subsurface soils in most cases have moisture values approaching the saturation percentage and rarely below the field capacity. Relatively high soil moisture percentages tend to alleviate some of the salinity hazards.

When occurring in unistratal communities, *S. depressa* is usually found in soils with a p^H ranging from 7.3 to 8.8 and averaging $8.0 \pm .2$. These data are based on over 100 soil samples taken in Kansas, Oklahoma, and Nebraska (Table 9). When *Suaeda* is present in bi- or multistratal communitites, the soil pH may range from 6.8 neutral to 10.0 highly alkaline. The pH of soil does not appear to limit this species distribution within its geographical range. Flowers (1934) in Utah and Keith (1958) in Canada have reported *S. depressa* in soils with a pH of 8.4-9.8 and 7.1-8.2. Gates et al. (1956), working in salt deserts of Utah, also found that pH had no effect on species distribution.

Salinity is by far the most important single factor effecting the distribution of *S. depressa* can grow in soils with salinities up to 4.9% on a dry weight basis and in non-saline soils with as little as 0.04% total salts. These data are based on a saturation extract percentage and with drying out even higher salinity percentages, possibly double these values will be reached (Table 9). Seed germination will take place in up to 4.0% sodium chloride, but it is doubtful that normal growth takes place when the maximum salinities are reached. Williams and Ungar (1972) report that *S. depressa* can complete its life cycle in up to 4.0% NaCl but growth is greatly reduced.

Keith (1958) has reported *S. depressa* growing in soils with salinities ranging from 1.0% to 2.4%; Kearney et al. (1914), 0.25% to 2.30%; Weaver

Table 9. Soil characteristics of the habitats of Suaeda depressa at 0 — 10 cm.

	SAMPLES	SOIL MOISTURE %	ORGANIC MATTER %	CHLORIDES	TOTAL SALT %	CONDUCTIVITY Mmhos./25½C	pH median
Great Salt Plains (Ungar, 1968b) Alfalfa County, Oklahoma							
Suaeda depressa Community	24	25.4	7.2	0.81	1.62	56.8	8.1
Distichlis-Suaeda Community	24	25.5	10.0	0.62	1.27	41.5	8.2
Distichlis-Hordeum Community	16	21.8	9.0	0.08	0.19	5.6	8.2
Big Salt Marsh (Ungar, 1965) Stafford County, Kansas							
Suaeda depressa Community	10	---	3.6	1.08	1.81	69.5	8.2
Distichlis-Suaeda Community	10	---	9.3	0.69	1.13	38.8	8.0
Distichlis-Hordeum Community	10	---	2.5	0.24	0.59	16.3	7.9
Lincoln Marshes (Ungar et al. 1969) Lincoln, Nebraska							
Suaeda depressa Community	15	31.9	---	1.75	3.59	45.3	7.7
Distichlis-Suaeda Community	10	25.4	---	2.06	4.00	64.0	7.8
Distichlis-Hordeum Community	10	37.4	---	0.30	0.76	12.9	7.5

and Clements (1938), 1.8%; and Flowers (1934) 0.2 to 4.0%. These data indicate a very wide salt tolerance and an ability to adapt to a broad range of nonsaline and saline conditions.

C. Physiology

Germination of the embryo is usually through a small beaklike projection of the seed, and usually the radicle emerges first. Rarely the plumule end of the embryo emerges before the radicle. Seeds are dormant in the laboratory and, unless some special treatment is given them, no germination will take place (Ungar and Capilupo 1969). Chapman (1947a, 1947b) has reported low germination in *Suaeda maritima* (L.) Dum. and *Suaeda fruticosa* Forsk. Binet (1960) has reported that acid treatment would break the dormancy of *Suaeda maritima*; and he also had success with stratification at 3 C.

Preliminary tests indicated that *S. depressa* seed would germinate if they were scarified in the beak area (Ungar 1962, 1965a). Sulfuric acid treatment for 10 minutes yielded 44% germination, mechanical scarification 46%, and stratification at 3 C for thirty days 42%. Stratification of seed at -2 C yielded 4% germination indicating temperatures below 0 C do not have the same stimulatory effect as 3 C (Ungar and Capilupo 1969). Alternating temperatures of 3 C and 25 C resulted in highest germination percentages. Preliminary results with *S. depressa* indicated high rates of water absorption after a relatively short period of time (Ungar and Capilupo 1969). This corroborated the results of Boucaud (1962) with other species of *Suaeda* indicating that water is taken up by the seed and the seed coats are not impermeable. Growth of the radicle appears to be impeded by the hard seed coat, preventing emergence of the seedling.

Seeds were germinated with and without a 3 C cold treatment and treated with gibberellic acid. These data indicate that high concentrations of gibberellic acid, 500 ppm, inhibit germination; however, greatest promotion of germination was found at 100 ppm (Ungar and Capilupo 1969). Our findings show that gibberellic acid does substitute for the cold treatment but it is not able to replace it entirely. Possibly two dormancies exist, one involving a hard seed coat which scarification or alternating temperatures would break, and a second embryonic dormancy which gibberellic acid alleviates.

Early germination experiments by Ungar (1965a) indicated an optimum germination in 0.5% NaCl at room temperature while experiments by Williams and Ungar (1972) indicated reductions in germination at 5-15 C temperatures with each salt increment. Ungar (1965b), Binet (1965a), Springfield (1966), and Malcolm (1964) have all found that germination of halophyte seed may be greater at some osmotic pressure higher than distilled water, and then at other temperatures the reverse would be true. The maximum salinity at which *S. depressa* seeds were found to germinate was 4.0% NaCl (Ungar and Capilupo 1969, Williams and Ungar 1972).

These data, as those of Ungar (1969) and Williams and Ungar (1972), indicate that the chief effect of high salinity is not lethal but in this case appears to be osmotic. There is also the possibility that the salt ions are inhibiting some metabolic processes involved in germination. An experiment was run with isotonic solutions of mannitol, sodium chloride, magnesium sulfate, and sodium carbonate, and these data corroborate that found by testing recovery of seed, indicating that the chief effect of excess salts is osmotic (Table 10).

Binet (1963) found that 1.0% NaCl solutions added to a nutrient solution produced better growth in *S. vulgaris* than in nutrient solution. Williams and Ungar (1972) investigations with *S. depressa* also indicated improved growth at 1.0% NaCl with adequate available nitrogen. The research of Ungar (1965b), Binet (1965b), and Williams and Ungar (1972) indicates that some species of *Suaeda* have very slow growth rates and this may explain why *S. depressa* is not able to compete and does occur in wet nonsaline areas.

Vegetative growth begins at the time of first seed germination in late March and early April in the field. Growth is slow and flowering takes place in late July, August, and September, and from preliminary experiments, appears to be stimulated by short days in the fall. The fruit develops in September, October, and early November. Large numbers of seed which germinate in the spring are produced by each plant but only a few seedlings survive due to the usual dryness and higher salinity of the soil in July and August. If heavy rains occur, more seed will germinate late in the season and these smaller plants will flower and fruit. Laboratory experiments also indicate that flowering is stimulated by short day photoperiods (Williams and Ungar 1972).

Allenrolfea occidentalis **Community**

A. Vegetation

Allenrolfea occidentalis (S. Wats.) Kuntze. is one of the most salt tolerant succulent shrub species occurring in salt pans. Distribution records indicate that it occurs from the California deserts east to Utah and in the southwest east to Texas (Mason 1957). It is more restricted in distribution than the other succulent species, and it has not migrated into the prairie region. Observations of *A. occidentalis* occurring in saline soil in the western states have been made by several researchers (Kearney et al. 1914; Flowers 1934; Billings 1945, 1949; Nichol 1962; Hunt and Durrell 1966; Branson, Miller, and McQueen 1967).

Kearney et al. (1914) and Vest (1962) describe the typical habitat for *A. occidentalis* as low hummocks on salt flats in Utah. It usually dominates these hummocks, forming almost pure stands, and is also apparently responsible for their growth and development. Rarely it can be found invading the open salt flats. Flowers (1934) feels that *A. occidentalis* is a secondary invader of saline soils after the species of *Salicornia*. Hummocks are formed from wind-blown sand caught by its vegetative parts. The distribution of *A. occidentalis* in a zone behind the annual *Salicornia* and *Suaeda* species indicates a lower degree of salt tolerance. He also found that *A. occidentalis* occurred in soils averaging one-half the salinity of the annual *Salicornia* communities. Vest (1962) reports that few other species invade hummocks occupied by *A.*

Table 10. Percentage germination of Suaeda depressa in osmotic solutions after 30 days. Five-day intervals of 3 C and 25 C. First five days 3 C. Five 25-seed replications used at each osmotic pressure (Ungar and Capilupo, 1969).

Days	Distilled Water	Sodium Chloride Atmosphere						Magnesium Sulfate Atmosphere						Sodium Bicarbonate Atmosphere						Mannitol Atmosphere					
		5	10	15	20	30	40	5	10	15	20	30	40	5	10	15	20	30	40	5	10	15	20	30	40
10	8	5	2	0	3	0	0	9	3	0	0	0	0	3	1	0	0	0	0	6	1	2	3	0[1]	0
20	28	13	4	3	4	0	0	16	5	5	1	0	0	13	6	6	0	0	0	13	3	3	3	0	0
30	35	30	13	3	4	0	0	30	24	5	1	0	0	27	20	6	0	0	0	27	11	3	5	0	0

1 Mannitol precipitates

occidentalis and in a quantitative study he found 503 plants in 16 random samples of 30 square meters; 92.4% of all plants counted were pickleweed, 7.4% *Atriplex nuttallii*, and 0.2% *Sarcobatus vermiculatus*. Kearney et al. (1914) report that *A. occidentalis* is associated with *Sarcobatus vermiculatus* and *Suaeda moquinii* on hummocks and with *Salicornia utahensis* in wetter depressions in Tooele Valley, Utah. In Nevada, *A. occidentalis* has been reported to occur in pure stands or in an herbaceous strata of *Distichlis stricta* between the bushes. Quantitative data indicated that *A. occidentalis* cover averaged 6.0% with 0.7 plants found per square meter. These soils were highly saline and had a pH of 8.8.

Marks (1950) in studies in the Lower Colorado Desert found *A. occidentalis* in pure dense stands on moist saline soils. He feels that the deeper rooted *A. occidentalis* indicates high salinity throughout the soil profile, while adjoining *Suaeda torreyana* communities indicate only surface soil salinity. It usually is an indicator of soils which are heavy textured ranging from silt loam to clay.

Studies in Death Valley, California, by Hunt and Durrell (1966) indicate that *A. occidentalis* forms nearly pure stands in an area ranging from ¼ of a mile to 2 miles wide bordering the west side of the salt pan. He describes mounds of two feet in height with a diameter of four or five feet. The spread of *A. occidentalis* vegetatively is due partly to underground runners, a few inches below soil surface, and partly by depressed branches on the soil surface. Plants in this area average less than a foot in height. Vest (1962) reports that some plants reach about 18 inches on the outer parts of hummocks which are scattered from 6 to 30 feet apart on the borders of the salt flats.

B. Soils

Salicornia stands usually occupy more poorly drained sites which are also more saline than those occupied by *A. occidentalis* (Flowers 1934). In the Great Salt Lake region he reports *Salicornia* maximum soil salinities of 6.0%, while the *Allenrolfea* maximum is 3.0% with optimum growth at 1.0% to 1.5% soil salinity for the latter. Hunt and Durrell (1966) report soil salinities as high as 3.0% with ground water salinities reaching as high as 10.0%. Thirty-five samples taken in soil below *A. occidentalis* plants by Marks (1950) had an average salinity of 1.9%, while *Suaeda* soils in this area average only 0.56% total salts. Kearney et al. (1914) report salinities ranging from 0.25% to 2.18% and an average surface soil salinity in the *Allenrolfea* community of 1.26% in Tooele Valley, Utah. The pH of these soils ranged from 6.5 to 8.2. In most areas studied in Death Valley, sodium chloride was the predominant salt but in some areas, sulfates were more abundant making up 1.2% of the total salts.

C. Physiological Data Concerning *Allenrolfea occidentalis*

Gold (1939) reported that *Allenrolfea occidentalis* seeds had high viability, and germination ranged from 85.0% to 90.0% in distilled water at room temperature. Scarified seeds had the high germination percentages reported above, whereas unscarified seed had only 1.0% to 2.0% germination. Cold treatment also stimulated 85.0% germination.

Salinity tests on germination indicated no or little retardation in germination in up to 1.5% NaCl solutions depending upon the temperature regime used (Gold 1939). They were placed in tap water and germination was not inhibited with seed in sodium sulphate solutions having 92.0% and those in Great Salt Lake brine 85.0% germination.

Seedling growth in Great Salt Lake dilutions equalling about 0.5% to 4.0% NaCl was not gradually reduced as one would expect from germination studies (Gold 1939). Seedline growth ranged from 1.8 cm at 4.0% to 2.9 at 3.0%.

Wiebe and Walter (1972) studied water soluble mineral ion concentrations in halophytes in the Curlew Valley, Utah. They found that *Allenrolfea occidentalis* had 1,044 mmoles/per liter tissue water sodium and 526 mmole chloride. Other halophytes in extremely saline soils including *Suaeda depressa, Salicornia rubra,* and *Distichlis stricta* had sodium concentrations of 1,093, 574, and 410/mmoles per liter and chloride values of 517, 535, and 364 mmoles per liter respectively. They felt that the lower sodium chloride content in *D. stricta* may be due to that species ability to excrete salts whereas the succulents stored sodium and chloride.

Bernstein (1967) has reported that *A. occidentalis* requires some salt increment for normal growth. Gold (1939) found that seedling growth was not reduced in tap water, but optimum growth occurred in solutions with a salt increment.

Distichlis stricta Community
A. Vegetation

This *Distichlis stricta* (Torr.) Rydb. [=*D. spicata* of early reports on inland soils and for inland soils by Penfound (1967)] community occurs in nearly pure stands in inland saline areas with a relatively high salinity and a high water table during at least part of the growing season (Aldous and Shantz 1924, Shantz and Piemeisel 1924, Daubenmire 1940, Marks 1950, Richards 1954, Weaver 1954). *D. stricta* is a widespread species in the western United States and Canada (Coupland 1950, Robinson 1958). Numerous authors including Harshberger (1911), Hilgard (1914), Kearney et al. (1914), Flowers (1934), Shreve (1942), Billings (1945), Bindschadler (1948), Bolen (1964), and Franklin and Dyrness (1969) have cited *D. stricta* as being one of the controlling species in large parts of inland salt pans and salt marshes. Bauer (1930) reports *D. stricta* as one of the dominants in the saline desert area of the Tehachapi Mountains, California.

In most of the areas in which it occurs, *D. stricta*, which is a shallow rooted rhizomatous perennial, is found in several plant communities or species combinations. It may be a co-dominant as it often is with *Hordeum jubatum, Puccinellia nuttalliana, Agropyron smithii,* and *Scirpus paludosus*.

This apparently depends a great deal on soil salinity concentrations, moisture relationships, availability of seed, and microtopography. At salinity extremes, where other species cannot survive, *D. stricta* becomes dominant. It also occurs in drier soils than some of its co-dominants. At lower salinities, averaging about 0.5% total salts, where most prairie species cannot survive, it may form monospecific unistratal stands. The number of species reported in *D. stricta* communities ranges from as low as two reported in Alberta, Canada, by Keith (1958) to as high as 26 reported by Dodd and Coupland (1966b) in Saskatchewan, Canada. The sampling of Dodd and Coupland (1966b) in Saskatchewan indicates that 22 of the 26 species reported from 28 different sample sites had a frequency of less than 10.0%, while *D. stricta* had a relative dominance value of over 84.0%. No other species in this community had a relative dominance value of over 6.0%. The more important associated species were: *Puccinellia nuttalliana, Hordeum jubatum, Chenopodium rubrum, Suaeda erecta (=S. depressa), Glaux maritima,* and *Aster ericoides*. Species found with lower frequency in Saskatchewan such as *Glycyrrhiza lepidota, Achillea lanulosa, Lepidium densiflorum* et al. were usually located in areas of lower salinity in more southern locations of the prairie studied by Ungar (1965a, 1970a). Keith (1958), in his study of ponds in Alberta, Canada, found it to be associated with *Chenopodium salinum, Suaeda depressa, Puccinellia nuttalliana,* and *Atriplex argentea* in soils with a salinity ranging from 1.0% to 2.4% total salts. In Stafford County, Kansas, *D. stricta* cover averaged 12.3% in nearly pure stands where soil salinity averaged about 0.5% and a high water table exists. A summary of frequency data for species occurring in various parts of the United States and Canada for the *D. stricta* community is included in Table 11. These data indicate that *D. stricta* when growing under most favorable conditions of salinity and moisture, form nearly pure stands (Bolen 1964, Ungar 1965a). Similar results were found by Dodd and Coupland (1966b) in saline soils of Saskatchewan in which it makes up 63% to 100% of the total graminoid cover. Total cover in this community type has been reported to range from a low of 3.0% in Lincoln, Nebraska, (Ungar, Hogan, and McClelland 1969) to as high as 20% in central Kansas and Saskatchewan, Canada (Ungar 1965a, Dodd and Coupland 1966).

Distichlis stricta occurs with prairie species and in other less saline communities than this, and also in extremely saline communities, the Dwarf *D. stricta*, and with *Salicornia rubra* and *Suaeda depressa*. In the Dwarf *D. stricta* areas the salinity averages 1.1% and *D. stricta* cover averages 1.2% (Ungar 1965a).

The species found in the *Distichlis-Hordeum* and *Hordeum* communities are often similar to those found in the *Distichlis stricta* community (Tables 11, 22). *Hordeum jubatum* is not as salt tolerant as *D. stricta* and in Lincoln, Nebraska, and Cloud County, Kansas, appears to characterize slightly wetter soils and a drop in microtopography. These three areas have been treated as a single unit by Ungar (1967b) where it is named a *Distichlis-Hordeum* complex. In this analysis of the marsh vegetation, *Hordeum* occurs in 91.0% and *Distichlis* in 47.0% of the one hundred .2 X .2 quadrats. Other species and their respective frequencies were *A. patula* var *hastata* 52.0%, *Iva annua*

Table 11. Percentage frequency of species in *Distichlis stricta* communities (+ = present, but not in quantitative data).

Total Salts - Location -	1.0% Alberta, Canada Keith (1958)	0.6% Stafford Co., Kansas Ungar (1965a)	0.7% Cloud Co., Kansas Ungar (1967b)	1.5% Juab Co., Utah Bolen (1964)
Species				
Distichlis stricta	100	100	100	100
Atriplex patula var. hastata	---	+	70	---
A. argentea	25	+	---	---
Suaeda depressa	5	+	---	---
S. intermedia	---	---	---	+
Hordeum jubatum	---	+	20	---
Poa arida, Poa sp.	---	10	---	---
Kochia scoparia	---	+	10	---
Polygonum ramosissimum	---	---	30	---
Aster ericoides	---	+	---	---
Aster exilis	---	+	---	---
Scirpus paludosus	---	30	---	---
Cordylanthus canescens	---	---	---	+

32.0%, *Suaeda depressa* 16.0%, *Poa arida* 11.0%, and *Polygonum ramosissimum* 3.0%. Rarer species occurring were *A. smithii, Aster exilis, Kochia scoparia, Rumex crispus,* and *Scirpus americanus*.

B. Soils

Distichlis stricta has been found growing in soils high in carbonates and bicarbonates (Tolstead 1942), high in chlorides (Ungar 1965a; Ungar, Hogan and McClelland 1969), and high in sulfates (Flowers 1934, Ungar 1970a) indicating that these specific ions are not limiting for the distribution of *D. stricta*. It has a very broad range of salt tolerance and has been found growing in soils ranging from 0.03% to 5.6% total salts (Hunt and Durrell 1966, Ungar, Hogan, and McClelland 1969) (Table 12).

Soil pH did not appear to be a controlling factor for the *D. stricta* community. It was usually alkaline between 7.5 and 8.5, but the extremes ranged from 6.8 to 9.2. This broad pH range indicates *D. stricta* can grow in all marsh communities and that its absence from certain areas cannot, therefore, be attributed to minor pH variations which occur in the salt marsh environment (Table 12).

Soil texture ranges from sand and gravel (Hunt and Durrell 1966) to clay (Ungar 1968b). It grows in depressions which generally accumulate finer soils and so it is commonly associated with this type. It is capable of growing in coarser soils in which moisture is available. In Oklahoma it was found growing in soils ranging from sand to clay in texture (Ungar 1967b).

Soil organic matter in the *D. stricta* community is made up of dead vascular plant material and blue-green algal crusts. It has been found to average from as low as 2.5% to as high as 14.7%. In wetter sites, thick algal crusts develop, whereas on drier sandy soil, these are usually not present. *Distichlis* has been found growing in sandy soils in Oklahoma with as little as 0.3% organic content (Ungar 1967b).

Soils in this *D. stricta* community have a relatively high water table (Robinson 1958) and, therefore, are generally close to field capacity in moisture content. Hunt and Durrell (1966) reports a depth of 1-2 ft to the water table in Death Valley. During drought periods in summer this may change with a sharp drop in the water table. This community is generally not found in as moist a soil as the *Scirpus paludosus* community nor in the drier soils occupied by *S. airoides* in Oklahoma. It is generally absent from prairie, plain, or desert communities throughout its range. Soil moisture percentages in studies by Ungar (1968c, 1970a) from Oklahoma to South Dakota averaged 22% to 46%, while saturation percentages range from 58% to 78%.

C. Physiology

The broad tolerance of *D. stricta* to saline conditions has been reported by Hilgard (1914), Harris (1920), Richards et al. (1954), and Ungar (1970a). It has been found growing in soils with salt concentrations as low as 0.03% to 5.6% total salts. In more saline zones it takes a stunted growth form and total cover decreases sharply (Schaffner 1898, Ungar 1965a) forming a dwarf *D.*

Table 12. Soil - Distichlis stricta community.

Location	pH Median	Saturation Percentage %	Soil Moisture %	Conductivity mmhos/cm	Na %	Cl %	SO$_4$ %	Total Salts %
North Dakota Hadley and Buccos (1967)	---	---	---	23	---	---	---	1.5
Canada Keith (1958)	8.2	---	---	5	---	---	---	1.7
South Dakota Ungar (1970a)	7.9	59	40	7	0.1	0.0	0.4	1.1
Nebraska Ungar, Hogan, McClelland (1969)	7.9	77	30	9	0.2	0.3	0.1	0.6
Kansas Ungar (1967b)	7.6	78	46	14	---	---	0.3	0.7
Oklahoma Ungar (1968b)	8.2	58	22	6	---	---	0.1	0.2
Utah Flowers (1934)	8.6	---	---	---	---	---	---	0.8
California Hunt (1966)	8.5	---	---	---	0.4	0.2	0.7	1.3

stricta type of community.

Nielson (1956) has found that a dormancy exists in seed of *D. stricta*. Sandpaper scarification yielded germination up to 72.0% in distilled water, and with scarification plus 1.0% NaC1-21% germination and a treatment with 3.5% NaC1 plus scarification yielded 23.0% germination. Controls without scarification had from 0.0% to 3.0% germination. He found from field growth studies that after establishment it spreads primarily by rhizomes. Variation in morphological characteristics occurred from different seed sources in Utah, South Dakota, Nevada, Washington, and California indicating broad genetic variability (Nielson 1956).

Studies of evapotranspiration rates of *D. stricta* indicate an increase in water loss with an increase in the average temperature during the growing season (Robinson 1958). Data indicate that evapotranspiration rates are 68% to 75% of pan evaporation.

Distichlis stricta chief adaptation to high salinities is its ability to maintain high osmotic pressure in the cell sap. Dodd and Coupland (1966a) found a mean cell sap osmotic pressure of 30.4 atm with a range from 21.7 to 47.8 atm in a saline meadow community. The soils in this community had osmotic pressures of 8 atm during the drought of August 1957. *Distichlis* is also able to excrete salts which may allow it to overcome toxic effects of excessive ionic accumulation (Frey-Wyssling 1935).

Puccinellia nuttalliana Community

A. Vegetation

Puccinellia nuttalliana (Schultes) Hitchc. is a widespread species occurring in saline soils in the western half of the United States and Canada (Flowers 1934, Hanson and Whitman 1938, Hitchcock 1950, Nielsen 1953, Love and Love 1954, Mason 1957, Keith 1958, Hadley and Buccos 1967, Ungar 1970a).

Puccinellia nuttalliana is a perennial bunch grass that begins growth in Canada in mid-May, flowers late in June, and matures by late July at Culm heights of 30 to 60 cm (Dodd and Coupland 1966b). Ungar (1970a) found a similar phenology in South Dakota with flowering in June and fruit maturing by July.

Dodd and Coupland (1966b) consider *Puccinellia nuttalliana* the most salt tolerant grass in saline soils of Saskatchewan with only the succulent forbs *Triglochin maritima, Salicornia rubra*, and *Suaeda depressa* occurring in more saline soils. *Puccinellia nuttalliana* was found growing in highly saline soils of South Dakota, ranging from 1.0% to 3.6% total salts (Ungar 1970a). Only *S. rubra* and *S. depressa* occurred in soils of higher mean salinity. Field investigations and laboratory experiments with *P. nuttalliana* indicates that it can grow in areas of low salinity and in nonsaline soils (Ungar 1970a, Macke and Ungar 1971). Why *P. nuttalliana* is not more common in nonsaline prairies of freshwater marshes is not known, but it is apparently due to an inability to compete with more robust prairie and fresh-water marsh species.

Puccinellia nuttalliana was a common species bordering saline pans in northern South Dakota (Ungar 1970a). It made up 75.1% relative cover in the *P. nuttalliana* community (Table 13). Associated with it were *Salicornia rubra, Distichlis stricta*, and *Suaeda depressa*, all having a frequency fo 50.0% or more but less than 10% relative cover. It also occurred occassionally in the more saline *Salicornia rubra* zone as well as in the less saline *Distichlis-Hordeum* areas. It is also common on saline soils of North Dakota (Hanson and Whitman 1937, 1938; Sloan 1970). Hanson and Whitman (1938) found *P. nuttalliana* dominating low stream terraces and depressions in which drainage was poor. Other graminoids frequently occurring with it were *Distichlis stricta, Agropyron smithii*, and *Hordeum jubatum*. Dodd and Coupland (1966b) found *P. nuttalliana* in seven saline communities. It dominated the *P. nuttalliana* community, 86.7% relative cover and the *Puccinellia-Distichlis* community, 46.5%. In the former community, *Salicornia rubra, Distichlis stricta, Triglochin maritima*, and *Hordeum jubatum* made up the remaining herbaceous cover with no species equalling 5.0% relative cover. In the *Puccinellia-Distichlis* community, *D. stricta* was the co-dominant making up 42.7% of the relative cover with other species mentioned above plus *Poa canbyi, Agropyron* spp., and *Muhlenbergia richardsonis* making up the remainder. Dodd and Coupland (1966b) found small percentage, lower than 5.1%, of *P. nuttalliana* in communities dominated by *S. rubra, T. maritima, H. jubatum, D. stricta* and *D. stricta-Agropyron*. This illustrates a broad tolerance to edaphic conditions and indicates that competition from other species at lower salinities and tolerance at the highest salinities were limiting to *P. nuttalliana* distribution.

Table 13. Percentage relative dominance and frequency of species in the *Puccinellia nuttalliana* community (+ = present, but not included in quantitative data). 1. Saskatchewan, Canada (Dodd and Coupland 1966b). 2. Codington County, South Dakota (Ungar 1970a). 3. Park County, Colorado.

	Locations		
	1.	2.	
Total Salts (mean)		2.8%	
Species	R. Do.	R. Do.	Freq.
Puccinellia nuttalliana	86.7	75.1	100.0
Salicornia rubra	4.7	11.2	100.0
Distichlis stricta	4.6	10.8	60.0
Hordeum jubatum	1.1	+	+
Triglochin maritima	2.9	-------	----------
Suaeda depressa	+	2.9	50.0
Atriplex patula var. *hastata*	-------	+	+

P. nuttalliana ranged from 81%-95% of the vegetation cover in the sites it dominated. A large number of forbs (22) occurred in the 23 sites sampled by Dodd and Coupland (1966b). The frequency of all but five of these was less than 4.0% *Chenopodium rubrum* occurred in 15.3%, *Suaeda erecta* in 18.8%, *Spergularia salina* in 41.1%, *Chenopodium salinum* in 7.7%, and *Crepis glauca* in 6.5%.

The overwhelming dominance of *P. nuttalliana* in the community it characterizes appears to be the chief factor delimiting this community. Throughout its range, as is the case for almost all other halophytic communities, associated species may vary, but the physiognomic and quantitative dominance of *P. nuttalliana* characterize this community.

B. Soils

Puccinellia nuttalliana is an important component of saline soils in the northern and western parts of its range (Flowers 1934; Dodd, Rennie, and Coupland 1964; Dodd and Coupland 1966b; Ungar 1970). In Kansas, Oklahoma, and Nebraska it is replaced by *D. stricta* in highly saline soils (Ungar 1966; Ungar, Hogan, and McClelland 1969). Dodd, Rennie, and Coupland (1964) report mean conductivities in Saskatchewan samples of 28.1 mmhos in 22 samples at the 0-6 cm level in *P. nuttalliana* stands. In four soil samples of a *Puccinellia-Distichlis* stand, surface salinity averaged 23.1 mmhos. Both of these values were obtained using the saturation extract technique while Ungar (1970a), using South Dakota material, found a mean value from 15 samples of 18.0 mmhos using the 1:1 soil extract procedure (Table 14). Converting these to field capacity values, with saturation values being about ½ and these 1:1 values 1/3 those expected at field capacity values of 56.2 mmhos, 46.2 mmhos, 54.0 mmhos are obtained for each case respectively. These soils contain about 50.0% sulfates with magnesium, calcium, and sodium being abundant in the cation fraction. Chlorides occur but make less than 0.2% of the total salts. Total salinity in the surface soils range from 1.0% to 6.0%, averaging 2.8%, while subsurface salinities range from 0.6% to 5.5% in the South Dakota Stink and Bitter lakes.

Dodd, Rennie, and Coupland (1964) report a median pH of 8.3 in Saskatchewan soils dominated by *Puccinellia nuttalliana*, with Hanson and Whitman reporting a range from 7.7 to 8.6 in saline North Dakota soils containing a *D. stricta-Puccinellia nuttalliana* community. In South Dakota soils studied by Ungar (1970a), soil pH ranged from 8.0 to 8.6. *P. nuttalliana* is associated with moderately to highly alkaline soil conditions.

Soil texture is quite variable in *P. nuttalliana* stands ranging from sandy loam to heavy clay (Dodd, Rennie, and Coupland 1964; Ungar 1970a) and appear not to strongly influence *P. nuttalliana* distribution.

Puccinellia occurs in several parts of marsh moisture gradients. It dominates in soils which average 25.2% moisture. Since these soils have saturation percentages of 50.0, those values appear to be close to the field capacity (Richards 1954, Ungar 1970a). Soil moisture varies throughout the growing season with values ranging from 15.3% to 75.4% in areas it

Table 14. *Puccinellia* community Soils.

Location	pH median	Saturation %	Soil Moisture %	Conductivity mmhos/cm	Na %	Cl %	SO$_4$ %	Total Salts %
Canada Dodd, Rennie and Coupland (1964)	8.3	--	--	28.1	290 meq/l	M	H	1.8
North Dakota Hanson and Whitman (1938)	7.7	--	--					
South Dakota Ungar (1970)	8.2	50.0	25.2	18.0	0.2	0.1	1.4	2.8
Colorado								

273

dominates, averaging 25.2% or about field capacity at Stink Lake, South Dakota (Ungar 1970a). As in other marsh communities, droughts and sporadic rainy periods greatly influence moisture relationships.

C. Physiology

Puccinellia nuttalliana can germinate at osmotic potentials of -16 bars in solutions of sodium sulfate, sodium bicarbonate, sodium chloride, and ethylene glycol (Macke and Ungar 1971). Germination percentages in distilled water, -4 and -8 bars, were approximately equal averaging 84.0 under these conditions. At -12 bars germination was sharply reduced to 35.0% with only 3.2% germinating at -16 bars. No seed germinated in -25 bars. Soaking seed for 25 days at -42 bars did not inhibit germination in any salt solution. After this treatment, 84.0% of the seed germinated in distilled water (Macke and Ungar 1971).

Seedling growth of *P. nuttalliana* reached an optimum at an osmotic potential of -4 bars with the greatest reduction in growth coming between -8 and -16 bars. Chapman (1960) has reported similar results for *P. maritima*.

Studies of photoperiod responses indicate that *P. nuttalliana* is stimulated to flower by long days in early summer (Macke and Ungar 1971). This correlates well with its behavior in the field with new growth beginning in May, flowering in June, and fruit ripening in July.

Triglochin maritima Community

A. Vegetation

Triglochin maritima L. is a widespread species occurring in saline and nonsaline soils having a high water table in the United States and Canada (Rydberg 1932, Rawson and Moore 1944, Dodd and Coupland 1966b, Ungar 1971). Baalman (1965) and Ungar (1968b) did not find *T. maritima* growing on the Great Salt Plains in northern Oklahoma nor has it been found in other salt marsh areas of the eastern prairie region (Hadley and Buccos 1967, Ungar 1970). It was not reported to occur in Kansas by Gates (1940) and Barkley (1968) nor was it recorded in a flora of Oklahoma by Waterfall (1952). Muenscher (1944) cites *T. maritima* as occurring in central Nebraska, South Dakota, and North Dakota.

Dodd and Coupland (1966b) found that *T. maritima* was the dominant near the saline depressions beyond the *Salicornia rubra* zone in saline marshes studied in Saskatchewan. They also found *T. maritima* bordering areas that were completely unvegetated. *Triglochin maritima* communities in Saskatchewan and south of the Antero Reservoir in Colorado covered 80.0% to 100.0% of the vegetated area (Dodd and Coupland 1966b, Ungar 1971, unpublished data). In the Saskatchewan salt pans, *T. maritima* cover averaged from 5.0% to 17.0%, while in Colorado salt pans studied by Ungar, cover averaged 13.3% in drier and 23.0% in wetter sites. In other communities studied in Colorado, including those dominated by *S. rubra* and *D. stricta*, *T. maritima* cover was less than 1.0% and in stands dominated by *P. nuttalliana* it averaged 2.0% (Table 15).

Triglochin maritima also occurred as a minor component, with less than

Table 15. Percentage relative dominance, cover and frequency of species in the *Triglochin maritima* community (+ = present). Locations: 1. Saskatchewan, Canada (Dodd and Coupland, 1966b); 2. Park County, Colorado (wet area); 3. Park County, Colorado (drier area).

	Locations				
	1.	2.		3.	
	R.Do.	R.Do.	Freq.	R.Do.	Freq.
Triglochin maritima	93.6	58.4	100.0	82.0	100.0
Puccinellia nuttalliana	3.5	22.2	100.0	11.1	54.0
Distichlis stricta	0.8	----	-----	0.3	8.0
Hordeum jubatum	0.8	----	-----	---	----
Salicornia rubra	0.7	2.1	50.0	5.5	86.0
Eleocharis palustris	0.3			---	----
Poa canbyi	0.3		-----	---	----
Scirpus americanus	---	3.0	24.0	---	----
Aster brachyactis	---	+	4.0	0.3	2.0
Ranunculus cymbalaria	---	14.2	86.0	0.8	4.0
Aster pauciflorus	---	+	2.0	+	2.0
Juncus bufonius	---	0.1	4.0	+	----

3.0% cover, in four other Saskatchewan communities including the *Salicornia rubra*, *Puccinellia nuttalliana*, *Puccinellia-Distichlis*, and *Distichlis stricta*. It was also found growing in these community types by Ungar (1971, unpublished data) in South Park, Colorado. It dominated on two sites, one with wet soil containing several inches of standing water and the other on drier pan bottoms with no standing water. In the former areas, *T. maritima* made robust growth, while in sites in which it was associated with *D. stricta* or *S. rubra*, growth appeared to be retarded and flowering specimens were not common. In wetter sites nearly all plants were flowering and most of these were in fruit by September 1971. Dodd and Coupland (1966b) reported that *T. maritima* initiates growth in May with plants reaching 30 cm tall by late June and 60 cm at maturity in September.

Stewart and Kantrud (1969) found that *T. maritima* was common in wet meadow zones of the glaciated prairie region of North Dakota with soil solution conductivities ranging from 3.5 to 70.0 mmhos. In Saskatchewan soils it dominated in soils averaging 31.8 mmhos. Dodd, Rennie, and Coupland (1964) and Ungar (1971, unpublished data) found *T. maritima* growing in soils with conductivities ranging from 16.0 mmhos to 72.0 mmhos and it dominated in soils averaging 24.0 mmhos.

Species of graminoids and succulents associated with *T. maritima* stands that it was the dominant species in Saskatchewan included *S. rubra, D. stricta, H. jubatum, Eleocharis palustris,* and *Poa canbyi*. All of these species had less than 1.0% relative dominance in these sites. *Puccinellia nuttalliana* had a relative frequency of 3.5% in this community compared to a value of 86.7% in the *Puccinellia* community. Forbs associated with *T. maritima* in these Saskatchewan locations having less than 9.0% frequency included *Chenopodium rubrum, Suaeda erecta, Atriplex hastata, Halerpestes cymbalaria, Glaux maritima, Juncus ater,* and *Sonchus uliginosus* (Dodd and Coupland 1966b). Ungar (1971, unpublished data) has found *T. maritima* dominating parts of several pans on the southern border of Antero Reservoir in South Park, Colorado. In areas of this type it made up 82.0% of the relative dominance and had a frequency occurrence of 100.0%. In wetter areas relative dominance equalled 58.5% and it was associated with *P. nuttalliana, Ranunculus cymbalaria, Salicornia rubra, Aster pauciflorus, A. brachyactus, Juncus bufonius,* and *Scirpus* spp. In drier areas no other species made up greater than 11.1% of the relative dominance (Table 15).

B. Soil

Surface horizons in the *Triglochin maritima* communities usually had higher salt concentrations than subsurface samples. Dodd, Rennie, and Coupland (1964) report mean conductivity values of 31.8, 19.7, 19.7, and 23.2 mmhos/cm at depths of 0-6 in., 6-12 in., 12-24 in., and 24-36 in. In pans south of Antero Reservoir it was found in soils with conductivities ranging from 16.0 to 32.0 mmhos/cm with total salt concentrations averaging 1.5%. Chlorides and sulfates were high in Saskatchewan soils and the cations sodium and magnesium were also high, 350 and 178 meq/l respectively (Table 16).

Table 16. *Triglochin* community soils.

	pH median	Soil moisture %	Conductivity mmhos/cm	Na %	Cl %	SO_4 %	Total Salts %
Saskatchewan, Canada Dodd, Rennie and Coupland (1964)	8.2	- - -	31.8	350 meq/l	high	high	2.1
Park County, Colorado Ungar (1971 unpublished)	8.1	212.3	13.0	0.6	1.8	- -	2.9

Soils had a pH median of 8.2 in Saskatchewan soils with only slight changes in the profile. Colorado soils tested had a median pH of 8.1 (Table 16).

Soils were often moist ranging from areas with standing water to apparently well drained. Soil moisture ranged from 20.5% to 376.2% in *T. maritima* stands south of Antero Reservoir. The wetter areas were typified by more robust growth of *T. maritima* and a greater species diversity.

Sporobolus airoides Community
  A. Vegetation

Sporobolus airoides Torr. grows in nonsaline prairies, but is also commonly found as a primary or secondary invader of saline soils of inland North America. Occurring with it are other salt tolerant species including *Suaeda depressa, Distichlis stricta,* and *Tamarix pentandra,* while in nonsaline soils it is associated with multitudinous prairie species. *S. airoides* may form dense, nearly pure stands on borders of saline pans where the mean salinity is 0.7%. Many researchers have reported it growing in saline soils of the western half of the United States and Canada (Kearney 1914; Hilgard 1914; Harris 1920; Aldous and Shantz 1924; Flowers 1934; Shreve 1925, 1942; Penfound 1953; Shields 1956; Bolen 1964; Baalman 1965; Hunt and Durrell 1966; Ungar 1965a, 1966, 1969). *Sporobolus airoides* is a common species in the southwest, growing in moist alkaline flats (Robinson 1958). Potter (1957), in his phytosociological study of the San Augustine Plains, New Mexico, reported that *S. airoides* was a common species on heavy alkaline soils.

The total range of salinity under which this species grew in Stafford County, Kansas, was 0.003% to 1.60%. In the *S. airoides* community where it was very abundant and had the dominant basal cover, the salinity ranged from 0.04% to 0.19%. The fact that *S. airoides* could establish itself on open salt flats indicates its tolerance for high salinities. Growth in prairies indicates that it has no requirement for excess salts and should be considered a facultative halophyte. In this area were only 11 species of vascular plants, all of which occurred in other areas of the Big Salt Marsh. Others flowering in early spring were *Poa arida*, later *Elymus canadensis*, followed by *Ambrosia psilostachya, Atriplex patula* var *hastata, Conyza canadensis, Desmanthus illinoense, Euphorbia marginata, Suaeda depressa,* and *Distichlis stricta* flowering later during the summer months. *Sporobolus* is common as a salt pan invader and appears to be very important in communities bordering salt pans. It represents the vegetation stage just prior to prairie in the successional series on marsh borders (Table 17).

Sporobolus airoides may invade the saline flats directly or follow a succulent stage. It appears to play a part in a cycle involving periods of decreased and increased salinity. With decreased salinity it will grow normally and produce hummocks with its dense root and rhizome system. Prairie species invade the hummocks and a successional pattern is completed.

Table 17. Percentage frequency of species in *Sporobolus airoides* communities.

1. Stafford County, Kansas (Ungar 1965a)
2. Alfalfa County, Oklahoma (Ungar 1968b)
3. Alfalfa County, Oklahoma (Ungar 1968b)
4. Juab County, Utah (Bolen 1964)

Total salts	0.1%	0.7%	0.0%	2.8%
Location	1.	2.	3.	4.
Sporobolus airoides	100.0	100.01	100.0	80.0
Distichlis stricta	40.0	+	90.0	45.0
Ambrisia psilostochya var. *coronopifolia*	40.0	---	---	---
Aster ericoides	30.0	---	---	---
Elymus canadensis	10.0	---	---	---
Sporobolus texanus	+	+	+	---
Tamarix pentandra	+	+	---	---
Kochia scoparia	+	---	0.2	---
Baccharis salicina	+	---	+	---
Chenopodium leptophyllum	---	---	+	---
Juncus balticus	---	---	---	20.0
Allenrolfea occidentalis	---	---	---	+

Commonly there is an increase in subsurface salinity which kills *S. airoides*. These areas are then reinvaded by succulent or *S. airoides* and the cyclical pattern continues. There are two possibilities then, one which is a repeating cycle, a second involving a lowering of the water table by increased soil accumulation and a successional process leading to the development of the prairie community (Ortenburger and Bird 1931, Baalman 1965, Ungar 1968b). Richards (1954) reports that soil salinity may range from 0.3% to 3.0%, but that best growth takes place in the range of 0.3% to 0.5%.

Vogl and McHargue (1966) reports that *S. airoides* can tolerate highly alkaline soils in California. It occurs here with the palm *Washingtonia filifera* (Table 18). Its average importance value in the seep oasis zone was 32.6, with a decline in areas with wash oases that were less stable and dependent on flood water. After fire at Willis Palms, *S. airoides* was an early invader and fire was also observed to apparently revitalize *S. airoides* clumps (Vogl and McHargue 1966). Bolen (1964) reports that *Sporobolus* was a co-dominant with *D. stricta*. Frequency of *S. airoides* was 80%, *Juncus balticus* 20%, and *D. stricta* 45%. It also is common in *Juncus balticus* dominated meadows with a frequency of 66%. These communities are among the most saline on the spring fed marshes, studied by Bolen (1962, 1964) in western Utah. Fanning, Thompson, and Isaacs (1965) found *S. airoides* occurring with *Trichloris crinita*, *Trichadne californica*, *Tridens eragrostoides*, and *T. albescens* in saline

Table 18. Prevalent and common species for California wash and seep oases with their average cover percentage and importance values (I.V.). Importance values were obtained for each species in each oasis by summing the percentage frequency, density, and cover (from Vogl and McHargue 1966).

Species	Seep Oases Average I.V.	Wash oases Average I.V.
Washingtonia filifera	52.6	36.9
Pluchea sericea	9.8	37.3
Sporobolus airoides	32.6	5.3
Juncus acutus	14.2	7.1
Distichlis spicata	7.3	1.9
Juncus mexicanus	5.0	--------
Prosopis pubescens	1.0	5.3

range soils in southwest Texas.

B. Soils

Sporobolus airoides grows in soils containing high NaCl concentrations (Ungar 1965a, 1968b) and is found in areas containing mixtures of salts including high bicarbonate and sulfate (Bolen 1964). Its tolerance to salinity appears to be broad and it has been found growing in soils with as low as 0.003% total salts to as high as 3.0% (Richards 1954, Ungar 1965a). It was commonly found in soils of central Kansas and northern Oklahoma with a mean salinity of 0.7%, while in Utah and other areas it could be found in soils whose mean salinity were between 2.0% to 3.0% total salts (Bolen 1964; Ungar 1965a, 1968) (Table 19.)

Table 19. *Sporobolus airoides* community soils.

	pH median	Saturation %	Soil Moisture %	Conductivity mmhos/cm	Na %	Cl %	SO_4 %	Total salts %
Kansas Ungar (1965a)	7.2	34.0	- -	0.6	--	0.0	- -	0.1
Oklahoma Ungar (1968b)	8.7	26.0	3.7	1.9	--	0.4	- -	0.7
Oklahoma Ungar (1968b)	8.1	34.0	11.4	35.8	--	0.0	- -	0.0
Utah Flowers (1934)	- -	- -	- -	- -	--	- -	- -	0.5

Soil pH in communities containing *S. airoides* ranged from 6.5 to 10.0 (Bolen 1964; Vogl and McHargue 1966; Ungar 1965a, 1968b). Bolen (1964) did not consider pH a significant factor determining halophyte distribution, and others have come to similar conclusions. The broad range of pH that *S. airoides* can occur within indicates that this is not a limiting factor in its distribution. (Table 19).

Soil texture in *S. airoides* stands ranges from sand to clay (Ungar 1968b). It is commonly found on sandier soils bordering salt pans. The tufts tend to accumulate blowing sand (Table 20).

Table 20.   Soil texture and loss on ignition.

	Sand %	Silt %	Clay %	Loss on Ignition %
Kansas Ungar (1965a)	85.9	5.0	9.1	2.2
Oklahoma Ungar (1968b)	90.7	1.2	8.1	0.3
Oklahoma Ungar (1968b)	65.5	10.0	24.5	3.3

Organic matter of the soil in this community is generally low. In central Kansas, values ranged from 1.1% to 2.2% based on loss on ignition. These were the lowest values on the Big Salt Marsh (Ungar 1965a). In northern Oklahoma, values were also low, ranging from 0.09% to 6.4%. Most values were between 0.2% and 0.6% in the *S. airoides* community. Bolen (1964) reports a low value, 4.0% to 4.6%, for his Utah study site.

Bolen (1964) found that *S. airoides* occurs in communities with the highest matric suction on the marsh. Osmotic pressure values of the soil solution tolerated were up to 15 atmospheres. Soil moisture in sandy soils of Oklahoma ranged from 0.0% to 8.7% moisture, while saturation percentages ranged from 23.0% to 29.0%. This indicates great matric suction during drought periods. The low water potential in these areas becomes a limiting factor to plants during mid-summer when periodic droughts occur. In one site in Oklahoma in June 1964, soil moisture equalled 14.9% and in July 1964, this dropped to 3.1%, rising to 19.1% in August. During July 1964 only 9.0% of normal rainfall occurred and soil stress caused many species to die off.

C. Physiology

Seed germination studies by Knipe (1970) with *S. airoides* indicate that large seed germinate more rapidly and had higher final germination percentages than small seed. Large seed average 83.0%, medium 59.0%, and small 26.0% germination. He felt that quicker rates of development in the larger seeds produce plants which are better adapted to harsh environments. Light exposures of 9 to 13 hours were found to delay germination 28 hours, while longer exposures of 13 hours delayed germination 72 hours. Germination was reduced from 96.0% in total darkness to 77.0% in 19 hours light exposures, to 59.0% in 24 hours continuous light (Knipe 1971a).

Seed of *S. airoides* germinate at constant temperatures ranging from 60F to 100F, while no germination occurs at 50F or 110F. Germination increased with each temperature increment from 60F, 70F, 80F, and 90F with germination values of 14.5% 22.3%, 71.8%, and 87.5% obtained in each case respectively (Knipe 1967).

Sporobolus airoides was found to be sensitive to moisture stress and seed germination percentages decrease with each salt increment (Knipe 1971b). Seeds were germinated at 0.0, 1.0, 4.0, 7.0, 10.0, 13.0, and 16.0 bars tension by Knipe (1968) and he found that germination percentages decreased with each increase in moisture tension yielding, respectively, 76.8%, 69.3%, 50.3%, 39.0%, 18.5%, 7.5%, and 3.0% germination. These data indicate that *S. airoides* does not require increased moisture stress for germination and may explain its occurrence in moister habitats.

Hordeum jubatum Community

A. Vegetation

Hordeum jubatum L. occurs in moderately saline communities throughout the western United States and Canada. In those stands where it is one of the characteristic species, *H. jubatum* may occur in relatively pure stands or with co-dominants *Distichlis stricta, Atriplex patula* var. *hastata,* or *Iva annua* (Keith 1958; Dodd and Coupland 1966b; Hadley and Buccos 1967; Ungar 1965a, 1967b; Ungar, Hogan, and McClelland 1969). It also occurs in nonsaline soils in open ground, dry sandy soils, prairies, meadows, and waste places (Hitchcock 1950; Rydberg 1932). *H. jubatum* is not cited in Fernald (1950), Hitchcock (1950), or Rydberg (1932) as occurring in saline soils, but data of Flowers (1934), Coupland (1950), Hadley and Buccos (1967), Ungar, Hogan, and McClelland (1969), and Dodd and Coupland (1966b) indicate that it is capable of growing in saline soils and may be an indicator of moderate salinities locally. It evidently has a very broad range of environmental tolerance since it occurs in climates ranging from forest to desert. Edaphically it occurs in soils ranging from fresh-water marsh to dry roadsides and prairie, as well as physiologically dry soils due to high ionic concentrations. It is generally limited to soils with salinities less than 1.0% total salts, averaging 0.6% (Keith 1958; Dodd and Coupland 1964; Ungar 1967b). Branson, Miller, and McQueen (1970) report that it is a dominant species in lowland communities of northeastern Montana which have total soil moisture stress values of 30 bars and occur over a range from 30 to 45 bars.

In some cases it is difficult to determine what physical factor is varying to produce pure *H. jubatum* stands rather than mixed stands with *D. stricta, I. annua,* or *A. patula* var. *hastata.* In vegetation-soil studies by Ungar (1967b) there appeared to be little difference in edaphic factors and yet a distinct *Hordeum-Iva* and *Hordeum-Atriplex* community developed. In the former, *Hordeum* had a frequency of 90.0%, *Distichlis*-50.0%, *I. annua*-80.0%, *Poa arida*-70.0%, and *Suaeda depressa*-0%, while the latter had a *H. jubatum* frequency of 100.0%, *Atriplex patula* var. *hastata*-90.0%, *D. stricta*-0.0%, *I.*

annua-30.0%, *P. arida*-10.0%, and *S. depressa*-60.0%. The *D. stricta* communities were usually at slightly higher salinities 0.7% while *Hordeum* dominated stands averaged 0.5% total salts. Chance distribution or biological interactions may play an essential part in determining community make-up in these areas. Species occurring with *Hordeum* vary from as few as 4 to as many as 14 in Alberta stands (Keith 1958). Dodd and Coupland (1966) report 31 species in a *Hordeum* community, but *H. jubatum* has a relative dominance of 94.2% of all graminoids and contributed 84% to 100% of the vegetative cover in sites sampled.

The species commonly occurring with *H. jubatum* in Kansas (Ungar 1967b) were *Suaeda depressa, D. stricta, A. patula* var. *hastata, Kochia scoparia, Rumex crispus, Iva annua* and *Poa arida*. In Alberta, Canada, other species commonly found were *Agropyron smithii, Carex* sp., *Sonchus arvensis* and *Aster ericoides* (Table 21). Kearney et al. (1914) and Flowers (1934) report that *H. jubatum* occurs with *D. stricta, Triglochin maritima, Puccinellia airoides, (=P. nuttalliana), Spartina gracilis, Sporobolus airoides, Melilotus alba, Suaeda erecta*, and others in grass meadows of the Great Salt Lake region in Utah. It is not cited as a dominant species in these areas. A summary of frequency data for species occurring in various parts of the United States and Canada for *Hordeum* and *Distichlis-Hordeum* communities is included in Tables 21 and 22.

Hadley and Buccos (1967, 1970) report *H. jubatum* commonly occurs with *D. stricta, Poa* spp., *Aster ericoides, Sonchus arvensis, Sphenopholis obtusata, Agropyron trachycaulon, Ambrosia artemisiifolia*, and *Grindelia squarrosa* in North Dakota communities studied. Working in the Oakville Prairie Grand Forks County, North Dakota, Redmann and Hulett (1964) found that *Sonchus arvensis, Aster ericoides, Grindelia squarrosa*, and *Ambrosia artemisiifolia* tended to occur in areas of higher salinity than other compositae.

Distributional data indicates *Hordeum jubatum* has a wide range of distribution in nonsaline communities and is limited to moderately saline soils. On the border of saline and nonsaline soils it occurs with many more species than in the saline communities. Prairies in which it occurs may contain 100 to 200 species, whereas the highest number of species found with it in saline environments is 31.

Other species occurring with *H. jubatum* may also have very wide distributions like *Atriplex patula* var. *hastata*, or may be limited to western United States and Canada like *D. stricta*. Some species such as *Iva annua* and *Agropyron dasystachyum* may even have more limited distributions (Fernald 1950). The chief characteristic linking the *H. jubatum* dominated communities in western United States and Canada is the great importance of *H. jubatum* in these areas and its very showy fruiting appearance in early summer. There is considerable variation in total species complement as can be seen in the data of Keith (1958) and in comparing studies in other areas (Table 21). Cusick (1970) has reported that *H. jubatum* occurs with

Table 21. Percentage frequency of species in the Hordeum-Community (+ = present but not in quantitative data). 1. Lincoln, Nebraska (Ungar, Hogan, McClelland, 1969); 2. Cloud Co., Kansas (Ungar, 1967b); 3, 4. Alberta, Canada (Keith, 1958).

Total Salts	0.7	0.5	0.5	0.3
Location	1.	2.	3.	4.
Species				
Hordeum jubatum	100	100	100	100
Iva annua	98	30	---	---
Atriplex patula var. hastata	1	90	---	---
Poa arida, Poa sp.	24	10	---	---
Scirpus paludosus	12	---	---	---
Eleocharis palustris	1	---	---	---
Rumex crispus	+	10	---	---
Spartina pectinata	+	---	---	---
Spartina gracilis	---	---	15	5
Distichlis stricta	---	---	45	---
Suaeda depressa	---	60	---	---
Kochia scoparia	---	30	---	---
Puccinellia nuttalliana	---	---	10	---
Aster ericoides	+	---	---	25
Agropyron smithii	---	---	---	60
Agrostis scabra	---	---	---	10
Muhlenbergia asperifolia	---	---	---	15
Carex spp.	---	---	---	75
Juncus balticus	---	---	---	10
Ranunculus cymbalaria	---	---	---	15
Taraxacum officinale	---	---	---	65
Sonchus arvensis	---	---	---	30

Table 22. Percentage frequency of species in the Distichlis-Hordeum community type (+ = present, but not in quantitative data). 1. Alfalfa Co., Oklahoma (Ungar, 1968); 2. Grand Forks Co., North Dakota (Hadley and Buccos, 1967); 3. Codington Co., South Dakota (Ungar, 1970a).

	1. Alfalfa Co., Oklahoma (Ungar, 1968)	2. Grand Forks Co., North Dakota Hadley and Buccos (1967)	3. Codington Co., South Dakota (Ungar, 1970a)
Total Salts	0.2	1.3	1.1
Species			
Distichlis stricta	100	100	100
Hordeum jubatum	80	90	68
Poa arida, Poa sp.	80	3	46
Aster exilis	10	---	---
A. ericoides	---	33	---
Atriplex patula var. hastata	---	---	1
A. argentea	+	---	---
Polygonum ramosissimum	+	---	---
Kochia scoparia	+	---	2
Suaeda depressa	---	33	1
Puccinellia nuttalliana	---	40	---
Spartina pectinata	---	10	---
Ambrosia artemisiifolia	---	37	5
Agropyron smithii	---	---	---
A. dasystachyum	---	33	---
Grindellia squarrosa	---	23	---
Sphenopholis obtusata	---	3	2
Sonchus arvensis	---	37	9
Scirpus paludosus	---	---	4

Salicornia rubra and *Atriplex patula* var. *hastata* in saline soils located in Rittman, Ohio.

There is a definite change in seasonal aspect in the *H. jubatum* communities. *H. jubatum* dominates the spring and early summer aspects, and in Kansas and Nebraska, *Iva annua* and *Atriplex patula* var. *hastata* are characteristic of the late summer and fall aspects. Other species evident at this time throughout the range of *H. jubatum* are *Chenopodium rubrum*, *Distichlis stricta, Atriplex argentea,* and *Suaeda depressa.* Growth of *Iva* and *Atriplex* begins in April but it is slow and the showy inflorescences of *H. jubatum* make it the dominant of the early seasonal aspects. It also matures early in the growing season while most of the other members of this community are late blooming species. Bolen (1964) reports that *Distichlis stricta* begins growth about April 14, but anthesis does not occur until June

10. In the case of *H. jubatum*, vegetative growth also begins in April, but by May flowering is observed.

B. Soils

Hordeum jubatum is commonly found in saline soils from Oklahoma north to Canada (Penfound 1953; Coupland 1950). It also occurs from the east coast of the United States to the west, but does not occur in southeastern United States (Hitchcock 1950). The specific ion of the soil is apparently not limiting to its distribution since it occurs in both chloride and sulfate dominated soils (Flowers 1934; Dodd and Coupland 1964; Ungar, Hogan, and McClelland 1969; Ungar 1970a). *H. jubatum* is not as salt tolerant as *D. stricta*, but does occur in soils with total salt concentrations ranging from 0.003% to 1.3% in Kansas and Oklahoma (Ungar 1965a, 1966) (Table 23). Median salinities for this community type is 0.5%. Hanson and Whitman (1938) have found *H. jubatum* growing in soils with salinities ranging up to 1.4% total salts in North Dakota.

Soil pH in Saskatchewan saline meadows ranged from 6.4 to 9.5, with a median value of 8.1 in the surface soils (Dodd, Rennie, and Coupland 1964). These values compare favorably to those for *H. jubatum* communities in Oklahoma, Kansas, Nebraska, and South Dakota: 8.2, 7.7, 7.5, 7.9, respectively (Ungar 1967b, 1968b, 1970a; Ungar, Hogan, and McClelland 1969). The extremes of pH found in these areas ranged from 6.9 in Kansas to 9.2 in Oklahoma. Flowers (1934) found *Hordeum* growing at a pH of 8.6 in Utah. These data indicate that *H. jubatum* has a broad tolerance to variations in pH and this may be one of the reasons it occurs in several different communities throughout its range (Table 23).

Soil texture in *Hordeum jubatum* communities is quite variable, but generally on the heavy side. Hanson and Whitman (1938) found it growing in textures ranging from loam to clay with clay content varying from 17% to 56%. Dodd, Rennie, and Coupland (1964) report saline soils ranging from sandy loam to clay. In studies in Nebraska and Oklahoma, Ungar (1967b) and Ungar, Hogan, and McClelland (1969) found *H. jubatum* growing in soils ranging from sandy loam to clay. Soil texture is probably not a limiting factor to *H. jubatum* distribution, but it may be eliminated from sandy soils due to the fact that the surfaces dry out rapidly and it cannot sustain long dry periods.

Organic matter measured by loss on ignition ranged from 2.6% to 7.0% in North Dakota (Hanson and Whitman 1938). In Kansas (Ungar 1965a, 1967b), loss on ignition in *Hordeum* stands ranged from 2.5% to 15.5%. Values for Oklahoma averaged 8.8% (Ungar 1968). The total organic matter content is not only made up of dead vascular plants in these areas but a good deal of it is composed of blue-green algal mats. The variable organic matter content in areas harboring *H. jubatum* may indicate it is not necessary for its establishment, but high organic content and algal crusts improves water relations and may favor seedling development and growth.

Table 23. Hordeum community soils.

	pH median	Saturation %	Soil Moisture %	Conductivity mmhos/cm	Na %	Cl %	SO$_4$ %	Total Salts %
North Dakota Hadley and Buccos (1967)	---	---	---	12.0	---	---	---	0.77
Canada Keith (1958) Dodd, Rennie, and Coupland (1964)	7.5 8.1	--- ---	--- ---	12.0 16.5	--- ---	--- ---	0.05 ---	0.43 1.03
South Dakota Ungar (1970a)	7.9	59.0	39.8	7.0	0.1	0.0	0.4	1.1
Nebraska Ungar, Hogan, and McClelland (1969)	7.5	77.0	30.3	12.7	0.2	0.4	0.1	0.7
Kansas Ungar (1967b)	7.8	74.0	38.2	11.8	---	0.2	---	0.5
Oklahoma Ungar (1968b)	8.2	58.0	21.8	5.5	---	0.1	---	0.2

C. Physiology

Hordeum jubatum is a short lived perennial bunch grass species (Cords 1960). Seeds usually germinate in late August or September and if conditions are favorable, the tufts will grow to a diameter of 5-10 cm the next year (Stevens 1963). Seed that do not germinate by late summer will overwinter and germinate the following spring.

In Nevada, other western states, and Canada, *H. jubatum* occurs in native pastures and meadows, which usually have high water tables and are frequently saline (Cords 1960, Wilson 1967). Cords (1960) found that storage of seeds for a year in the laboratory caused a drop from 50% initial germination to 20%. After four weeks dry storage, 22% was more deleterious to seed than wet storage at 35°F. Using alternating temperatures of OC-10 hrs. 20C-14 hrs, 74% germination was obtained while temperatures of 5C-20C yielded 73% after a 60 day germination period (Ungar, unpublished data). These seed were collected at lincoln, Nebraska, July 1967 and tested October 1967. During this storage, seed were stored dry at 5C. Reduction in water table caused a great decrease in root and shoot growth, with greatest decrease at an 18 in. water table. At all water table depths, 0-18 in., plants grown in saline conditions (10 mmhos cond, pH 8.8) produced poorer growth than those grown in nonsaline soils (0.5 mmhos, pH 6.8). With surface water tables, *Eleocharis* sp. offered enough competition to reduce *H. jubatum* while under lower water table conditions, *Festuca* and tall wheat grass were able to reduce its prevalence (Cords 1960). Wilson (1967) has found that *H. jubatum*, when growing alone, was hindered by high salinity, but that high soil moisture favored its growth. In combination with other species *Agropyron desertorium* (Fisch.) Schult. *Agropyron elongatum* (Host.) P.B., *Phalaris arundinacea* Schreb., and *Dactylis glomerata* L., *H. jubatum* did not generally grow as well. Only with *D. glomerata* under wet-saline soil conditions where orchard grass grew poorly did *H. jubatum* develop well. Under wet-nonsaline conditions it had its second best development, but *D. glomerata* grew well here and offered more intensive competition. All other species bettered *H. jubatum* growth under saline and nonsaline conditions. This ability of *H. jubatum* to compete under wet-saline conditions as illustrated by the experiments with *D. glomerata* may explain its distribution in wet saline meadows.

Hordeum jubatum has been found growing in soils that are nonsaline, 0.003% total salts to soils which contain as much as 1.5% total salts. It is usually not found invading the more highly saline pans with over 1.0% total salts. Seed germination studies indicate no reduction in percentage germination in up to 1.0% NaCl, but a regular decrease in germination with salinity increases between 1.0% and 2.0% which appears to be the maximum salinity it can tolerate. Seedlings were a more sensitive indicator of salinity stress, and with each increase in stress there was a corresponding decrease in growth. Plants could survive in up to 2.0% NaCl solutions but growth did not occur.

Iva annua (I. ciliata) occurs in saline soils from Oklahoma north to a single saline location in North Dakota (Stevens 1963). It occurs in saline soils averaging about 0.6% total salts. Field growth appears to be best in moist nonsaline soils (Ungar, Hogan, and McClelland 1969). Plants growing in 0.1% total salts reached 73 cm tall, while adjacent plants growing in soils with 0.7% total salts were only 37 cm tall by August 1969. These data indicate a salt tolerance rather than a requirement for excess salts in the case of *I. annua*. Seed germination experiments indicate that germination will take place in up to 1.5% NaCl solutions and that the optimum temperature for germination lies between 15 to 20 C.

Additions of salt to the germination media slowed the rate of germination as well as decreasing the final percentages (Ungar and Hogan 1970). Again, as in studies with other species, recovery after soaking in high salinities occurs. Seeds were soaked for ten days in up to 23% NaCl solutions and germination averaged 57% at 5 C and 51% at 5 C when seeds were returned to distilled water at 20 C. Seeds placed at 35 C were permanently injured and only 3% germination occurred (Ungar 1970b).

Growth studies by Hogan (1968) indicate that plants will maintain themselves in media with up to 0.75% NaCl added. This fits in well with the total soil salinities of the natural areas in which it is found. Growth of *Iva* took place in up to 0.75% NaCl and Hoagland and Arnon number 2 solutions. Seedling L.D. 50's were 28 days at 0.5%, 7 days at 100%, and 7 days at 1.5%.

Croft (1930) has demonstrated that *Atriplex patula* var. *hastata* can adjust the ionic concentration of its cell to increased osmoticum in the soil solution. He found that at 0.8%, 2.0%, 4.5%, and 8.5% soil salinity concentrations, plant extracts contained 5.5%, 7.6%, 8.7%, and 10.0% total salts.

Atriplex patula var. *hastata* is dimorphic in its seed production. A single plant may have large light colored seed averaging 2.0 mm in diameter while the black small seed average 1.4 mm (Ungar 1971). Initially the large seed have germination percentages of 43% without scarification, while the small seed require scarification and maintain about 73% germination. After storage for three years both types of seed require scarification to germinate. There is no loss of viability after eight years refrigerated storage. Seeds apparently germinate in the field either enclosed in their fruiting bract or they become released from the bract and then germinate (Drysdale, Benner, and Ungar 1971, unpublished data).

Scirpus paludosus Community
A. Vegetation

Scirpus paludosus Nels. occurs on the borders of saline lakes and in wet depressions and shallow ponds throughout the western half of the United States (Schaffner 1898; Flowers 1934; Muenscher 1944; Fernald 1950; Penfound 1953; Ungar 1965a, 1967b, 1968b, 1970a). It generally occurs in nearly pure stands, but may be found in mixed assemblages (Flowers 1934;

Rawson and Moore 1944; Ungar, Hogan and McClelland 1969; Ungar 1970a). Dix and Smeins (1967) indicate that *S. paludosus* is a common emergent in Nelson County, North Dakota. It was found to mix only infrequently with other species and was limited to saline soils although it showed a broad tolerance to other environmental factors. Other species commonly occuring with *S. paludosus* include *Distichlis stricta, Scirpus americanus, Suaeda depressa, Atriplex patula* var. *hastata*, and *Hordeum jubatum*. Its distribution appears to be narrowly limited by two factors. Salinities above 1.5% are detrimental to *S. paludosus* growth, and dwarfed and non-fruiting forms are found at salinities beyond its tolerance. Unavailability of soil moisture is a second factor playing an important part in determining available sites. It does not occur in prairie soils or other well-drained areas bordering saline marshes.

Robust growth of *S. paludosus* occurs at salinities averaging 0.5%, but it has been found growing over a much broader range, from 0.03% to 3.7% total salts in the surface soils (Ungar 1965a, 1967b, 1968b). Biological interactions may eliminate it from nonsaline ponds since *Scirpus validus, S. acutus, Typha latifolia,* and *T. angustifolia* are more robust plants which would tend to crowd out *S. paludosus* under optimum conditions for their growth. Bolen (1964) found that *S. paludosus* dominated a single stand with invasions from dominants of other communities including *S. olneyi, S. acutus, Distichlis stricta,* and *Eleocharis rostellata. S. paludosus* was not common at sites studied by Bolen (1964) and he felt that the Fish Springs stand represented a relict community. Though *S. paludosus* may have been n more abundant in the past, it appears to have been replaced by *S. olneyi*. It presently occurs in this one location at lower salinities, o.2%, than it is usually found in, however, other emergent species occupy more saline sites.

Sampling by Ungar (1968b) in northern Oklahoma indicates that *S. paludosus* can grow in soils reaching a mean total salinity of 1.5%. Growth in this stand was not good and the plants appeared weak. A stand analyzed on the Great Salt Plains of northern Oklahoma, containing well over a thousand culms in 1964, was almost completely killed due to raised soil salinities. In 1965 only three or four green culms were present. Surface salinity rose from 2.5% to 3.7% during the period between June and August 1964. Heavy rains in September reduced salinities to 0.1%, but the extreme conditions earlier apparently was lethal to *S. paludosus*. In Cloud County, Kansas, Ungar (1967b) has found *S. paludosus* dominating a shallow pond community; associated with it were *Suaeda depressa, Atriplex patula* var. *hastata, Hordeum jubatum, Polygonum ramosissimum, Eleocharis palustris,* and *Loptochloa fascicularis* (Table 24). Rawson and Moore (1944) report that *S. paludosus* was common at or near waters edge of Saskatchewan saline lakes.

Scirpus paludosus appears to be an indicator species for saline lakes of South Dakota. It was found growing in pure stands or accompanied by *Scirpus americanus* around Stink and Bitter lakes (Ungar 1970a). Only during the 1968 growing season, which was dry, did other species including *Kochia scoparia, Chenopodium rubrum, A. patula* var. *hastata, Suaeda depressa,* and

Table 24. Percentage frequency of species in Scirpus paludosus communities. (+ = present, but not represented in the quantitative data.) 1. Alfalfa County, Oklahoma (Ungar (1968b); 2. Stafford County, Kansas (Ungar 1965a); 3. Lancaster County, Nebraska, Ungar, Hogan, and McClelland 1969); 4. Codington County, Ungar (1970a); 5. Juab County, Utah (Bolen 1964).

Location Total Salts	1. 1.6	2. 0.5	3. 1.9	4. 1.4	5. 0.3
Species					
Scirpus paludosus	98.6	100.0	100.0	100.0	100.0
Hordeum jubatum	- -	+	8.0	- -	- -
Distichlis stricta	- -	100.0	- -	- -	66.0
Polygonum ramosissimum	1.4	60.0	- -	- -	- -
P. aviculare	- -	- -	+	- -	- -
Poa arida	- -	+	- -	- -	- -
Scirpus americanus	- -	- -	- -	+	- -
S. olneyi	- -	- -	- -	- -	50.0
S. acutus	- -	- -	- -	- -	11.0
Eleocharis rostellata	- -	- -	- -	- -	+
E. palustris	- -	- -	+	- -	- -
Suaeda depressa	- -	10.0	24.0	- -	- -
Atriplex patula var. hastata	- -	+	62.0	- -	- -
A. argentea	- -	+	- -	- -	- -
Heliotropium curassavicum	- -	+	- -	- -	- -

Salicornia rubra invade pans produced by lake margin regression inside the *S. paludosus* zone. Sloan (1970a) reports that *S. paludosus* is the dominant emergent species in saline prairie potholes of North Dakota. These saline potholes range from 0.3% to 3.5%+ in salinity. In brackish ponds, ranging from 0.1% to 1.5% salts, it is the co-dominant with *Scirpus acutus*.

Jones and Peterson (1970) report that *S. paludosus* dominates a high water table area in Regina, Saskatchewan. This community had the lowest diversity of the communities studied along a moisture gradient. It contained 23 species. Four to seven were encountered at each one meter square quadrat, with a mean value being 5.5. Other species found in this area include *Eleocharis palustris, E. pauciflora, Scirpus americanus, S. validus, Carex aquatilis, C. lanuginosa, C. sartwellii, C. praegracilis, Hordeum jubatum,* and others.

Prairie potholes described by Sloan (1970b) are water holding depressions of glacial origin found in South Dakota, Iowa, North Dakota, Minnesota in the United States, and Manitoba, Saskatchewan, and Alberta, Canada. Studies in North Dakota indicate that the configuration of the water table surrounding the pothole has a marked influence on their permanence and water salinity through its influence on inflow and outflow seepage. If there is no outflow saline, permanent ponds develop, especially when water table slopes toward the pothole, while when the water table slopes down away from the pothole and outflow high the fresh water, temporary ponds develop (Sloan 1970b).

B. Soils

Soil texture varies from sand to clay in areas occupied by *S. paludosus*. The soils in central Kansas had 64.3% sand, 8.8% silt, and 26.9% clay, while those of northern Kansas had 28.4% sand, 21.6% silt, and 50.0% clay. Other areas had soils which generally fell somewhere between these two.

Soil moisture appears to be extremely important in determining *S. paludosus* distribution. Median saturation percentages in the prairie regions studied was 64.0% (Table 25). Soil moisture percentages determined by the gravimetric method had a median value of 44.0%, well above the field capacity value. *S. paludosus* grows in soils which are saturated for a good portion of the growing season. It does not occur in well-drained prairie sites or in other sites of raised microtopography.

Scirpus paludosus appears to occur in about the center of the salinity gradient in the areas studied. It fills the more or less open niche for primary invaders of emergent habitats in areas of high salinity. It is absent from nonsaline sedge-meadows, but it has been found in with as low as 0.03% total salts (Table 25). The limiting factors in this case appears to be competition from more robust fresh-water marsh species. Highly saline flats are sometimes invaded by *Scirpus paludosus*, but in these habitats dwarf non-fruiting forms are evident. Rising salinities due to excess salts leaching into its area or droughts eliminate *S. paludosus* (Ungar 1968b). Kaushik (1963) found that salt concentrations of 1.5% were about the limit of *S. paludosus* tolerance.

Table 25. Scirpus paludosus community soils.

	pH median	Saturation %	Soil Moisture %	Conductivity mmhos/cm	Na %	Cl %	SO$_4$ %	Total Salts %
South Dakota Ungar (1970a)	8.4	50.0	40.2	10.0	0.1	0.0	0.6	1.4
Nebraska Ungar et al (1969)	6.5	77.0	48.4	28.8	0.7	0.8	0.1	1.9
Kansas Ungar (1967b)	8.9	74.0	44.0	8.9	--	0.3	--	0.5
Kansas Ungar (1965a)	8.4	47.0	--	16.8	--	0.2	--	0.5
Oklahoma Ungar (1968b)	7.7	64.0	43.1	48.9	--	0.7	--	1.6
Utah Flowers (1934)	8.6	--	22.8	--	--	--	--	1.4
Utah Bolen (1964)	8.2	--	--	5.0	--	--	--	0.3

Field data indicate that, at least for short periods, surface salinities of up to 3.0% may be tolerated.

Median pH values ranged from a low of 6.5 in northern Kansas to a high of 8.9 in Lincoln, Nebraska (Ungar 1967b; Ungar, Hogan, and McClelland 1969). The median value for five study areas sampled from northern Oklahoma to South Dakota was 8.4. These data indicate that soil pH is variable and it has little influence on *S. paludosus* distribution.

C. Physiology

Seed germination studies by Isely (1944) indicate that *S. paludosus* seed require a pre-treatment before they will germinate. Sulfuric acid for 18 minutes treatment produced 30% germination. Continuous illumination at 30-32 C facilitated germination and values of 51% and 95% germination obtained; seeds in the dark had no germination or 1%. Wet storage facilitated germination, while dry storage reduced percentages.

Field studies indicate that *S. paludosus* can grow in a broad range of saline soils throughout western North America. Kaushik (1963) has found that *S. paludosus* can germinate in salinities as high as 1.5% (Table 26). As with many other salt tolerant species, maximum germination occurred in distilled water. The first significant drop in germination percentage occurred at 0.7% salt concentrations. Adult plants grew at all salinities up to 1.5%, but at the latter concentration, mortality was about 70.0% (Table 28). Table 27 indicates the decrease in height of plants, and dry weight decreases from 0 to 180 meq/1 NaCl solutions. This broad tolerance at both the germination stage and during growth explains why *S. paludosus* persists in shallow ephemeral saline ponds throughout the prairie region and in more western areas.

Table 26. Seed germination of *Scirpus paludosus* (from Kaushik 1963) 100 seeds at each concentration.

Treatment meq/1	Days							
	3	4	5	6	7	8	9	10
0	7.5	23.5	37.0	46.0	61.0	62.5	67.5	70.5
30	4.0	16.5	21.0	38.5	59.0	62.0	66.5	70.0
60	0.0	0.0	11.5	28.0	50.5	53.5	53.5	56.0
90	0.0	0.0	4.5	20.5	42.0	45.5	45.5	46.0
120	0.0	0.0	3.0	7.0	16.0	18.5	18.5	22.0

Table 27. Growth of *Scirpus paludosus* under various salinity treatments (from Kaushik 1963)

Treatment meq/l	Total length cm	Shoot wet gm	Shoot dry gm	Root wet gm	Root dry gm
0	76.1	41.7	2.1	18.9	0.9
90	76.4	41.2	2.2	18.7	0.9
120	73.6	37.4	2.1	14.8	0.8
150	61.7	28.9	1.9	12.5	0.7
180	44.5	22.2	1.4	8.3	0.6

Table 28. Survival rates of *Scirpus paludosus* at high salinities (from Kaushik 1963).

meq/l	Percentage Survival	Mean Days to Death
150	100	0
180	80	49
220	50	10
240	34	8

Tamarix pentandra Community
  A. Vegetation

Tamarix pentandra Pall. (*T. gallica* L. of early reports) is an introduced shrub species occurring in saline marshes as far north as northern Kansas and more commonly in the southwestern United States (Fernald 1950, Robinson 1958). Ortenburger and Bird (1931) did not list *T. pentandra* as occurring in saline soils of Oklahoma nor was it listed by other early researchers in their studies of saline prairie soils (Schaffner 1898). This species is currently an important element in the flora of the Great Salt Plains and has been found by Ungar (1966) and Morley (1964) in northern Kansas marshes.

Tamarix pentandra is the only small tree occurring in the inland saline marsh communities. It is an introduced species which is native to southern Europe and western Asia. This probably accounts for its absence in earlier floristic studies in the prairie and southwest desert regions. Robinson (1958) reports that the first recorded introduction in the western states was in Texas in 1884. It is now an agressive bottom land shrub invading saline and nonsaline soils in the southwest. An indication of its aggressiveness is the report of its complete absence from the Pecos River basin before 1912, while by 1915 *Tamarix* covered 600 acres of this area and 36,270 acres by 1953 (Robinson 1958). This equals about a 1,000 acre spread per year from 1912 when the first seedlings were observed until 1953, and makes *Tamarix* an important problem phreatophyte in the dry southwest region of the United States.

This community represents one of the few situations in inland saline marsh soils that has a definite shrub stratum with an herbaceous undergrowth dominated by grasses. *Tamarix pentandra* appears to be a primary invader either by means of seed dispersal or root sprouts in highly saline soils on the Great Salt Plains, Oklahoma. This occupation of highly saline conditions may only be temporary and represents a cyclic invasion pattern involving establishment on salt pans and a dying off of the colony. New shoots would develop on highly saline soils which would succumb to excess salts during the dry summer months or would be killed due to flooding by highly saline streams.

Tamarix pentandra communities in Oklahoma had their lower strata dominated by *D. stricta* (Ungar 1968b). Other species occurring with *T. pentandra* on salt pans in the prairie include *Sesuvium verrucosum*, *Suaeda depressa*, *Sporobolus airoides*, *Polygonum ramosissimum*, and *Poa arida*. These were generally scattered and always had less than 1.0% relative cover. These Oklahoma *T. pentandra* communities generally had no other tree or shrub species, however, occasionally an individual specimen of *Eleagnus angustifolia* or *Baccharis salicina* could be found growing on the least saline sites. Campbell and Dick-Peddie (1964) report that *T. pentandra* accounts for 19.7% to 55.6% of the cover on hydric and mesic sites on the Rio Grande in New Mexico, while in xeric locations, maximum cover was 17.8%. Other trees and shrubs occurring at more than five sites in this community include *Baccharis glutinosa*, *Prosopis pubescens*, *Populus fremontii*, *Lycium andersonii*, *Suaeda frutescens*, *Salix gooddingii*, and *Eleagnus angustifolia*. *Distichlis stricta* and *Sporobolus airoides* are the most common grasses in the herbaceous stratum in these shrub communities (Campbell and Dick-Peddie 1964). Marks (1950) describes a *Tamarix-Pluchea* community as the dominant in the lowest bottom lands of the lower Colorado Desert of California, Arizona, and Mexico. The dominants were never found bordering these areas on the upper terraces. Species reported by Marks (1950) that were also found in New Mexico associations include *Populus fremontii*, *Salix gooddingii*, and *Pluchea sericea*.

B. Soils

Marks (1950) reports that *Tamarix* occurrence is completely independent of soil textures and that if subsurface moisture is available, it can occur on soils ranging from sand to silty clay loam. Depth of soil does not appear to be a determining factor either. In the prairie region, *T. pentandra* was found growing in a wide variety of soil types ranging from loamy sand to clay in texture (Ungar 1968b).

Distribution data indicate that *T. pentandra* is not an obligate halophyte but that it is tolerant of moderately saline conditions. Marks (1950) indicates that the majority of sites containing *Tamarix* had only 0.14% total salts, but that it was able to grow in more saline soils. In Kansas and Oklahoma, *T. pentandra* was found growing in soils ranging from a low of 0.14% to 3.7% total salts with a mean of 1.7% for the 44 sites sampled in highly saline soils (Ungar 1966).

Campbell and Dick-Peddie (1964) could find no correlation between *T. pentandra* distribution and pH or total salts of soils in the phreatophyte communities they studied. The pH values in the Oklahoma and Kansas prairie study sites ranged from 7.3 to 9.5 and this factor did not appear to be a good indicator for *T. pentandra* distribution (Ungar 1967b; Ungar 1968b).

C. Physiology

Merkel and Hopkins (1957) report that *T. pentandra* seeds have high viability and that germination can take place within 24 hours if soil conditions are favorable. Viability of seeds is high and by the end of five days germination values reached 98.0% (Merkel and Hopkins 1957, Robinson 1958).

Flowering occurs from late May through October (Merkel and Hopkins 1957). Seed production takes place throughout the summer and seed viability remains high, 40.0% through September. A $10°C$ storage temperature retained high viability in seed, higher storage temperatures were less favorable for germination.

Seed germination studies by Tomanek (unpublished report) indicate that *T. pentandra* can germinate as well in salinities of 0.5% to 1.0% as it does in nonsaline solutions. This ability to germinate at high salinities explains its occurrence in saline soils. Germination studies by Ungar (1967b) corroborate the findings of Tomanek and indicate that seed can germinate in up to 5.0% NaCl solutions. Hopkins and Tomanek (1957) report high mortality for *Tamarix* seedlings from July 9 to September 15, 1956, with 72.0% of the new seedlings not surviving. Summer droughts and accompanying increased soil osmotic stress are responsible for eliminating *Tamarix* seedlings and other halophytes from pans due to the greatly increased total soil moisture stress.

Literature Cited

Adams, D. A. 1963. Factors influencing vascular plant zonation in North Carolina salt marshes. Ecology 44:445-456.

Aldous, A. E., and H. L. Shantz. 1924. Types of vegetation in the semiarid portion of the United States and their economic significance. J. Agr. Res. 28:99-128.

Anderson, G. C. 1958. Seasonal characteristics of two saline lakes in Washington. Limnol. Oceanogr. 3:51-68.

──────────. 1958. Some limnological features of a shallow saline meromictic lake. Limnol. Oceanogr. 3:259-270.

Axelrod, D. I. 1950. Studies in late Tertiary paleobotany VI. Evolution of desert vegetation in western North America. Carnegie Inst. Wash. Publ. 590:215-306.

Baalman, R. J. 1965. Vegetation of the Salt Plains Wildlife Refuge, Jet, Oklahoma. Univ. of Oklahoma, Ph.D. thesis. 129 pp.

Barbour, M. G. 1970. Is any angiosperm an obligate halophyte. Amer. Midl. Natur. 84:105-120.

──────────, and C. B. Davis. 1970. Salt tolerance of five California salt marsh plants. Amer. Midl. Natur. 84:262-265.

Barkley, T. M. 1968. A manual of the flowering plants of Kansas. Kansas State Univ. Manhattan, Kansas. 402 pp.

Bauer, H. L. 1930. Vegetation of the Tehachapi Mountains, California. Ecology 11:263-280.

Baumeister, W., and L. Schmidt. 1962. Uber de Rolle des Natriums im pflanzlichen Stoffwechsel. Flora Allg. Bot. Zeitung 152:24-56.

Bernstein, L. 1967. Plants and the supersaline habitat. Contr. Marine Science 12:242-248.

Billings, W. D. 1945. The plant associations of the Carson desert region, Western Nevada. Butler Univ. Botan. Studies 7:89-123.

──────────. 1949. The shadscale vegetation zone of Nevada and eastern California in relation to climate and soils. Amer. Midl. Natur. 42:87-109.

Bindschadler, H. 1948. Native vegetation in relation to soil in parts of Wyoming. Wyoming Range Manage. 3:1-4.

Binet, P. 1960. La dormance des semences de *Suaeda vulgaris* Moq. et de *Suaeda macrocarpa* Moq. Bull de la Soc. Bot. Fr. 107:159-162.

_____ 1963. Etude de quelques aspects de la croissance et du metabolisme chez le genre *Suaeda.* Ann. Sci. Nat. Bot. 4:539-556.

_____ 1965a. Action de la temperature et de la salinite sur la germination des graines de *Cochlearia anglica* L. Rev. Gen. Bot. 72:221-236.

_____ 1965b. Vue d'emsemble la croissance et la nutrition minerale de *Suaeda maritima* Dum. Bull. Soc. Fr. Physiol. Veg. 10:227-241.

Bolen, E. G. 1962. Ecology of spring fed marshes. Utah State Univ., M. S. thesis. 124 pp.

_____ 1964. Plant ecology of spring fed marshes in western Utah. Ecol. Monogr. 34:143-166.

Boucaud, J. 1962. Etude morphologique et ecophysiologique de la germination de trois varietes de *Suaeda maritima* Dum. Bull. Soc. Linn. Norm. 3:63-74.

Bourn, W. S. 1935. Sea water tolerance of *Ruppia maritima* L. Boyce Thompson Institute Contrib. 7:249-255.

_____ 1932. Ecological and physiological studies in certain aquatic angiosperms. Boyce Thompson Institute Contrib. 4:425-496.

Branson, F. A., R. F. Miller, and I. S. McQueen. 1967. Geographic distribution and factors affecting the distribution of salt desert shrubs in the United States. J. Range Manage. 20:287-296.

_____, _____, and _____ 1970. Plant communities and associated soil and water factors on shale-derived soils in northeastern Montana. Ecology 51:391-407.

Brown, R. W. 1971. Distribution of plant communities in southeastern Montana Badlands. Amer. Midl. Natur. 85:458-477.

Campbell, C. J., and W. A. Dick-Peddie. 1964. Comparison of phreatophyte communities of the Rio Grande in New Mexico. Ecology 45:492-502.

Carpelan, L. H. 1958. The Salton Sea: Physical and chemical characteristics. Limnol. Oceanogr. 3:373-386.

Chapman, V. J. 1960. Salt marshes and salt deserts of the world. Interscience Publ., New York. 392 pp.

_____ 1947a. *Suaeda fruticos* Forsk. J. Ecology 35:303-310.

_____ 1947b. *Suaeda maritima* L. Dum. J. Ecology 35:293-302.

Chouduri, G. N. 1968. Effect of soil salinity on germination and survival of some steppe plants in Washington. Ecology 49:465-471.

Christensen, E. M. 1967. Bibliography of Utah botany and wildland conservation. Brigham Young Univ. Sci. Bull. 9:1-136.

Cords, H. P. 1960. Factors affecting the competitive ability of foxtail barley (*Hordeum jubatum*). Weeds 8:636-644.

Coupland, R. T. 1950. Ecology of mixed prairie in Canada. Ecol. Monogr. 20:271-315.

Croft, A. R. 1930. Some osmotic phenomena in *Atriplex hastata* L. Utah Acad. Sci. Proc. 7:52-54.

Cusick, A. W. 1970. An assemblage of halophytes in Northern Ohio. Rhodora

72:285.

Daubenmire, R. F. 1940. Contributions to the ecology of the Big Bend Area of Washington. II. Indicator significance of the natural plant communities in the region of the Grand Coulee Irrigation project. Northwest Sci. 14:8-10.

Dix, R. L., and F. E. Smeins. 1967. The prairie, meadow, and marsh vegetation of Nelson County, North Dakota. Can. J. Botany 45:21-59.

Dodd, J. D., and R. T. Coupland. 1966a. Osmotic pressures of native plants of saline soil in Saskatchewan. Can. J. Plant Sci. 46:479-485.

――――――, and ――――――. 1966b. Vegetation of saline areas in Saskatchewan. Ecology 47:958-968.

Dodd, J. D., D. A. Rennie, and R. T. Coupland. 1964. The nature and distribution of salts in uncultivated saline soils in Saskatchewan. Can. J. Soil Sci. 44:165-175.

Fanning, C. D., C. M. Thompson, and D. Isaacs. 1965. Properties of saline soils of the Rio Grande Plains. J. Range Manage. 18:190-194.

Fernald, M. L. 1950. Gray's manual of botany. American Book Co., New York. 1632 pp.

Fireman, M., and H. E. Hayward. 1952. Indicator significance of some shrubs in the Escalante Desert, Utah. Botan. Gaz. 114:143-155.

Flowers, S. 1934. Vegetation of the Great Salt Lake region. Botan. Gaz. 95:353-418.

――――――, and F. R. Evans. 1966. The flora and fauna of the Great Salt Lake region, Utah. In H. Boyko. Salinity and aridity. W. Junk Publishers, Hague. pp. 367-393.

Foshag, W. F. 1926. Saline lakes of the Mohave desert region, California. Economic Geology 21:56-64.

Franklin, J., and C. T. Dyrness. 1969. Vegetation of Oregon and Washington. U.S.D.A. Forest Serv. Res. Paper PNW-80:1-216.

Frey-Wyssling, A. 1935. Die Stoffauschiedung der hoheren Pflanzen. Berlin.

Ganong, W. F. 1903. The vegetation of the Bay of Fundy salt and diked marshes. Botan. Gaz. 36:161-186, 280-302, 349-367, 429-455.

Gates, D. H., L. A. Stoddart, and C. W. Cook. 1956. Soil as a factor influencing plant distribution on salt deserts of Utah. Ecol. Monogr. 26:155-175.

Gates, F. C. 1940. Flora of Kansas. Cont. No. 291. Dept. of Botany, Kansas State College, Manhattan, Kansas. 226 pp.

Gold, H. 1939. A preliminary study of salt effects in the germination of *Allenrolfea occidentalis*. Univ. of Utah, M. A. thesis. 60 pp.

Good, R. E. 1965. Salt marsh vegetation, Cape May, New Jersey. Bull. N. J. Acad. Sci. 10:1-11.

Gould, C. N. 1900. The Oklahoma salt plains. Trans. Kansas Acad. Sci. 17:181-184.

Hadley, E. B. 1970. Net productivity and burning responses of native eastern North Dakota prairie communities. Amer. Midl. Natur. 84:121-135.

_____, and R. P. Buccos. Plant community composition and net primary production within a native eastern North Dakota prairie. Amer. Midl. Natur. 77:116-127.

Halket, A. C. 1915. The effect of salt on the growth of *Salicornia*. Ann. Bot. 29:143-154.

Hanson, H. C., and W. Whitman. 1937. Plant succession on solonetz soils in western North Dakota. Ecology 18:516-522.

_____, and _____. 1938. Characteristics of major grassland types in western North Dakota. Ecol. Monogr. 8:57-114.

Harrington, H. D. 1954. Manual of the plants of Colorado. Sage Books, Denver, Colorado. 666 pp.

Harris, F. S. 1920. Soil Alkali. John Wiley and Sons, New York. 258 pp.

Harshberger, J. W. 1911. An hydrometric investigation of the influence of sea water on the distribution of salt marsh and estuarine plants. Proc. Amer. Phil. Soc. 50:457-496.

Harward, M. R., and I. McNulty. 1965. Seasonal changes in ionic balance in *Salicornia rubra*. Proc. Utah Acad. Sci. Arts. Lect. 42:65-69.

Hay, R. 1890. Notes on some Kansas salt marshes. Trans. Kansas Acad. Sci. 12:97-100.

Hilgard, E. W. 1914. Soils, their formations, properties, composition, and relations to climate and plant growth. McMillan Co., New York. 593 pp.

Hinde, H. P. 1954. Vertical distribution of salt marsh phanerogams in relation to tide level. Ecol. Monogr. 24:209-225.

Hitchcock, A. S. 1898. Ecological plant geography of Kansas. Trans. Acad. Sci. St. Louis 8:55-69.

_____ 1950. Manual of the grasses of the United States. U.S. Govt. Printing Office, Wash. Misc. Pub. 200:1-1051.

Hitchcock, C. L., A. Cronquist, M. Ownbey, and J. W. Thompson. 1964. Vascular plants of the Pacific Northwest. Univ. Washington Press, Seattle. 597 pp.

Hogan, W. C. 1968. The effects of salinity on the germination and the growth of two inland halophytes. Ohio Univ., M.S. thesis. 38 pp.

Hopkins, H. H., and G. W. Tomanke. 1957. A study of the woody vegetation at Cedar Bluff Reservoir. Trans. Kansas Acad. Sci. 60:351-359.

Hunt, C. B., and L. W. Durrell. 1966. Plant ecology of Death Valley. U.S. Geol. Surv. Prof. Paper 509:1-68.

Isely, D. 1944. A study of conditions that effect the germination of *Scirpus* seeds. Cornell Univ. Agr. Expt. Sta. Mem. 257:1-27.

Jewell, M. E. 1927. Aquatic biology of the prairie. Ecology 8:289-298.

Johnson, D. S., and H. H. York. 1915. The relation of plants to tide level. Carnegie Inst. Wash. Publ. 206:1-162.

Jones, G. J., and E. B. Peterson. 1970. Plant species diversity in a woodland-meadow ecotone near Regina, Saskatchewan. Can. J. Bot. 48:591-601.

Kaushik, D. K. 1963. The influences of salinity on the growth and reproduction of marsh plants. Utah State Univ., Ph.D. thesis. 122 pp.

Kearney, T. H., L. J. Briggs, H. L. Shantz, J. W. McLane, and R. L. Piemeisel. 1914. Indicator significance of vegetation in Tooele Valley, Utah. J. Agr. Res. 1:365-417.

Keith, L. B. 1958. Some effects of increasing soil salinity on plant communities. Can. J. Bot. 36:79-89.

Keller, B. 1925. Halophyten and xerophyten - studien. J. Ecology 13:224-261.

Knight, J. B. 1934. A salt marsh study. Amer. J. Sci. 28:161-181.

Knipe, O. D. 1967. Influence of temperature on the germination of some range grasses. J. Range Manage. 20:298-299.

_____ 1968. Effects of moisture stress on germination of alkali sacaton, galleta, and blue grama. J. Range Manage. 21:3-4.

_____ 1970. Large seeds produce more, better alkali sacaton plants. J. Range Manage. 23:369-371.

_____ 1971a. Light delays germination in alkali sacaton. J. Range Manage. 24:152-154.

_____ 1971b. Effect of different osmotica on germination of alkali sacaton (Sporobolus airoides Torr.) at various moisture stresses. Botan. Gaz. 132:109-112.

Kurz, H., and K. Wagner. 1957. Tidal marshes of the Gulf and Atlantic coasts of northern Florida and Charleston, South Carolina. Florida State Univ. Stud. 24:1-168.

Langbein, W. B. 1961. Salinity and hydrology of closed lakes. U.S.G.S. Prof. Paper 412:1-20.

Latta, B. F. 1950. Geology and ground water resources of Barton and Stafford Counties, Kansas. Univ. Publ. St. Geol. Sur. Kansas 88:1-228.

Love, A., and D. Love. 1954. Vegetation of a prairie marsh. Bull. Torr. Bot. Club 81:16-34.

MacDougall, D. T. 1914. The Salton Sea. Carnegie Inst. Wash. Publ. 193:1-182.

Macke, A. J., and I. A. Ungar. 1971. The effects of salinity on germination and early growth of *Puccinellia nuttalliana*. Can. J. Bot. 49:515-520.

Malcolm, C. V. 1964. Effects of salt, temperature and seed scarification on germination of two varieties of *Arthrocnemum halocnemoides*. J. Roy Soc. West. Aust. 47:72-74.

Maples, R. S. 1968. An investigation of anamolous secondary growth in selected species of Amaranthaceae and Chenopodiaceae with special reference to the effect of salinity. Univ. Nebraska, Lincoln, Ph.D. dissertation. 102 pp.

Marks, J. B. 1950. Vegetation and soil relations in the Lower Colorado desert. Ecology 31:176-193.

Mason, H. L. 1957. A flora of the marshes of California. Univ. California Press, Berkeley and Los Angeles. 878 pp.

McCarraher, D. B. 1962. Northern Pike, *Esox lucius* in alkaline lakes of

Nebraska. Trans. Amer. Fish Soc. 91:326-329.

――――――, O. E. Orr, C. P. Agee, G. R. Foster, and M. O. Stern. 1961. Sandhills Lake survey. Nebraska Game and Forest and Parks Commission Job number 2. Lincoln, Nebraska. 83 pp.

McGregor, D. J. 1964. Ground water supply for the city of Langford. South Dakota State Geol. Sur. Spec. Report 29:1-27.

McKay, E. M. 1934. Salt tolerance of *Ruppia maritima* L. in lakes of high magnesium sulfate content. Wash. State College, Pullman, Wash., Ph.D. dissertation. 41 pp.

McMillan, C., and F. N. Moseley. 1967. Salinity tolerances of five marine spermotophytes of Redfish Bay, Texas. Ecology 48:503-506.

Merkel, D. L., and H. H. Hopkins. 1957. Life history of salt cedar (*Tamarix gallica* L.). Trans. Kansas Acad. Sci. 60:360-369.

Miller, W. R., and F. E. Egler. 1950. Vegetation of the Wequetequock-Pawcatuck tidal marshes, Connecticut. Ecol. Monogr. 20:143-172.

Morley, G. E. 1964. A floristic study of Republic County, Kansas. Trans. Kansas Acad. Sci. 67:716-746.

Muenscher, W. C. 1944. Aquatic plants of the United States. Comstock Publ. Co., Inc., Ithaca, New York. 374 pp.

Munz, P. A., and D. D. Keck. 1965. A California flora. Univ. California Press, Berkeley. 1,681 pp.

Nelson, N. F. 1955. Ecology of the Great Salt Lake Marshes. Utah Acad. Proc. 32:37-40 (1955).

Nichol, A. A. 1952. The natural vegetation of Arizona. Univ. Ariz. Agric. Expt. Sta. Tech. Bull. 68:189-230.

Nielson, E. L. 1953. Revegetation of alkali flood plains adjoining the North Platte River, Garden County, Nebraska. Amer. Midl. Natur. 49:915-919.

Nielson, A. K. 1956. A study of the variability of *Distichlis stricta* selections from several geographical locations in the western United States. Utah State Univ., Logan, Utah, M.S. thesis. 46 pp.

Ortenburger, A. I., and R. D. Bird. 1931. The ecology of the western Oklahoma salt plains. Okla. Biol. Survey Bull. 3:49-64.

Peattie, D. C. 1922. The Atlantic coastal plain element in the flora of the Great Lakes. Rhodora 24:57-70, 80-88.

Penfound, W. T. 1953. Plant communities of Oklahoma lakes. Ecology 34:561-583.

――――――. 1967. A physiognomic classification of vegetation in conterminous United States. Botan. Rev. 33:289-326.

――――――, and E. S. Hathaway. 1938. Plant communities in the marshlands of southeastern Louisiana. Ecol. Monogr. 8:1-56.

Potter, L. D. 1957. Phytosociological study of San Augustin Plains, New Mexico. Ecol. Monogr. 27:113-136.

Pound, R., and F. E. Clements. 1898. The phytogeography of Nebraska. Jacob North and Co., Lincoln, Nebraska. 415 pp.

Rawson, D. S., and J. E. Moore. 1944. The saline lakes of Saskatchewan. Can. J. Res. 22D:141-201.

Redmann, R. E., and G. Hulett. 1964. Factors affecting the distribution of certain species of compositae on an eastern North Dakota prairie. Proc. N. D. Acad. Sci. 18:10-21.

Richards, L. A. 1954. Diagnosis and improvement of saline and alkali soils. USDA Handbook 60:1-160.

Robinson, T. W. 1958. Phreatophytes. U. S. Geol. Surv. Water Supply Paper 1423:1-84.

Rydberg, P. A. 1932. Flora of the prairie and plains of central North America. New York Botan. Garden, New York. 969 pp.

Schaffner, J. H. 1898. Notes on the salt marsh plants of northern Kansas. Botan. Gaz. 25:255-260.

Schofield, W. B. 1959. The salt marsh vegetation of Churchill, Manitoba and its phytogeographic implications. Natl. Mus. Canada Bull. 160:107-132.

Shantz, H. L., and R. L. Piemeisel. 1940. Types of vegetation in Escalante Valley, Utah, as indicators of soil conditions. USDA Tech. Bull. 713:1-46.

——————, and ——————. 1924. Indicator significance of the natural vegetation of the southwestern desert. J. Agr. Res. 28:721-802.

——————, and R. Zon. 1924. The natural vegetation of the United States. U.S.D.A. Atlas Amer. Agr. Pt. 1, Sect. E. 29 pp.

Shields, L. M. 1956. Zonation of vegetation within the Tularosa Basin, New Mexico. Southwest. Natur. 1:49-68.

Shirk, C. J. 1924. An ecological study of the vegetation of an inland saline area. Ph.D. dissertation, Univ. of Nebraska, Lincoln. 126 pp.

Shjeflo, J. B. 1968. Evapotranspiration and the water budget of prairie potholes in North Dakota. U. S. Geol. Survey Prof. Paper 585-B:1-49.

Shreve, F. 1925. Ecological aspects of the deserts of California. Ecology 6:93-103.

——————. 1942. The desert vegetation of North America. Botan. Rev. 8:195-246.

Sloan, C. E. 1970a. Biotic and hydrologic variables in prairie potholes in North Dakota. J. Range Manage. 23:260-263.

——————. 1970b. Prairie potholes and the water table. U. S. Geol. Survey Prof. Paper 700-B:227-231.

Springfield, H. W. 1966. Germination of fourwing saltbush seeds at different moisture stress. Agr. J. 58:149-150.

Stevens, O. A. 1963. Handbook of North Dakota plants. North Dakota Institute for regional studies, Fargo, N. D. 324 pp.

Stevenson, R. E., and K. O. Emery. 1958. Marshlands at Newport Bay, California. Allan Hancock Foundation Publications 20:1-109.

Stewart, R. E., and H. A. Kantrud. 1969. Proposed classification of potholes in the glaciated prairie region, in small water areas in the prairie pothole region. Can. Wildlife Service Report 6:57-69.

St. John, H., and W. D. Courtney. 1924. The flora of Epsom Lake. Amer. J.

Bot. 11:100-107.

Steiner, M. 1934. Zur okologie der salzmarschen der nordostlichen vereinigten staaten von nordamerika. Jahrb. Wiss. Bot. 81:94-202.

Strogonov, B. P. 1964. Physiological basis of salt tolerance of plants. Israel Program Scientific Translations, Jerusalem. 366 pp.

_____, and L. P. Laping. 1964. A possible method for the separate study of toxic and osmotic action of salt in plants. Fiziol. Rest. 11:674-680.

Svenson, H. K. 1927. Effects of post pleistocene marine submergence in eastern North America. Rhodora 29:41-48, 57-72, 87-93.

Taylor, N. 1938. A preliminary report on the salt marsh vegetation of Long Island, New York. New York State Museum Bull. 316:21-84.

_____ 1939. Salt tolerance of Long Island salt marsh plants. New York State Mus. Circ. 23:1-42.

Teeter, J. W. 1963. The influence of sodium chloride on the growth and reproduction of the sago pondweed (*Potamogeton pectinatus*). M.A. thesis, Utah State Univ., Logan, Utah. 73 pp.

Tolstead, W. L. 1942. Vegetation of the northern part of Cherry County, Nebraska. Ecol. Monogr. 12:255-292.

Ungar, I. A. 1962. Influence of salinity on seed germination in succulent halophytes. Ecology 43:763-764.

_____ 1964. A phytosociological analysis of the Big Salt Marsh, Stafford County, Kansas. Trans. Kansas Acad. Sci. 67:50-64.

_____ 1965a. An ecological study of the vegetation of the Big Salt Marsh, Stafford County, Kansas. Univ. Kansas Sci. Bull. 46:1-98.

_____ 1965b. Soil-vegetation relationships in saline soils of Kansas and Oklahoma. Bull. Ecol. Soc. Amer. 46:90 (abstract).

_____ 1966. Salt tolerance of plants growing in saline areas of Kansas and Oklahoma. Ecology 47:154-155.

_____ 1967a. Influence of salinity and temperature on seed germination. Ohio J. Sci. 67:120-123.

_____ 1967b. Vegetation-soil relationships on saline soils in northern Kansas. Amer. Midl. Natur. 78:98-120.

_____ 1968a. Ecology of *Suaeda depressa* (Pursh) Wats. in inland saline soils. Bull. Ecol. Soc. Amer. 48:107-108 (abstract).

_____ 1968b. Species-soil relationships on the Great Salt Plains of northern Oklahoma. Amer. Midl. Natur. 80:392-406.

_____ 1968c. *Sporobolus texanus* Vasey in Lincoln, Nebraska. Rhodora 70:450-451.

_____ 1969. Species-soil relationships of halophytes in Nebraska and South Dakota. XI International Botanical Congress 223 (abstract).

_____ 1970a. Species-soil relationships on sulfate dominated soils in South Dakota. Amer. Midl. Natur. 83:343-357.

_____ 1970b. Salinity and temperature effects on *Iva annua* L. establishment. Bull. Ecol. Soc. Amer. 51:25 (abstract).

_____ 1971. *Atriplex patula* var *hastata* L. Gray seed dimorphism. Rhodora 73:548-551.

_____, and F. Capilupo. 1969. An ecological life study of *Suaeda depressa* Pursh Wats. Adv. Front. Plant Sci. 23:137-158.

_____, W. Hogan, and M. McClelland. 1969. Plant communities of saline soils at Lincoln, Nebraska. Amer. Midl. Natur. 82:564-577.

_____, and W. C. Hogan. 1970. Seed germination in *Iva annua* L. Ecology 51:150-154.

Vest, E. D. 1962. Biotic communities in the Great Salt Lake Desert. Inst. Env. Biol. Res. Ecology and Epizoology Series 73:1-122.

Vogl, R. J., and L. T. McHargue. 1966. Vegetation of California Fan Palm Oases on the San Andreas Fault. Ecology 47:532-540.

Walker, B. H., and R. T. Coupland. 1970. Herbaceous wetland vegetation in the aspen grove and grassland regions of Saskatchewan. Can. J. Bot. 48:1861-1878.

Waterfall, U. T. 1952. A catologue of the flora of Oklahoma. Oklahoma A. and M. College, Stillwater, Oklahoma. 243 pp.

Weaver, J. E. 1918. The quadrat method in teaching ecology. Plant World 21:267-283.

_____. 1954. North American Prairie. Johnson Publ. Co., Lincoln, Nebraska. 348 pp.

_____, and F. E. Clements. 1938. Plant ecology. McGraw Hill Book Co., New York. 601 pp.

Webb, K. L. 1966. NaCl effects on growth and transpiration in Salicornia bigelovii, a salt marsh halophyte. Pl. Soil 24:261-268.

West, N. E. 1968. Ecology and management of salt desert shrub ranges. Utah Agric. Expt. Sta. Series 505:1-30.

_____, and Ibrahim. 1968. Siol-vegetation relationships in the shadscale zone of southeastern Utah. Ecology 49:445-456.

Wiebe, H. W., and H. Walter. 1972. Mineral ion composition of halophytic species from northern Utah. Amer. Midl. Natur. 87:241-245.

Williams, M. D., and I. A. Ungar. 1972. The effect of environmental parameters on the germination, growth, and development of *Suaeda depressa* (Pursh) Wats. Amer. J. Bot. (in press).

Wilson, J. N. 1958. The limnology of certain prairie lakes in Minnesota. Amer. Midl. Natur. 59:418-437.

Wilson, D. B. 1967. Growth of *Hordeum jabatum* under various soil conditions and degrees of plant competition. Can. J. Plant Sci. 47:405-412.

Young, R. T. 1923. Resistance of fish to salts and alkalinity. Amer. J. Physiol. 65:373-388.

A REVIEW OF STRUCTURE IN SEVERAL NORTH CAROLINA SALT MARSH PLANTS

Charles E. Anderson

Department of Botany
North Carolina State University
Raleigh, North Carolina 27607

Abstract

The objective of this paper is to present the basic structural features of seven plants which occur in North Carolina salt marshes. These plants are: *Spartina alterniflora* Loisel., *Spartina patens* (Ait) Muhl., *Distichlis spicata* (L.) Greene, *Aster tenuiflolius* L., *Juncus roemerianus* Scheele, *Salicornia virginica* L., and *Limonium spp.* Literature on related species will be cited where appropriate. It is hoped that this information may serve as a background for future developmental, physiological, and ecological studies.

Materials and Methods

All plants studied were collected from North Carolina salt marshes. Marshes such as the one at Sunset Beach are very useful because they contain all the species studied within a very limited area.

Some material was fixed in formalin-acetic acid-alcohol, dehydrated in tertiary butyl alcohol, and infiltrated with Paraplast. Other material was frozen for cryostat sectioning. Staining was primarily done with safranin and fast green.

The cytoplasmic vacuoles spoken about in the text are best seen with fresh or frozen sections. Normal fixation seems to cause distortion which makes them hard to recognize. Only *Distichlis spicata* provided an occasional exception to this.

Both *Limonium carolinianum* and *Limonium Nashii* occur in North Carolina salt marshes. At the time this work was done, it was impossible to distinguish between these two species.

Species Studied

Spartina alterniflora

Probably the most important low marsh plant in North Carolina is *Spartina alterniflora*. Much has been published about the genus, including work by Sutherland and Eastwood (1916) on the structure of *S. townsendii* and papers by Skelding and Winterbotham (1939) and Levering and Thomson (1971) on salt glands in *S. townsendii* and *S. foliosa*, respectively.

Leaf. The leaves of *S. alterniflora* are long and thin with sheathing bases. The adaxial surface of the leaf is composed of ridges and furrows which conform to the series of vascular bundles and surrounding mesophyll tissue which make up the leaf. The abaxial surface is only slightly undulating. Both epidermal layers of *S. alterniflora* are lignified and cutinized. Small papillae characterize the adaxial surface, but are generally lacking on the abaxial epidermis (Fig. 1). Parallel rows of long and short cells make up both epidermal layers (Fig. 2). Sutherland and Eastwood (1916) suggested that the short cells of *S. townsendii* are of two types, saddle cells and silica cells. There appear to be silica bodies in some small cells of *S. alterniflora*.

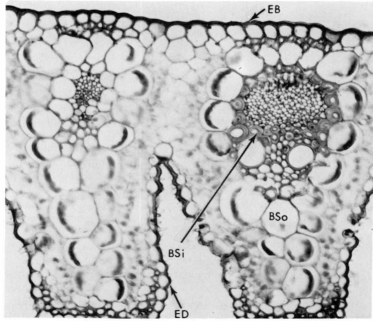

Fig. 1. *S. alterniflora* leaf x.s. ED — adaxial epidermis; EB — abaxial epidermis; BSo — outer bundle sheath; BSi — inner bundle sheath; M — Mesophyll chlorenchyma (800x).

Stomates occur in rows most protusely in the furrows of the adaxial surface of the leaf. They are also common on the abaxial surface, especially between the veins. Some stomates occur on the bases of the ridges of the adaxial surface. Accessory cells associated with stomates, particularly those in the furrows, typically possess branched papillae which extend over the stomatal aperture (Fig. 3).

Salt glands occur in both epidermal layers, but are most profuse in the furrows of the adaxial epidermis. The glands are composed of two cells, a basal cell which is large and lies beneath the epidermis and a cap cell which is smaller and protrudes through the epidermis (Fig. 3). The ultrastructure and

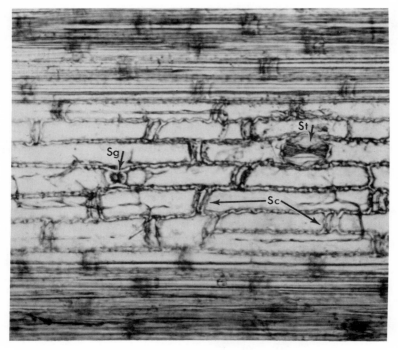

Fig. 2. **S. alterniflora** leaf abaxial epidermis surface view. St — stomate; Sg — salt gland; LC — long cell; Sc — saddle and silica cells (400x).

proposed mechanism of function for these glands in *S. foliosa* is described by Levering and Thomson (1971). If light microscope observations are any indication, the glands of *S. alterniflora* and those described by Skelding and Winterbotham (1939) for *S. townsendii* are probably very similar in structure and function.

The internal structure of the leaf blade consists of varying sizes of vascular bundles surrounded by two bundle sheaths (Fig. 1). In large bundles the inner sheath is composed of small highly lignified sclerenchyma type cells. The xylem and phloem appear to be separated by these cells as well. In smaller bundles the inner sheath cells have very little lignification. The outer bundle sheath in all bundle sizes is composed of large highly vacuolated cells with large chloroplasts concentrated along their outer walls. Cytoplasmic vacuoles are also present in the outer sheath cells, but are visible only in fresh section (Fig. 4). The outer bundle sheath extends the length of the adaxial ridge, ending at the layer of fibers which is just beneath the adaxial epidermis at the tip of the ridge. A similar layer of fibers occurs between the abaxial epidermis and the outer bundle sheath. The mesophyll cells are radially elongated and slightly branched, developing some intercellular spaces. However, compared to the rest of the plant the leaf blade is tightly packed. The chloroplasts of

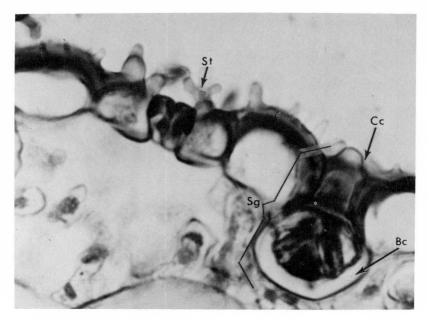

Fig. 3. S. alterniflora leaf x.s. St – stomate; Sg – salt gland; Bc – basal cell; Cc – cap cell (3600x).

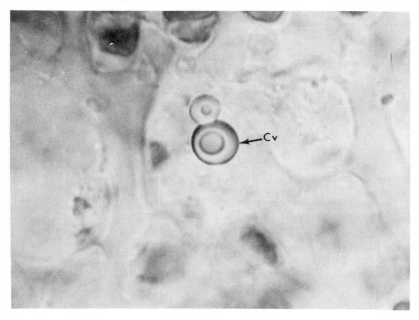

Fig. 4. S. alterniflora leaf. Cv – cytoplasmic vacuole (2000x).

the mesophyll cells are somewhat smaller than those of the bundle sheath. There are cytoplasmic vacuoles found also in the mesophyll.

The leaf sheath is similar in structure to the blade with a few exceptions. Large lacuna occupy the space between the bundles in the sheath. In *S. townsendii* these lacuna are present in the leaf blade as well (Sutherland and Eastwood 1916). The sheath epidermis contains fewer and smaller papillae, and is less lignified. The chlorenchymous mesophyll is primarily on the adaxial side of the leaf. Even the outer bundle sheath is nearly devoid of chloroplasts on the abaxial side (Fig. 5).

Stem. The culm of *S. alterniflora* is short and compact until floral initiation. The lower portions of the stem are quite aerenchymous, allowing for direct air flow with the lacuna of the leaf sheath. The internodes of the stem retain active intercalary meristems and ultimately elongate this way during flowering (Fig. 6).

The rhizome is a hollow cylinder with slight ridges (Fig. 7). The epidermis is lignified as is the hypodermal layer. A scalloped band of sclerenchyma touches the hypodermal layer and dips inward enclosing thin walled parenchyma cells which form lysogenous cavities in some rhizome tissue. Scattered collateral vascular bundles occur in the parenchymous ground tissue.

Root. The root system is adventitious. It originates primarily from the nodes of the rhizome and from the base of the culm. There are two types of roots, one larger, which Sutherland and Eastwood (1916) call fixing roots, and a smaller type, called absorbing roots by Sutherland and Eastwood. The larger roots grow straight down into the marsh as a rule and have few

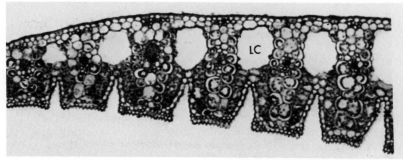

Fig. 5. *S. alterniflora* leaf sheath x.s. L — lysogenous cavity (160x).

branches. Their root hairs, when occurring, are very short. In transection the epidermis is subtended by four or five hypodermal layers. The hypodermal cells are small and compact. The cortex in young root tissue consists of tightly packed radial rows of cells, but during maturation many of these cells break down leaving radial files of cells and cell walls and extensive intercellular space. The inner cortical layer remains intact surrounding an endodermis which is lignified on its radial and inner tangential walls. A

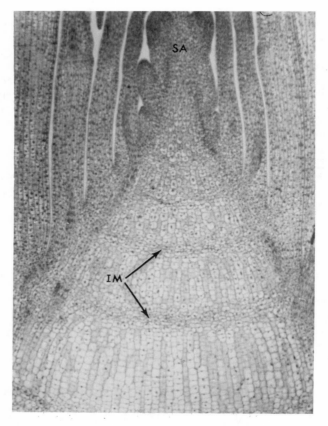

Fig. 6. **S. alterniflora** culm apex l.s. SA — shoot apex; IM — intercalary meristem (160x).

pericycle occurs inside the endodermis enclosing the phloem and highly lignified polyarched xylem (Fig. 8).

The smaller roots (absorbing roots) have the same general tissue pattern as the larger ones. There is less stele tissue, fewer layers of cortical tissue, and fewer hypodermal layers. The smallest roots observed have a single xylem vessel and essentially no lysogenous cavities in the cortex which is composed of only two to three layers. Root hairs are much larger and abundant in the absorbing roots. Extensive branching occurs, forming a highly diffuse root system.

Spartina patens

Spartina patens, like *S. alterniflora*, has not been investigated broadly from a structural or developmental viewpoint. Prat (1934) compared leaf epidermal features of *S. patens* and *S. alterniflora*, and Metcalfe (1960) mentions *S. patens*, but offers no description.

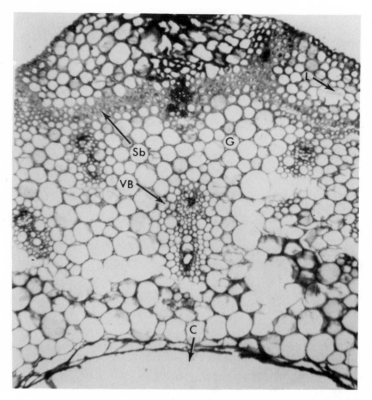

Fig. 7. **S. alterniflora** rhizome x.s. C — center cavity; G — ground tissue; Sb — sclerenchyma band; L — lysogenous cavity (400x).

Leaf. The leaves of *S. patens* are composed of ridges and furrows on the adaxial side as in *S. alterniflora* (Fig. 9). Both epidermal layers are composed of short and long cells in rows. Some of the short cells appear to contain silica bodies. The surface of the abaxial epidermis has some papillae, but is relatively smooth as compared to the adaxial epidermis which has large, often branched, thick walled papillae. Prat (1934) pointed out that the papillae of *S. patens* are much larger than those of *S. Alterniflora*, but less numerous. In addition to the papillae of the adaxial surface, larger pointed trichomes occur which typically have six small basal cells surrounding them. The only stomates observed occur in the furrows of the adaxial epidermis. Branched papillae extend over the stomates. Protruding through the epidermis and also occurring primarily in the furrows are salt glands. These glands appear to be of the same general structure as those of *S. alterniflora*, with a larger basal cell and a smaller extended cap cell (Fig. 10).

The leaf is composed of a series of vascular bundles and a relatively small

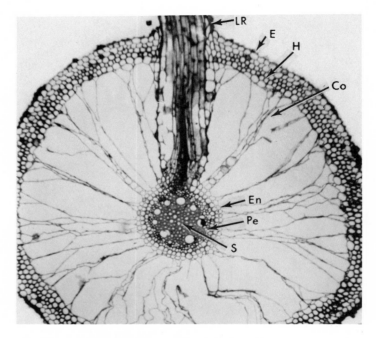

Fig. 8. **S. alterniflora** root x.s. E – epidermis; H – hypodermis; Co – cortex; LR – lateral root; En – endodermis; Pe – pericycle; S – stele (48x).

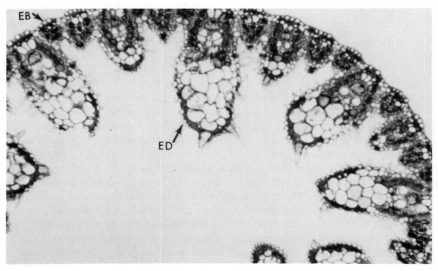

Fig. 9. **S. patens** leaf x.s. ED – adaxial epidermis; EB – abaxial epidermis (160x).

ECOLOGY OF HALOPHYTES

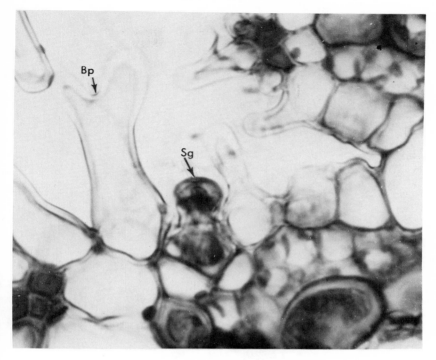

Fig. 10. *S. patens* leaf x.s. Sg — salt gland; Bp — branched papillae (2000x).

amount of mesophyll tissue (Fig. 11). All bundles contain a double sheath. The inner sheath is composed of small highly lignified sclerenchyma cells in larger veins and less lignified cells in smaller veins. The outer sheath is composed of large, rather thin walled cells with large chloroplasts. The adaxial side of the outer bundle sheath expands forming a mass of achlorophyllous parenchyma cells. This tissue is much reduced or lacking in smaller veins. Subtending the epidermis, both above and below the veins, are one to two layers of sclerenchyma tissue. The remainder of the leaf tissue is composed of radially elongated chlorenchyma cells which surround the vascular bundles (Fig. 11).

Stem. The culm and rhizome system of *S. patens* collected from North Carolina is consistently quite compact. This is unlike the more trailing system found in *S. alterniflora*. The tissues of the rhizome are much more compact also, showing little aerenchymous tendency. In transection the rhizome consists of an outer fleshy cortex with an active periderm and leaf traces. Inside this outer layer is the stele composed of scattered collateral vascular bundles and ground tissue. A pericycle occurs on the outer edge of the stele. The center cells of the stele are morphologically distinct and might be considered pith-like, although such an interpretation may be questioned in

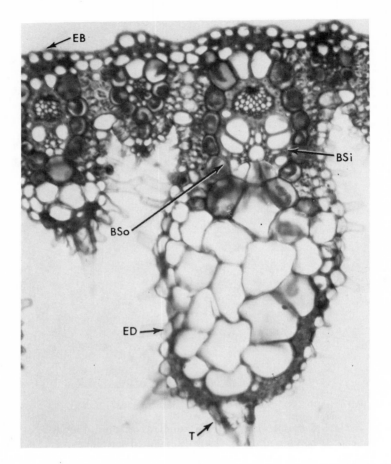

Fig. 11. **S. patens** leaf x.s. ED — adaxial epidermis; EB — abaxial epidermis; BSo — outer bundle sheath; BSi — inner bundle sheath; T — trichome (400x).

grasses (Fig. 12).

Root. The root system of *S. patens* is entirely adventitious. The larger roots have the characteristic pattern of deteriorating cortex found in so many aquatic plants. The epidermis and three to four hypodermal layers surround the cortex and may be lignified. Root hairs are not numerous and, when present, those observed from field collections were very small. The two to four cortical layers just outside the stele may be highly lignified. Steles in the larger roots exhibit some secondary growth. The polyarched xylem is highly lignified and frequently possesses a relatively large lacuna in the center. The root may possess an active pericycle and an endodermis which is thickened on its radial and inner tangential walls. In many instances the pericycle appears lignified (Fig. 13).

Limonium spp.

Several species of *Limonium* have received attention in the literature. The

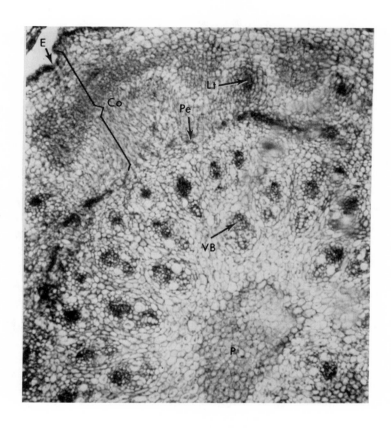

Fig. 12. **S. patens** rhizome x.s. E – epidermis; Co – cortex; Pe – pericycle; Lt – leaf trace; VB – vascular bundle; P - pith (48x).

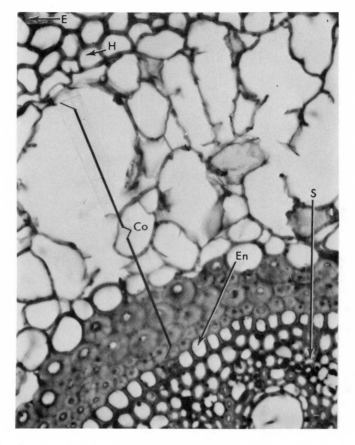

Fig. 13. **S. patens** root x.s. E – epidermis; H – hypodermis; Co – cortex; En – endodermis; S – stele (160x).

salt glands of *L. vulgare* have been investigated by Ziegler and Luttge (1966). Vegetative reproduction was considered for *L. mayeri* and *L. latifolium* by Katina (1962). A review of the structure of several species of *Limonium* and related genera is covered by Metcalfe and Chalk (1950).

Leaf. The leaves of *Limonium* occurring in North Carolina are somewhat fleshy and tightly packed (Fig. 14). The thickness of the cuticle is quite variable. Scattered stomates and salt glands occur on both surfaces of the leaf. As described in the literature, the salt glands of *Limonium* are composed of at least two inner collecting cells which accumulate small vesicles of salt through large plasmodesmata. In addition, there are usually six or more secreting cells which are not covered with a cuticle. This physical description seems to fit the North Carolina *Limonium* species as well (Fig. 15).

The vascular bundles in *Limonium* are small but fairly numerous. The mesophyll is not highly differentiated. Palisade layer cells are not very elongated and comprise two to three cell layers of the leaf. The cells of the

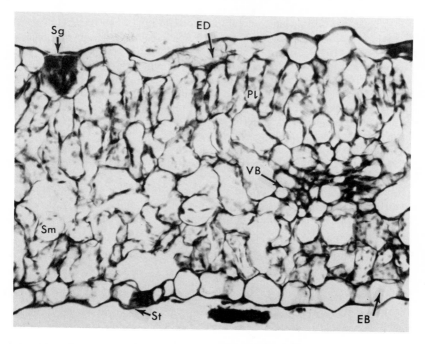

Fig. 14. **Limonium** leaf x.s. ED — adaxial epidermis; EB — abaxial epidermis; VB — vascular bundle; Pl — palisade layer; Sm — spongy mesophyll; Sg — salt gland; St — stomate (800x).

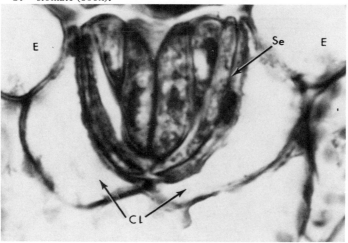

Fig. 15. **Limonium** salt gland x.s. Cl — collecting cells; Se — secreting cells; E — epidermis (2000x).

spongy layer are somewhat more elongated than might be expected. The cells of the inner mesophyll are larger and seem to have fewer chloroplasts, possibly due to their larger vacuole. The lack of more definite tissue differentiation may be related to the usual upright position of the leaf in the marsh. Cytoplasmic vacuoles have been observed in both epidermal layer cells and the mesophyll cells. Considerable variability occurs both in the size and occurrence of these vacuoles. It is not obvious how these large vacuoles might relate to the smaller ones which supposedly pass into the collecting cells of the salt glands in accumulating salt.

The petiole of *Limonium* is quite distinct. Aerenchymous tissue forms an extensive network of air passages. The scattered vascular bundles are much larger than in the lamina (Fig. 16).

Stem. New shoots are produced each year from a perennial stem. This stem is surrounded by an extensive periderm with abundant cork. The cortex is aerenchymous as is the pith. Secondary growth occurs, producing very small

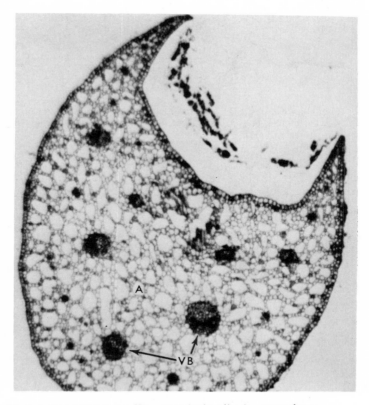

Fig. 16. **Limonium** leaf petiole x.s. VB — vascular bundle; A — aerenchyma.

xylary elements and a small amount of phloem (Fig. 17). Of all the stems examined, this one was by far the easiest to section, probably indicating little lignification.

Root. The root system of *Limonium* is not very extensive and it is entirely

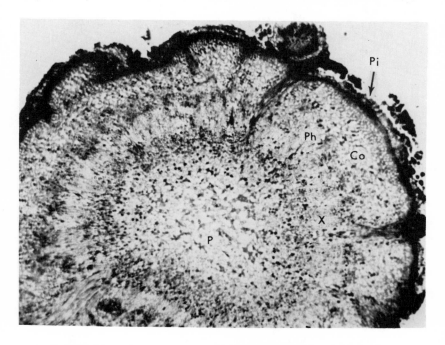

Fig. 17. **Limonium** stem x.s. Pi – periderm; Co – cortex; Ph – phloem; X – xylem; P – pith (48x).

adventitious, arising from a pericycle just outside the phloem of the stem. The older roots sectioned all exhibited some secondary growth around a tri or tetrarch xylem. The cortex is aerenchymous with curving radial files of cells. The outer cortical tissue was found to contain cork. In the oldest roots some corky periderm was present (Fig. 18). This general tissue pattern was present as well in the smaller branched roots except for the corky tissue and secondary growth.

Distichlis spicata

Work on this genus seems restricted to the older literature. Holm (1891, 1901-2) described several features of plants in the genus including leaf tissue. Prat (1936) also contributed some observations to the literature, but Metcalfe (1960) suggested his observations regarding the presence of double bundle sheaths were incorrect.

Leaf. *Distichlis spicata* differs from *Spartina* grasses in several ways,

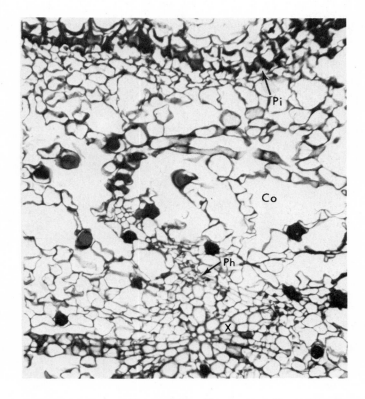

Fig. 18. **Limonium** root x.s. Pi – periderm; Co – cortex; X – xylem; Ph – phloem (200x).

although the basic leaf structure is similar (Fig. 19). The adaxial epidermis contains numerous papillae of the size similar to *S. patens*. The stomates on the adaxial surface are limited to the furrows in the furrow and ridge system. Papillae extend over the stomatal aperture. The cell arrangement in both epidermal layers is composed of long and short cells in rows with very irregular walls. Stomates on the abaxial surface are much more numerous, also occurring in rows. Salt glands occur in the furrows of the adaxial surface and between bundles on the abaxial surface (Fig. 20). In limited observations, these glands appear to be composed of a single bar bell shaped cell with a single nucleus and very dense cytoplasm. Numerous cytoplasmic strands pass from one end of the cell to the other, being constricted in the isthmus of the cell. The nucleus may be seen in either end of the cell in any given leaf section. The evidence for these cells being salt glands is the presence of liquid droplets and salt crystals on leaves in the greenhouse.

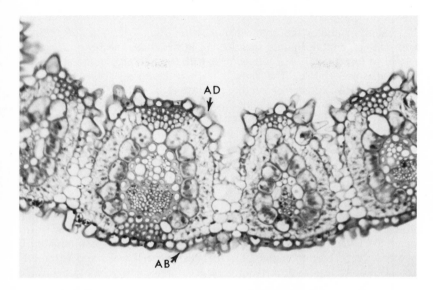

Fig. 19. **Distichlis spicata** leaf x.s. AD — adaxial epidermis; AB — abaxial epidermis (400x).

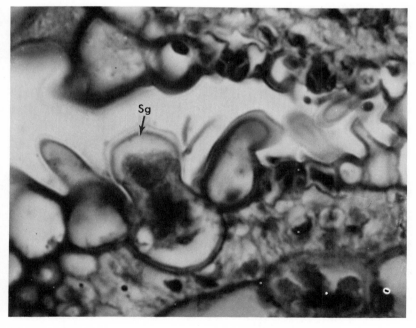

Fig. 20. **Distichlis spicata** salt gland x.s. Sg — salt gland (2000x).

The vascular bundles are somewhat variable (Fig. 21). The larger ones contain distinct double bundle sheaths. As in *Spartina*, the outer sheath is composed of large, highly vacuolated cells with large chloroplasts and at least one cytoplasmic vacuole. The inner sheath is composed of small highly lignified living cells. In smaller bundles the extent of lignification in the inner sheath becomes limited to the abaxial side. It is difficult to distinguish sheath cells on the adaxial side, but the presence of living nucleated cells there tends to support the existence of a continuous sheath even in the smallest veins.

The remainder of the mesophyll is composed of radially oriented chlorenchyma cells and some sclerenchyma above and below the bundles. In the larger veins the sclerenchyma cells are continuous with the bundle sheath and should be considered bundle sheath extensions. Bulliform cells separate each bundle segment of the leaf.

Stem. The vegetative culm of *D. spicata* is much more extensive than in the *Spartinas* described. In transection, the epidermal layer, plus two to four hypodermal layers, is highly lignified. The cortex is composed of deteriorating cells leaving radial files of cells and cell walls and extensive cavities. The inner layers of cortex are intact and typically highly lignified. This lignification continues into the outer stele layers and gradually decreases toward the inside of the stem. Collateral vascular bundles are scattered

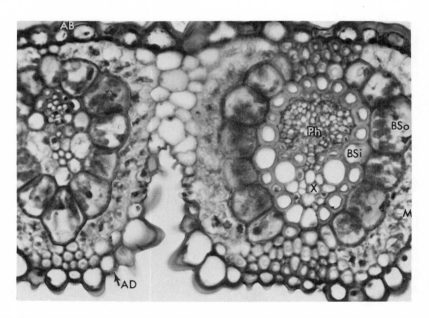

Fig. 21. **Distichlis spicata** leaf x.s. BSo — outer bundle sheath; BSi — inner bundle sheath; M — mesophyll chlorenchyma; AD — adaxial epidermis; AB — abaxial epidermis; X — xylem; Ph — phloem (800x).

throughout the stele (Fig. 22).

The rhizome is very similar to the culm in structure (Fig. 23). The hypodermal layer is not as distinct as in the stem. Cells of the hypodermis are small, more numerous than in the stem, and tend to deteriorate as do those of the cortex. The inner cortex outer stele region is again a band of small highly lignified cells. A pericycle is present, originating lateral roots. Scattered collateral vascular bundles are found in the stele surrounded by relatively thin-walled parenchyma cells. Each vascular bundle has a single large vessel which contains a tan colored material in fresh section. No attempt was made to identify this material but it is retained throughout the usual paraffin slide-making process. Starch grains are more numerous in this rhizome than in any other organ of any plant described in this paper.

Root. The root system of *D. spicata* is adventitious. The epidermal layer contains the longest root hairs found in larger roots of this general type (Fig. 24). One to three hypodermal layers subtend the epidermis. The cortex is composed of radial files of deteriorating cell walls and an occasional file of

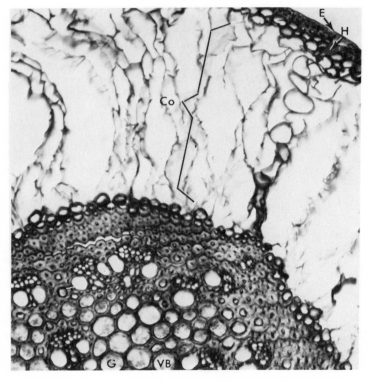

Fig. 22. **Distichlis spicata** stem x.s. E — epidermis; H — hypodermis; Co — cortex; VB — vascular bundles; G — ground tissue (200x).

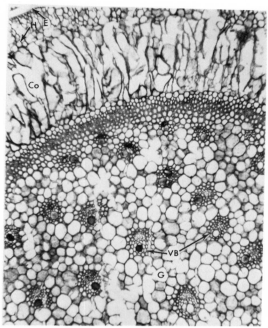

Fig. 23. **Distichlis spicata** rhizome x.s. E − epidermis; H − hypodermis; Co − cortex; VB − vascular bundle; G − ground tissue (100x).

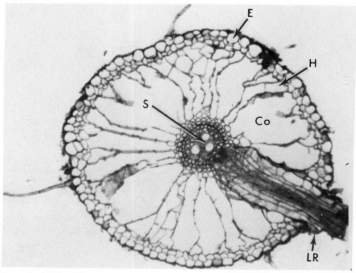

Fig. 24. **Distichlis spicata** root x.s. E − epidermis; H − hypodermis; Co − cortex; S − stele; LR − lateral root (48x).

intact cells. The polyarched stele is highly lignified. Lateral branching is common, being initiated by a pericycle. As in other species described, the endodermis is thickened on its radial and inner tangential walls. Also, as in other marsh species, the smaller roots contain a smaller stele with fewer xylem archs and a restricted amount of cortical tissue.

Salicornia virginica

The genus of *Salicornia* has received much attention from a structural point of view. Morphological studies have been performed on various species by DeBary (1884), Ganong (1903), Cross (1909), DeFraine (1912), Peck (1941), Munz (1959), Fahn and Arzee (1959), James and Kyhos (1961), and Anderson (1971). The descriptions here relate specifically to *S. virginica*.

Shoot. The so-called leafless shoot of *S. virginica* is, in fact, a stem with two appressed leaves surrounding each internode (Fig. 25). In transection there is a single layered somewhat papillate epidermis composed of small cells in longitudinal rows. The external walls are highly cutinized. Frequent and randomly distributed stomates occur (Fig. 26). Beneath the epidermis is a

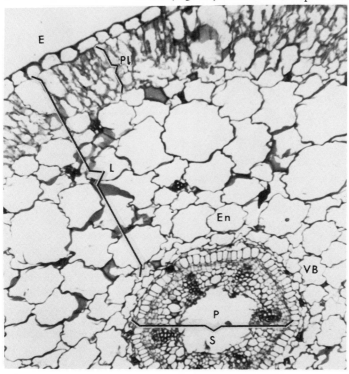

Fig. 25. **Salicornia virginica** shoot x.s. L — leaf; Pl — palisade layer; E — epidermis; En — endodermis; VB — vascular bundles; P — pith; S — stele or stem tissue (48x).

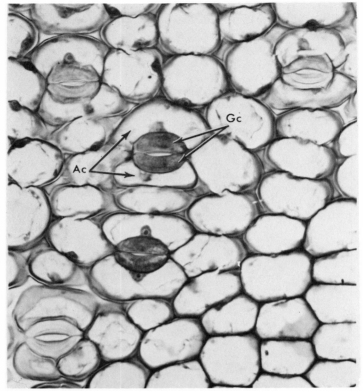

Fig. 26. **Salicornia virginica** leaf surface section. Gc — guard cell; Ac — accessory cell (2000x).

palisade tissue composed of two to three cell layers. Interspursed within the palisade cells are clumps of tracheoidioblasts (Fig. 27). These large spirally thickened cells possess a plasmolemma, and are considered by some to be involved in the maintenance of water balance. They should not be considered salt glands, since *S. virginica* is not known to secrete salt, and these cells do not demonstrate high salt content with histochemical tests. The remainder of the leaf tissue is primarily parenchymous, with large intercellular spaces. Scattered fibers occur near the inner regions of the tissue.

Inside the leaf tissue, in what is considered stem tissue, occurs a tissue layer which is thickened on its radial and inner tangential walls. Casparian strips have not been demonstrated in this layer, but it does have the other structural characteristics of an endodermis. Beneath the endodermis is a pericycle. This pericycle not only forms adventitious roots, but gives rise to the vascular and cork cambia. Six collateral bundles make up the primary vascular tissue. Two leaf traces, one for each leaf, may also be seen near the nodes. The pith area

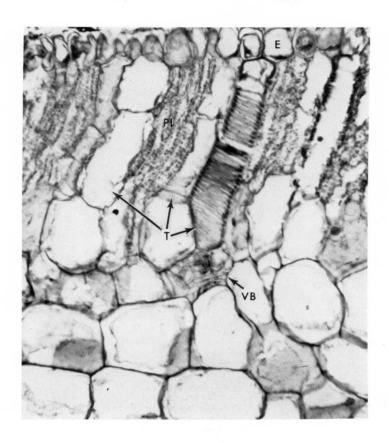

Fig. 27. **Salicornia virginica** leaf x.s. E – epidermis; Pl – palisade layer; T – tracheoidioblasts; VB – vascular bundle (2000x).

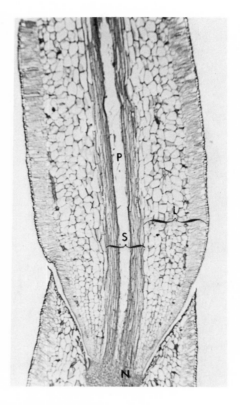

Fig. 28. **Salicornia virginica** shoot l.s. P − pith; S − stem or stele; L − leaf; N − node (24x).

of the stem is hollow at the internodes and solid at the nodes (Fig. 28).

Secondary growth starts rather quickly in *S. virginica*. The pericycle forms several cell layers. The outer layer gives rise to the cork cambium, while the inner layer functions as the vascular cambium. The cork cambium develops a typical periderm which becomes most profuse as the leaf tissue sloughs off. The vascular cambium forms vascular bundles and parenchyma tissue to the inside (Fig. 29) and aerenchymous tissue to the outside. The inner parenchyma tissue becomes highly lignified but remains alive (Figs. 30, 31).

Root. The root system of *S. virginica* is not very extensive. It is represented

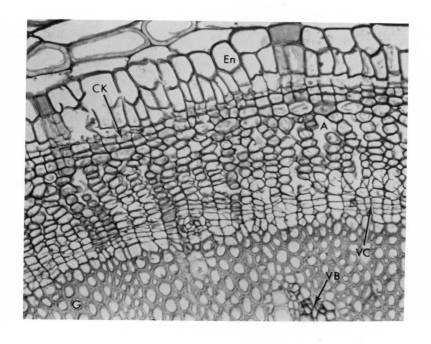

Fig. 29. **Salicornia virginica** stem x.s. VC – vascular cambium; CK – cork cambium; A – aerenchyma; VB – vascular bundles; G – ground tissue; En – endodermis (400x).

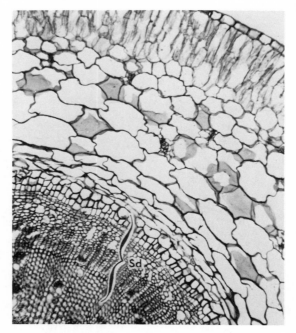

Fig. 30. **Salicornia virginica** shoot x.s. Sd — secondary growth (160x).

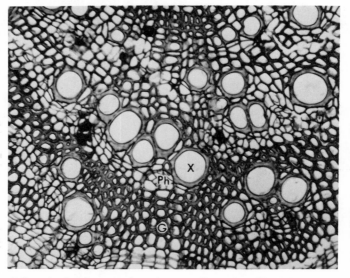

Fig. 31. **Salicornia virginica** secondary tissue x.s. X — xylem; Ph — phloem; G — ground tissue (800x).

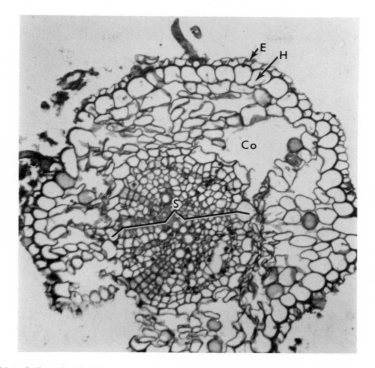

Fig. 32. Salicornia virginica root x.s. E — epidermis; Co — cortex; H — hypodermis; S — stele (48x).

by both a seminal and adventitious system. Structurally the root is quite similar to the older stem except that it contains xylem in the center. There are few root hairs in the epidermal layer. A single hypodermal layer occurs as well as a cortex composed of radial files of cells. Large intercellular spaces develop in the cortex. The stele is usually diarch or triarch. Secondary growth is not extensive (Fig. 32).

Juncus roemerianus

A detailed study of the development of *Juncus roemerianus* has been reported by Seibert (1969). Other studies involve different species. General leaf structure was considered by Adamson (1925) and Thielke (1948). European species of *Juncus* were reported on by Buchenau (1906). More recently Cutler (1969) reviewed the entire Juncaceae, but did not mention *J. roemerianus*.

<u>Leaf.</u> The leaf of *J. roemerianus* is long, pointed, and radially symmetrical (Fig. 33). Its growth is produced by an intercalary meristem at the base of the blade. The leaf sheath is bifacial. Structurally, there is a single layered

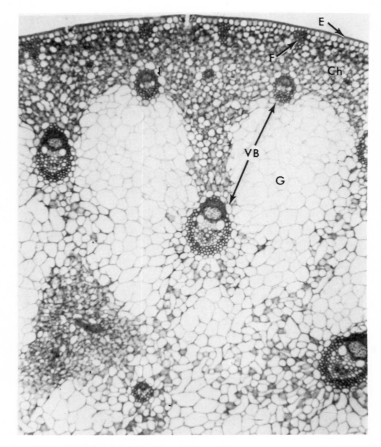

Fig. 33. **Juncus roemerianus** leaf x.s. E – epidermis; Ch – chlorenchyma layer; VB – vascular bundle; G – ground tissue; F – fibers (160x).

epidermis with lignified and cutinized outer walls. Stomates occur in parallel rows and are quite numerous. Beneath the epidermis lie five to ten cell layers of palisade tissue which alternate with wedge shaped groups of sclerenchyma tissue. The remainder of the leaf is composed of large highly vacuolated parenchyma cells and large intercellular spaces. There are also bands of smaller cells with dense cytoplasm which occur between vascular bundles. Collateral vascular bundles are scattered throughout the leaf tissue. These have transverse branches which connect the perpendicular bundles. Each vascular bundle is surrounded by a double bundle sheath. The outer layer is parenchymous and often contains starch. The inner bundle is composed of one to nine layers of sclerenchyma cells and is highly lignified.

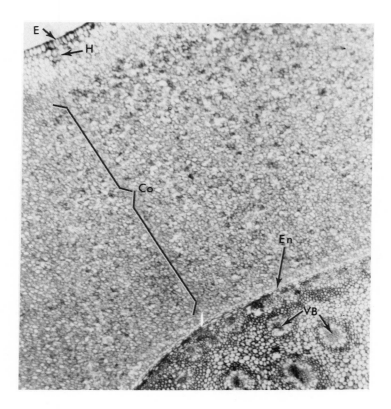

Fig. 34. **Juncus roemerianus** rhizome x.s. E – epidermis; H – hypodermis; Co – cortex; VB – vascular bundles; En – endodermis (48x).

Stem. The culm of *J. roemerianus* is generally short and aerenchymous except following floral initiation. The tissues are quite woody. The rhizome is formed as a lateral branch from one of the lower leaves of the culm. It produces cataphylls, usually about nine, before forming foliar leaves. A rhizome continuation bud then forms in a sympodial branching pattern.

Structurally the rhizome has a single epidermis layer and three to five hypodermal layers which usually become lignified in older tissue (Fig. 34). Further inside occurs a cortical region composed of loosely packed parenchyma tissue. An endodermal layer indicated by thickening on the radial and inner tangential walls occurs next, followed by two to three layers of pericycle and the other stele tissue. Scattered amphivasal bundles embedded

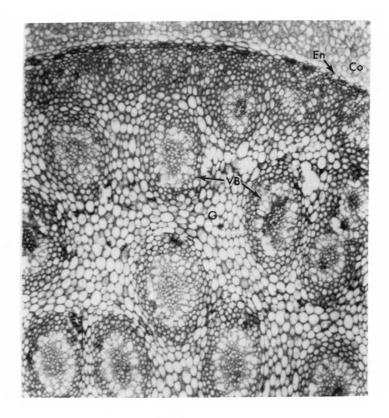

Fig. 35. **Juncus roemerianus** rhizome x.s. Co — cortex; En — endodermis; VB — collateral vascular bundles; G — ground tissue (160x).

in parenchyma tissue make up the remainder of the stele (Fig. 35). The whole tissue system is quite porous.

The adventitious root system is initiated by the rhizome pericycle primarily. Two types of roots occur as in *Spartina alterniflora.* The larger roots comprise a deep penetrating system with very few, short root hairs and infrequent lateral roots. These larger roots tend to grow straight down. Smaller roots originate from around the rhizome. These roots branch profusely forming a very fine fibrous system. Structurally the difference in

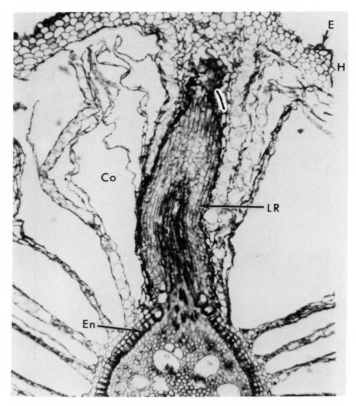

Fig. 36. Juncus roemerianus root x.s. E — epidermis; H — hypodermis; Co — cortex; LR — lateral root; En — endodermis (160x).

the two root systems is merely quantitative (Fig. 36). The larger roots have an epidermal layer subtended by five to six hypodermal layers. The bulk of the cortex is composed of radial files of cells, cell walls, and intercellular spaces. Two to three cortical layers are intact surrounding the endodermis which is thickened on its radial and inner tangential walls. The pericycle comprises two to three cell layers inside the endodermis which encloses the polyarched stele. Extensive long root hairs, fewer hypodermal layers, and a smaller cortex and stele are characteristic of smaller roots. All roots are somewhat aerenchymous.

Aster tenuiflolius

Of all the plants reported on in this study, *Aster tenuiflolius* seems the least likely to be found in a salt marsh habitat from a structural point of view.

Leaf. The leaf surface of *A. tenuiflolius* is relatively smooth. The abaxial epidermal cells have a pattern of very small ridges on their surface. The

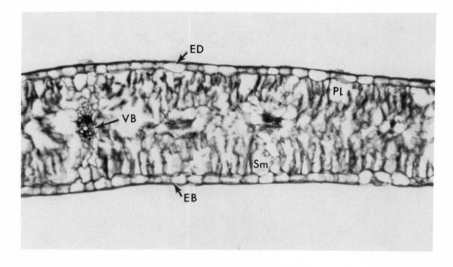

Fig. 37. **Aster tenuiflolius** leaf x.s. ED — adaxial epidermis; EB — abaxial epidermis; VB — vascular bundle; Pl — palisade layer; Sm — spongy mesophyll (160x).

epidermal cells of both surfaces are relatively large and irregular in arrangement. Stomates occur on both surfaces, also in an irregular pattern. The cuticle is well-developed in plants examined.

The mesophyll is more typical of a mesophyte (Fig. 37). Three to four cell layers make up the palisade layer with a similar number in the spongy mesophyll. In some tissue, particularly around the edges of the leaves, the spongy layer is organized in a manner more characteristic of a palisade layer. This may be due in part to the upright orientation the leaf usually assumes. The vascular bundles are small and surrounded by a bundle sheath. While the leaf does have several cell layers, these cells cannot be considered very tightly packed, nor is the leaf very succulent. About the only unique feature of the leaf is the presence of a single cytoplasmic vacuole in each of the mesophyll cells.

Stem. The stem of *A. tenuiflolius* elongates throughout the growing season. It is green due to a palisade type chlorenchymous layer which makes up about one-half of the cortex. The epidermis contains stomates and like the leaf is not highly lignified. The outer cortex is quite aerenchymous while the inner portion is much more tightly packed (Fig. 38).

The stele of the stem is composed of collateral vascular bundles, medulary rays, and an inner pith. Phloem caps occur on the outer edges of the bundles representing the only highly lignified tissue present.

Root. The root material examined is much more typical of aquatic plants

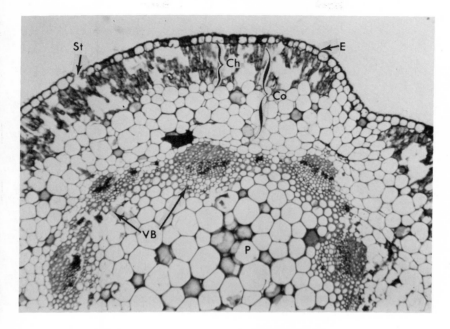

Fig. 38. **Aster tenuiflolius** stem x.s. E — epidermis; Ch — chlorenchyma; Co — cortex; VB — vascular bundle; P — pith; St — stomate (160x).

than the rest of the plant (Fig. 39). A cutinized epidermal layer with three to four hypodermal layers makes up the outside of the root in transection. The central cortical region is composed of radial files of cell fragments. The triarch stele, is surrounded by several intact cortical layers. There is no obvious endodermis, although casparian strips were not stained for. The presence of a pericycle is evident through lateral root production.

Summary

Few generalizations can be made about salt marsh plants. They are all small, probably due to water stress providing a long term selecting action. Root systems are routinely aerenchymous. Since the mode of action in dealing with salinity is so variable in different plants, structural features are also quite variable. For example, there is the presence of salt glands in some plants and not others. The tendency for succulence is also variable. In *Salicornia virginica* succulence may increase up to a certain percent salinity and then decrease above that point. Lignification is quite variable, but in certain plants seems to fluxuate with salinity. Cytoplasmic vacuoles are not present in all salt marsh plants.

Structure can give a clue as to the mode of dealing with salinity. Extensive starch in rhizomes and roots may provide sugar to help cope with salinity fluxuations. Lignin may offer wall strength to deal with high internal salinity.

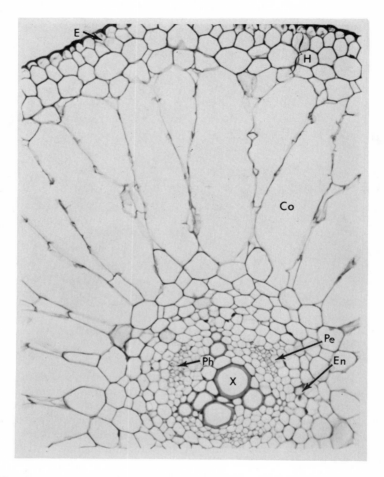

Fig. 39. **Aster tenuiflolius** root x.s. E — epidermis; H — hypodermis; Co — cortex; X — xylem; Ph — phloem; Pe — pericycle; En — endodermis (400x).

Salt glands certainly suggest a means for maintaining water balance. Well-developed vascular tissue, thin cuticles, and abundant stomates may suggest high transpiration rates as a mechanism for dealing with salt stress.

None of these suggestions is intended to imply that dealing with salt water is not primarily a physiological problem. Nor is it suggested that these structural features are caused by high salinity, since most develop in the greenhouse under fresh-water conditions as well. Many are manipulated in degree by salinity. Many do indicate the sort of physiological pathways various plants use in dealing with the salt marsh environment.

Key to figures

ED — adaxial epidermis
EB — abaxial epidermis
BSo — outer bundle sheath
BSi — inner bundle sheath
St — stomate
Sg — salt gland
LC — long cell
Sc — saddle cell and silica cell
Bc — basal cell
Cc — cap cell
CV — cytoplasmic vacuole
L — lysogenous cavity
IM — intercalary meristem
SA — shoot apex
C — center cavity
VB — vascular bundle
M — mesophyll chlorenchyma
G — ground tissue
E — epidermis
Co — cortex
H — hypodermis
LR — lateral root
En — endodermis
Pe — pericycle
S — stele
Bp — branched papillae
P — pith
Cl — collecting cells
SE — secreting cells
Pi — periderm
Ph — phloem
X — xylem
Pl — palisade layer
Sm — spongy mesophyll
T — tracheoidioblasts
N — node
L — leaf
A — aerenchyma
VC — vascular cambium

Ck — cork cambium
Sd — secondary growth
F — fibers
Ch — chlorenchyma
Sb — sclerenchyma band
Lt — leaf trace
Gc — guard cell
Ac — accessory cell

Literature Cited

Adamson, R. S. 1925. On the leaf structure of *Juncus*. Ann. Bot. 39(155):599-612.

Anderson, C. E. 1971. Shoot Development in *Salicornia virginica*. Assoc. SE. Biol. Bull. 18(2):25 (Abstract).

Buchenau, F. 1906. Juncaceae. Pages 1-284 *in* Das Pflanzenreich. Edited by A. Engler. H. R. Engelman. Weinheim, Germany.

Cross, B. D. 1909. Observations on some New Zealand Halophytes. Trans. and Proc. New Zealand Institute 42:545-574.

Cutler, D. F. 1969. Anatomy of the Monocotyledons. IV. Juncales. Clarendon Press. Oxford.

DeBary, A. 1884. Comparative anatomy of the vegetative organs of the phanerogams and ferns. Clarendon Press. Oxford.

DeFraine, Ethyl. 1912. The anatomy of the genus *Salicornia*. Jour. Linn. Soc. Bot. 41:317-348.

Fahn, A., and T. Arzee. 1959. Vascularization of articulated Chenopodiaceae and the nature of their fleshy cortex. Amer. Jour. Bot. 46(5):330-338.

Ganong, W. F. 1903. The vegetation of the Bay of Fundy salt and diked marshes. An ecological study. Bot. Gaz. 36:349-367.

Holm, T. 1891. A study of some anatomical characters of North American Gramineae. III. *Distichlis* and *Pleuropogon*. Bot. Gaz. 16:275-281.

Holm, T. 1901-1902. Some new anatomical characters for certain Gramineae. Beih. bot. Zbl. 11:101-133.

James, L. E., and D. W. Kyhos. 1961. The nature of the fleshy shoot of *Allenrolfea* and allied genera. Amer. Jour. Bot. 48(2):101-108.

Katina, Z. F. 1962. The biology of reproduction in *Limonium spp*. Bot. Zhur. 47(5):693-697.

Levering, C. A., and W. W. Thomson. 1971. Salt Glands of *Spartina foliosa*. Planta (Berl.) 97:183-196.

Metcalfe, C. R. 1960. Anatomy of the Monocotyledons. I. Gramineae. Clarendon Press. Oxford.

Metcalfe, C. R., and L. Chalk. 1950. Anatomy of the Dicotyledons. Clarendon Press. Oxford.

Munz, P. A. 1959. A California flora. University of California Press. Berkeley and Los Angeles.

Peck, M. E. 1941. A manual of the higher plants of Oregon. Bingords and Mort Publishers. Portland.

Prat, H. 1934. Contribution a '1' etude systematique et histologique de chloridees. Bull. Soc. Bot. Fr. 81:475-491.

Prat, H. 1936. La Systematique des Graminees. Ann. Sci. Nat. Bot. Ser. 10,

18:165-258.

Seibert, R. W. 1969. Flowering Patterns, Germination, and Seed and Seedling Development of *Juncus roemerianus*. Ph.D. dissertation. North Carolina State University.

Skelding, A. D., and J. Winterbotham. 1939. The Structure and Development of the Hydathodes of *Spartina townsendii* Groves. New Phytol. 38:69-79.

Sutherland, G. K., and A. Eastwood. 1916. The Physiological Anatomy of *Spartina townsendii*. Ann. Bot. 30:333-351.

Thielke, C. 1948. Beilrage zur Entwicklungsqesichichte unifazialer Blatter. Planta. 36:154-177.

Ziegler, H., and U. Luttge. 1966. Die Salzdrusen von *Limonium vulgare*. I. [Milleilung:] Die Feinstruktur. Planta (Berl.) 70:193-206.

PHYSIOLOGY OF COASTAL HALOPHYTES

William H. Queen

Chesapeake Research Consortium

University of Maryland, College Park, Maryland 20742

Introduction

A number of very fine publications have appeared during the past several years that contain reviews of much of the literature that is pertinent to the physiology of coastal halophytes. Foremost among these are: "Physiological Basis of Salt Tolerance of Plants" (Strogonov 1964), "Salt transport by plants in relation to salinity" (Rains 1972). "The Biology of Halophytes" (Waisel 1972), "Ecology of Salt Marshes and Sand Dunes" (Ranwell 1973). "Is any angiosperm an obligate halophyte?" (Barbour 1971). Also, other papers in this volume contain relevant information: "Physiology of desert halophytes" (Caldwell), "Mangroves: a review" (Walsh), "Salt tolerance of mangroves and submerged rooted aquatics" (McMillan), "Nutrient limitation in salt marsh vegetation" (Valiela and Teal), and "The potential economic uses of halophytes" (Mudie) all have material relating, either directly or indirectly, to the physiology of coastal halophytes. For these reasons, a comprehensive review of the literature does not appear in this paper, and all physiological mechanisms possessed by coastal halophytes for coping with high soil salt concentrations are not described. Instead, this article will have a rather narrow focus. Selective ion absorption and cellular tolerance of high ionic concentrations, two key factors in the adaptation of vascular plants to the highly saline coastal environment, will be examined.

Generally, the higher salinity region of the coastal zone is a hostile environment for most vascular plant species. Many of the problems faced by plants in this zone are related to characteristics of sea water. Of these characteristics, the most significant are: (1) a high osmotic pressure, and (2) an ionic mix substantially different from that of the soil solution encountered by most terrestrial plants.

Plants cannot acquire water from an external medium unless the osmotic pressure within the plant exceeds that of the external medium. If the external pressure is higher than the internal pressure, water will flow from the plant to the medium. For years, many investigators were of the opinion that the stress imposed on plants by high soil salinities was that of a water deficit (Bernstein and Hayward 1958). Other workers (Slatyer 1961, Greenway 1962) doubted the validity of this explanation. They were of the opinion that adequate quantities of water are acquired under saline conditions because plants respond to high external osmotic pressures by increasing the internal pressure to a level higher than that of the external medium by accumulating additional salt from the external medium. The latter view was later confirmed (Bernstein 1961). Thus, for those species able to develop and tolerate high internal osmotic pressures, the acquisition of water is not a factor in preventing their occurrence in a highly saline coastal habitat.

However, plant adaptation to a saline environment is not assured by the accumulation of additional salt. In fact, the additional salt, while necessary for the continued acquisition of water, creates other problems. Among these are: (1) ionic imbalances that are potentially toxic, and (2) internal salt concentrations that are sufficiently high to be toxic to most plant species. Cells are highly intricate metabolic machines and many of their critical processes can occur only within very narrow ionic concentration ranges. Halophytes [generally defined as plants able to withstand soil salt concentration of 0.5% and higher] differ from glycophytes in two important respects: (1) they possess "mechanisms" that prevent the development of toxic ionic imbalances, and (2) their cells are able to tolerate the higher internal salt concentrations. Most of the more well known adaptive mechanisms of halophytes function to prevent the development of toxic ionic imbalances. Salt glands and vesicular hairs certainly belong to this group. Succulence, increased growth, and salt excretion from the roots, while reducing the internal ionic concentration, also, are probably of importance to halophytes in controlling ion balance. Their utility in reducing the internal ion concentration is limited because of the previously discussed requirement plants growing in a saline medium have for maintaining a high internal osmotic pressure. Recent work concerned with the above mentioned halophyte mechanisms is discussed in the following paper by Caldwell and will not be reviewed here. Rather, selective ion absorption, which has great potential importance in the regulation of internal salt content will be examined. Also, recent studies relating to cellular tolerance of halophyte tissue to high salt concentrations will be reviewed. Before proceeding to these topics, it should be noted that a clear distinction cannot always be made between halophytes and glycophytes. Many plant genera have species belonging to both categories. Moreover, many typical halophyte features, e.g. succulence, are characteristic of plant groups that are not particularly salt tolerant. Salt tolerance, in many instances, has resulted from a number of specific adaptations and these are often quantitatively rather than qualitatively different from glycophyte characteristics.

Selective Ion Absorption

Absorption of ions from the external medium is regulated by active transport mechanisms located in cellular membranes (Rains 1972). However, control is not absolute. If potentially toxic ions are present in the external medium, they will be absorbed (Steward and Sutcliffe 1959). Also, less discrimination over ion absorption occurs under highly saline conditions (ibid). In a coastal environment, considerable quantities of sodium and chloride will be absorbed even though these ions are required only in exceedingly small quantities (Rains and Epstein 1967). Although these "inadequacies" occur, selective absorption may be as important in the regulation of internal ionic balance as salt glands, vesicular hairs, succulence, and the other well known adaptive structures of halophytes.

Epstein et al. (1963) reported that potassium ions are absorbed by two separate mechanisms, designated 1 and 2. Subsequently, the dual mechanism has been demonstrated for other ions, and in a wide variety of plant tissues. For these reasons, it can be considered to occur almost universally in higher plant tissues (Epstein 1969). Significant functional differences exist between the two mechanisms. Ions which occur in the external medium in exceedingly small concentrations (approximately 1 mM and below) are absorbed by mechanism 1. Those occurring in higher concentrations (1 mM and above) are absorbed, for the most part, by mechanism 2. Mechanism 1 has a very high ion affinity and is inhibited only slightly by the presence of competing ions. Absorption of potassium by *Avicennia marina*, a mangrove, was found to be inhibited by the presence of sodium only when the ratio, Na/K substantially exceeded 100 to 1 (Rains and Epstein 1967). The rate of absorption by mechanism 1 increases with increases in the external concentration of the required ion from exceedingly low to moderately low levels. However, potassium absorption by roots of tall wheatgrass. *Agropyron elongatum*, via mechanism 2, does not increase appreciably in rate with increases in the potassium concentration of the external medium above approximately 0.2 mM (ibid.). Generally, mechanism 2 which is inhibited to a greater degree by the presence of chemically similar ions than is mechanism 1, is inactive at extremely low concentrations. Absorption of potassium occurs primarily via mechanism 2 with a high concentration of potassium in the external medium (ibid.).

Rains and Epstein (1967) investigated the absorption of potassium in the presence of high sodium chloride concentrations by leaf tissue of *Avicennia marina*, a mangrove. They found that the type 1 mechanism for potassium absorption of this halophytic species had a weaker affinity for potassium than the type 1 mechanism of barley, which is not salt tolerant. However, the type 2 mechanism of *Avicennia* was found to be inhibited less by high sodium chloride concentrations than the mechanism 2 of barley. As Rains and Epstein noted, these findings have significance relating to plant adaptation to various soil salinities. Terrestrial plants growing on nonsaline soils must frequently abosrb ions of required elements from a soil solution in which the required ions are found in exceeding low concentrations. Furthermore, some of the required ions, such as potassium, must be built up to high internal concentrations. An absorptive mechanism with a high affinity for the essential ion is highly useful, if not essential, if this requirement is to be met. Because of the ionic composition of sea water, coastal halophytes, such as *Avicennia marina*, rarely face the problem of building up high internal ion concentrations from a medium having an exceedingly low concentration of the required ion. Thus, the presence in *Avicennia* of a type 1 mechanism for potassium absorption that has a weaker ion affinity than that of glycophytes should not be considered as a hindrance to halophyte growth in a highly saline environment. With the type 2 mechanism, the situation is reversed. Because non-saline soils usually do not have high concentrations of competing ions (i.e. non-essential ions which compete with the essential ions for the same active transport mechanism), the high sensitivity of the type 2 mechanism does not represent a serious problem for non-halophytes. On the other hand, a type 2 mechanism that is only moderately sensitive to the presence of competing ions is highly advantageous to coastal halophytes because of the presence of high concentrations of competing ions in sea water.

A dual mechanism for ion absorption has been demonstrated in other halophytes (Black 1960, Osmond 1966). If the ion affinities and sensitivities reported by Rains and Epstein (1969) for the dual absorptive mechanisms of *Avicennia marina* are found for other halophytes, then selective ion absorption must be considered of primary importance in the adaptation of halophytes to high salinity environments.

Cellular Adaptation

The activity of many plant enzymes is affected by ion concentration (Dixon and Webb 1958, Evans and Sorger 1966). Miller and Evans (1956) reported that both univalent and divalent cations markedly stimulate the activity of particulate cytochrome oxidase from roots of higher plants. The activity of glucose-6-phosphate dehydrogenase from yeast was found to be increased by as much as six-fold by low concentrations of certain ions (Rutter 1957). Salt-sensitive enzymes of halophilic bacteria actually require high salt concentrations for optimal activity (Baxter and Gibbons 1957). Maximal activity was obtained at a salt concentration of 1.0 M for malate dehydrogenase of *Halobacterium solinarium* (Holms and Halvorson 1965, Hockstein 1970).

Indirect evidence of the effects of ions on plant enzymes is also available. Reports by Robertson (1951) and Lundegardh (1955) that respiration rates of plant tissues containing low concentrations of salts are markedly stimulated by the addition of salts to the bathing medium can be interpreted as evidence of the effects of ions on enzyme activity. Similar interpretations can be made in regard to results of other investigators (Hasson-Porath and Poljakoff-Mayber 1970, Lee 1971) who have reported increases in the activity level of enzymes of plants that had been subjected to increasing soil salinities. However, results such as these must be interpreted with care because factors other than ionic strength of the soil solution could be responsible for the changes in enzyme activity. Stewart et. al. (1971) reported that nitrate reductase activity of *Suaeda maritima* decreased in going from the lower to the upper marsh. This variation was found to be related to nitrate availability; not to salinity.

Because of these and numerous other studies concerning the effects of ion concentration on enzyme activity, and the general awareness that cellular ion concentration increases with increases in the ion concentration of the external medium, various investigators have speculated that the salt-sensitive enzymes of halophytes may be more tolerant of high cellular ion concentrations than those of glycophytes (Adriani 1958, Greenway 1968). Data to either support or refute this hypothesis is both scanty and contradictory. The contradictory nature of the results is particularly perplexing because most of the investigations (Lee 1971, Greenway and Osmond 1972, Yopp, In Press) have involved the same enzyme, malate dehydrogenase, which is highly salt sensitive (Hiatt and Evans 1960).

Lee (1971) studied the effects of salt concentration on the activity of malate dehydrogenase of two highly salt tolerant coastal halophytes, *Spartina alterniflora* and *Salicornia virginica*. Extracts of leaf material from both species was found to be affected by the salt concentration of the assay medium. Increases in the NaCl concentration up to 0.16 M for *Spartina* and 0.20 M for *Salicornia* resulted in greater malate dehydrogenase activity. Further increases in NaCl concentration were found to be inhibitory, but in no case, over the range tested, did the activity drop below that observed in the absence of NaCl. The effects of NH_4Cl and KCl concentrations on malate dehydrogenase of *Spartina alterniflora* were comparable to those of NaCl. Similar and simultaneous studies by Lee with corn produced strikingly different results. Maximal malate dehydrogenase activity was obtained at the much lower concentration of 0.06 M.

Yopp (In Press), working with *Salicornia pacifica* has obtained results that are comparable to those of Lee. Malate dehydrogenase was maximally stimulated at a NaCl concentration of 0.2 M. Also, the enzyme showed considerable activity at a NaCl concentration of 1.5 M, which is considerably higher than the maximal concentration of 0.3 M reported by Weimberg (1967) for a glycophytic tissue. Yopp speculated that the greater enzyme activity at high NaCl concentrations may reflect some structural adaptations in the malate dehydrogenase of *Salicornia*.

Results reported by Lee and Yopp concerning the effects of salt concentration on the activity of malate dehydrogenase from halophytes differs from those obtained for the same enzyme from glycophytes. Hiatt and

Evans (1960) reported that the activity of malate dehydrogenase from spinach was stimulated by the presence of sodium chloride in the assay medium up to a concentration of 0.04 N. The enzyme responded to other salts in the same quantitative manner. Weimberg (1967) found malate dehydrogenase of pea seeds to be very sensitive to the ionic concentration of the assay medium. Activity was stimulated by increases in the concentration of NaCl up to 0.02 M. Concentrations above 0.02 M inhibited activity in relation to maximal activity. At NaCl concentrations of 0.3 M, malate dehydrogenase was completely inactive. Also, as reported above, maximal activity of malate dehydrogenase from corn was obtained at a sodium chloride concentration of 0.06 M (Lee 1971).

The above reported differences between halophytes and glycophytes relating to the effects of salt concentration on malate dehydrogenase activity suggest that salt-sensitive enzymes of halophytes are more tolerant of high salt concentrations than those of glycophytes. However, this conclusion is not supported by results obtained by other investigators. Greenway and Osmond (1972) studied the effects of salt concentration on the *in vitro* activity of four salt-sensitive enzymes (malate dehydrogenase, aspartate transaminase, glucose-6-phosphate dehydrogenase, and isocitrate dehydrogenase) of two salt tolerant species, *Atriplex spongiosa* and *Salicornia australia*. While the response varied with each enzyme, extracts fo salt-tolerant *Atriplex* and *Salicornia* showed only marginally higher stimulation of enzyme activity by sodium chloride than those of salt-sensitive *Phaseolus*. Comparable results were obtained with KCl. Also, *Atriplex* and *Phaseolus* seedlings were grown in culture solutions containing up to 400 mM NaCl. *Atriplex* showed substantially enhanced growth in cultures containing up to 200 mM Nacl. Phaseolus grew poorly with as little as 150 mM NaCl in the culture solution. However, these culture treatments neither induced major changes in the specific activity of the four enzymes studied, nor altered the response of these enzymes to added NaCl during assay. From these investigations, Greenway and Osmond concluded that there are no significant differences in salt sensitivity between enzymes of salt-tolerant and salt-sensitive species.

Contradictory results concerning the response of salt-sensitive enzymes of halophytes to high salt concentrations should not be surprising in view of the limited number of investigations. However, failure to conclusively demonstrate a greater tolerance of high salt concentrations by salt-tolerant species than by salt-sensitive species has resulted in increased interest in the idea that critical cellular mechanisms of salt-tolerant species are protected by ion compartmentalization (Rains and Epstein 1967, Greenway and Osmond 1972, Osmond and Greenway 1972). Recent studies (Larkum 1968, Larkum and Hill 1970) suggest that ion concentration varies with cytoplasmic component.

Summary

Recent studies of the dual mechanism of ion absorption have provided significant new insights concerning the adaptation of vascular plants to their highly saline environment. Comparable success has not been achieved in the few investigations of the *in vitro* response of salt-sensitive enzymes of halophytes to high salt concentrations. Data both supporting and refuting the idea that enzymes of halophytes are more tolerant of high ion concentrations than those of glycophytes have been reported. Results of other studies suggest salt-sensitive mechanisms may be protected by cellular ion compartmentalization.

Literature Cited

Adriani, M. J. 1958. Halophytes. In: W. Ruhland, ed., Handbuch des Pflanzenphysiologie, Vol. 4. Springer, Berlin. pp. 709-736.

Barbour, Michael G. 1970. Is any angiosperm an obligate halophyte? Amer. Midl. Natur. 80:105-120.

Baxter, R. M. and N. E. Gibbons. 1956. Effects of sodium and potassium chloride on certain enzymes of *Micrococcus halodentrificans* and *Pseudomonas salinaria*. Can. J. Bot. 2:599-605.

Bernstein, L. 1961. Osmotic adjustment of plants to saline media. 1. Steady state. Amer. J. Bot. 50:360-370.

Bernstein, L. and H. E. Hayward. 1958. Physiology of salt tolerance. Ann. Rev. Plant Physiol. 9:25-46.

Black, R. F. 1960. Effects of NaCl on the ion uptake and growth of *Atriplex vesicaria* Heward. Aust. J. Biol. Sci. 13:249-266.

Dixon, M. and E. C. Webb. 1958. Enzymes. Longmans Green and Co., London.

Epstein, F. 1969. Mineral metabolism in halophytes. *In* "Ecological Aspects of the Mineral Nutrition of Plants" (I. H. Rorison, ed.). pp. 345-355., Blackwell, Oxford.

Epstein, E., D. W. Rains and O. E. Elzam. 1963. Resolution of dual mechanisms of potassium absorption by barley roots. Proc. Natn. Acad. Sci. U. S. 49:684-692.

Evans, H. J. and G. J. Sorger. 1966. Role of mineral elements with emphasis on the univalent cation. Ann. Rev. Plant Physiol. 17:47-112.

Greenway, H. 1962. Plant response to saline substrates. I. Growth and ion uptake of several varieties of *Hordeum* during and after sodium chloride treatment. Aust. J. Biol. Sci. 15:16-38.

Greenway, H. 1968. Growth stimulation by high chloride concentrations in halophytes. Israel J. Bot. 17:169-177.

Greenway, H. and C. B. Osmond. 1972. Salt responses of enzymes from species differing in salt tolerance. Plant Physiol. 49:256-259.

Hasson-Porath, E. and A. Poljakoff-Mayber. 1969. The effect of salinity on the malic dehydrogenase of pea roots. Plant Physiol. 44:1031-1034.

Hasson-Porath, E. and A. Poljakoff-Mayber. 1970. Lactic acid content and formation in pea roots exposed to salinity. Plant and Cell Physiol. 11:891-897.

Hiatt, A. T., and H. J. Evans. 1960. Influence of salts on activity of malic dehydrogenase from spinach leaves. Plant Physiol. 35:662-672.

Holms, P. K. and H. Halvorson. 1965. Purification of a salt-requiring enzyme from an obligately halophilic bacterium. J. Bacteriol. 90:312-316.

Hochstein, L. I. 1970. Studies on a halophilic malic dehydrogenase. In: Extreme Environments. p. 42. Ames Research Center. California: N.A.S.A.

Lee, J. 1971. Studies of malate dehydrogenase from selected salt marsh plants. MS Thesis. University of South Carolina. Columbia, S. C.

Larkum, A. W. D. 1968. Ionic relations of chloroplasts *in vivo*. Nature 218:447-449.

Larkum, A. W. D. and A. E. Hill. 1970. Ion and water transport in *Limonium*. V. The ionic status of chloroplasts in the leaf of *Limonium vulgare* in relation to the activity of the salt glands. Biochem. Biophys. Acta 203:133-138.

Lundegardh, H. 1955. Mechanism of absorption, transport, accumulation, and secretion of ions. Ann. Rev. Plant Physiol. 6:1-24.

Miller, G., and H. J. Evans. 1957. The influence of salts on pyruvate kinase from tissues of higher plants. Plant Physiol. 32:346-354.

Osmond, C. B. 1966. Divalent cation absorption and interaction in *Atriplex*. Aust. J. biol. Sci. 19:37-48.

Osmond, C. B. and H. Greenway. 1972. Salt responses of carboxylation enzymes from species differing in salt tolerance. Plant Physiol. 49:260-263.

Rains, D. W. 1972. Salt transport by plants in relation to salinity. Ann. Rev. Plant Physiol. 23:51-72.

Rains, D. W. and E. Epstein. 1967. Preferential absorption of potassium by leaf tissue of the mangrove *Avicennia marina*: an aspect of halophytic competence in coping with salt. Aust. J. biol. Sci. 20:847-857.

Ranwell, D. S. 1973. Ecology of Salt Marshes and Sand Dunes. Chapman and Hall. London, England.

Robertson, R. N. 1951. Mechanism of absorption and transport of inorganic nutrients in plants. Ann. Rev. Plant Physiol. 2:1-24.

Rutter, W. J. 1957. The effect of ions on the catalytic activity of enzymes: yeast glucose-6-phosphate dehydrogenase. Acta Chem. Scand. II. 1576-1586.

Slatyer, R. O. 1961. Effects of several osmotic substrates on the water relations of tomato. Aust. J. biol. Sic. 14:519-540.

Steward, F. C., and J. F. Sutcliffe. 1959. Plants in relation to inorganic salts. *In* "Plant Physiology, A Treatise" (F. C. Stewart, ed.), Vol. 2, pp. 253-478. Academic Press, New York.

Stewart, G. R., J. A. Lee and T. O. Orebamjo. 1972. Nitrogen metabolism in halophytes. I. Nitrate reductase activity in *Suaeda maritima*. New Phytol. 71:263-267.

Strogonov, B. P. 1964. "Physiological Basis of Salt Tolerance of Plants (as affected by various types of salinity)". Akad. Nauk. SSSR. Translated from Russian. Israel Progr. Sci. Transl. Jerusalem.

Waisel, Y. 1972. Biology of Halophytes. Academic Press. New York. 395 pp.

Weimberg, R. 1967. Effect of sodium chloride on the activity of a soluble malate dehydrogenase from pea seeds. J. Biol. Chem. 242:3000-3006.

Yopp, J. In Press. Effect of low water potential on the activity of mitochondrial, chloroplast and supernatent malic dehydrogenase from the halophyte, *Salicornia pacific*.

PHYSIOLOGY OF DESERT HALOPHYTES

Martyn M. Caldwell

Utah State University

As stated in the preceding paper, salinity, water, and halophytes have received limited attention in physiological literature. However, much less attention has been paid to adaptations of halophytes to other environmental factors. For this reason, the special case of desert halophytes deserves particular attention.

In addition to salinity, non-phreatophytic desert plants must endure very low soil moisture potentials which not only increase the effective soluble salt concentrations, but also present a severe total soil moisture stress. Low atmospheric humidity also increases the total water stress. Since most desert areas are continental, great temperature extremes can be expected during the course of the year. Although deserts are known for high summer temperatures, the low winter temperatures of many arid areas should not be overlooked. The time available during the year for active growth and carbon uptake is usually quite limited by combinations of extreme temperatures and lack of moisture. Furthermore, the sporadic nature of precipitation and temperature fluctuations in most desert areas greatly adds to the hazards of existence. Thus, desert halophytes must be well adapted so as to avoid imprudent advances in growth or reproductive activity when the probability of immediate drought or temperature extremes exists. In addition to salinity, other edaphic factors may present additional problems. Low available soil nitrogen is characteristic of most desert areas. Other elements such as phosphorus cab also be limiting. Furthermore, elements such as boron can be excessively high in concentration.

This review will concentrate on literature of the last 15 years since the earlier literature has been well covered in reviews by Adriani (1956–1958) and Chapman (1960). Also, much of the material covered in recent articles by Jennings (1968) and Barbour (1970) will not be reviewed again here. The recent reviews of Kylin (in press) and Waisal (1972) were not available at the time this review was written.

Halophyte Categories

Definitions and classifications of halophytes have absorbed much attention in the earlier literature. Several such schemes are discussed by Chapman (1960) and Adriani (1958). Most of these classifications have been based upon plant tolerance of various external salt concentrations, stimulation of growth by certain external salinity levels, or requirements of plants for significant quantities of salts. Barbour (1970) suggests that terms such as "facultative" or "obligate" halophyte should only be defined on a relative basis. He points out that salt tolerance of plants is often quite variable, and species restricted to saline soils are very rare. Furthermore, salinity tolerance often appears to reflect conditions of the experiment as well as the imposed salinity level. Gale, Naaman, and Poljakoff-Mayber (1970) demonstrated this convincingly for *Atriplex halimus* L. Under low humidity conditions growth and yield were maximal at low salinity concentrations of approximately -5 atm. However, under a high humidity environment, optimal growth occurred in the non-saline control solution. Such experiments indicate the equivocal nature of halophyte definitions which hinge on stimulation of plant growth by certain salinity concentrations in the growing medium.

Furthermore, as Barbour points out, species which require sodium as an essential micronutrient require such small quantities of sodium that they can hardly be classed as "obligate" halophytes. Ecological salt tolerance should only be defined by the ability of a species to compete and reproduce in a particular environment. Physiological definitions of salt tolerance must be carefully formulated so as to prescribe exactly the cultural conditions for determing salinity tolerance or requirements.

In this review only the relative terms halophyte and glycophyte will be employed.

Salt Accumulation and Osmotic Adjustments

Salt accumulation in the tissues of desert halophytes is basically no different than that of other halophytic species. Several recent reports have documented ion concentrations in various plant parts of desert halophytes (Chatterton and McKell 1969, Chatterton et al. 1970, Wiebe and Walter 1972, Wallace and Romney 1972, Moore, Breckle, and Caldwell 1972). Preferential accumulation of various cations and anions is often found in plant tissues from different species growing in the same soil (Walter 1972). Walter felt the tendency for preferential accumulation of various ions was so pronounced that classification of different halophytes as "chloride-halophytes" or "sulfate-halophytes" might be justified. However, even though many species do demonstrate pronounced affinity for accumulation of prominent cations or anions, many are intermediate in selectivity for particular ions (Wiebe and Walter 1972, Moore et al. 1972). Ion specificity and uptake processes by plant membranes are becoming reasonably well understood and probably differ very little between glycophytes and halophytes in basic mechanism (Epstein 1969; Osmond 1968; Jeschke 1970, 1972).

Uptake of reasonably rapidly permeating electrolyte solutes is a well-known phenomenon in both glycophytes and halophytes. Increased cellular concentrations of such solutes will allow the plant to maintain turgor in the face of decreased water potentials in the external medium. This so-called osmotic adjustment in halophytes appears to differ in magnitude more than principle from glycophytes (Slatyer 1961; Boyer 1965; Meiri and Poljakoff-Mayber 1969; Meiri, Kamburoff, and Poljakoff-Mayber 1971; McNulty 1969; Chatterton et al. 1970, Greenway, Gunn, and Thomas 1966; Jennings 1968; Waisel and Pollack 1969; Gale and Poljakoff-Mayber 1970). Both McNulty (1969), working with *Sarcobatus vermiculatus* Hook, and Black (1960), working with *Atriplex vesicaria* Heward, found salinity tolerance and active growth at concentrations of 1 M NaCl (-44 atm) in solution cultures. This tolerance was accomplished by a slow stepwise osmotic adjustment in each case. The maintenance of a suitable water potential gradient between the plant and the external solution appears qualitatively similar for glycophytes and halophytes.

Recently Mozafar and Goodin (1970) reported a lack of osmotic adjustment in *Atriplex halimus* L. even though the experimental plants presumably continued to thrive and grow. This conclusion was based on determinations of the electrical conductivity of leaf sap, which was found to be less than that of the culture medium. Although not explicitly stated in their paper, the plants were presumably still actively growing under these conditions. Also unexplained were large discrepancies between the stated concentrations of the solution cultures and the reported conductivity of these solutions. Other details of the experiment are also difficult to reconstruct from the paper. Further skepticism is warrented by the results of Gale and Poljakoff-Mayber (1970) who worked with the same species and found that osmotic potentials of the expressed leaf sap were always more negative than those of the culture solutions throughout a range of culture osmotic potentials from 0 to -15 atm. These results were obtained by salinization using either NaCl or Na_2SO_4, whereas Mozafar and Goodin (1970) used equal mixtures of KCl and NaCl in their experiments. However, this difference in salts is unlikely to account for the lack of osmotic adjustment reported by Mozafar and Goodin. Apart from this questionable exception, the apparent universality of osmotic adjustment in halophytes continues to be maintained by the recent literature.

Salt Tolerance in Halophyte Tissues

Passive water uptake in response to the water potential gradient established by active electrolyte accumulation in halophytes, i.e., osmotic adjustment, seems to be a reasonable mechanism for extraction of water from soils of low water potential. However, electrolyte accumulation in plant protoplasm must be limited even in halophytes and avoidance or tolerance mechanisms are necessary. Protoplasmic salt resistance was determined for a variety of plants by Repp, McAllister, and Wiebe (1959). *Halogeton glomeratus* (M. Bieb.) C. A. Mey and *Salsoli kali* L., the two halophytes selected for comparison with a number of agricultural species, clearly exhibited a much higher protoplasmic salt resistance. These tests were based mainly on ability of cells when bathed in saline solutions to undergo normal plasmolysis and to maintain cytoplasm and membrane integrity as determined by microscopic observation. Other physiological parameters, however, were not measured. Slatyer (1967) briefly discussed some of the metabolic implications of increased electrolyte concentrations in plant protoplasm. Changes in protein hydration and conformation, reduced enzyme activity, altered nucleic acid metabolism, and stimulation of respiration were mentioned as consequences of salt build up in plant tissues of glycophytes. Whether these are the result of decreased tissue water potential or specific ion effects is not always clear. Although there are a number of mechanisms to avoid high electrolyte concentrations in plant tissues, as will be discussed later, halophytes must invariably endure higher tissue salt concentrations than glycophytes if osmotic adjustment is to be accomplished when these plants are growing in saline media. This must again lead to the question of how halophytes might be especially adapted to cope with high plant tissue salinity. To date, new answers to this old question are still seriously wanting.

Maintenance of enzyme activity under high tissue salt concentrations would seem to be a reasonable attribute of halophytes. This would certainly indicate that severe conformational changes in important proteins or other serious disarrangements had not occurred. For halophilic bacteria, salt resistance of enzymes and even high salt requirements for optimal enzyme activity have been reported (Baxter and Gibbons 1957). Recent reports of Greenway and Osmond (1972) and Osmond and Greenway (1972) have, however, failed to support this notion for higher plants. Several enzymes including carboxylating enzymes were isolated from species of varying salt tolerance including the halophyte *Atriplex spongiosa* F.v.M. Enzymes from halophytic species were not more salt resistant under saline assay conditions in vitro than their counterparts from glycophytes. In the case of phosphoenolpyruvate carboxylase from species which fix carbon dioxide via the C_4 dicarboxylic acid pathway, enzymes extracted from glycophytes were even somewhat more salt resistant than those from halophytes. Furthermore, culturing two desert halophytes, *Atriplex spongiosa* and *A. nummularia*, in various culture solutions of NaCl up to 400 mM did not result in any change in specific activity of malate dehydrogenase, glocuse-6-P dehydrogenase, or isocitrate dehydrogenase as compared to plants cultured in 1 mM NaCl solutions. Salt sensitivity of these enzymes also remained the same. Although negative results must always be considered with some reservation, the failure to demonstrate differences in enzyme salt sensitivity in vitro for halophytes versus glycophytes is indeed perplexing. If future research continues to fail to demonstrate a greater salt resistance for enzymes of halophytic or salinized plants, hypotheses concerning ion compartmentalization within cells will certainly become more attractive.

Salt Avoidance Mechanisms

Mechanisms for avoidance of high salt concentrations within plant cells are well known. Cellular ion concentrations may be reduced by increased water uptake by cells, often termed as increased succulence, increased plant growth, ion excretion via salt glands, or exclusion of salts by root membranes.

Increased succulence in both glycophytes and halophytes is well known. Succulence is usually taken to mean increased thickness of plant organs. Increased cellular volume/surface ratios and cellular water content are associated with increased plant succulence, although total plant water content may not change significantly. Mendoza (1971) reported that *Atriplex hastata* L., when grown in solutions of varying degrees of salinity, exhibited increased succulence with increased salinization. However, this succulence was apparent in total plant volume/surface ratio but not in plant water content measured either as g water/g fresh weight or g water/cm^3. Mendoza also documented substantial changes in leaf anatomy which included a description of increased mesophyll cell size in salinized plants. This would result in greater cell water content per cell surface area which would serve to dilute electrolyte concentrations in the cell protoplasm.

Recent literature also supports the original hypothesis of Keller (1925) that Na^+ has a greater tendency to promote succulence that K^+ and Cl^- has a greater tendency than SO_4^- (Jennings 1968, McNulty 1969, Gale and Poljakoff-Mayber 1970).

Apart from increased succulence, growth itself will alter electrolyte concentrations within cells and tissues. Greenway and Thomas (1965) have proposed a simple negative feedback scheme whereby growth reduces cellular solute concentrations which in turn lessens the water potential gradient between plant and soil, and thereby reduces growth. To complete the loop, reduced growth would allow increased accumulation of solute concentrations which in turn would accelerate growth again. Such a simple feedback scheme would appear to apply to most plants, both halophytes and glycophytes.

Regulation of internal ion concentrations can also be at least partially effected through ion exclusion by root membranes. Scholander and colleagues (Scholander et al. 1966, Scholander 1968) have demonstrated this for mangroves and other halophytes. Xylem sap was always very low in solute content, usually greater than -0.5 atm osmotic potential. One exception was the desert halophyte *Atriplex polycarpa* (Torr.) S. Wats. with xylem sap osmotic potentials ranging between -.8 and -1.8 atm.

Leaf sap osmotic potentials of halophytes, meanwhile, remained quite low, often 100 fold less than in the xylem sap (Scholander et al. 1966). Scholander (1968) suggested that a non-metabolic ultrafiltration process is employed by these halophytes for removal of salt in the root membranes. Extreme temperatures and metabolic inhibitors did not disrupt this filtration capacity. Xylem sap tensions, which are presumably in equilibrium with leaf protoplasmic water potentials, are often extremely low in many desert halophytes. Minimal tension values of -80 atm have been reported for the halophytes *Eurotia lanata* (Pursh.) Moq. and *Atriplex confertifolia* (Torr. & Frem.) wats. in the field (Love and West 1972, Moore and Caldwell 1972) and approaching -120 atm under laboratory conditions for rooted plants in soil (Moore, White, and Caldwell 1972). Detling (1969) measured minima of -85 atm for *Suaeda fruticosa* (L.) Forsk. and *S. depressa* (Pursh) S. Wats., and Branson, Miller, and McQueen (1969) reported a minimal value of -128 atm for *Atriplex obovata* in the field.

Although this ultrafiltration process in halophyte roots seems to be widespread among different species, the efficacy of filtration appears to vary. For example, as has been shown in mangroves, *Rhizophora mangle* seems to be much more effective in salt exclusion than other species such as Avicennia nitida (Scholander et al. 1966). Therefore, excretion of salt in the leaves by special salt glands in *Avicennia* appears to be a compensating mechanism for ion regulation apart from growth and succulence. The limited evidence for *Atriplex* species also suggests that ultrafiltration may not be quite as effective as in other halophytes (Scholander et al. 1966, Mazafar and Goodin 1970) and, therefore, the dependence upon vesicular hairs or trichomes for salt excretion from the leaves appears to be important (Osmond et al. 1969, Luttge 1971).

Salt Excretion

Salt excreting glands have been described in several halophytic genera including *Tamarix* (Thomson, Berry, and Liu 1969), *Limonium* (Arisz et al. 1955), and *Atriplex* (Luttge 1971). Since the *Atriplex* genus is well represented among desert halophytes, the salt excreting vesicular hairs of this genus will be discussed here. Recently, the structure and function of these vesicular hairs have been well elucidated (Osmond et al. 1969, Luttge et al. 1970, Luttge and Osmond 1970, Pallaghy 1969, Mazafar and Goodin 1970). These trichomes appear as large spherical balloons supported by intermediate stalk cells on the leaf surface. Mozafar and Goodin (1970) reported extremely high electrlyte concentrations, mostly NaCl, with osmotic potentials on the order of -500 to -700 atm in these bladders. Although the exact magnitude may be somewhat questionable, the extremely high salt concentrations in these cells was clearly demonstrated. Although the stalk and bladder cells contained chloroplasts, these are apparently not very active (Luttge and Osmond 1970). Two mechanisms are apparently operating in the active ion pump responsible for these high concentrations in the bladder cells. One mechanism operates at the same rate in the light and dark and is thought to involve a potassium-sodium exchange mechanism (Luttge et al. 1970). Chloride accumulation in the bladder is strongly stimulated by light suggesting another independent mechanism which is thought to be driven by mesophyll chloroplast photosynthesis (Luttge and Osmond 1970). Although the bladder and stalk cells themselves contain chloroplasts, these are not found to be very active in photosynthesis. Instead, electron flow or reducing power from mesophyll photosynthesis is apparently the active driving force for the chloride pump. Symplasmatic transport between bladder and mesophyll cells has been suggested by Luttge and Osmond (1970) to explain how electron transport in chloroplasts of one cell can be responsible for chloride transport in the membranes of another cell. Eventually these vesicular bladders rupture with subsequent deposition of salts on the leaf surface (Mozafar and Goodin 1970).

The notion that these trichomes or vesicular bladders might be important in atmospheric moisture uptake was refuted by Mozafar and Goodin (1970). They demonstrated that these bladders in the intact state were quite impermeable to water, and water uptake could not be induced even when the bladders were immersed in liquid water. Once the bladders have ruptured and salts have been deposited on the leaf surface, this surface becomes very hygroscopic and water uptake on the leaf surface may be expected. However, it is still doubtful that the living mesophyll tissue can benefit substantially from the hygroscopic nature of the leaf surface. Water uptake from the atmosphere has been demonstrated by increases in weight of the entire leaf (Bjerregaard, unpublished data), though this certainly does not provide any evidence of water uptake by the living portion of the leaf.

There is little doubt that the trichomes of *Atriplex* species function as a salt excreting mechanism which can compensate for salt accumulation in these species. However, as was mentioned earlier, these trichomes are effectively external to the leaf in terms of water potential. Therefore, maintenance of suitable water potentials in the leaf still would depend on reasonably high electrolyte concentrations in the mesophyll tissues. These electrolyte concentrations must result in osmotic pressures sufficient to balance the extreme hydrostatic tensions that develop during the day in many desert halophytes as was discussed earlier in this review (Scholander et al. 1966).

Oxalate Accumulation

Not all electrolytes in the shoot tissues of halophytes are mineral ions. Organic acid salts, particularly oxalates, have been demonstrated to comprise a sizeable portion of the total electrolytes in the tissues of many halophytes, such as **Halogeton glomeratus**. Oxalate accumulation has been documented in halophytes which fix CO_2 via the C_4 dicarboxylic acid pathway as well as in those which carry on normal C_3 photosynthesis (Osmond 1963, Mendoze 1971, Williams 1960). As a functional anion, oxalate has been thought to replace inorganic anions such as chloride in several species and the propensity for oxalate synthesis has been suggested to be inversely correlated with chloride tolerance among various halophyte species (Osmond 1963). Although this correlation might suggest some adaptive advantage of oxalate synthesis and accumulation in some halophytes, the distinct benefit of oxalate production and storage is still obscure, if any exists. The occasional though dramatic incidence of domestic sheep death from oxalate poisoning following the comsumption of great quantities of certain halophytes would suggest a possible advantage of oxalate accumulation as a biotic defense agent. However, other primary herbivores such as the black-tailed jack rabbit are apparently not affected by high oxalate ingestion (Saul and Westoby, unpublished data).

Energy Requirements of Salt Metabolism

Salt deserts are severe environments for plant growth and productivity. It is, therefore, appropriate to consider the energy and carbon that halophytes must expend for osmotic adjustment and salt excretion. In the same context, oxalate production should be considered. This is hardly a novel consideration. As is discussed by Adriani (1958), Lundegardh and his colleagues had considered this problem some 40 years ago. Much of this is summarized in Lundegardh's classic text (Lundegardh 1954). He suggested that root respiration is comprised of two components—a basal component and a component closely linked with anion accumulation. The relationship between moles of CO_2 respired and moles of anion accumulated seemed to vary with both the anion and the plant species. Although very few data exist, the earlier work of van Eijk (1938) suggested that the halophyte *Aster tripolium* possessed a much lower CO_2 respiration requirement for anion uptake than glycophytes such as wheat and tobacco. Though much progress has recently been made in elucidating the mechanisms of ion pumps in plant tissues, there seems to be little further work reported on quantifying energy costs of anion uptake.

Although the concept of salt respiration is concerned with root respiration, ion pumps in above-ground portions of plants would also require an expenditure of energy. As was discussed earlier, there exists in most halophytes a striking difference in electrolyte concentrations between the cell sap and the nearly pure water of the xylem stream. These osmotic gradients may often be on the order of 60 to 80 atm in many halophytic species.

Although cations might be transported passively due to the fact that the interior of most plant cells is negatively charged with respect to the exterior, cation uptake will obviously be limited by the active uptake of anions, or in the case of some species, the synthesis of oxalates or related compounds (Pallaghy 1969, Osmond and Avadhaniani 1968). Therefore, the establishment and maintenance of sizeable electrolyte gradients in plant tissues of both roots and shoots must always require an expenditure of energy.

Though salt excretion via vesicular hairs in *Atriplex* most likely does not involve ATP, either electron transport or reducing power are required (Luttge and Osmond 1970). The quantitative energy requirements per mole of anions secreted into the bladders have not as yet been established. However, if the determinations of Mozafar and Goodin (1970) for the osmotic potential of bladder cells in *Atriplex* are in the correct order of magnitude (-550 to -700 atm), a considerable energy investment for this ion transport would seem to be requisite.

Finally, for those halophytes which substitute organic acid salts for inorganic ions in the total electrolyte balance of the cell sap, a carbon and energy investment in these organic anions is required.

In *Atriplex spongiosa*, oxalate synthesis is apparently linked to oxidation of glyoxylate derived from the TCA cycle in dark acid metabolism (Osmond and Avadhani 1968). Though oxalate may be a subsequently metabolized into CO_2, most of it apparently remains in a very slowly exchangeable pool, and a sizeable component may often be in the form of oxalate crystals (Osmond and Avadhani 1968, Mendoza 1971). Since oxalates are very low energy compounds, they would represent an inefficient means of energy storage which would not be readily available. When species such as *Halogeton glomeratus* accumulate as much as 38% oxalate (of total tissue dry weight in the shoots, Williams 1960) a considerable investment of energy must indeed be recognized. The comparative energy invested in mineral anion accumulation versus oxalate production is still, however, quite open to question.

Stimulation of Halophyte Growth by Salts

As stated above, halophytes are distinguished from glycophytes by their remarkable tolerance of high salinities. In addition, it has long been known that a certain quantity of salts in a growing medium may actually stimulate halophyte growth. This stimulation has led to a variety of conclusions concerning halophyte physiology and definitions which would distinguish different categories of species such as "facultative" or "obligate" halophytes (Adriani 1958, Barbour 1970). The recent literature continues to verify the stimulation of halophyte growth by certain quantities of salts, particularly NaCl, in the growing medium (Williams 1960, McNulty 1969, Chatterton and McKell 1969, Greenway 1968, Wallace and Romney 1972, Mendoza 1971) for desert halophytes. Although sodium has been shown to be an essential micronutrient for the growth of some halophytes (Brownell 1965, Brownell and Crossland 1972), sodium requirements for nutrition can be satisfied by very small quantities (0.1 mM, Brownell and Crossland 1972). Optimal growth of many halophytes can be evidenced with salinity concentrations as high as 100 to 200 mM (Greenway 1968, Chatterton and McKell 1969, Mendoza 1971).

As Barbour (1970) has cautioned, rigid categorizations of halophytes must be reassessed. Although there appears to be marked stimulation of growth in many halophytes by salts in excess of that needed for micro-nutrient requirements, this cannot be viewed as a salt requirement. Furthermore, the conditions under which such determinations of salt stimulation have been made must be carefully considered. Gale et al. (1970) have shown for *Atriplex halimus* L. that under low relative humidity (27%), maximum growth was attained with a NaCl solution of -5 atm (approx. 110 mM), but under a higher relative humidity of 65%, optimal growth occurred under the nonsaline control solution. This would suggest that under low humidity conditions, optimal growth in the -5 atm solution occurred because salts were available for accumulation by the plant which in turn facilitated water uptake necessary to meet the imposed transpirational stress. Concentrations of salts greater than this -5 atm solution only contributed to greater plant water stress and decreased growth. At high humidities, total plant water stress was reduced, ion accumulation to facilitate water uptake was not so necessary, and even low concentrations of salts were harmful to the plant. Therefore, optimal growth took place in the nonsaline solution. For plants growing in soils, an analogous situation should be expected. If soil water potentials are low, there should be an advantage for a halophyte to have salt available in the soil solution for uptake and osmotic adjustment. If, on the other hand, soil water potentials are not particularly low and atmostpheric moisture stresses are also minimal, there would be no advantage to a halophyte if salts were available in the soil solution. Under these circumstances, salts in the soil solution would only decrease the soil moisture potential. Evidence to support this hypothesis is only indirect since most of the physiological studies have been carried out using culture solutions. Waisel and Pollak (1969) have shown that the equilibration points (approximate wilting points) of two halophytes, *Aeluropus litoralis* (Willd.) Parl. and *Suaeda monoica* Forsk., occurred at about the same soil water potential (approximately -15 atm) as sunflower when these halophytes were grown under nonsaline soils in the laboratory. When grown in the field in saline soils, both halophytes exhibited equilibration points at much lower soil water potentials, e.g. -55 atm for *Aeluropus*.

From the foregoing discussion, it would seem that if salts are merely a convenient rapidly permeating electrolyte for osmotic adjustment by halophytes, growth stimulation by salts in the culture medium of halophytes would be primarily a matter of providing the permeating electrolytes necessary to meet the osmotic adjustment demands of the total atmospheric and soil moisture stresses presented to the plant. It would certainly be well to consider many of the earlier physiological studies in this context in order to critically review the reports of salt stimulation or even salt requirements of halophytes. Nevertheless, the possibilities for growth stimulation by Na^+ or Cl^-, apart from micronutrient requirements or merely increased succulence, certainly cannot be ruled out (Greenway 1968, Jennings 1968). Also, the notable increase in succulence resulting from Cl^- or Na^+ ions, although initially only altering the dimensions and water content of cells, should often result in increased dry matter accumulation for the plant due to increased leaf area or height of the plant and a greater display of chlorophyll for photosynthesis.

Evaluating Salinity Tolerance Limits of Desert Halophytes

Studies of salinity tolerance, whether conducted in the laboratory or based on soil data from the field, must be interpreted with caution, particularly for desert halophytes. As has just been discussed, other factors in laboratory experiments besides salinity of the culture solution will alter plant response to salinity. There is also the problem of extrapolating results from culture solutions to soils. Field soil data, if properly interpreted, can provide much information about actual salinities tolerated by halophytes in desert areas. However, soil salinities are often reported as merely percent dry weight of salts in the soil or as conductivity of a saturated soil paste. In either case, this is difficult to relate to the soil salinity in a partially wetted soil. As soils dry, salinity obviously increases in the remaining soil solution. For example, Moore and Caldwell (unpublished data) measured total soluble salt percentages in a saturated paste of approximately 1% at 90 cm depth in a mixed *Eurotia lanata, Atriplex confertifolia* community in northern Utah. Yet, by psychrometric determination, a minimum soil osmotic potential of -35 atm was evident for this depth (Moore and Caldwell 1972) which would yield a soluble salt percentage of at least 4.7% assuming most of the salts to be NaCl. Under these particular conditions, total soil moisture potential was between -45 and -50 atm. Without actual measurement, it is difficult to estimate the osmotic potentials of soil solutions since not all salts of the soil solution will remain in solution as the soil dries. Therefore, much caution is warranted in relating laboratory data in which culture solutions were used to field conditions.

Halophytes and Desert Soils

In addition to salinity, desert halophytes must tolerate other edaphic hardships. Low available soil nitrogen is a common feature of desert areas due to the very low organic matter content of most desert soils. Yet, many desert shrubs, particularly in the Chenopodiaceae, have very high leaf protein contents (Cook 1971, Dwyer and Wolde-Yohannis 1972, Goodin and McKell 1971). Nitrogen concentration in many *Atriplex* species is similar to that in harbaceous nitrogen-fixing legumes (Beadle, Whalley, and Gibson 1957).

Surprisingly, little is known about the nitrogen metabolism of desert halophytes. In the coastal halophyte *Suaeda maritima* L. Dunn. Stewart, Lee, and Orebamjo (1972) found a close correlation between plant nitrogen content and nitrate reductase activity in this species. Both in the laboratory and in the field, nitrate reductase activity in the plant apparently responded to available soil nitrogen levels and did not seem to be dependent upon soil salinity within the gradient measured in their study. Goodman and Caldwell (1971) determined nitrate reductase activity for several desert halophytes, some of which exhibited very high activity levels, e.g. *Eurotia lanata*. Considerable genetic variation, however, was evident in samples of the same species taken from nearby communitites.

Nitrogen fertilization experiments for desert halophytes have also been very limited. Goodman (unpublished data), and Wallace and Romney (1970) both found that fertilization of desert perennial shrubs produced no immediate apparent effect on plant biomass. However, for annuals, including halophytes such as *Halogeton glomeratus*, there was a pronounced response to nitrogen fertilization.

Chatterton et al. (1971) reported that nitrogen in excess of 28 ppm in a nutrient solution culture had little effect on the growth of *Atriplex polycarpa*, a perennial salt desert shrub. Apparently this species is very efficient in absorption of low levels of soil nitrogen. At low substrate nitrogen levels, most of the plant nitrogen was in protein form. At substrate nitrogen levels approaching 1000 ppm, growth was reduced somewhat at both high and low salinity levels. However, non-protein nitrogen in the plant usually remained low. Strogonov (1962) has suggested that in salt intolerant species, damage from high salt concentrations was due to disruption of protein metabolism and accumulation of non-protein nitrogen compounds, particularly ammonia and various free amino acids. Amino acid accumulation seems to vary with plant species, types of nitrogen available in the soil, and plant organs, e.g. roots versus shoots.

Although Strogonov (1962) reviews a considerable amount of intensive work on the nitrogen metabolism of glycophytes, little is mentioned concerning the physiological adaptations of halophytes in terms of nitrogen metabolism.

The work of Chatterton et al. (1971) suggests that the halophyte *Atriplex polycarpa* is able to minimize non-protein nitrogen accumulation in plant tissues even at high substrate nitrogen concentrations and at high salinity levels. From the available information, it is also apparent that desert halophytes are able to maintain reasonably high tissue nitrogen levels despite very low available soil nitrogen concentrations. There must be a limit, however, as Medina (1971) has shown in growing *Atriplex patula* spp. *hastata* Hall & Clem. in a culture solution of only 14 ppm available nitrogen. Under these conditions, leaf nitrogen, soluble protein, and carboxydismutase levels were greatly reduced. Furthermore, a concomitant reduction in photosynthesis was closely correlated with the reduction in carboxydismutase activities. He suggested that the lack of nitrogen reduced the quantity of carboxydismutase rather than simply the activity of this key photosynthetic enzyme. The minimum substrate nitrogen levels which these plants require for optimal development and photosynthetic activity was not determined in his study. Perhaps at substrate levels above 25 ppm available nitrogen, there would be no indication of nitrogen limitation as Chatterton et al. (1971) found.

Although the picture is far from complete, desert halophytes appear to be well adapted to low available soil nitrogen conditions through efficient nitrogen uptake and metabolism. Furthermore, high salinities apparently do not exert the same disruptions of protein metabolism that occur in glycophytes.

Apart from salinity, other soil elements in desert areas may often be approaching what would be considered as toxic levels for most species. A particular case in point is boron, concentrations of which are often much higher than would ever be tolerated by agronomic species. Southard (unpublished data), for example, reports concentrations of 24 ppm boron in the upper soil layers in some sites occupied by *Atriplex nuttallii* spp. *tridentata* (Kuntze) H & C in northern Utah. Goodman (unpublished data) found that this species was little affected by boron concentrations exceeding 400 ppm. Wallace and Romney (1972) reported only slight reduction in growth of *Franseria dumosa* Gray and *Atriplex hymenelytra* (Torr.) Wats in solutions containing up to 100 ppm boron.

Chatterton et al. (1970) noted that growth reduction in *Atriplex polycarpa* was only effected by boron concentrations of at least 80 ppm in the culture solution. However, at these concentrations, boron in the shoot reached 600 ppm although root tissue boron levels remained quite low at approximately 70 ppm.

Root Growth and Productivity

The dynamics of halophyte root systems have been largely neglected.

Although underground plant biomass in some halophytic shrubs can be quite substantial, e.g. approaching 80% of total plant biomass (Bjerregaard 1971), annual root productivity remains largely unknown for most desert halophytes. Although it might seem logical that the periods of root growth activity during the year should correspond with shoot growth, this is not necessarily the case. Preliminary experiments (Fernandez and Caldwell, unpublished) have shown that root growth of *Atriplex confertifolia* in the field can take place long after shoot growth has terminated for the year. This root extension was active at soil moisture potentials less than -50 atm. In laboratory experiments with *Atriplex vesicaria* Hew., Cowling (1969) has shown the Cl^{36} uptake was dependent on new root growth and activity. He also found that roots could continue to grow and take up Cl^{36} in dry soil up to a period of 60 weeks, as long as another portion of the root system was located in moist soil. Although the greatest abundance of nutrients is usually in the upper horizons of desert soils (Cowling 1969, Bjerregaard 1971), these layers are also often the driest regions of the soil profile. Therefore, in these perennial shrubs, if lower roots are in areas of reasonably sterile but moist soil, this can sustain root growth and active nutrient uptake in the upper soil layers even though these upper regions of the soil profile may be very dry.

Gas Exchange, Salinity and Drought

Desert halophytes enjoy only a limited period during the year when moisture and temperature conditions are truly suitable for active metabolic activity, including photosynthesis. These plants must possess the ability to carry on photosynthesis and other aspects of metabolism as long as possible in the relatively brief periods of the year when conditions permit. Maximizing carbon gain, when growing in salty desert soils that are usually at least partially dry, is essential for survival.

Photosynthesis and dark respiration have been studied in halophytes which have been cultured in solutions of varying salinity (Anderson 1962, McNulty 1969, Gale and Poljakoff-Mayber 1970). Nieman (1962) reported that respiration rates for several glycophytes were stimulated by low concentrations of NaCl in the nutrient culture medium. Although only a limited amount of evidence is available, this does not appear to be the case for desert halophytes such as *Sarcobatus vermiculatus*, except for the very young leaves (Anderson 1962, McNulty 1969). Detling (1969) found a slight decrease in dark respiration of three desert halophytes, *Sarcobatus vermiculatus*, *Suaeda fruticosa*, and *Distichlis spicata*, as plant moisture potentials decreased to -50 atm. However, this decrease in dark respiration activity was only very slight even at the extreme plant moisture potentials.

As with glycophytes, halophytes respond to increasingly salinized culture media with decreasing net photosynthetic rates, though the depression is not so pronounced as with glycophytes. Gale and Poljakoff-Mayber (1970) found that *Atriplex halimus* exhibited only a slight depression in photosynthetic activity as plants were grown in media of increasing NaCl concentration up to 440 mM (-20 atm). Anderson (1962) found that photosynthesis of *Sarcobatus vermiculatus* was greatly depressed at NaCl concentrations in excess of 450 mM. Detling (1969), working with the same three halophytes mentioned above, found a marked decrease in net photosynthesis as total plant moisture potential approached -50 atm. However, soil moisture potentials or salinities were not determined in his study. Some CO_2 uptake was measured in all three species at plant moisture potentials of -50 atm. Moore (1971) reported CO_2 uptake at soil moisture potentials less than -75 atm in *Atriplex confertifolia* and *Eurotia lanata*. Under these conditions total plant moisture potential approached -90 to -120 atm. In these soils the osmotic component of the total soil moisture potential comprised approximately one-third of the total.

Measurements of transpiration in the studies by Moore, White, and Caldwell (1972) and Gale and Poljakoff-Mayber (1970) indicated that transpiration also decreased as water potential of the growing medium decreased. Stomatal diffusion resistance calculated by Gale and Poljakoff-Mayber for *Atriplex halimus* continued to increase as plants were grown in increasingly salinized media up to 440 mM (-20 atm). Calculations of mesophyll diffusion resistance showed an initial decrease as plants were cultured in salinized media of approximately -8 atm (184 mM). When plants were grown at higher salinities, this mesophyll resistance began to increase. Therefore, only at the greater salinities when both stomatal diffusion resistance and mesophyll resistance increased was there a noticeable decrease in photosynthetic rate. The decrease in mesophyll resistance which occurred when plants were grown under only moderate salinities, coupled with the increased stomatal diffusion resistance, resulted in more favorable photosynthesis/transpiration ratios for this species. Whether or not these relationships apply when *Atriplex halimus* is growing in saline soils in the field is, however, still open to speculation. The influence of salinized culture solutions on photosynthesis and transpiration may be quite dissimilar to the influence of decreasing total soil moisture potential for plants growing in the field.

The interaction between anatomical characteristics produced by salinization of the growing medium such as changes in stomate size and number, leaf succulence, surface area development, and mesophyll structural changes with plant gas exchange need further elucidation. Changes in leaf area certainly alter total plant assimilation capacity even if net photosynthetic capacity per unit leaf material remains constant. Changes in leaf anatomy should be expected to alter at least the stomatal and mesophyll diffusion resistances for photosynthesis and transpiration. Certainly, many effects of salinization on plant gas exchange may act through these changes in leaf anatomy and size.

Effects of specific ions such as Cl⁻ on photosynthesis and respiration, apart from decreasing osmotic potentials, have been suggested (Chapman 1960, McNulty 1969), but too little information is yet available to assess the influence of specific ions on halophyte gas exchange rates.

Since desert halophytes are able to continue photosynthesis as soil moisture potentials decrease to extreme dryness and respiration is not stimulated under conditions of high salinity or low moisture potentials, they would seem quite well equipped to enjoy at least some level of positive carbon gain during much of the year when temperature conditions permit.
permit.

The gas exchange response of desert halophytes to temperature and irradiation will be next discussed as species possessing C_4 and C_3 photosynthesis are compared.

C_4 Photosynthesis in Desert Halophytes

The C_4 dicarboxylic acid pathway of photosynthesis is well represented among desert halophytes (Tregunna and Downton 1967, Osmond 1970, Welkie and Caldwell 1970). However, plants possessing the normal C_3 photosynthetic pathway also thrive quite well in salt desert environments. Since the discovery of the C_4 pathway, a great deal of research has been centered on elucidation of the biochemical and anatomical features of C_4 species. Much less has been done to unravel the ecological perspectives of this rather elaborate photosynthetic pathway. Based on laboratory evidence primarily with agronomic species, several characteristics have been consistently associated with C_4 species which would suggest they should be well adapted to warm arid environments. Most C_4 species possess high rates of photosynthesis which has led some investigators to label these as "efficient" plants (Black 1971). In addition, they undergo light saturation of photosynthesis only at very high irradiation levels, have high temperature optima for photosynthesis, i.e. 35–40°C, comparatively high photosynthesis/transpiration ratios, and lack apparent photorespiration (Black 1971). Many other anatomical and biochemical features are associated with C_4 plants; however, the ones just mentioned are those which appear most relevant to this discussion.

Some C_4 desert shrubs do indeed appear to possess characteristics similar to those of agronomic C_4 speices. When comparative laboratory studies were carried out on pairs of *Atriplex* species such as *A. rosea* L. (C_4) and *A. patula* spp. *hastata* Hall & Clem. (C_3) (Bjorkman 1971) and *A. spongiosa* F. v.M. (C_4) and *A. hastata* L. (C_3) (Osmond et al. 1969, Slatyer 1970), the C_4 species in both cases did possess higher rates of photosynthesis on a leaf area basis and exhibited higher temperature optima for photosynthesis, high light saturation intensities, and greater photosynthesis/transpiration ratios. Jones, Hodgkinson, and Rixon (1969) also reported high photosynthetic rates at high temperatures along with a high temperature optimum for photosynthesis in *Atriplex nummularia* Lindl., a C_4 species common to many desert areas in Australia. But, on a leaf area basis, this species was still far inferior in photosynthetic rate to corn. Such comparative studies based primarily on gas exchange characteristics of single leaves in C_4 halophytes present the C_4 species in a favorable light when compared to their C_3 counterparts, although net photosynthetic rates in the C_4 halophytes may still be much less than reported for C_4 agronomic species such as corn and sugar cane. The question must still be raised as to how these C_4 halophyte species perform in their natural habitat, and whether or not they are indeed superior to the C_3 halophyte species.

Crop species possessing the C_4 pathway have been traditionally acclaimed for very high productivity and also for very favorable productivity/water-use ratios (Shantz and Piemeisel 1927, Black 1971). Slatyer (1970) analyzed growth of *Atriplex spongiosa* (C_4) and *A. hastata* (C_3) over a 23-day period. His results indicated that the grwoth performance of the C_4 species was not always superior to that of the C_3 species. Net photosynthetic rates per unit leaf area of *A. spongiosa* were indeed higher and never less than those of *A. hastata*. However, though total leaf area was initially greater in *A. spongiosa*, towards the end of the experiment total leaf area per plant was much less than in *A. hastata*. Therefore, the daily growth rate of *A. spongiosa* actually fell below that of *A. hastata* in the latter part of the experiment. Although Slatyer's experiment with these young potted greenhouse-grown plants does lend some clues as to what might happen under field conditions, it is hard to extrapolate to field situations where high degrees of mutual leaf shading, moisture stress, increased soil salinity, and variability of gas exchange rates throughout the different seasons of the year are to be expected. It is also not clear as to whether or not Slatyer took into account underground productivity of these two species in his growth analysis. Although Hofstra and Hesketh (1969) found net photosynthesis per unit leaf area of single leaves of *A. nummularia* to be quite inferior to corn at several temperatures, Jones et al. (1969) found *A. nummularia* to be superior to corn when net photosynthetic rates per unit leaf area were measured for whole plants under field conditions with optimal moisutre and nutrient conditions. Apparently *A. nummularia* could carry on photosynthesis at approximately 67% of the maximum rate for individual leaves whereas *Zea mays* was functioning in the field at only approximately 35% of maximum leaf rates. In a 500-day growth analysis study, Jones et at. (1969) grew *A. nummularia* under optimal soil moisture and nutrient conditions. They found that growth rates computed over this period of time for this C_4 shrub were in the same range of magnitude as highly productive C_4 crop species such as sugar cane and corn (Warren Wilson 1967). These growth analyses included underground plant productivity, which is seldom included in most productivity studies.

Goodin and McKell (1971) reported that when C_4 *Atriplex* species such as *A. polycarpa*, *A. lentiformis*, and *A. canescens* were grown under non-irrigated agronomic conditions, dry weight yields as high as 10,000 kg per ha per growing season could be realized in southern California. This, when compared to forage sorghum (*Sorghum vulgare*) at 40,000 kg per ha per season in California (Loomis and Williams 1963), is certainly quite respectable. It was not, however, clear as to whether underground plant parts were included in this productivity estimate.

In a greenhouse experiment, Dwyer and DeGarmo (1970) found that total shoot plus root production for the C_4 shrub *Atriplex canescens* (Pursh) Nutt. was much higher than that of C_3 species such as *Larrea tridentata* (D. C.) Coville, *Prosopis juliflora* (Swartz) D. C., and *Gutierrezia sarothrae* (Pursh) Britt. & Rusby. Furthermore, the water use efficiency, i.e. water used per total plant dry weight produced, was much lower for *Atriplex canescens* than for the other three species. Wallace (1970) and Dwyer and Wolde-Yohannis (1972) found that water use efficiency of *Salsoli kali* L., a C_4 halophyte, to be highly superior to that of C_3 species in greenhouse experiments. Field determinations of photosynthesis/transpiration ratios for *Atriplex confertifolia*, a C_4 halophyte, indicated these ratios to be much greater throughout the growing season than for *Eurotia lanata*, a C_3 halophyte (White, Moore, and Caldwell, unpublished data; Caldwell 1972). These species were growing in their natural saline habitat during these experiments.

Except for the last experiment mentioned, most of the comparative studies dealing with production, photosynthesis, and water-use efficiency of C_4 and C_3 halophytes have been carried out on plants grown in nonsaline media, as far as it is possible to discern from the descriptions of methodologies. Under saline conditions it would be reasonable to expect lower rates of net photosynthesis and lower productivities than under non saline conditions. As was mentioned earlier, under most situations, photosynthesis of halophytes decreases with increasing salinization of the growing media. Also, as was discussed earlier in this review, various aspects of salt metabolism including active anion transport, salt excretion by vesicular hairs, oxalate production, etc. are energy requiring processes and must ultimately divert energy that might otherwise go into increased production.

Detling (1969) measured net photosynthesis in three halophytes growing in saline soils in the laboratory. The C_3 species *Sarcobatus vermiculatus* was intermediate in net photosynthesis between two C_4 halophytes, *Distichlis spicata* which possessed much lower rates of net photosynthesis and *Suaeda fruticosa* which possessed higher rates. This applied for a wide range of plant moisture potentials and at temperatures of both 25 and 30°C.

In situ field measurements of photosynthesis and transpiration in *Atriplex confertifolia* (C_4) and *Eurotia lanata* (C_3) (Moore 1971; White, Moore, and Caldwell, unpublished data; Caldwell 1972) revealed that throughout the season neither species possessed particularly high rates of net photosynthesis. Both species exhibited low rates of net photosynthesis never exceeding 18 mg CO_2 dm^{-2} hr^{-1} even when these plants were exposed to saturating irradiation intensities from a xenon arc lamp in the laboratory. *Atriplex confertifolia* the C_4 species, never exceeded *Eurotia lanata* (C_3) in net photosynthetic activity. Maximal photosynthetic rates for the two species were roughly only one-third of the maximal rates reported for C_4 species of agronomic use (Bull 1969) and a C_3 warm desert glycophytic shrub *Encelia farinosa* (Cunningham and Strain 1969). However, these maximal rates are in the same range of magnitude as those reported for a variety of other arid land shrubs in Israel by Whiteman and Koller (1967), and in western Australia by Hellmuth (1971a). Both *Atriplex* and *Eurotia* were able to carry out photosynthesis during the spring months at leaf temperatures of -5°C and exhibited low temperature optima for photosynthesis (approximately 10–15°C) at that time. For the C_4 species, this certainly represents an anomaly when compared with the reported characteristics of other C_4 species. Later in the season, temperature optima for net photosynthesis of *Atriplex confertifolia* had shifted to values around 30°C.

If the C_4 species did have an advantage over the C_3 species, it was not in higher rates of net photosynthesis but rather in a longer period of photosynthesis during the course of the year. *Atriplex* exhibited positive CO_2 uptake from early March to late October; *Eurotia lanata* exhibited positive CO_2 uptake only from the early spring to approximately late July or August. *Atriplex* also possessed a consistantly more favorable photosynthesis/transpiration ratio throughout the entire season. Bjerregaard (1971) reported that although *Atriplex confertifolia* exhibited approximately 20% higher shoot productivity on an annual basis, total plant biomass per ground area in adjacent pure communities of *Atriplex* and *Eurotia lanata* were approximately the same.

Certainly much more remains to be unraveled concerning the comparative characteristics of C_4 and C_3 halophytic species growing in salt desert areas. For example, the interaction of salinized soils with photosynthetic and respiratory characteristics of C_4 and C_3 species still remains largely in question. For example, the recent findings by Brownell and Crossland (1972) indicate that among species surveyed, C_4 species including many halophytes possessed a sodium requirement as a micronutritional element which was satisfied by a very low quantity of sodium (0.1 mM). None of the C_3 species tested possessed this sodium micronutrient requirement. There is nothing in the comparative biochemistry of C_4 and C_3 species that would immediately indicate why C_4 species might require sodium as a micronutrient. Furthermore, the nature of salt accumulation, excretion, and oxalate synthesis does not appear to be fundamentally different between C_3 and C_4 halophytes covered in this review article.

Many of the generalizations currently in the literature, though pertinent to many agronomic species, should be carefully considered in the context of wild land species, especially desert halophytes. Furthermore, the notation of C_4 species as "efficient" (Black 1971) should be abandoned. As has been discussed, many C_4 species are not particularly efficient in net photosynthetic activity and many c_3 species are known to have very high rates of net photosynthesis (e.g. McNaughton and Fullem 1970). Since both the C_4 and C_3 photosynthetic pathways have been successful strategies for desert halophytes throughout the world, clear-cut advantages of one system over the other is questionable.

Heat and Low Temperature Tolerance

Desert halophytes must in the course of the year experience great temperature extremes since they normally occur in continental habitats. Little attention, however, has been paid to the heat or low temperature resistance of these species. Working in western Australia, Hellmuth (1971b) studied the heat tolerance of several desert species including *Arthrocnemum bidens* and *Kochia planifolia*. Heat resistance of both species varied with soil moisture and plant osmotic potentials. *Arthrocnemum* could endure temperatures of 54°C under optimal soil moisture but 58°C under moisture stress. Osmotic potentials of the plant sap increased from approximately -60 atm under optimal soil moisture to -80 atm under moisture stress. Whether osmotic potential was directly related to heat resistance is unclear. Similarly for *Kochia* under optimal soil moisture, 50°C could be endured, but under moisture stress, this species could tolerate temperatures up to 56°.

As has been pointed out by Kappen (1969a), the natural halophyte communities in the cool temperature latitudes of the world are rather poor in numbers of species compared to similar communities in milder and warmer climates. Furthermore, very few halophytes growing in cold regions retain their leaves throughout the winter. Those species which do, certainly provide interesting subjects for physiological investigations. It has long been assumed that as the salt concentration (NaCe) of plant tissues increases, the frost resistance would be decreased (Kappen 1969 a & b). Kappen's (1969 a & b) studies with halophytes in the northern coastal areas of Europe indicated that salt concentrations in the tissues decreased during the winter months, while sugar concentrations increased. Species such as *Halimione portulacoides* (L.) Aellen tolerated -20°C during February. Not only did total sugars increase during the winter months, but also the kinds of sugars changed appreciably. However, Kappen was able to demonstrate under laboratory conditions that the frost resistance of halophytes could be greatly increased even when the leaves contained high quantities of salt. Furthermore, the difference in frost resistance between leaves with high and low salt concentrations was very small (Kappen 1969 a & b) and these halophytes could develop cold tolerance as efficiently as any glycophytes growing in the same areas. The hypothesis was put forward by Kappen that there might be partitioning of salts within the halophyte cells so that most of the salts were concentrated in the vacuole so as not to interfere with the protoplasmic frost resistance.

Much remains to be done in the elucidation of frost hardiness mechanisms in halophytes. Many halophytes in temperate northern latitudes in desert areas retain their leaves during the winter and must endure temperatures in the range of -30 to -40°C.

Literature Cited

Adriani, M. J. 1956. Page 902 in "Handbuch der Pflanzenphysiologie" Vol. 3. Edited by W. Ruhland. Springer-Verlag, Berlin.

Adriani, M. J. 1958. Page 709. in "Handbuch der Pflanzenphysiologie". Vol. 4. Edited by W. Ruhland. Springer-Verlag, Berlin.

Anderson, N. W. 1962. "The Effect of NaCl and KCl on the Respiration and Photosynthesis of Greasewood, *Sarcobatus vermiculatus* (Hook.) Torr." M. S. Thesis, Univ. Utah, Salt Lake City.

Arisz, W. H., I. J. Camphuis, H. Heikens, and A. J. van Hodan. 1955. Acta. Bot. Neer. 4:322.

Barbour, M. G. 1970. Amer. Midl. Natural. 84:105.

Baxter, R. M. and N. E. Gibbons. 1957. Can. J. Microbiol. 3:461.

Beadle, N. C. W., R. D. B. Whalley, and J. B. Gibson. 1957. Ecology. 38:340.

Bjerregaard, R. S. 1971. "The Nitrogen Budget of Two Salt Desert Shrub Plant Communities of Western Utah." Ph.D. Thesis, Utah State Univ., Logan, Utah.

Bjorkman, O. 1971. Page 18. in "Photosynthesis and Photorespiration." Edited by M. D. Hatch, C. B. Osmond, and R. O. Slatyer. Wiley, New York.

Black, C. C. 1971. Adv. Ecol. Res. 7:87.

Black, R. F. 1960. Austr. J. Biol. Sci. 13:249.

Boyer, J. S. 1965. Plant Physiol. 40:229.

Branson, F. A., R. F. Miller, and J. F. McQueen. 1969. Abstract in "International Conference on Arid Lands in a Changing World." Univ. Ariz. Press, Tucson.

Brownell, P. F. 1965. Plant Physiol. 40:460.

Brownell, P. F. and C. J. Crossland. 1972. Plant Physiol. 49:794.

Bull, T. A. 1969. Crop Sci. 9:726.

Caldwell, M. M. 1972. Page 27 in "Eco-Physiological Foundation of Ecosystems Productivity in Arid Zones." Edited by L. E. Rodin. USSR Acad. Sci., Leningrad.

Chapman, V. J. 1960. "Salt Marshes and Salt Deserts of the World." Interscience, New York.

Chatterton, N. J., J. R. Goodin, and C. Duncan. 1971. Agron. J. 63:271.

Chatterton, N. J., and C. M. McKell. 1969. Agron. J. 61:448.

Chatterton, N. J., C. M. McKell, F. T. Bingham, and W. J. Clawson. 1970. Agron. J. 62:351.

Cook, C. W. 1971. Utah Agric. Exper. Sta. Bull. 483.

Cowling, B. W. 1969. "A Study of Vegetation Activity Patterns in a Semi-Arid Environment." Ph.D. Thesis, Univ. New England, N.S.W. Australia.

Cunningham, G. L., and B. R. Strain. 1969. Photosynthetica. 3:69.

Detling, J. K. 1969. "Photosynthetic and Respiratory Responses of Several

Halophytes to Moisture Stress." Ph.D. Thesis, Univ. Utah, Salt Lake City.
Dwyer, D. D., and H. C. DeGarmo. 1970. New Mex. State Univ. Agric. Exper. Sta. Bull. 570.
Dwyer, D. D., and K. Wolde-Yohannis. 1972. Agron. J. 64:52.
Eijk, M. van. 1938. Proc. Kon. Ned. Acad. V. Wetensch. 41:1115.
Epstein, E. 1969. Page 345 in "Ecological Aspects of the Mineral Nutrition of Plants." Edited by I. H. Rorison. British Ecol. Soc. Symp. No. 9. Blackwell, Oxford.
Gale, J., R. Naaman, and A. Poljakoff-Mayber. 1970. Austr. J. Biol. Sci. 23:947.
Gale, J., and A. Poljakoff-Mayber. 1970. Austr. J. Biol. Sci. 23:937.
Goodin, J.R., and C. M. McKell. 1971. Page 235 in "Food, Fiber and the Arid Lands." Edited by W. G. McGinnies, B. J. Goldman, and P. Paylord. Univ. Ariz. Press, Tucson.
Goodman, P. J. and M. M. Caldwell. 1971. Nature 232:571.
Greenway, H. 1968. Israel J. Bot. 17:169.
Greenway, H., A. Gunn, and D. A. Thomas. 1966. Austr. J. Biol. Sci. 19:741.
Greenway, H., and C. B. Osmond. 1972. Plant Physiol. 49:256.
Greenway, H., and D. Thomas. 1965. Austr. J. Biol. Sci. 18:505.
Hellmuth, E. O. 1971a. J. Ecol. 59:225.
Hellmuth, E. O. 1971b. J. Ecol. 59:365.
Hofstra, G., and J. D. Hesketh. 1969. Planta 85:228.
Jennings, D. H. 1968. New Phytol. 67:899.
Jeschke, W. D. 1970. Planta 94:240.
Jeschke, W. D. 1972. Planta 106:73.
Jones, R., K. C. Hodgkinson, and A. J. Rixon. 1969. Page 31 in "The Biology of *Atriplex*." Edited by R. Jones. C.S.I.R.O., Canberra.
Kappen, L. 1969a. Ber. Dtsch. Bot. Ges. 82:103.
Kappen, L. 1969b. Flora 158:232.
Keller, B. 1925. J. Ecol. 13:224.
Loomis, R. S. and W. A. Williams. 1963. Crop Sci. 3:67.
Love, L. D., and N. E. West. 1972. Northwest Sci. 46:44.
Lundegardh, H. 1954. "Klima und Boden." VEB Gustav Fischer Verlag, Jena.
Luttge, U. 1971. Ann. Rev. Plant Physiol. 22:23.
Luttge, U., and C. B. Osmond. 1970. Ausrr. J. Biol. Sci. 23:17.
Luttge, U., C. K. Pallaghy, and C. B. Osmond. 1970. J. Membr. Biol. 2:17.
McNaughton, S. J., and L. W. Fullem. 1970. Plant Physiol. 45:703.
McNulty, I. 1969. Page 255 in "Physiological Systems in Semi-Arid Environments". Edited by C. C. Hoff and M. L. Piedesel. Univ. New Mexico Press, Albequerque.
Medina, E. 1971. Carnegie inst. Wash. Yearbook 70:551.
Meir, A., J. Kamburoff, and A. Poljakoff-Mayber. 1971. Ann. Bot. 35:837.
Meiri, A. and A. Poljakoff-Mayber. 1969. Israel J. Bot. 18:99.
Mendoza, M. M. 1971. "The Effects of NaCl on Atomical and Physiological Processes in *Atriplex hastata* L." M. S. Thesis. Univ. Utah, Salt Lake City.

Moore, R. T. 1971. "Transpiration of *Atriplex confertifolia* and *Eurotia lanata* in Relation to Soil, Plant, and Atmospheric Moisture Stresses." Ph.D. Thesis, Utah State Univ. Logan, Utah.
Moore, R. T., S. W. Breckle, and M. M. Caldwell. 1972. Oecologia (in press).
Moore, R. T., and M. M. Caldwell. 1972. *In* "Psychometry in Water Relations Research." Edited by R. W. Brown and B. P. van Haveren. Utah Agric. Exper. Sta. (in press).
Moore, R. T., R. S. White, and M. M. Caldwell. 1972. Can. J. Bot. (in press).
Mozafar, L. A. and J. R. Goodin. 1970. Plant Physiol. 45:62.
Nieman, R. H. 1962. Bot. Gaz. 123:279.
Osmond, C. B. 1963. Nature 198:503.
Osmond, C. B. 1968. Austr. J. Biol. Sci. 21:1119.
Osmond, C. B. 1970. Z. Pflanzenphysiol. 62:129.
Osmond, C. B. and P. N. Avadhani. 1968. Austr. J. Biol. Sci. 21:917.
Osmond, C. B. and H. Greenway. 1972. Plant Physiol. 49:260.
Osmond, C. B., U. Luttge, K. R. West, C. K. Pallaghy, and B. Shacher-Hill. 1969. Austr. J. Biol. Sci. 22:797.
Osmond, C. B., J. H. Troughton, and D. J. Goodchild. 1969. Z. Pflanzenphysiol. 61:218.
Pallaghy, C. K. 1969. Page 57 *in* "The Biology of *Atriplex*." Edited by R. Jones. C.S.I.R.O., Canberra.
Repp, G. I., D. R. McAllister, and H. H. Wiebe. 1959. Agron. J. 51:311.
Scholander, P. F. 1968. Physiol. Plant. 21:251.
Scholander, P. F., E. D. Bradstreet, H. T. Hammel, and E. A. Hemmingsen. 1966. Plant Physiol. 41:529.
Shantz, H. L. and L. N. Piemeisel. 1927. J. Agric. Res. 34:1093.
Slatyer, R. O. 1961. Austr. J. Biol. Sci. 14:519.
Slatyer, R. O. 1967. "Plant-Water Relationships." Academic Press, New York.
Slatyer, R. D. 1970. Planta 93:175.
Stewart, G. R., J. A. Lee, and T. O. Orebamjo. 1972. New Phytol. 71:263.
Strogonov, B. P. 1962. "Physiological Basis of Salt Tolerance of Plants." Edited and translated by Poljakoff-Mayber and A. M. Mayer. 1964. Monson, Jerusalem.
Thomson, W. W., W. L. Berry, and L. L. Liu. 1969. Proc. Nat. Acad. Sci. 63:310.
Tregunna, E. B. and J. Downton. 1967. Can. J. Bot. 45:2385.
Waisel, Y. 1972. "Biology of Halophytes." Academic Press, New York.
Waisel, Y. and G. Pollak. 1969. J. Ecol. 57:789.
Wallace, A. 1970. Soil Sci. 110:146.
Wallace, A. and E. M. Romney. 1972. "Radioecology and Ecophysiology of Desert Plants at the Nevada Test Site." U.S. Atom. Energy Comm., Oak Ridge, Tennessee.
Walter, H. 1970. "Vegetationszonen und Klima." E. Ulmer, Stuttgart.
Warren Wilson, J. 1967. Page 53 *in* "The Collection and Processing of Field Data." Edited by E. F. Bradley. Interscience, New York.

Welkie, G. W., and M. M. Caldwell. 1970. Can. J. Bot. 48:2135.
Whiteman, P. C. and D. Koller. 1967. J. Appl. Ecol. 4:363.
Wiebe, H. H. and H. Walter. 1972. Amer. Midl. Natural. 87:241.
Williams, M. C. 1960. Plant Physiol. 35:500.

SALT TOLERANCE OF MANGROVES AND SUBMERGED AQUATIC PLANTS

Calvin McMillan

The Department of Botany and The Plant Ecology Research Laboratory
The University of Texas at Austin, Austin, Texas 78712

Abstract

Mangroves and submerged aquatic plants, the seagrasses, share a niche attribute of tolerance to broad and rapid changes in salinity. Among five seagrasses studied, *Halodule*, which often occurs as the sole occupant of shallower and more hypersaline bays, has the greatest tolerance of salinity. *Cymodocea* and *Halophila* have the narrowest tolerance ranges, and *Ruppia* is the only one that can survive for extended periods in nonsaline conditions. In order of their decreasing tolerance to low salinity, the five seagrasses are *Ruppia*, *Halodule*, *Thalassia*, *Cymodocea* and *Halophila*. The mangroves also survived rapid changes in salinity but among three species studied, narrower tolerances were shown by *Rhizophora* and *Laguncularia* than by *Avicennia*. All survive for indefinite periods in nonsaline conditions, but *Avicennia* withstands rapid changes to salinities over 118 0/00. Among *Avicennia* plants, those of various age or stage of development showed different salt tolerance. Seedlings and younger plants had greater tolerance to hypersaline conditions. Although salt tolerance has been explored in these estuarine plants, many questions remain concerning the mechanisms underlying the diverse tolerance ranges of the seagrasses and of different ages and stages of development in mangroves.

Introduction

Mangroves and submerged aquatic plants, the seagrasses, share a niche attribute of tolerance for changing salinity. Each has special niche relations to a particular range of salinity, but the limits of tolerance are difficult to establish because of a common tolerance for salinities near that of seawater, and a physiological system that reacts slowly to changes in salinity. The niche relations which permit survival in an estuarine ecosystem also make salt tolerance in these plants difficult to evaluate.

Seagrasses and their habitats have been observed in detail by Phillips (1960) in Florida, Humm (1956) and Simmons (1957) in Texas and Lot-Helgueras (1968) in Mexico, among others. Experimental evaluations of seagrasses include those of Marmelstein et al. (1968) and Fuss and Kelley (1969) in Florida, and McMahan (1968) and McMillan and Moseley (1967) in Texas. The seagrasses include members of five or six plant families. Salt tolerance has been investigated primarily in turtle-grass, *Thalassia*, and in *Halophila* of the Hydrocharitaceae; in Widgeongrass, *Ruppia*, of the Ruppiaceae; in shoalgrass, *Halodule (Diplanthera)* and in manateegrass, *Cymodocea (Syringodium)*, of the Zannichelliaceae.

Mangroves include members of diverse plant families, but the experimental evaluations of salt tolerance have been largely with *Avicennia*, in the Verbenaceae or Avicenniaceae, and *Rhizophora* of the Rhizophoraceae. Of the mangroves, *Avicennia* has received major attention because of its accessibility to temperate stations and because of its wide distribution. The black mangrove, *A. germinans (A. nitida)*, extends into the Gulf Coast of North America to near 30°N latitude. The gray mangrove, *A. marina*, extends to Auckland, New Zealand (38°S) and along the Red Sea (30°N). The plants of western Africa are treated as *A. germinans (A. africana)* and those of eastern Africa as *A. marina*.

Seagrasses

Flowering in *Ruppia* is influenced by salinity. The studies by Setchell (1924) in California and by Bourn (1935) and Phillips (1960) in Florida suggest that flowering is confined to lower salinities, and that seed set occurs at a salinity of 28 0/00 or less. In unpublished studies by this author, California plants from Mendocino Co. and Texas plants from Redfish Bay flowered in tap water. Of the seagrasses, only *Ruppia* can be maintained indefinitely in tap water.

McMahan (1968) studied the salt tolerance of *Halodule* and *Cymodocea* from the lower Laguna Madre of Texas. He grew plants of *Halodule* in tap water, and in a range of salinity, 3.5-87.0 0/00. After a six-week test, plants were still surviving from 3.5 to 52.5 0/00. *Cymodocea* plants kept at 35.0, 44.0 and 52.5 0/00 died after three weeks at the highest salinity and were rated as not vigorous at 44.0. His tests showed a broad salinity tolerance for *Halodule* and narrower tolerance for *Cymodocea*.

McMillan and Moseley (1967) compared the responses of five seagrass species from Redfish Bay, Texas, to increasing salinity using tanks, under out-of-doors conditions at Port Aransas and under controlled photoperiod-temperature conditions. During a 55-day test, *Halodule* leaves showed continuous height increase as the salinity rose to 72 0/00. *Thalassia* leaves showed no further height increase after the salinity reached 60 0/00. *Cymodocea* reached its ultimate height at approximately 40 0/00. *Ruppia* showed a tolerance that was intermediate between *Thalassia* and *Cymodocea*.

Halophila had sporadic survival and its salt tolerance was not determined.

Subsequent unpublished studies by McMillan and Krause have shown that *Halodule* survives at salinities above 72 0/00. In containers of artificial seawater that were allowed to evaporate, *Halodule* retained green leaves in salinities of 70, 71, 73, and 80 0/00. As salinities exceeded 90 0/00 most plants died, but a few had green tissue at 95 0/00. These plants were grown in river sand and artificial seawater (Instant Ocean) under low light intensities in the greenhouse.

In other unpublished studies, *Halodule* and *Halophila* from Redfish Bay were compared under fixed salinities of 23, 37, 50, and 60 0/00. The plants were transferred to pint containers of river sand in late March and kept in battery jars of artificial seawater for three weeks prior to the test. As indicated in Table 1, plants of both *Halodule* and *Halophila* survived for 13 weeks at 23 and 37 0/00. In *Halophila* four of six plants survived two weeks at 60 0/00, but the six were dead by the fourth week. *Halophila* plants that had survived at 37 0/00 in the greenhouse for four months lost most of their chlorophyll within 24 hours after being placed in tap water and dilute sea water, 5 0/00. Those transferred to 10 0/00 had a trace of green tissue for five days, but those at 13 and 18 0/00 had some green leaves after a week. *Halodule* transferred from 37 0/00 retained green tissue in tap water and at 5 0/00 for the two-week test.

Halophila that was transplanted in January flowered profusely after two months in artificial sea water (37 0/00) cultures. The transplanted clones had appeared morphologically different in leaf form and these differences were correlated with staminate and pistillate plants. The January transplants were kept only under one photoperiod, 14 hr. No flowering was recorded among late March transplants under 12-, 14- or 16-hr. photoperiods.

Thalassia and *Cymodocea* were compared in fixed salinities of 5, 10, 13, 18, 23, 37, 50, 60 0/00. The plants had been transplanted to pint containers of river sand and placed in artificial sea water in late March, three months prior to the test. The *Cymodocea* from Port Isabel, Texas, was flowering at the time of transplanting in March. *Thalassia* from Port Isabel flowered (staminate only) during the month after transplanting to 14- and 16-hr conditions and 37 0/00. This is probably the first report for flowering of *Thalassia* in Texas. *Thalassia* from Redfish Bay had no flowering among plants that were transplanted in mid-Jaunary or in late March.

Neither *Thalassia* nor *Cymodocea* plants retained green tissue in a salinity of5 0/00 for more than a few days, but both had some green tissue at 10 0/00. The plants at various salinities 10-50 0/00 survived for two weeks or more but those at the highest salinity, 60 0/00 did not survive or had only a trace of green tissue. In plants at 50 0/00 that were shifted to 5 0/00, the leaves of *Thalassia* remained green for several days but those of *Cymodocea* became brown within 24 hours. The salinity tolerances of *Thalassia* and *Cymodocea* were similar, but the leaves of *Thalassia* reacted more slowly to a rapid change in salinity.

Table 1. Comparison of *Halodule* and *Halophila* plants from Redfish Bay, Texas, under four salinities..[a]

	Salinity (%oo)							
Time (Weeks)	23	37	50	60	23	37	50	60
	Halodule				*Halophila*			
4	V	V	S	S	V	V	S	X
5	V	S	BS	BS	V	V	BS	X
8	S	S	BS-X	X	S	S	BS	X
13	S	S	BS-X	X	BS-S	S	X	X

a. Plants were transplanted to river sand and artificial seawater (Instant Ocean) in a greenhouse at Austin, Texas, on 27 March. They were placed in the graded salinity series under three photo-periods, 12, 14, and 16 hours on 15 April. The plants showed no day length correlations and the results include plants from the three conditions. The symbols are V, vigorous; S, surviving; BS, surviving but not as much green tissue as in S; and X, no green tissue.

The tests of salt tolerance of the seagrasses can be described as exploratory, but they suggest certain natural history facts that correlate with distributional data. The broad salt tolerance of *Halodule* was suggested by its broad distribution. The fact that it is often the sole occupant of shallow estuaries where salinity fluctuates broadly (Simmons, 1957) and the fact that it occurs in mixed stands with the other seagrasses suggests its broad tolerance (Humm, 1956; Simmons, 1957; Phillips, 1960). Its restriction from some areas may be due to factors of sedimentation. We have noticed in our cultures that *Halodule* leaves are easily buried by settling of sand particles. Heavy settling of sand particles covers the short stems and leaves of *Halophila*, but does not affect *Thalassia* or *Cymodocea*. The scarcity of *Cymodocea* in the estuaries of the central Texas coast where salinities fluctuate widely and rapidly and its abundance in the lower Laguna Madre where salinities are more constantly near that of seawater (McMahan, 1968) may result from its greater sensitivity to changes in salinity.

The seagrasses show relatively slow responses to broad changes in salinities above that of seawater (McMillan and Moseley, 1967; McMahan, 1968), but some of them, such as *Halophila* and *Cymodocea*, react rapidly to changes in salinity between 5-13 0/00. The studies under gradually increasing salinity suggest tolerances that exceed those demonstrated in studies of fixed salinity. By combining the results from fixed and varying salinities, *Halodule* showed the broadest salinity tolerances. *Halophila* showed the narrowest range of tolerance, and *Cymodocea* and *Thalassia* were intermediate but slightly different from each other. *Ruppia* has a broad tolerance that ranges from nonsaline to hypersaline conditions.

Whether different populations of a given seagrass differ in their salt tolerance has not been determined. It seems most likely that a cosmopolitan species, such as *Ruppia maritima*, has niche differentiation with respect to a range of habitat conditions, including salinity. In *Halophila*, which shows morphological differentiation even with a local population, niche variation between populations may also be found. The rarity of flowering in *Halodule* and the sporatic flowering of *Thalassia* suggests that the chances for populational differentiation are perhaps least. In these latter two seagrasses, extensive areas, particularly near their northern limits, may be vegetative propogation of a type with relatively uniform niche capacities.

The survival of the seagrasses in artificial sea water and river sand in a greenhouse 200 miles from the Gulf of Mexico suggests that adaptation to estuarine ecosystems involves broad tolerance. The Austin tap water, however, is derived from the Colorado River which eventually empties into Gulf Coast estuaries and the highly calcareous river sand with a pH of 8-8.5 is not unlike the depositions along the Texas coast.

Mangroves

Salt tolerance of mangroves is complicated by the differentiation among them with respect to salt secretion. The salt-secreting group that includes *Avicennia, Aegiceras* and *Aegialitis* has a high xylem sap tension (Scholander, 1968). The non-secreting group that includes *Rhizophora, Bruguiera, Ceriops*, and *Sonneratia*, among others, has a lower xylem sap tension (Scholander, 1968). The former group secretes salt from special glands on the leaves and the ionic composition is similar to that of sea water (Scholander et al., 1962). Salt concentration in sap of the non-secreting group is approximately 10 times less than that of the secreting group.

Various studies have suggested an altered relationship to salinity between the seedling and the parent plant of the two groups. Walter and Steiner (1936) showed that the chloride content differed in various parts of the seedlings and the parent plants. *Rhizophora*, in the non-secreting group, showed less variation in chloride content between seedlings and parent plants, however, the chloride content was considerably lower in the seedlings. In the secreting group, *Avicennia* showed a wide range of chloride among different stages of development. Seedlings of *Avicennia* that were attached to the parent plant had an extraordinarily low chloride level. Walter and Steiner reported that the chloride content rises rapidly as the seedling is rooted but that the chloride level is not very high compared with that of the parent plant. They concluded from their examinations of the field-collected samples that seedlings remove water from sea water.

Scholander (1968) demonstrated that seedlings of *Avicennia* in the salt-secreting group and *Ceriops* in the non-secreting group could remove water from sea water. Using decapitated seedlings, he collected and tested sap using a pressure device. By various treatments he demonstrated that the removal of almost pure water from sea water was accomplished by an ultrafiltration process that continued in dead tissue. Scholander was primarily concerned with the physiological mechanisms and did not emphasize the obvious shift in niche relations to salt from the seedling to the adult stage in *Avicennia*.

There have been various other reports of differences in salt concentration in seedling and parent tissue. Chapman (1962) reported that chloride ions in parent tissue differed from that in seedlings. Lotscherts (1968) suggested that chloride enters seedlings before they drop from the parent plant and he suggested that this is the way in which a young plant is adapted to its new habitat. Genkel (1962) reported that the chloride content increased in sprouts in proportion to size and he concluded that salt tolerance of the plant increased with an increase in size. He also reported that an excess of chlorides in the soil delayed the abscission of the fruit. Chapman (1939) reported that the chloride content in the leaves of *Avicennia germinans* was greater in the highest leaves on the tree, and that chloride concentration varied from leaf to leaf.

Numerous studies have demonstrated that various types of mangroves can be grown almost indefinitely in non-saline conditions. Winkler (1931) reported that *Rhizophora, Avicennia, Ceriops,* among others, thrived in stagnant fresh water, some for up to 15 years. Stern and Voigt (1959) cultured *Rhizophora mangle* from Florida in tap water and various dilutions of artificial sea water. They reported that *Rhizophora* showed its greatest growth in the salinity closest to that of sea water. Conner (1969) cultured *Avicennia marina* in various modifications of Hoagland's solution. He reported that all levels of KCl and $CaCl_2$ suppressed growth, but a positive growth response to $NaCl$ occurred in the range of 10-20 0/00. Although Stern and Voigt (1959) suggested that their highest salinity may have been twice that of sea water as a result of evaporation, most studies have not dealt with hypersaline conditions.

McMillan (1971) tested the effects of salinity, water turbulence, water depth, and temperature on the establishment of *Avicennia germinans* from the central coast of Texas. Salinity was not the chief factor limiting seedling establishment because seedlings rooted in distilled water and in salinities approaching twice the concentration of sea water. Water turbulence, either of distilled or sea water, inhibited root and seedling development. Seedlings tumbled for as much as 12 weeks, but showed rapid root development when stabilized. Various water depths promoted extensive root systems, but seedlings did not become established until water depth was reduced to 5 cm or less. High temperature treatment was lethal to stemless seedlings but not to seedlings with stems and leaves. McMillan (1971) concluded that seedling establishment is probably phenologically timed by environmental relations during winter and early spring as a protection against the lethal effects of high temperature rather than high salinity.

In unpublished studies of *A. germinans* from the central Texas coast, differences in salt tolerance have been demonstrated for plants of different age and stage of development. Plants that were 15 months from rooting in fine sandy loam were severely damaged after 48 hours at salinities 4-5 times greater than sea water (Table 2). Seedlings 3 months from rooting showed no observable effects of the hypersaline condition.

Table 2. Percentage of Avicennia germinans plants from Harbor Island, Texas, with leaf damage and surviving after exposure to various hypersaline conditions.[a]

	Number of Months from Rooting[b]			
	15		3	
	Leaf Damage	Survival[c]	Leaf Damage	Survival[c]
4 Days at 145 /oo	100	33	0	100
2 Days at 170 /oo	66	33	0	100
4 Days at 170 /oo	100	0	0	100
2 Days at 190 /oo	100	0	0	100

a. Plants potted in fine sandy loam were exposed to hypersaline conditions in sets of three plants. The pots were submerged so that only the stem and leaves were above the hypersaline seawater prepared from artificial sea salts (Instant Ocean). The leaf damage and survival was recorded during the exposure to seawater and during the subsequent period in nonsaline conditions.
b. Although the seedlings were derived from Harbor Island, they were beach collections and, therefore, represent mixed progenies.
c. Seedlings planted in river sand were severely damaged at all salinities and none survived.

Another comparison under varying hypersaline conditions showed similar results with *Avicennia* (Table 3). Plants that were 18 months from rooting in fine sandy loam showed severe leaf damage after 6-8 days in salinities above 150 0/00 and none of them survived. Plants 6 months from rooting showed slight chlorosis after 10 days at the high salinities and showed increasing chlorosis subsequently under nonsaline conditions. Among 2-month plants, neither chlorosis nor death of leaf tissues was evident during 10 days under hypersaline conditions. However, some showed slight to severe wilting of leaves and bending of stems. After two weeks in nonsaline conditions, most surviving seedlings regained their normal appearance.

Avicennia plants rooted in fine sandy loam and in sand were compared at various salinities. Those in the loam showed the above tolerances but those in the sand had no added protection against hypersaline conditions.

Macnae (1963), in his studies in South Africa, mentioned that mangroves usually grew in mud and he suggested that there might be a protective influence by the highly calcareous muds. While this may offer a partial explanation, the Colorado River sand that was used in the above comparison is largely composed of limestone fragments. This calcium sand showed no protective influence in a comparison with plants placed directly into hypersaline conditions without sand or soil.

In an exploratory comparison, *Rhizophora*, *Laguncularia*, and *Avicennia* differed in response to hypersaline conditions. The year-old plants rooted in fine sandy loam were moved from nonsaline conditions to 118, 135 and 152 0/00 for 48 hours and then returned to nonsaline conditions. *Laguncularia* plants from Florida showed wilting of leaves and stem tips under hypersaline conditions. Leaves began dropping 36 hours after the return th nonsaline conditions. *Rhizophora* plants from Florida showed wilting of the leaves and branches 36 hours after their return to nonsaline conditions. After an additional five days all leaves dropped from *Rhizophora* plants, the apices shriveled, and none of the plants survived. The *Avicennia* plants from Texas showed slight leaf mottling at the highest salinity, but all survived and maintained their original leaves.

The colonizing ability of various mangroves varies significantly and their patterns of distribution probably suggests a relationship to salt tolerance. Mangroves such as *Rhizophora* which tend to occupy an outer zone are more constantly exposed to salinities near that of sea water. An inability to cope with hypersaline conditions may restrict adult plants to sites which are regularly covered by sea water. *Avicennia* occupies various sites and is often intermixed with *Rhizophora* as well as *Laguncularia*. It is the sole occupant of sites such as those along the coast of Texas where salinities fluctuate broadly. The salt tolerance of *Avicennia* seedlings suggests extremely broad occurrence and the tolerance of *Avicennia* adults permits survival at hypersalinities that are potentially damaging to *Rhizophora* and *Laguncularia*.

The studies of Atkinson et al. (1967) on salt regulation provide insight into salt tolerance of mangroves. They demonstrated in *Aegialitis*, a salt-secreting

Table 3. Percentage of <u>Avicennia germinans</u> plants with leaf damage and surviving after exposure to two increasing hypersaline conditions for 10 days.[a]

Number of Months From Rooting[b]	Date of Leaf Damage or Wilting	Salinity (°/oo)			
		150-165		155-170	
		Leaf Damage	Survival	Leaf Damage	Survival
18	2	0		0	
	4	100A		100A	
	6	-	0	50B	0
	8	100B		100B	
	10	-		-	
6	2	0		0	
	4	0		0	
	6	0	100[c]	0	100[c]
	8	0		0	
	10	0		66A	
3	2	0		0	
	4	0		0	
	6	30W	100	60W	66
	8	30W		60W	
	10	30W		60W	

a. Plants were potted in fine sandy loam. The pots were submerged so that only the stem and leaves were above the hypersaline seawater. The date for the first appearance of leaf discoloration is denoted by A, and the date for total disappearance of green tissue in the leaves is denoted by B. The wilting of plant without the loss of green tissue is indicated by W. The survival rating is for the number of plants with green leaves 14 days after their return to nonsaline conditions.

b. The 18-month and 6-month plants were from the beach of the estuary delta, Harbor Island. The 3-month plants were from a Gulf of Mexico beach collection on Mustang Island.

c. All of the plants 6 months from rooting became very chlorotic after being returned to nonsaline conditions. Two plants (66%) from the 155-170 ppt treatment lost most of the leaves after 2 weeks in nonsaline conditions.

plant, that the ratio of Na/K in leaves was 3, in xylem sap 8 and in sea water 50. They noted that most of the salt secretion was NaCl and as a result there was a buildup of divalent ions in the leaves, particularly Mg which was five times greater. They showed in *Rhizophora mucronata* that the Na/K ratio was lower in younger leaves (2.5-3.6) than in older leaves (7-14). Their study suggested that *Rhizophora* leaves maintained a constant level of chloride through exclusion and that *Aegialitis* balanced the input of chloride by secretion.

A salt-secreting plant such as *Avicennia* under hypersaline conditions increases secretion up to a tolerance level. The seedlings of this plant, however, using an exclusion mechanism described by Scholander (1968) can survive beyond the tolerance level of the salt-secreting adults. The chlorosis that results in the seedlings, however, suggests some difficulty in ionic uptake possibly involving Mg or Fe under hypersaline conditions. The wilting of the youngest seedlings in hypersaline conditions suggests that the ultrafiltration process in very small stems cannot extract water at a sufficiently high rate to meet demands. The slow recovery from wilting after being returned to nonsaline conditions indicates that some damage to the system may have resulted from the exposure to hypersaline conditions. In adult plants the gradual necrosis of leaves, first involving leaves near the stem apex, suggests that leaves on the same plant may differ in their response to hypersaline conditions. The niche relations in *Avicennia* that vary from seedling to adult suggest adaptive strategies that have permitted this type of mangrove to exceed all others in latitudinal distribution and salinity tolerance.

The niches of mangroves and the submerged aquatic plants, the seagrasses, have been explored for salt tolerance. At this point it is evident that more laboratory evaluations of niche relations are needed. As indicated by Connor (1969) in his *Avicennia* study, it is difficult to conclude anything about the physiological response of mangroves from observations of their distribution. The salinity at any specific site can vary widely from one day to the next, particularly in estuaries at the higher latitudes. Also, reports of the salinity for the day of observation and the occurrence of a particular plant may not provide clues to niche relations to salinity. The studies of Walter and Steiner (1936), however, are an example of the useful information that can be obtained in the field by careful observers. Their measurements of chloride levels in the water and soil as well as in diverse parts of the mangrove plants provide remarkable insight into the different salt relations of seedling and adult plants. The fact that estuarine plants can be grown in the laboratory with relative ease as a result of their broad niche capacity should provide impetus to more studies of specific niche relations of mangroves and seagrasses.

Literature Cited

Atkinson, M. R., G. P. Findlay, A. B. Hope, M. G. Pitman, H. D. Saddler, and K. R. West. 1967. Salt regulation in the mangroves *Rhizophora mucronata* Lam. and *Aegialitis annulata* R. Br. Austral. J. Biol. Sci. 20:589-599.

Bourn, W. S. 1935. Sea-water tolerance of *Ruppia maritima* L. Boyce Thompson Inst. Contri. 7:249-255.

Chapman, V. J. 1939. Cambridge University Expedition to Jamaica. Part 2. A study of the environment of *Avicennia nitida* Jacq. in Jamaica. Journ. Linn. Soc. (Bot.) 52:448-485.

Chapman, V. J. 1962. Respiration in mangrove seedlings. Bull. Mar. Sci. Gulf and Carib. 12:137-167.

Connor, D. J. 1969. Growth of grey mangrove (*Avicennia marina*) in nutrient culture. Biotropica 1:36-40.

Fuss, C. M., and Kelly, J. A. 1969. Survival and growth of seagrasses transplanted under artificial conditions. Bull. Mar. Sci. Gulf and Carib. 19:351-365.

Genkel, P. 1962. On the ecology of the mangrove swamp. Akad. Nauk. S.S.S.R. (Moscow) 223-232. (Abstract from Referat. Zhur. Bio. 1962. 7V151.)

Humm, H. J. 1956. Seagrasses of the northern Gulf coast. Bull. Mar. Sci. Gulf and Carib. 6:305-308.

Lot-Helgueras, A. 1968. Estudios sobre fanerogamas marinas en las cercanias de Veracruz, Ver. Biologo Tesis. Universidad Nacional Autonoma de Mexico, D. F.

Lotschert, W. 1968. Speichern die keimlinge von Mangrovepflanzen Salz? Umsch. wiss. Tech. 68:20-21.

Macnae, W. 1963. Mangrove swamps in South Africa. J. Ecol. 51:1-25.

Marmelstein, A. D., P. W. Morgan, and W. E. Pequegnat. 1968. Photoperiodism and related ecology in *Thalassia testudinum*. Bot. Gaz. 129:63-67.

McMahan, C. A. 1968. Biomass and salinity tolerance of shoalgrass and manteegrass in Lower Laguna Madre, Texas. J. Wildl. Manag. 32:501-506.

McMillan, C. 1971. Environmental factors affecting seedling establishment of the black mangrove on the Central Texas coast. Ecology 52:927-930.

McMillan, C. and F. N. Moseley. 1967. Salinity tolerances of five marine spermatophytes of Redfish Bay, Texas. Ecology 48:503-506.

Phillips, R. C. 1960. Observations on the ecology and distribution of the Florida seagrasses. Professional Papers Series,Fla. Board Conserv. 2:1-72.

Scholander, P. F. 1968. How mangroves desalinate water. Physiol. Plant 21:251-261.

Scholander, P. F., H. T. Hammel, E. Hemmingsen, and W. Carey. 1962. Salt balance in mangroves. Plant Physio. 37:722-729.

Setchell, W. A. 1924. *Ruppia* and its environmental factors. Proc. Nat. Acad. Sci. 10:286-288.
Simmons, E. G. 1957. An ecological survey of the upper Laguna Madre of Texas. Publ. Inst. Mar. Sci. Univ. Tex. 4:156-200.
Stern, W. L. and G. K. Voigt. 1959. Effect of salt concentration on growth of red mangrove in culture. Bot. Gaz. 121:36-39.
Winkler, H. 1931. Einige Bemerkungen uber Mangrove-Pflanzen und den Amorphophallus Titanum im Hamburger botanischen Garten. Ber. deut. bot. Ges. 49:87-102.
Walter, H. and M. Steiner. 1936. Die Okologie der Ost-Afrikanischen Mangroven. Zeit. Bot. 30:65-193.

PART III. HABITAT ASSOCIATIONS OF HALOPHYTES

MATHEMATICAL MODELING — *SPARTINA*[1]

Robert J. Reimold

The University of Georgia Marine Institute
Sapelo Island, Georgia 31327

Introduction

The translation of physical, chemical, and biological ecosystem components into a set of mathematical relationships may yield an abstraction of the ecosystem. The resultant abstraction or model attempts to represent the structure and function of the real world system. Models are constructed by using a number, or series of numbers, to represent each part or component of the system. Two rather separate strategies have evolved in the development of ecological models: compartmental and experimental.

The compartmental approach emphasizes the quantitative properties of the compartments of the ecosystem. These models are usually expressed as a system of fairly simple differential equations. This is the type of approach to ecological modeling utilized by Patten (1971), Odum (1972), and Van Dyne (1969). Contrasted with this is an approach which focuses on the detailed analysis of ecological processes. Here emphasis is placed on the interactions and systems equations rather than on identification of quantitative measurements to describe the state of the system. This experimental component approach has been proposed by Holling (1966).

Some of the more recent ecological models have integrated the attributes of both types of models. In these models, emphasis is not only placed on prediction of the systems reaction of the system to manipulations and perturbations, but also on data description and summary.

Van Dyne (1969) has presented an integrated model of the grasslands ecosystem which was designed to evaluate the major aspects of energy flow. This integrated type of model demonstrates a rational balance between realism, generality, and precision. Recently, several comprehensive reviews of mathematical models in ecology have been prepared by Walters (1971) and Kadlec (1971).

One of the few attempts to incorporate halophytes and their surrounding fauna and flora into a compartmental model has been made by Pomeroy *et al.* (1972). A model was developed from interactions between real world research and simulation by a variety of digital and hybrid methods, starting

[1] Contribution No. 257 from The University of Georgia Marine Institute.

This research has been supported in part by U. S. Atomic Energy Commission Contract AT 38-1-(639).

with a linear approach and concluding with an introduction into a totally nonlinear model. In the nonlinear system, the transfer coefficients are proportional to the amount of material in the compartment suffering the loss. The model utilized the solution of a Taylor series in two variables on the IBM 360-65, an application previously described by Glass (1967) and Conway et al. (1970). This Taylor series expansion is often used in classical solutions of differential equations. Here, if f (x,y) is an analytic function, the successive derivatives of y (x) may be obtained, and the series for y (x) written out in standard Taylor format. The subroutine is based on the historical method of Euler which employs a discrete set of y_k values for the arguments x_k, using the differential equation

$$y_{k+1} = y_k + h\, f(x_k, y_k)$$

where $h = x_{k+1} - x_k$. This theorem guarantees the existence of a unique solution of the classical problem under the hypothesis on f (x,y).

The purpose of this paper is to present detailed information on a five compartment model which is sensitive to discrete changes in *Spartina alterniflora*, the salt marsh cordgrass. A portion of the research focused on mathematical manipulation of the model, the other on field validation of the mathematical manipulation.

Methods and Materials

The basic mathematical construction of the nonlinear model has been described by Pomeroy et al. (1972) (Fig. 1). The nonlinear model of five compartments (X_1 = water, X_2 = sediments, X_3 = *Spartina*, X_4 = detritus and microflora, and X_5 = detritus feeders) is represented by the following five differential equations:

$$dx_1/dt = k_1 * x_4^2 / x_4(0)^2 - (k_1 + k_2) * x_1 * x_4 / (x_1(0) * x_4(0)) \\ + k_2 * x_4 * x_5 / (x_4(0) * x_5(0)) + R * x_3 / x_3(0) \quad (1)$$

$$dx_2/dt = (k_1 + k_2) * x_1 * x_4 / (x_1(0) * x_4(0)) - C * x_2 / x_2(0) \\ - R * x_2 / x_2(0) \quad (2)$$

$$dx_3/dt = C * x_2 / x_2(0) - D * x_3 / x_3(0) + R * x_2 / x_2(0) - R * x_3 / x_3(0) \quad (3)$$

$$dx_4/dt = D * x_3 / x_3(0) - (k_1 + k_2) * x_4^2 / x_4(0)^2 \quad (4)$$

$$dx_5/dt = k_2 * x_4^2 / x_4(0)^2 - k_2 * x_4 * x_5 / (x_4(0) * x_5(0)) \quad (5)$$

Here C, D, and R are oscillatory transfer functions with a period equal to one year. The values of C and D are taken from Odum and Smalley (1959), and R is from Reimold (1972).

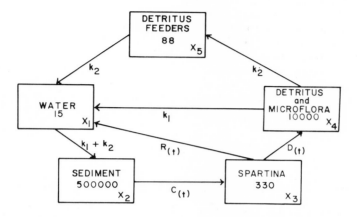

Figure 1. Nonlinear model of the flux of phosphorus through the salt marsh ecosystem.

A number of the mathematical manipulations have used a leak term equal to:

$$-0.1 * (x_1 - 3.0)$$

inserted in the right hand side of the first equation (1). This term represents a supposed loss of phosphorus to the sea whenever the concentration in the marsh water exceeds 3 mg m^{-2}.

The solution of the five differential equations is done by a two-term Taylor series with a one-day interval. The final output, covering 1,500 days, is produced by a Calcomp plotter at 30-day intervals with traces for compartments $x_1, ..., x_5$ appearing on the plot. Because the vertical scaling for the five phosphorus concentrations is usually different, it is necessary to run three distinct computer operations. Stage one solves the differential equations, stage two is designed to introduce the necessary scaling, headings, and footnotes for the plot, and stage three plots the complete resultant figure with headings, scaling, etc.

To simulate the removal of *Spartina* by harvesting, a three phase program was developed in which the normal seasonal change of phosphorus in the marsh grass began on day 1 (1 January) and at a specified day the grass was reduced to 0.1 which mathematically simulated the mowing of the grass. Regrowth of the grass was programmed using data from Smalley (1958). Mathematical manipulation of the "cutting and regrowth" phase of *Spartina* was evaluated for response of *Spartina* harvesting for day 90 (30 March), 120 (29 April), 150 (29 May), 180 (28 June), and 210 (28 July), and its resultant effect on phosphorus concentrations in each of the five compartments for a four-year period.

Since some compartments are more easily manipulated and measured (in the field), X_1 (water) and X_3 (*Spartina*) were initially chosen for comparisons between the model and the real world. In order to accomplish this field testing, four 0.4 hectare research plots were constructed in a pure stand *Spartina alterniflora* salt marsh by constructing dikes. The perimeter of the plot was diked with the exception of a 20 m opening (Fig. 2). Normal tidal influence was maintained by natural tributary creeks entering through these openings. Foot bridges were constructed from high ground to the middle of the opening in each plot. Automatic samplers, which collect 500 ml water samples at hourly intervals for 24 hours, were installed on each bridge.

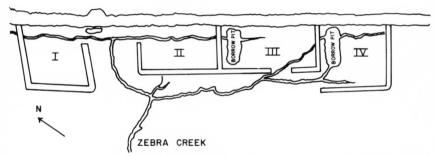

Figure 2. Diagram of four 0.4 hectare plots depicting dike surrounding all but 20 m perimeter of each plot. (Note two plots have borrow pits within.)

Color infrared photographic remote sensing of salt marsh productivity (Reimold *et al*. 1972, Gallagher *et al* 1972) was utilized six times during the past year to evaluate any abnormal short term changes in the production of *Spartina*. Low level precision photography was also employed as a novel method for determination of volume in each of the experimental plots. After establishing four level control points in the harvested plots (using a three wire level network tied into a primary U.S.G.S. bench mark), a stereo pair of aerial black and white infrared photographs was compiled in a stereo plotter to provide the x (width), y (length), and z (depth) coordinates. These were quantified with an H. Dell Foster quantilizer to yield watershed volumes per increment of tide height. Coupled with this system was a tide level recorder which allowed accurate volume computation each day of water sample collection. Phosphorus data, volume, and tide data were recorded on IBM cards, and a program written to compute transport of phosphorus in and out of the experimental plots.

Phosphorus concentrations of the marsh water were monitored using a

Technicon II autoanalyzer and the phosphate routine according to Technicon Instruments Corp. (1971). The chemistry was based on that introduced by Murphy and Riley (1962) and the autoanalysis methodology as modified by Cherry (1972) and the author. The modifications included: (1) expansion of the range to 0 to 20 ug at P/1, (2) use of 5.5:1 cam on the sampler, and (3) use of 4.0 ml Levoir IV per 1000 ml of water diluent. Organic matter was digested by the persulfate technique of Menzel and Corwin (1965) or the ultraviolet digestion technique of Armstrong and Tibbitts (1968). The ultraviolet technique was modified by limiting air circulation around the UV lamp until the lamp reached 285 volts. Four drops of 30% hydrogen peroxide were added to each sample prior to the two-hour digestion period (Thomas 1971). All dissolved phosphorus samples were filtered through a Reeve Angel® Ultra glass filter with 50 mm Hg vacuum. Each field water sample was analyzed for dissolved inorganic phosphorus, dissolved total phosphorus, particulate phosphorus, and total phosphorus.

In early spring 1972, a field program was initiated to compare phosphorus concentrations in each of the four research plots. In order to field validate mathematical perturbations, it was necessary to remove the *Spartina* from the salt marsh and follow changes in phosphorus concentration of the marsh water. On 29 May 1972 (day 150), all the aerial portions of *Spartina* were harvested from two of the four experimental plots. A Gravely® tractor equipped with dual wheels and sickle bar mower was used to mechanically harvest the marsh grass. Where the substrate was too soft for mechanical harvesting hedge clippers were employed. The harvested *Spartina* was left on the research plots. A high spring tide and southeasterly breeze that evening carried all the grass out of one plot. In the other harvested plot, the wind and tide piled the grass along the side away from the opening. This fortuitous act resulted in one of the two harvested plots being cleared of all cut grass, and the other experimental plot being "enriched" with potential detritus. The two control plots remained unaltered. Following the harvesting of *Spartina*, water samples were collected at hourly intervals for a 24-hour period every three days.

Results

The initial results of the five compartment nonlinear model over four years (Fig. 3) demonstrate that there is a periodicity in each of the compartments and that it assumes a logical sequence (i.e. increase in phosphorus in *Spartina* (X_3), followed by an increase in the detritus (X_4) and water (X_1) compartments). The sediment compartment (X_2) contains most of the standing stock of phosphorus and undoubtedly gives the great stability to the rest of the system.

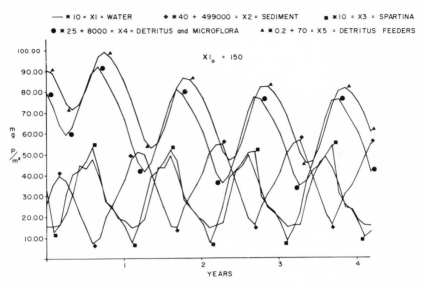

Figure 3. Four year plot of phosphorus in each of the five compartments. This represents the solution resulting from the nonlinear, undisturbed initial conditions.

The growth and regrowth of *Spartina* has been a major concern. Using a mathematical representation of growth and regrowth after clipping, a deterministic growth function was incorporated into the differential equations. This growth, clipping, regrowth function based on data from Smalley (1958) demonstrated that *Spartina* would grow back differently depending on what time of year it was harvested. Since *Spartina* is the most significant primary producer, it helps drive the flux of phosphorus through the rest of the ecosystem (Pomeroy *et al.* 1969; Reimold, in press). Fig. 4 depicts the predicted effect of cutting the marsh grass on day 120. Mathematical simulation of the cutting of the grass reduced the resultant concentrations of phosphorus in the water during the remainder of the first year. By the end of the first year the flux of phosphorus through the *Spartina* component nearly returned to its normal level.

The effect of cutting the *Spartina* on day 120 was also evaluated with the addition of a leak function in the water (X_1) (Fig. 5). The leak (proportional to X_1 and positive when X_1 was greater than 3.0 and negative when X_1 was less than 3.0) was designed to simulate input into the ocean during periods of high run-off and consequently a net seaward transfer. The result of this leak over a four-year period was a damping effect on the oscillation of phosphorus in the water (X_1) (Fig. 5). A very small decrease in the phosphorus content of the detritus and microflora was suggested for the four-year period. The sediment compartment (X_2) remained unchanged.

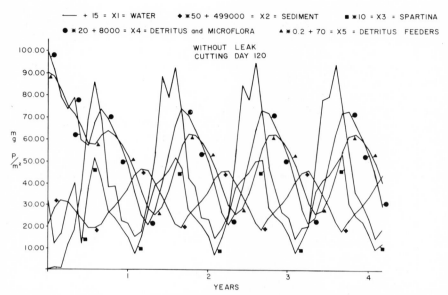

Figure 4. Four year plot of phosphorus in the five compartments resulting from mathematical simulation of harvesting the *Spartina* at day 120.

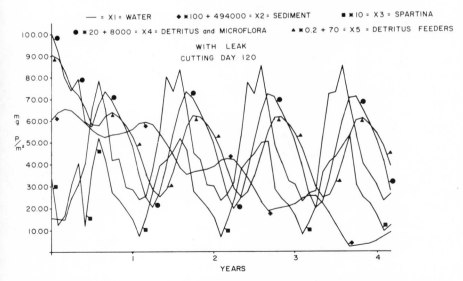

Figure 5. Four year plot of phosphorus in the five compartments resulting from mathematical simulation of harvesting the *Spartina* at day 120. This also incorporates a leak of phosphorus out of the systems from X_1 (water).

The predicted effect of cutting the *Spartina* on day 150 is presented in Fig. 6. This appeared to produce more pronounced changes in the water compartment than any of the other theoretical cutting days tested; however, it is also incorporated the leak in X_1 (water). To contrast this, the mathematical prediction of harvesting on day 150 was again run; however, the leak function in the water was omitted (Fig. 7). This resulted in less decrease in standing stock in all of the compartments and represented a more steady state solution. Since the addition of the leak did not appear to alter the overall response of the model significantly, further manipulation was conducted without the leak in the water compartment.

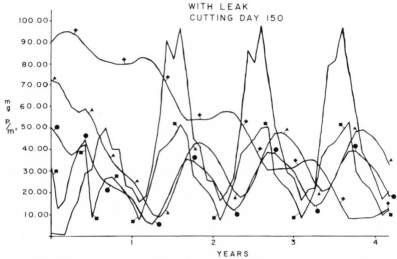

Figure 6. Four year plot of phosphorus in the five compartments resulting from mathematical simulation of harvesting the *Spartina* at day 150.

The field program initiated in early spring 1972 to compare the phosphorus concentrations in the four research plots suggested that there was initially no difference in phosphorus concentrations in the water in the four plots. The *Spartina* was harvested from plots I and IV (Fig. 2) on day 150. Aerial photography (for remote sensing productivity) taken one month after harvesting revealed very little regrowth of *Spartina* in the two experimental plots and confirmed the presence of the dead, cut grass stacked alongside the edge of the one experimental plot. The photography also indicated that the two control plots were unaltered and remained similar in "remote sensed productivity" to the surrounding undisturbed salt marsh. To date, there has been no significant change in production that can be related to the construction of the dikes. Furthermore, each of the plots and dikes has

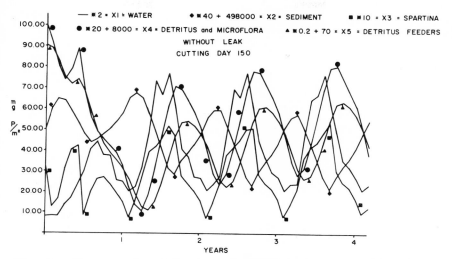

Figure 7. Four year plot of phosphorus in the five compartments resulting from mathematical simulation of harvesting the *Spartina* at day 150. This also incorporates a leak of phosphorus out of the system from X_1 (water).

stabilized and has withstood storm tides and spring tides.

The early results of phosphorus analyses suggest that the water phosphorus concentrations are similar to those predicted by the model. The diel fluctuations in phosphorus concentration in the water follow a pattern of high concentrations at low tide and low concentrations at high tide. The total phosphorus concentration of water in experimental plot I was contrasted to control plot II (Table 1). This comparison was made on the basis of samples collected at low water (corrected to volume per unit area). These observations are also contrasted to the mathematical prediction of phosphorus content in the control and experimental plots. Generally, the field concentrations are higher than the predicted concentrations but the concentrations of phosphorus in the experimental plots (both from field measurements and from mathematical predictions) are higher than the control plots.

Discussion

This model has been founded on field research following the flux of phosphorus through the marsh ecosystem (Pomeroy *et al.* 1969) including the salt marsh cord grass (Reimold 1972). The model has followed a formidable mathematical development and analytical validation (Pomeroy *et al.* 1972) suggesting that this is a relatively simple but stable ecosystem. The computer manipulation has permitted testing the effect of many perturbations. By selecting (from the resultant mathematical plotback) perturbations which produced demonstrable changes in the system, field experiments were implemented to test the mathematical model. The early results of the field

Table 1. Comparison of predicted phosphorus concentration* for control and experimental plots contrasted with field measured concentrations. D O Y = day of year

EXPERIMENTAL			CONTROL	
computer prediction	field result	D O Y	field result	computer prediction
150		001		150
168		030		156
168		060		152
272		090		262
344		120		329
	840	123	148	
480		150		461
	1559	151	855	
616		180		440
	1990	184	1094	
856		210		454
	3670	213	2058	
896		240		375
768		270		242
760		300		245
520		330		173
512		360		165

* Phosphorus concentration mg P/ M^2

testing appear to agree with the model's predictions. The differences between the two results will be utilized to refine the mathematical model.

The model described above gives a predictive insight into the real world and extends our intuitive understanding of the flux of phosphorus through the system. McRoy (1970) has similarly demonstrated the effect of long term climatic changes on the distribution of eel grass. Williams and Murdock (1972) have modeled the production of needlerush (*Juncus roemerianus*) in North Carolina marshes and contrasted the results of the model with field observations on the undisturbed system. Teal (1962) made an early effort to assemble the components of the *Spartina* salt marsh into a working model. His model was based on the flux of energy. Odum (1970) assembled a model of the flux of energy through a south Florida estuary. This system was based on productivity of phytoplankton and mangroves.

Another approach to modeling involves economic considerations of *Spartina*. Since the *Spartina* interfaces with both terrestrial and marine ecosystems, it has multiple use values. It is an inseparable part of the estuarine ecosystem since estuarine waters depend on the export of detritus from the *Spartina* for sustained productivity (Odum and de la Cruz 1967). Another use of the coastal marshes in England and the United States has been to supplement the grazing food chain with *Spartina*. Ranwell (1967) reports that *Spartina* is palatable and digestible for sheep and beef cattle. Taschdjian (1954) has demonstrated that microbial conversion of *Spartina* can result in a significant increase in protein concentration. Kirby (1971) recently demonstrated that *Spartina* could be converted into a protein rich microbial concentrate (50% protein) with an efficiency of nearly 50%.

Economic considerations of an acre of salt marsh are summarized in Table 2. Most of these economic values are in some way dependent upon the *Spartina*. Since the oysters are filter feeders and, consequently, dependent on the *Spartina* detritus, they rely on a certain finite input of the detritus for their existence. Microbial conversion of the detritus to single cell protein and the water quality improvement both depend on *Spartina* for substrate.

The sum of all these estimates results in an economic model of an acre of *Spartina alterniflora* situated in coastal Georgia. If this first approximation of estimate of worth is equated over 20 years, its value would be one hundred thousand dollars. This still is far from being a good estimate of worth because it needs better validation. Nonetheless, it serves as an illustration of an economic model.

The models cited above have been established on a variety of data bases. The common denominators of these systems are the first and second laws of thermodynamics. Regardless of whether we consider the biosphere, the ecosystem, or the organism, they all possess the same thermodynamic characteristics which allow the direct comparison of descriptive quantitative measurements (calories, phosphorus, carbon, etc.).

In another example of ecosystem modeling, Forrester (1971) presented a simulation of the world based on a five compartment system with five first order differential equations. His system was devised to employ the computer to deduce consequences of the whole system from a set of relationships which were not intuitive. In several ways his model was similar to the one considered herein. Boyd (1972) considered that Forrester's assumption was valid only if the computer solution was insensitive to different sets of plausible assumptions. If the results of the model are not what the technological optimist would anticipate, then the model would indeed be of merit according to Boyd's view (1972). The *Spartina* ecological model also meets the requirements of the results being different from what intuition alone would suggest.

There are few examples where modeling has been utilized to solve basic or applied ecological problems. Most all the technology used in systems ecology has been borrowed from other disciplines. There remains a need to develop methodology that is appropriate for ecology and to field test the results in the real world. When this has been done, systems analysis may add as much to

Table 2. Estimates of potential annual value of an acre of marsh.

	Annual Value
Aquaculture--moderate culture level [1]	
Oyster meat production	1800 lb.
Dockside value @35¢	$630
Value added during processing (70% dockside value)	440
Total	$1070
Assume 4 acres marsh per acre oyster rafts	
Net Value	$262 per acre marsh
Aquaculture--intensive raft culture of oysters [2]	
Oyster meat production when stocked at 1/10 water surface	4500 lb.
Dockside value @35¢	$1575
Value added during processing (70% dockside value)	1100
Total	$2675
Assume 4 acres marsh per acre oyster rafts	
Net Value	$670 per acre marsh
Grazing by beef cattle	
7500 lb. dry wt. grass x 10% conversion [3]	1000 lb. animal tissue
Wholesale value @35¢	$350
Value added in processing	115
	$465
Intensive microbial conversion to single-cell protein (50% protein)	
7500 lb. x 0.4 = raw high protein product	3000 lb.
Value @15¢ [4]	$450
Water quality improvement (waste assimilation)	
19 lb. BOD removed per day @4¢ per lb. for incremental secondary treatment [5]	$247
Economic contribution of allowing 1 mg./m reduction in minimum level of dissolved oxygen from 4 mg./l to 3 mg./l [6] (increasing BOD loadings by 4.3 lb./day)	$360
"Life support" value [7]	
Based on net primary production of 4×10^6 gm. dry wt.	$1600
Based on gross production (assuming net production x 2)	$3200
Total	$4054 to $5654

[1]/ from Advisory Committee on Mineral Leasing, 1968.
[2]/ see J. E. Bardach, 1968.
[3]/ see Ranwell, 1967.
[4]/ see Callihan and Dunlap, 1971.
[5]/ present BOD "nature" removes in 5 mid-Atlantic estuaries (Sweet, 1971).
[6]/ data for Delaware estuary (Sweet, 1971).
[7]/ see Odum and Odum, 1972.

fundamental ecological understanding as radio-nuclide tracers did in the past 20 years.

Literature Cited

Advisory Committee on Mineral Leasing. 1968. A report on the proposed leasing of state-owned lands for phosphate mining. University of Georgia.

Armstrong, F. A. J., and S. Tibbitts. 1968. Photochemical combustion of organic matter in sea water for nitrogen, phosphorus and carbon determination. J. Mar. Biol. Assoc. U. K. 48:143-152.

Bardach, J. E. 1968. Science 161:1102.

Boyd, R. 1972. World Dynamics: A note. Science 117(4048):516-519.

Callihan, C. D., and C. E. Dunlap. 1971. Construction of a chemical-microbial pilot plant for production of single-cell protein from cellulosic wastes. U. S. Environmental Protection Agency. SW-24C.

Cherry, B. 1972. Personal communication.

Conway, G. R., N. R. Glass, and J. W. Wilcox. 1970. Fitting nonlinear models to biological data by Marquardt's algorithm. Ecology 51:503-507.

Forrester, J. W. 1971. World Dynamics. Wright-Allen Press. Cambridge, Mass.

Gallagher, J. L., R. J. Reimold, and D. E. Thompson. 1972. Remote sensing and salt marsh productivity. Proceedings 38th Annual Meeting. American Society of Photogrammetry. Washington, D. C. Pages 338-348.

Glass, N. R. 1967. A technique for fitting nonlinear models to biological data. Ecology 48:1010-1013.

Holling, C. S. 1966. The strategy of building models of complex ecological systems. Pages 195-214 in Systems Analysis in Ecology. Edited by K. E. F. Watt. Academic Press, New York. 276 pp.

Kadlec, J. A. 1971. A partial annotated bibliography of mathematical models in ecology. Analysis of Ecosystems. International Biological Program. School of Natural Resources, The University of Michigan, Ann Arbor.

Kirby, C. J. 1971. The annual net primary production and decomposition of the salt marsh grass *Spartina alterniflora* Loisel. in the Barataria Bay estuary of Louisiana. Doctoral dissertation, Louisiana State University. 74 pp.

McRoy, C. P. 1970. On the biology of eelgrass in Alaska. Doctoral dissertation, University of Alaska.

Menzel, D. W., and N. Corwin. 1965. The measurement of total phosphorus in seawater based on the liberation of organically bound fractions by persulfate oxidation. Limnol. Oceanogr. 10(2):280-282.

Murphy, J., and J. P. Riley. 1962. A modified single solution method for the determination of phosphate in natural waters. Anal. Chim. Acta 27:30.

Odum, E. P., and A. A. de la Cruz. 1967. Particulate organic detritus in a Georgia salt marsh estuarine ecosystem. Pages 383-388 in Estuaries. Edited by Lauff. AAAS Publ. No. 53.

Odum, E. P., and H. T. Odum. 1972. Natural areas as necessary components of man's total environment. North American Wildlife Conf. (in press).

Odum, W. E. 1970. Pathways of energy flow in a south Florida estuary. Dissertation. University of Florida. 180 pp.

Patten, B. C. (ed.). 1971. Systems Analysis and Simulation in Ecology. Academic Press, New York.

Pomeroy, L. R., R. E. Johannes, E. P. Odum, and B. Roffman. 1969. The phosphorus and zinc cycles and productivity of a salt marsh. Symposium of Radioecology. Edited by D. J. Nelson and F. C. Evans. U.S.A.E.C. TID-4500. Pages 412-419.

Pomeroy, L. R., L. R. Shenton, R. D. H. Jones, and R. J. Reimold. 1972. Nutrient flux in estuaries. Pages 274-293 in Nutrients and Eutrophication. Edited by G. E. Likens. American Society of Limnology and Oceanography. Special Symposia Volume 1.

Ranwell, D. S. 1967. World resources of *Spartina townsendii* and economic use of *Spartina* marshland. J. Applied Ecol. 4(1):239-256.

Reimold, R. J. 1972. The movement of phosphorus through the salt marsh cord grass, *Spartina alterniflora* Loisel. Limnol. Oceanogr.

Reimold, R. J., J. L. Gallagher, and D. E. Thompson. 1972. Coastal mapping with remote sensors. Proceedings of Coastal Mapping Symposium. American Society of Photogrammetry, Falls Church, Va. Pages 99-112.

Smalley, A. E. 1958. The role of two invertebrate populations, *Littorina irrorata* and *Orchelimum fidicimum*, in the energy flow of a salt marsh ecosystem. Doctoral dissertation. Zoology Department, University of Georgia, Athens.

Sweet, D. C. 1971. The Economic and Social Importance of Estuaries. Environmental Protection Agency, Water Quality Office, Washington, D. C.

Taschdjian, E. 1954. A note on *Spartina* protein. Econ. Bot. 8:164.

Teal, J. M. 1962. Energy flow in the salt marsh ecosystem of Georgia. Ecology 43:614-624.

Technicon Instruments Corp. 1971. Orthophosphate in sea water. Autoanalyzer II Industrial Method No. 155-71W. Technicon Instruments Corp., Tarrytown, New York. 2 pp.

Thomas, J. P. 1971. Release of dissolved inorganic matter from natural populations of marine phytoplankton. Mar. Biol. 11(4):311-323.

Van Dyne, G. M. 1969. Grassland management, research and teaching viewed in a systems context. Range Science Department. Science Series No. 3. Colorado State University. 39 pp.

Walters, C. J. 1971. Systems Ecology: the systems approach and mathematical models in ecology. Pages 276-292 in Fundamentals of Ecology. 3rd ed. Edited by E. P. Odum. W. B. Saunders Co., Philadelphia. 574 pp.

Williams, R. B., and M. B. Murdock. 1972. Compartmental analysis of the production of *Juncus roemerianus* in a North Carolina salt marsh. Chesapeake Science 13(2):69-79.

THE ROLE OF OVERWASH AND INLET DYNAMICS IN THE FORMATION OF SALT MARSHES ON NORTH CAROLINA BARRIER ISLANDS

Paul J. Godfrey and Melinda M. Godfrey

Department of Botany/U. S. National Park Service
University of Massachusetts
Amherst, Massachusetts

Introduction

Although the roles played by oceanic overwash and inlets in building salt marshes on barrier islands have been mentioned in geological publications (Fisher 1962, Kraft 1971), there is little information available on the ecological significance of these processes. Godfrey (1970) discussed some general aspects of overwash ecology. This paper will describe studies in the vicinity of Core Banks, a low barrier island where overwash and inlet closure are the major ways in which new salt marshes are formed.

Core Banks, the largest island of Cape Lookout National Seashore, lies between 76°10'–76°33'W and 34°35'–35°N in Carteret County, North Carolina; Cape Lookout forms its southern end. The island is about 50 km long and from 0.5 to 2 km wide including marsh islands, and is broken by Drum Inlet. Swash Inlet, now closed, once divided Core Banks from Portsmouth Island, and Barden Inlet separates Core Banks from Shackleford Banks, also part of the National Seashore. The islands are undeveloped except for scattered fishing camps, and a few more substantial buildings on Cape Lookout. In contrast to other barrier islands to the north and west, the study area is in a relatively natural state. The berm crest is up to 2 m above mean sea level; the interior of the island is about 1 m above mean sea level, with dunes rising about a meter or two higher. A wide berm (up to 150 m) is characteristic of the island, with low, scattered dunes behind it, formed primarily by *Spartina patens*, along with *Uniola paniculata* and other dune species. Back of the dunes, flat grasslands and shrub thickets grade into salt marshes dominated by *Spartina patens* and *Juncus roemerianus* at the higher levels, and *Spartina alterniflora* at the lower. Core Sound, a shallow lagoon with broad underwater beds of *Ruppia maritima* and *Zostera marina*, separates Core Banks from the mainland. Normal tidal range is about 1 m on the ocean side, and somewhat less in the sounds. Sound tides are greatly influenced by winds and by distance from a major inlet. Figure 1 is a general map of Cape Lookout National Seashore; Figure 2 is an aerial view of the primary overwash study area at Codds Creek on Core Banks.

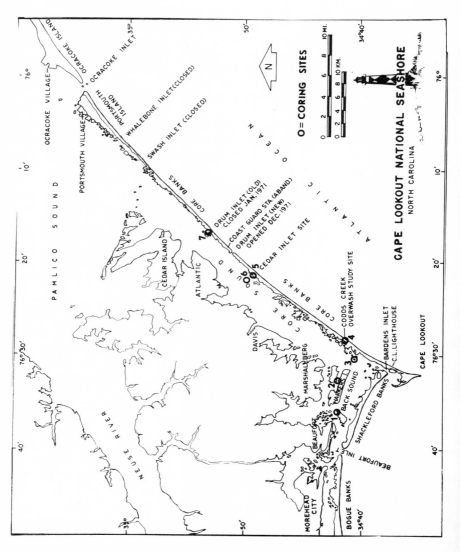

Figure 1. Map of Cape Lookout National Seashore; numbered circles refer to coring sites.

Figure 2. Core Banks at the Codds Creek study area. Dotted lines refer to the cross-sectional profiles shown in Figures 4 and 5.

Methods

Changes in the physiography of the island were documented by studying aerial photographs from 1939 to the present, and by examining old maps and records of former inlets. The stratigraphy of Core Banks and nearby marsh islands was determined from piston cores whose sites are shown in Figure 1. Peat and sand samples were retained from each core for further analysis. Determination of above ground standing crop, expressed as oven dry weight, were taken as the average of three randomly placed 0.25 m^2 quadrats. Topographic profiles across Core Banks were determined with a transit, using elevations from bench marks established by the U. S. Army Corps of Engineers in 1960. Carbon 14 analysis was used to date one peat sample (U.S. Geol. Survey Sample W-2307).

Results and Discussion: Overwash

Changes in salt marsh borders at Codds Creek, determined from aerial photographs taken in 1939 and 1967, are shown in Figure 3. In 1939 the creek water extended about 150 meters farther into the land than it does now. Storms in the 1950's washed beach sand onto the marsh bisected by Transect 1 and into the creek to the left of the transect. The dotted line that crosses the transect represents the boundary between the marsh and the open sand of the berm and dunes in 1939. Today the filled parts of the creek and former marsh support salt marsh, grassland, and dune vegetation. There was also a net retreat of the beach from its position in the 1939 photograph. Line 2 is a transect through a new salt marsh which has grown up during the last decade on the overwash fill.

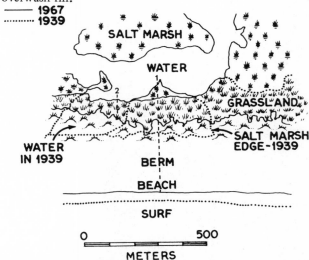

Figure 3. Changes at Codds Creek as determined from aerial photographs taken in 1939 and 1967. Dotted lines represent borders of the major zones in 1939, solid lines are the borders in 1967. Dashed lines indicate transects and profiles described in the text.

Figure 4 shows a stratigraphic profile of Transect 1, as interpreted from cores at 30 meter intervals, and a generalized vegetation sequence as it relates to the topography of the island. Core No. 4 was taken on the bare berm and shows upper layers of beach sand and shell. About a meter below the surface is a stratum of *Juncus roemerianus–Spartina patens* peat, the surface of a former high salt marsh. Below the peat are more layers of beach sand and oceanic shells such as *Spisula, Donax, Mactra,* and *Anatina*. Core No. 5, from the overwash pass between the dunes, shows strata similar to those in No. 4, including the high marsh peat. At position No. 6 the sand-shell strata are 95 cm thick and overlie a zone of organic matter and dark gray sand, which appears to be a continuation of the organic layers in No. 4 and No. 5. Below the gray sand are two layers of *Spartina alterniflora* peat from a low salt marsh. The first peat layer was marsh in the 1940's, according to the aerial photographs. The edge of this marsh is about where the dotted line parallel to the beach in Figure 3 crosses Transect 1. Below the first layer of peat in Core No. 6 are another zone of beach sand and shells, a second layer of peat, and then a zone of fine gray sand. The lowest peat layer represents a marsh which apparently predates the marshes of this century; it was buried by overwash deposits and then a new marsh developed on top of the older one.

Cores No. 7 to 10 show successively thinner upper layers of beach sand and shells going toward the existing salt marsh. All these cores contained peat at the same level as the first peat layer in Core No. 6; the surface of this older marsh is below the level of the present marsh, although the two marshes are continuous and are presumably of the same initial age. The present marsh undoubtly continued to accumulate organic matter vertically after the overwashed part of the marsh was buried; part of the depression is also probably due to compaction of the peat by the overlying sand. Underlying the peat in Cores No. 7 to 10 are zones of mixed peat and sand, and then homogeneous fine gray sand; the overwash deposits under the peat of No. 6 did not get as far as No. 7.

The material overlying the old surfaces now supports five vegetation types. Dunes forming on the overwash deposits are dominated by *Spartina patens* and *Uniola paniculata*, with other dune species present. The overwash passes and fans, which constitute the most recent terraces, support a sparse grassland of *S. patens* and *Solidago sempervirens* with low standing crop (0.047 kg/m^2). Toward the back of the overwash terraces is a closed grassland dominated by *S. patens* and *Fimbristylis spadicea*, with a number of associated species, having the highest standing crop (1.475 kg/m^2) on this transect. Where the overwash deposits are flooded by spring tides, there is a high salt marsh with dense stands of *Spartina patens, Fimbristylis,* and *Juncus roemerianus*. On the outer edge of the overwash are salt pannes containing *Salicornia* spp. and *Limonium carolinianum*.

The overwash sediment that filled part of Codds Creek was bare in 1958, but now supports salt marsh (Figure 3). Figure 5 contrasts a new marsh

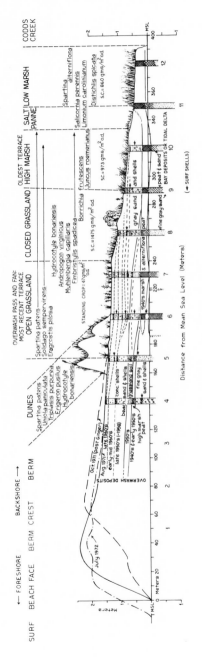

Figure 4. Stratigraphic and vegetation profile across Core Banks at Codds Creek. Dated lines connecting various strata in the profile are probable surfaces as interpreted from aerial photographs and core data. Arrows indicate locations from which oceanic shells were identified. The solid line shows the profile as it was in August 1971, when vegetation data and cores were obtained. The dashed line shows the profile after Hurricane Ginger on 30 September 1971. The third profile is the beach as it appeared in July 1972. The surface has been lowered from the post-Ginger overwash profile by wind removal of fine sands. Standing crop data were obtained in 1970.

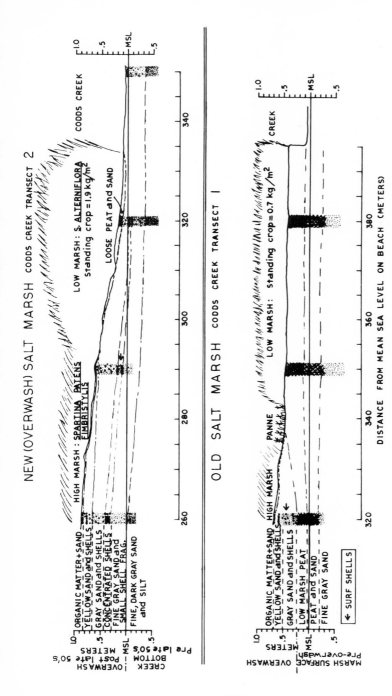

Figure 5. Comparative transects and stratigraphy across a salt marsh that developed on late 1950's overwash material, and an older marsh nearby, not completely covered by overwash. Standing crop data were collected in 1971.

profile (Transect 2) with an older marsh profile (Transect 1). It is on Transect 2 that the role of overwash in creating new marshes is evident. Cores from this transect show upper layers of overwash material, with the fine gray sand underneath. In the 14 years since 1958, *Spartina alterniflora* has colonized the new material and attained a standing crop of 1.9 kg/m^2. The grass grows in its tall form, over 2 meters high, and has built up about 10 cm of new peat on top of the overwash sand. The overwash material slopes gradually into the water, providing a surface on which the grass can advance into the creek. Frequent tidal flooding on this slope benefits the marsh, and the sound ecosystem as well, since the water carries detrital nutrients away to the estuary and allows marine organisms to feed and take shelter among the *Spartina* stems.

The older marsh of Transect 1 (and much of the sound side of Core Banks) is bordered by an erosion scarp. The marsh can no longer advance into the creek, and waves are actually cutting away the peat. The marsh can go on growing vertically; but since the surface is now so high and the edge eroding, the grass roots cannot readily get down to the creek bottom. There will be no more significant lateral growth unless overwash brings in sand to provide new substrate. The *Spartina* that grows here on the old, eroding peat is of the short form and has a standing crop of only 0.7 kg/m^2; not only is productivity lower, but the organic matter reaches the estuary less regularly since tidal flooding is less frequent. Overwash is thus one process which creates the salt marsh fringe and keeps the system highly productive.

Aerial photographs show overwash deposition all along the Shackleford Banks–Core Banks–Portsmouth Island chain, much of it during the past decade. Most of the sand that was washed back into the sounds during the 1950's today supports well developed *Spartina alterniflora* marsh. Figures 6 and 7 show changes during recent years at the site of the Atlantic Coast Guard Station on Core Banks. Figure 6 was taken in 1963, soon after the station was decommissioned. The harbor was then still usable by boats; note the mooring posts and sea wall. There was a great deal of bare sand due to storms and grazing. (Livestock was removed in the late 1950's.) Figure 7, taken in 1970, shows a great increase in grass cover, most of it on sand dropped by several overwashes. An active overwash channel has emptied a large fan of new sediment, now being colonized by *Spartina alterniflora*, into the sound at the lower right corner of the photograph. Salt marsh grass has invaded the new sand where deep water was maintained in the harbor until about 1958, and the marsh fringe has increased all along the back of the island. The arrow indicates the same mooring post as in Figure 6 (the station burned in 1968).

Spartina alterniflora was transplanted into a 2m by 5m plot on overwash sand at the Atlantic Coast Guard Station site (Figures 8 and 9). Individual plants were taken from the marsh nearby and spaced 0.5 m apart.

After two growing seasons, the grass had increased until the planted plot was indistinguishable from the surrounding natural marsh. Under favorable

Figure 6. The Atlantic Coast Guard Station on Core Banks in 1963. Note mooring posts and sea wall.

Figure 7. The Atlantic Coast Guard Station on 22 January 1970. The harbor has been partially filled in with overwash sand and new salt marsh. A recent overwash fan can be seen on the right. The small square dark patch near the marked mooring post is the marsh planting shown in the ground views in Figures 8 and 9. Ice covers the water of Core Sound.

Figure 8. **Spartina alterniflora** planted on 8 April 1969 at the site of Atlantic Coast Guard Station.

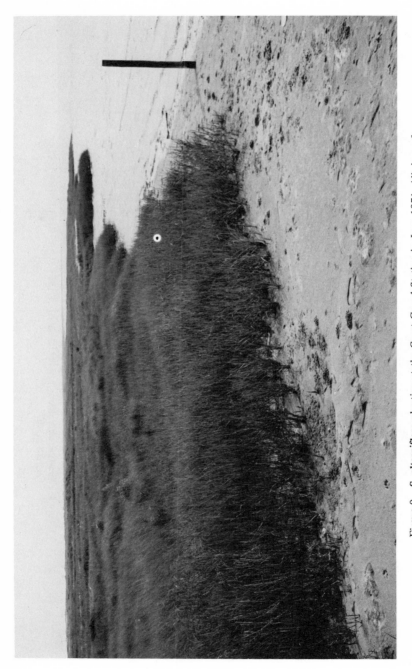

Figure 9. **S. alterniflora** planting at the Coast Guard Station in June 1971. All signs of artificial planting have been obliterated. Note the increase of the natural marsh in the background compared to Figure 8.

conditions, marsh grass invades new sediments at a rate of about a meter/year, mainly by vegetative growth; the rhizomes begin to advance outward from existing marsh as early as April. *Spartina* is now being used experimentally in coastal management for stabilizing dredge spoil and beaches on the sound side of barrier islands (Woodhouse, Seneca and Broome 1972); the technique can be highly successful if the substrate is at about mean sea level. In some cases, such reclamation may help compensate for loss of marshes through dredging and filling.

Submergence of Uplands

Rising sea level has caused the conversion of low woodlands to salt marsh at some places on the Outer Banks, as well as elsewhere in North Carolina (Adams 1963). The east end of Shackleford Banks was well wooded in the late 1800's, but one now finds scattered stumps in salt marshes where low lying forests once existed. Salt marshes between wooded dune lines on the western half of Shackleford also contain stumps of drowned woodlands. Chapman (1960) describes the flooding of former uplands in New England and the creation of salt marshes as sea level rises.

Inlet Dynamics

Fisher (1962) described the geomorphology of relic inlets on the Outer Banks and located evidence for 14 such inlets along Core Banks alone. He pointed out such features as former inlet channels extending up into the banks, and salt marsh islands that apparently developed on former tidal delta shoals behind inlets that are now closed.

During the course of our study, Drum Inlet completed the closure pattern described by Fisher. This inlet opened during the 1933 hurricane after having been closed for a number of years. Extensive tidal deltas built up behind the inlet, and in the late 1950's the updrift side began migrating southward in the general direction of the littoral current. The downdrift side eroded, exposing layers of peat under the berm and dunes, which shows that the sand was an overwash deposit on top of salt marsh. As the shoals behind the inlet built up enough to be dry at low tide, they began to be colonized by *Spartina alterniflora*. By January 1971, the updrift side of the inlet had migrated 2 km south of its 1958 position; a month later the inlet closed. For nearly a year, the sound behind the former inlet was affected more by wind rather than by lunar tides. This meant that the shoals would be out of the water during a northeast or east wind. Figure 10 shows the changes at Drum Inlet from 1958 on, as seen from aerial photographs; with the inlet closed, *Spartina alterniflora* continued to invade the shoals. When surveyed in 1971, the grass had covered small intertidal shoals and the outer edges of the larger ones. The standing crop of this new marsh grass was as much as 0.4 kg/m^2 in the denser stands. Low *Spartina patens* dunes had begun to develop on some of the exposed shoals, and *Ruppia maritima* and *Zostera marina* were colonizing the permanently submerged ones. Thick beds of *Zostera* were found near the letter "h" in "channel" in Figure 10.

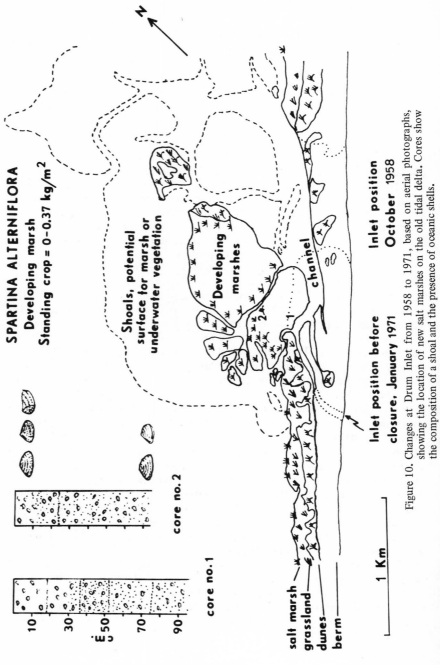

Figure 10. Changes at Drum Inlet from 1958 to 1971, based on aerial photographs, showing the location of new salt marshes on the old tidal delta. Cores show the composition of a shoal and the presence of oceanic shells.

Figure 10 also illustrates the contents and locations of two cores from the Drum Inlet area. Beach sand and oceanic shells were rather homogeneously mixed except for ilmenite bands, shown as dotted lines in the core profile.

Fisher determined that Cedar Inlet was open for about 100 years in the 1700's and 1800's. Early records place it about 10 km south of Drum Inlet, and maps of 1850 labeled its location even though the inlet was closed by then. Its site can be located readily today by matching the marsh islands on the 1850 maps of Core Sound with those on modern maps. The inlet site and three cores from the marshes behind are shown in Figure 11. The peat layer in Core No. 1 was about 20 cm thick, with a peat-sand mixture going down to 50 cm. Below this was fine gray sand with oceanic shell fragments. Core No. 2 showed about 10 cm of very fine white sand over marsh peat, which went down to 50 cm and contained silt and fine sand throughout. Below the organic layers were fine gray sand and oceanic shells as in Core No. 2. Core No. 3 was very similar to the other two.

An inlet is usually the result of storm winds driving water over the island and piling it up in the sound, often weakening the island at some narrow point. As the storm passes, the water may cut open an inlet as it rushes back over the island to the sea; the opening is more accurately an outlet than an inlet. Clearly, the opening and closing of such a passage is an important event on a barrier island. While the inlet is open, it allows a biologically vital exchange of water masses, nutrients, and organisms between the sound and the sea. When it closes, the marshes which appear on the delta shoals add to the productivity of the sounds. These marshes are eventually connected to the rest of the barrier by overwashes, significantly widening the narrow island. The salt marsh fringe and marsh islands so characteristic of the low lying Outer Banks are largely due to inlet dynamics and overwash deposits in the lagoons behind.

An experiment in inlet management is now underway on Core Banks. When Drum Inlet closed in January, 1971, the Corps of Engineers cut a new channel 4.5 km to the south, where the island was narrow and had no shoals behind it. Most of the dredge spoil was spread out behind the island at the proper elevation to be planted with salt marsh grass. The pattern taken by sand movement and deposition was similar to that seen at natural inlets. Re-establishment of tidal flow in Core Sound should aid the growth of the natural marshes at old Drum Inlet as well as that of the man-made ones at new Drum Inlet.

Further evidence for the youth of the Outer Banks salt marshes comes from cores taken near the mainland. The positions of four cores from behind the barrier islands (Nos. 1, 2, 3, and 6) are shown in Figure 1. The greatest depth of peat turned up in Core No. 2, from a marsh behind Harkers Island, an area of Pleistocene age (Hoyt and Henry 1971). Here pure peat was nearly 90 cm deep, with peat, sand, and silt mixtures throughout the rest of the 135 cm core length.

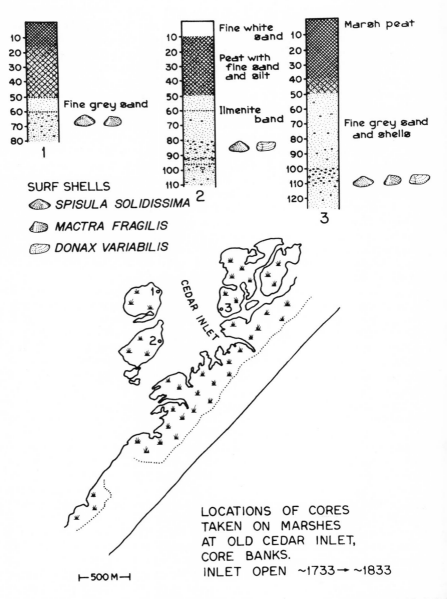

Figure 11. Map of the former Cedar Inlet, with cores taken in the marshes behind the barrier island.

The next deepest peat layer (Core No. 1) came from Middle Marsh, a salt marsh complex in Back Sound behind Shackleford Banks. Here the peat was 80 cm deep, underlain by fine gray sand without shells. This second oldest marsh could have grown up on a tidal delta that formed at a time when Beufort Inlet may have been east of its present position. Sea level was then at least 80 cm below its present stand. Wooded curving dune lines on the western half of Shackleford Banks, opposite Middle Marsh, suggest that this whole section of the island formed as a westward growing spit. It seems unlikely that Middle Marsh was formed on shoals built of sediments brought into Back Sound by North River (east of Beaufort), since this river is short and carries little sand. The lack of shells in Core No. 1 might seem to point to a nonoceanic source for the sediment, but it is hard to imagine what other source could have supplied so much material. In the opinions of Dolan (1971) and Pilkey (1972), the fine shell fragments that would have reached the site of the core have dissolved in the highly acid conditions under the marsh since the inlet was in this region. This may also explain the lack of shells under the marshes at Codds Creek.

Core No. 3, containing the next deepest peat stratum, came from Cockle Marsh Island, a salt marsh 1.75 km southeast of Harker's Island and 0.75 km northwest of Core Banks; that is the most distant marsh island which can be associated with the Core Banks system. Peat was 60 cm deep in this core, with a peat-sand mixture down to 80 cm and traces of organic matter as deep as 100 cm. Again, no shells were found under the peat.

Finally, Core No. 6 came from a *Ruppia-Zostera* bed in the shallow water of Core Sound about 2 km back from the beach at the Cedar Inlet site. This core was 40 cm long and consisted entirely of fine gray sand with no shells, the same material as that under Middle Marsh and the older marshes on Core Banks.

Carbon 14 dates obtained by Newman and Munsart (1968) show marsh peat deposits in Virginia 2.8–2.9 m below the present marsh surface to be about 2,500 years old. Interpolation on their submergence curve gives ages of about 800 years for peat 1 m thick, and approximately 300 years for a 0.5 m thick layer. If Cedar Inlet closed about 140 years ago, as reported by Fisher (1962), then the 0.2 m depth of pure peat found there would fit into Newman and Munsart's curve, giving a submergence rate of 7–8 cm/100 years. This would put the older Codds Creek marshes, now being covered by overwash, in the 300 year old class. (An inlet could have existed here prior to that time, when sea level was about 25 cm below its present position.) The peat and sand mixture underlying the pure peat strata probably represents the root zone at the time when the marshes were developing on the sand substrate. This means that the marshes are somewhat older than the peat layer alone would indicate, since pure peat is deposited on a mature marsh surface. However, the beginning of the pure peat stratum is probably a more reliable, although somewhat conservative, point from which to date a marsh. Attempts to date peat from our cores and from outcrops on the beach were abandoned when a sample from the beach on Shackleford, which also contained stumps, turned out to be less than 200 years old, too young to yield reliable C-14 dates.

As one goes from the mainland to the Outer Banks, the depth of the salt marsh peat, and hence the age of the marshes, decreases, with the youngest marshes growing on overwash sediments. Most marshes appear to be related in some way to the location of former inlets, and continuous overwash is slowly moving the barrier beach and marsh fringe over the older tidal delta marshes.

Summary

Analysis of time sequence aerial photographs, and field studies at Codds Creek, Cedar Inlet and Drum Inlet, suggest that the present pattern of salt marshes behind the Outer Banks is the result of overwash deposition and inlet dynamics. Oceanic overwash provides sand from the beach during every severe storm, while inlet closure forms new marshes at much longer intervals. Once the inlet is closed, however, marsh formation proceeds rapidly. Both processes are of prime importance and complement each other; submergence of uplands by rising sea level also leads to new salt marshes, but this is a more important process on the mainland than on the Outer Banks.

Figure 12 summarizes the processes involved in marsh formation and barrier island retreat. All stages shown can be found on the modern Outer Banks, so they do not necessarily have to occur at the same time. The stage in which a given section of the Outer Banks may be, at a particular time, depends on factors such as original elevation, orientation, exposure to storms, sand supply, and human impact.

Stage 1 is a hypothetical barrier island, which may have existed anywhere from present time to several thousand years ago, with a well developed dune line, or series of dunes, and a forest behind. In stage 2, the sea level has risen slightly and storms have knocked the dune barrier back into the woodlands. By stage 3, much of the barrier island has been overwashed and the dunes pushed back. The marsh has grown vertically, and has undergone some erosion, and former uplands are now salt marsh as a result of sea level rise. In stage 4, the barrier has retreated considerably from its original position. Dune and overwash sand has moved completely over the old forest, which is now exposed on the ocean side. Marshes near the island interior have been covered as well. Further retreat places sand completely over the original marsh surface and into the lagoon behind, where new marshes form. At stage 6, an inlet has opened and a typical tidal delta has appeared behind it. The inlet has closed in stage 7, and the tidal delta now supports salt marsh and low dunes. Overwashes have tied the marsh islands to the main barrier, and have filled in the old channels by stage 8. The salt marshes are now well developed on the old tidal delta, woods have grown up on the low dunes on these marsh islands, the salt marsh fringe behind the barrier is expanding, and on the barrier itself new dune lines and vegetation have appeared where only a short time ago there was water.

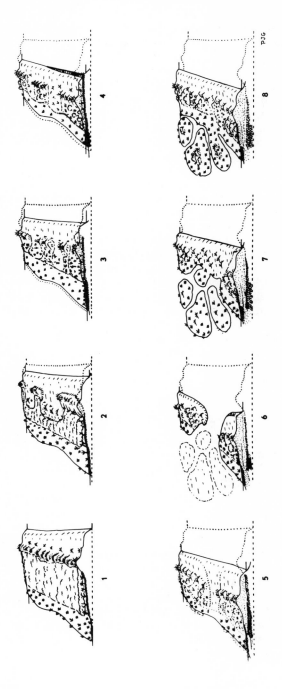

Figure 12. Generalized summary of barrier island dynamics, illustrating response of the island to rising sea level and storms, the formation of marshes by overwash, the development of an inlet, and the appearance of new marshes following closure of the inlet. Vertical scale and sea level rise are exaggerated.

Dune lines on the barrier help protect the marshes behind from excessive and continuous overwashes which might overwhelm the ability of the marshes to recover, but they do not stop recession. Thus, all parts of the barrier island system—beach, berm, dunes, grasslands, marshes and underwater vegetation—are interrelated and dependent on each other in some way. In manipulating systems of this kind, these interrelationships are too often overlooked. The processes of overwash and inlet dynamics are vital features of the environment on low barrier islands, and provide the substrate for the development of their eco-systems under present conditions. Any plans for management of barrier islands of this type must recognize and respect these natural processes.

By stage 8, the barrier island has completed one "roll over" from its original position as sea level continues to rise. (Should sea level stabilize, there might be some recession as a result of severe storms, but it would occur more slowly, if at all. Falling sea level would cause a seaward widening of the islands and drying of the present marsh system.) While the original barrier has been obliterated, nothing has really been lost since a new system has developed in response to changing conditions and estuarine productivity has been maintained. By such distinctive adaptations to this rolling over, barrier island ecosystems can persist through time in a kind of homoeostasis. The salt marshes we see today on the Outer Banks are young and are derived almost exclusively from overwashes and inlets.

Acknowledgements

This research was supported by the Office of Natural Science, U. S. National Park Service, Washington, D. C. and the Cape Lookout National Seashore, Beaufort, North Carolina. We thank G. A. Godfrey for helpful comments on the manuscript.

References

Adams, D. A. 1963. Factors influencing vascular plant zonation in North Carolina salt marshes. Ecol. 44(3):445–456.

Chapman, V. J. 1960. Salt Marshes and Salt Deserts of the World. Leonard Hill (Books) Ltd., London. 392 pp.

Fisher, J. J. 1962. Geomorphic expression of former inlets along the Outer Banks of North Carolina. M. A. Thesis, Univ. of North Carolina, Chapel Hill. 120 pp.

Dolan, R. 1971. Personal communication.

Godfrey, P. J. 1970. Oceanic Overwash and its Ecological Implications on the Outer Banks of North Carolina. Office of Natural Science, National Park Service, Washington, D. C. 37 pp.

Hoyt, J. H. and V. J. Henry, Jr. 1971. Origin of capes and shoals along the southeast coast of the U. S. Geol. Soc. Am. Bull. 82:59-60.

Kraft, J. C. 1971. Sedimentary facies patterns and geologic history of a holocene marine transgression. Geol. Soc. Am. Bull. 82:2132-2158.

Newman, W. S. and C. A. Munsart. 1968. Holocene geology of the Wachapreague Lagoon, eastern shore peninsula, Virginia. Mar. Geol. 6:81–105.

Pilkey, O. 1972. Personal communication.

U. S. Army, Corps of Engineers. 1964. Ocracoke Inlet to Beaufort Inlet, North Carolina Combined Hurricane Survey and Beach Erosion Report. U. S. Army Eng. Dist., Wilmington, N. C.

Woodhouse, W. W., Jr., E. D. Seneca, and S. W. Broome. 1972. Marsh building with dredge spoil in North Carolina. Bull. 445, Agr. Exp. Stat., N. C. State Univ., Raleigh. 28 pp.

PROBABLE AGENTS FOR THE FORMATION OF DETRITUS FROM THE HALOPHYTE, *SPARTINA ALTERNIFLORA*

Mallory S. May, III

Brunswick Junior College
Brunswick, Georgia

The trophic significance of detritus [ecologically defined by Odum and de la Cruz (1963) and Darnell (1967a)] in the food webs of terrestrial and marine communities is well established (Darnell, 1958, 1967b; Teal, 1962; E. P. Odum and de la Cruz, 1967; Heald and W. E. Odum, 1970; and E. P. Odum, 1971). Halophytes, such as *Spartina, Juncus, Distichlis, Salicornia* and *Rhizophora*, of the intertidal marsh and marine communities are major contributors of detritus (Macnae, 1956, 1957; Teal, 1962; Odum and de la Cruz, 1967; and Heald, 1969). There is a paucity of information available about the detritus contributions of *Salicornia, Allenrolfea, Sarcobatus* and *Atriplex*, primary-producer halophytes of the salt desert communities (Chapman 1960).

Microbial decomposition has been generally accepted as the major cause of dead-plant degradation to detritus (Burkholder and Bornside, 1957; Teal, 1962; de la Cruz, 1965; Darnell, 1967; and E. P. Odum and de la Cruz, 1967). Recently, Heald (1969) has discovered that several macroinvertebrates such as amphipods *Melita nitida* and *Corophium lacustre* and a xanthid crab, *Rithropanopus harrisii* are major degraders of red mangrove, *Rhizophora mangle*. Fenchel (1970) discovered that an amphiopod, *Parhyallela whelpleyi*, assisted in the degradation of turtle grass, *Thalassis testudinum*.

The purpose of this work was to investigate probable agents of detritus formation from smooth cordgrass, *Spartina alterniflora*, the dominant grass of the intertidal marshes of Georgia. The work included: a determination of the feasibility of SEM (scanning electron microscopy) as a tool for observing *in situ* microbial organisms during *Spartina alterniflora* decomposition; a SEM examination of sequential degradation of *Spartina alterniflora* culms in litter bags placed in the marsh; and an examination of the detritus producing role of *Cleantis planicauda*, an estuarine isopod, and several other macroinvertebrates associated with dead *Spartina alterniflora*.

METHODS

In December 1971 live smooth cordgrass, *Spartina alterniflora* used for the two SEM studies was cut into 18 centimeter sections, air dried, weighed, placed in 1.5 mm^2 PVC (polyvinyl chloride) screen litter "envelops" sealed with PVC solvent and placed in three locations (tidal creek, tidal creek slope and surface of high marsh) in a typical Georgia *Spartina* marsh. At various intervals through May 1972, litter bags were removed from their marsh locations. The decomposing *Spartina alterniflora* was air dried, weighed and prepared for SEM examination following the methods of Todd, Humphreys, and E. P. Odum (1972).

Cleantis planicauda with dead *Spartina alterniflora* culms were collected from the Georgia estuarine system at the confluence of the Frederica and MacKay rivers with a twenty-foot otter trawl. *Cleantis planicauda* were placed in 15-centimeter diameter culture dishes containing membraned filtered sea water (1.2 u Millipore membrane filter) and short lengths of *Spartina alterniflora* culms. Control culture dishes were without *Cleantis planicauda*. At 48 h intervals the isopod and culm were removed, the water was filtered through a preweighed, 1.2 u membrane filter (Millipore), air dried, and weighed. Twelve replicas for each of two experiments were done.

Cleantis planicauda and several macroinvertebrates found in the litter bags were fixed in Heidenhain's Suza, sectioned and stained with Hematoxylin and counter stained with Eosin. The gastro-intestinal tracts of these animals were examined for plant fragments.

OBSERVATIONS AND RESULTS

Figure 1, 2, and 3 are SEM photomicrographs. Figure 1 shows fungal hyphae in the *Spartina* leaf air passages and conducting vessels. Figure 2 is a set of photomicrographs showing *in situ* bacteria. Figure 3 is a "time lapse" study of the decomposition of *Spartina* culms from the litter bags.

The upper left photomicrograph of Figure 1 shows a single fungal hypha spreading into an air passage. The lower right photomicrograph of Figure 1 is of a different culm exposed in the marsh two months longer than the culm section shown in the upper left. Note the increase in density of fungal hyphae in the lower right compared to the upper left photomicrograph. The intervening photomicrographs are of culms of shorter durations of marsh exposure.

The bacteria seen in the upper left photomicrograph of Figure 2 were "seeded" on the sample. Prior to SEM examination, the *Spartina* culm sample was "rolled" on a Petri dish culture inoculated with decomposing *Spartina*. All of the other photomicrographs were of "nonseeded" culms. The lower right photomicrograph also shows a single, collapsed fungal hypha. Note the ridges perpendicular to the length of the hypha.

The upper left photomicrograph of Figure 3 shows only the epidermal surface of a *Spartina* leaf and the ridges of the leaf vascular system. The upper right, lower left and lower right photomicrograph of Figure 3 show epidermal surfaces of leaves exposed in the marsh for progressively longer periods of time. Note the density of the fungal hyphae and the "frayed" appearance of the epidermis in the lower right photomicrograph.

From the laboratory studies of *Cleantis planicauda* maintained in culture

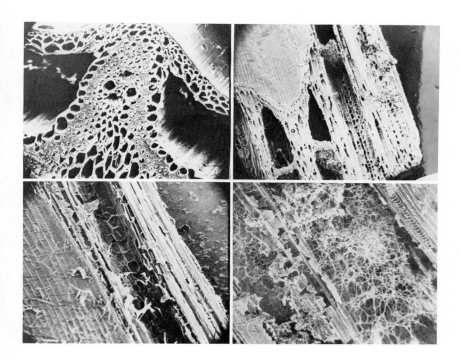

Figure 1: Upper left. Cross-section of Spartina alterniflora leaf with several fungal hyphae in right air passage; 280 x. Upper right and lower left. Top view of an oblique section of Spartina culms showing hyphae in air passages after 1-month marsh exposure; 100 x and 58 x. Lower right. Top view of an oblique section of 2-month marsh Spartina culm; 110 x.

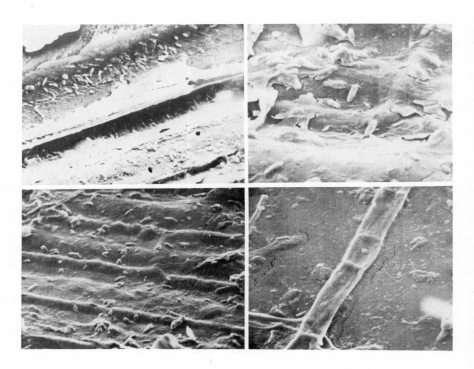

Figure 2: Upper left. Spartina alterniflora culm section with "seeded" bacteria; 2700 x. Upper right and lower left. Bacteria on leaf epidermal surface; 2190 x and 1090 x. Lower right. Collapsed fungal hypha revealing cell walls; 2000 x.

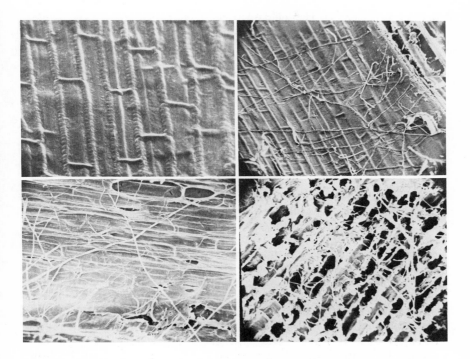

Figure 3: Upper left. Spartina alterniflora leaf before fungal invasion and leaf epidermis after 1-month marsh exposure; 200 x. Lower left. Fungal invasion and leaf epidermis after 1-month marsh exposure; 210 x. Lower right. Fungal invasion and leaf epidermis after 5-month marsh exposure; 210 x.

dishes with dead *Spartina alterniflora*, it was observed that *Cleantis* ate *Spartina*. In the upper left and upper right photomicrograph of Figure 4 plant debris is seen in a section of *Cleantis* intestine. In the upper right photomicrograph is a small, circular structure similar to the "barrel-hoop-like" support structures found in *Spartina*. These sclerenchymal rings are shown *in situ* in the SEM photomicrographs of *Spartina* (lower right and left of Figure 4). The middle left and middle right photomicrographs of Figure 4 are light microscopy sections of a *Spartina* culm. Note the ring-like structure of the vascular system in middle right photomicrograph. Gut analysis of gammarid amphipods associated with the litter bag culms revealed plant debris of much smaller size and less identifiable structure. The gut analysis of a polychaete, *Nereis succinea*, also found associated with the litter bag culms, revealed no identifiable plant debris.

The mean weight of detrital materials produced by the laboratory *Cleantis* was 0.012 g/animal/48 hr (s=0.005) and 0.007 g/animal/48 hr (s=0.002) for experiment one and two respectively. $\pm$ - distribution tests were calculated and both experiments were significantly larger than the controls at the 99% confidence interval.

DISCUSSION

SEM may be used as a research tool for *in situ* observations of bacteria and fungi on *Spartina alterniflora*. Although the work of Todd, Humphreys, and Odum (1972) did not reveal *in situ* microorganisms associated with *Spartina alterniflora*, they did predict the application of SEM as a research tool for ecological investigations. In this investigation bacteria and fungi were found by SEM to be associated with decomposing *Spartina alterniflora*. However, the SEM photomicrographs revealed a paucity of bacteria. Bacterial colonies were not found on any of the samples. These findings do not agree with those of Burkholder and Bornside (1957) and E. P. Odum and de la Cruz (1967), who explained *Spartina* degradation by bacterial decomposition. Dr. William J. Wiebe, a microbial ecologist at the University of Georgia, has consistently claimed to be unable to find *in situ* bacteria in large numbers on *Spartina* or other detrital particles (unpublished). In Odum (1971) Wiebe recommends caution in determining the ecological activity of bacteria by laboratory culture studies. The SEM photomicrographs demonstrating small numbers of bacteria are in agreement with Wiebe's findings. It does appear that the ecological importance of the bacteria is not in the decomposition of *Spartina*.

Fungal hyphae invade the air passages of the leaves and inter-leaf zone of the culms of *Spartina alterniflora* first. Invasion may also include the vascular conduction system of the leaves. In these areas cell walls do not obstruct the growth of the fungi. Later the fungal hyphae are found within the cellular pockets such as shown in the upper right of Figure 3. After five months of exposure in the litter bags, a large quantity of dead plant material had not been decomposed. This is consistent with the findings of E. P. Odum and de la Cruz (1967). The photomicrographs reveal the decomposition of the epidermal layer of the leaf (Figure 3) but much of the vascular system and

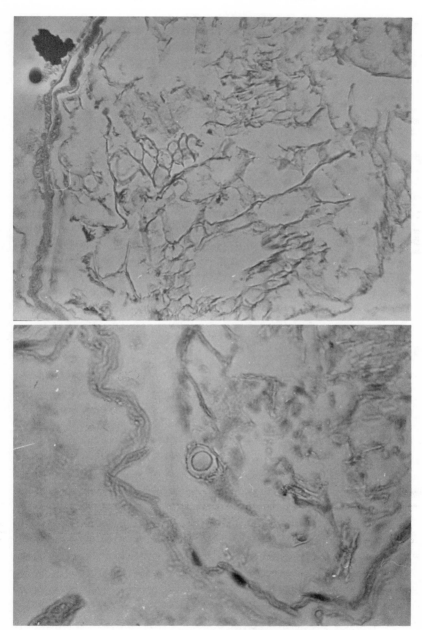

Figure 4a. Upper. Plant debris in gastro-intestinal tract of Cleantis planicauda; 430 x. Lower, sclerenchymal ring in gastro-intestinal tract of Cleantis planicauda 430 x.

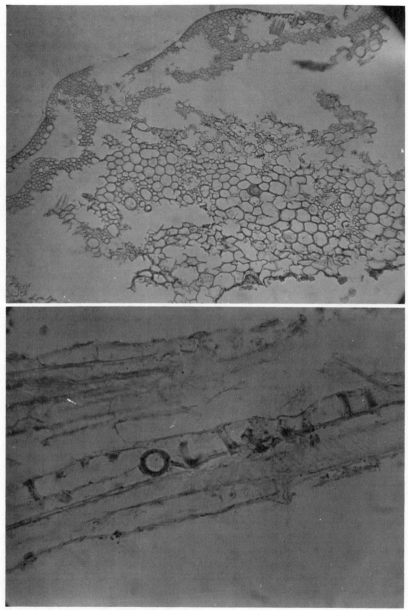

Figure 4b. Upper. Cross-section portion of Spartina culm; 430 x. Lower, Sclerenchymal ring in conduction system of Spartina leaf; 430 x.

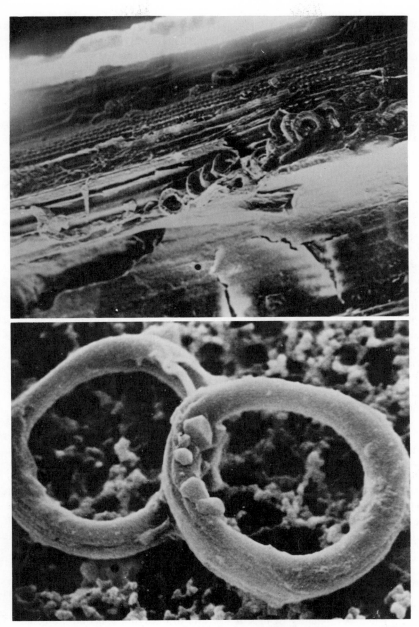

Figure 4c. SEM photomicrographs of sclerenchymal rings. Upper, 270 x; Lower, 2425 x.

inner leaf structure appear to be intact. This is supported by parallel light microscopy studies. Heald (1969) and Fenchel (1970) report similar findings with *Rhizophora mangle* and *Thalassia testudinum*, respectively. In these studies the investigators attributed much of the degradation of the dead plants to amphipods.

Gammarid amphipods were found associated with the dead *Spartina* in the litter bags. Examination of the gastro-intestinal tracts of several amphipods taken from the litter bags revealed small, rod-shaped structures similar to fragments of plant cell walls. The amphipods may be very active in *Spartina* decomposition. Further study of the feeding patterns of the amphipods is necessary before this can be concluded.

Undoubtedly, the plant debris observed in the cross sections of the gastro-intestinal tracts of *Cleantis planicauda* was *Spartina alterniflora*. These animals were in culture dishes containing only *Spartina alterniflora*. Also, the ring structures observed in several of the animals appear very similar to the sclerenchymal rings observed in *Spartina alterniflora* (lower right SEM photomicrograph of Figure 3, and lower right and left SEM photomicrograph of Figure 4). With the laboratory results of the degradation of *Spartina alterniflora* by *Cleantis planicauda* and the observations of *Spartina alterniflora* fragments in the gastro-intestinal tracts of *Cleantis planicauda* one may conclude that *Cleantis planicauda* is a major degrader of dead *Spartina alterniflora* to detritus. Since field animals have not been examined, further study is necessary for confirmation of this conclusion in the natural environment. The sclerenchymal rings may be the means by which this problem can be studied in the field. If *Spartina alterniflora* is the major or only producer of these sclerenchymal rings in the Georgia estuarine system, the rings may serve as naturally occurring tags for identifying *Spartina alterniflora* consumers in the Georgia estuarine system. Further study is necessary before this can be determined.

Similarly to the conclusions of W. E. Odum (1970), Fenchel (1970), and E. P. Odum (1971) regarding the action of macroinvertebrates in other dead-plant degradations, *Spartina alterniflora* may be the food for *Cleantis planicauda* or the *Cleantis planicauda* may remove other organisms as the *Spartina alterniflora* passes through the gut. Protozoa (Borror, 1965; Kirby, 1934; and Fenchel, 1969), nematodes (Teal and Wieser, 1966), copepods (Ruber, 1968), fungi, and other organisms are associated with marine marsh communities and, in particular, *Spartina alterniflora*. The answer to this must await further study of the feeding habits of *Cleantis planicauda* in the laboratory and natural environment.

CONCLUSIONS

1. The SEM is an adequate research tool for *in situ* observation of bacteria and fungi on *Spartina alterniflora*.

2. There is a paucity of *in situ* bacteria in decaying *Spartina* culms and a dense growth of fungi in air passages and the inter-leaf zone of culms.

3. Microbial decomposition of dead *Spartina* to detritus does occur and is probably due to fungi.

4. Macroinvertebrates, especially *Cleantis planicauda* are major degraders of dead *Spartina* to detritus.

ACKNOWLEDGEMENTS

This work was supported (in part) by a grant from the G. E. Foundation to Dr. E. P. Odum of the University of Georgia. The *Cleantis planicauda* were obtained from trawls made with the R/V *Driftwood* the estuarine research vessel of Emory University. I wish to thank Dr. R. L. Todd of the University of Georgia for assistance with SEM methodology; Dr. William Carlton of the University of Georgia for identification of the sclerenchymal rings and Dr. R. J. Reimold of the Marine Institute of the University of Georgia for suggestions and reading of the manuscript.

LITERATURE CITATION

Borror, Arthur C., 1965. New and little-known tidal marsh ciliates. Trans. Amer. Microscop. Soc., 84:550-565.

Chapman, V. J., 1960. Salt marshes and salt deserts of the world. Leonard Hill (Books) Limited, London. pp. 392.

Darnell, Rezneat M., 1958. Food habits of fishes and large invertebrates of Lake Pontchartrain, Louisiana, an estuarine community. Publ. Inst. Mar. Sci., Univ. Texas, 5:353-416.

Darnell, Rezneat M., 1967a. Organic detritus problem. *In* Estuaries, G. Lauff (ed.) Am. Assoc. Advant. Sci. Publ. No. 83, pp. 374-375.

Darnell, Rezneat M., 1967b. Organic detritus in relation to the estuarine ecosystem, pp. 376-382. *In* Estuaries, G. Lauff (ed). Am. Assoc. Adv. Sci. Publ. No. 83.

Fenchell, T., 1969. The ecology of marine microbenthos. IV. Structure and function of the benthic ecosystem, its chemical and physical factors and the microfauna communities with special reference to the ciliated protozoa. Ophelia, 6:1-182.

Heald, E. J. and W. E. Odum, 1970. The contribution of mangrove swamps to Florida fisheries. Proc. Gulf. and Carib. Fish. Inst., 22:130-135.

Kirby, Jr. Harold, 1934. Some ciliates from salt marshes in California. Arch. Protistenk., 82:114-133.

Macnae, W., 1956. Aspects of life on muddy shores in South Africa. S. African J. Sci., 53:40-43.

Macnae, W. 1957. The ecology of the plants and animals in the intertidal regions of the Zwartkops Estuary near Port Elizabeth, South Africa. J. Ecol., 45:113-131, 361-387.

Odum, E. P., 1971. *Fundamentals of Ecology*. Saunders, Philadelphia. 574 pp.

Odum, W. E., 1970. Pathways of energy flow in a South Florida estuary. Dissertation. University of Miami, 180 pp.

Odum, Eugene P. and Armando A. de la Cruz, 1963. Detritus as a major component of ecosystems. AIBS. Bull., 13:39-40.

Odum, E. P. and Armando A. de la Cruz, 1967. Particulate organic detritus in a Georgia salt marsh-estuarine ecosystem, pp. 383-388. *In* Estuaries, G. Lauff (ed). Am. Assoc. Adv. Sci. Publ. No. 83.

Ruber, Ernest, 1968. Description of a salt marsh copepod *Cyclops* (Apocyclops) *Spartinus* n. sp. and a comparison with closely related species. Trans. Amer. Microsco. Soc. 87:368-375.

Teal, John M., 1962. Energy flow in the salt marsh ecosystem of Georgia. Ecology, 43:614-624.

Teal, John M. and Wolfgang Wieser, 1966. The distribution and ecology of nematodes in a Georgia salt marsh. Limnol. and Ocean. 11:217-222.

Todd, R. L., W. J. Humphreys, and E. P. Odum, 1972. The application of scanning electron microscopy to estuarine microbial research. *In* Estuarine Microbial Ecology (ed) L. H. Stevenson and R. R. Colwell. Belle W. Baruch Library in Marine Science, Vol. 1. University of S. Carolina Press.

MARSH SOILS OF THE ATLANTIC COAST

LEO J. COTNOIR

Plant Science Department
College of Agriculture Science
University of Delaware

Marsh Soils of the Atlantic Coast

Old maps of the Atlantic Coast invariably delineate marsh areas with a common symbol suggesting that all marsh areas are essentially the same. Likewise, the Soil Survey maps published by the Bureau of Soils, USDA, starting at the turn of this century and later by the Division of Soil Surveys, Bureau of Plant Industries, USDA, make no attempt to separate different kinds of marshes. Until about 1935, the scale for most of the published maps was about one inch to the mile. At that time larger scales started to be used, and since the reorganization of the National Cooperative Soil Survey in 1952, most soil surveys have been published at a scale of three to four inches per mile. The use of larger mapping scales prompted some attempts in the Atlantic coastal counties to make soil delineations within marsh areas. The most commonly proposed separation was between "low marsh" and "high marsh." Actually, this criteria was never used in any of the soil surveys of these counties, perhaps fortunately, as there does not appear to be any basic soil boundary which corresponds to this separation. Two mapping units were used in the soil surveys of Washington and Kent counties (Roberts et al, 1939) and Newport and Bristol counties (Shearin et al, 1942), Rhode Island: (1) Tidal peat and (2) Peat, Tidal Marsh Phase. These units were based primarily on the depth of the organic tier.

The lack of information concerning marsh soils is apparently in part due to a past lack of interest by soil scientists, and to the difficulties involved in studying and mapping marsh soils. However, there is a growing interest in marsh soils. Hill and Shearin (1970) investigated a number of marsh areas along the Connecticut and Rhode Island coasts. They examined soil profiles in each of the four recognized marsh types: Deep Coastal Marshes, Very Shallow Coastal Marshes, Shallow Coastal Marshes, and Estuarine Marshes. Harmon and Tedrow (1969) conducted a study of the Hackensack Meadows. Thirty-six sites in this area were examined and some interesting pedologic information developed. They were mainly concerned with methods of establishing riparian ownership rather than developing criteria for classification or mapping. Soils along several transects in the marshes of Delaware have been described in our work, but we have no plans at the present for any intensive mapping efforts. Some mapping work was recently started in Rhode Island. A fairly ambitious program of marsh soil characterization is planned in Maryland. Interest has been expressed in soils work in the near future in the other Atlantic Coastal States, but little work is actually in progress at this time.

Classification of Organic Soils

The early soil classification schemes in the United States attest to the lack of attention given to marsh soils. In the classification system developed by Marbut (1935) and revised by Baldwin, Kellogg, and Thorp (1938) and by Thorp and Smith (1949), marsh soils were classified as intrazonal soils, suborder hydromorphic soils, and the Bog Great Soil Group. Soils of the Bog Soil Group were divided into peats and mucks, with peats defined as soils containing between 20% and 50% organic matter and mucks as those with over 50% organic matter. These criteria were not successfully used. Some marsh soils would probably also have been classified as Alluvial soils. This system was highly unsatisfactory for marsh soil classification. It lacked sufficient differentiation for marsh soils, for example, differentiation between salt marsh, fresh-water tidal marshes and swamps, and other types of wetlands. The system did not permit further separation at the lower levels of abstraction. The criteria were either poorly defined or lacking.

In 1960, a new comprehensive system of soil classification was introduced by the Soil Survey staff of the Division of Soil Surveys, USDA (Soil Survey 1960). This system, still under revision, appears to hold much more promise as a system for the classification of marsh soils. The highest level of abstraction consists of ten soil orders. Of these, two hold interest as orders in which marsh soils may be classified: Entosols and Histosols. Entosols are soils either without natural genetic horizons or with only the beginnings of horizons. At one extreme in age, they consist of recent Alluvium or other recent depositions; at the other extreme, they consist of high quartz soils in place for many thousands of years. In either case, because of time or because of conditions unfavorable to soil formation, soil development has been minimal. Some marsh soils on recent Alluvium with relatively low or thin organic tiers are Entosols. The exact criteria for Histosols is somewhat more complex, but useful criteria can be given (Soil Survey 1968).

The general rule is that a soil is classed as a Histosol if more than half of the upper 80 cm is organic, and it is classed as a Histosol without regard to thickness of organic matter if the soil rests on rock or fragmental material with interstices filled with organic materials. Organic soil materials have been defined as those that are either:

a. saturated with water for prolonged periods, or, artificially drained, and have 30% or more organic matter if the mineral fraction is 50% or more clay, or 20% or more organic matter if the mineral fraction is intermediate, or,

b. are never saturated with water for more than a few days, and have 35% or more organic matter.

The order Histosol is further divided into four suborders on the basis of the nature of the organic matter and the length of time they are saturated with water. These are:
1. Folist: never saturated with water for more than a few days and have:
   a. a lithic contact less than 1 m from the surface
   b. less than 75% of the organic materials are fibric spahagnum moss
2. normally saturated with water
   2.1 Fibrist: the organic material is predominantly fibric
   2.2 Hemist: the organic material is predominantly hemic
   2.3 Saprist: the organic material is predominantly sapric

The terms fibric, hemic, and sapric have been defined on the basis of the degree of decomposition of the organic material. At one extreme, fibric materials are those which contain a very high amount of fiber which is well preserved and readily identifiable as to botanical origin. At the other extreme, sapric materials are those representing the highest degree of decomposition. The bulk density of sapric materials is higher and the saturation water content is lower than for fibric or hemic materials. Hemic materials represent an intermediate state of decomposition. More detail definitions of these materials have been given, and field and laboratory criteria for them are being developed.

Provisions have been made at the lower levels of abstraction for other important paramaters such as depth, temperature, differences in hydraulic regime, salinity, presence of sulfur compounds, etc.

While the concept embodied in the classification system reflects field observations, as any satisfactory classification system must, there is need for a great deal more field testing, and revisions will very likely be needed. A work group of the Northeast Soil Survey Planning Conference (Northeast Soil Survey, 1972) have compiled information on over fifty profiles of marsh soils from Maryland to Maine. No profiles were reported that could be classified as Folist. Ten profiles were classified as Fibrist and there were about equally distributed between Sphagnofibrist, Borofibrist, and Medifibrist. Twenty-two were Hemist and included Sulfihemist, Borohemist, and Medihemist in approximately equal numbers. Nineteen were Saprist, of which twelve were Medisaprist, and the other seven were Borosaprist. A tour by a study group through the marshes of Maryland, Delaware, and Virginia indicated that the marshes in this area appear to be predominantly Saprist, especially Medisaprist. In both the Workgroup report and the Midatlantic tour, some Entosols were observed. These were primarily Hydraquent.

Scanty as the information may be, the evidence indicates that the Comprehensive System is a useful and workable classification. There is need for further field testing. Many of the criteria needed to classify soils in this system need further study. Nevertheless, the availability of a satisfactory classification system holds promise of much greater progress in marsh soil studies in the coming years.

Problems in Marsh Soil Studies

A number of problems are unique to the study of marsh soils. The lack of easily observed landscape features and the difficulty of traveling through the marshes are obvious. The classical approach to soil surveying in which one observes the landscape for clues of possible soil differences and then confirms these by auger borings is very difficult in the marsh environment. The landscape features are either lacking or very difficult, both because of accessibility and because of the inadequacy of presently available augers, for marsh work.

The soil profile description is the basic unit of data necessary for soil classification. The difficulties of making pits for marsh soil observations are obvious.

The well-established laboratory procedures, which are essential to provide the necessary chemical and physical data of soils, need modification for use with marsh soils. More specifically, the following are some of the problems encountered:

Organic Matter: The wet combustion methods (Jackson 1958), which have been extensively used to determine the soil carbon, give low values on marsh soil samples compared to the carbon furnace method. Loss on ignition on the other hand gives values which are too high.

Fiber Content: The classification discussed above places considerable emphasis on the degree of decomposition of the organic matter. Practical field estimates can be made by estimating the amount of fiber which breaks down when the material is rubbed between the fingers. This appears to offer a good rapid field estimate of degradation, but further correlation with laboratory sieve determinations are needed. A quick test in which the material is extracted with sodium pyrophosphate and the color of the extract compared to color chips of the Munsell Soil Color Book also offers promise.

Reaction: The pH of soil materials as determined by glass electrode in a 1:1 soil:water suspension is subject to variations with salt concentration. Because of the highly variable salt concentrations of marsh soils, standard procedures need to be established which take into consideration this variation in salt concentration.

Cation Exchange Capacity and Exchangeable Cations: The determination of cation exchange capacity and exchangeable cations is complicated by the fact that most of the exchange sites are due to organic groups and to the high concentration of cations in the solution phase. A separate determination of the cations in the solution phase after leaching with water, before replacement of the exchangeable cations, appears reasonable. Possible changes in the charge sites due to oxidation of the sample when it is exposed to the atmosphere needs to be investigated.

Sulfur: Many marsh soils contain large amounts of sulfur which is oxidized to sulfates when exposed to air. These marsh soils, called "cat clays," are of considerable interest. Pederson has developed a procedure involving the oxidation of the sample with H_2O_2 which appears to hold promise as a rapid method for the detection of this sulfur.

Soil Density: The usual core procedures for the determination of soil bulk density are difficult to use in marsh soils because of the tough mat of roots. Some rapid method for the estimation of this parameter is needed.

These are some of the major problems encountered in marsh soil studies. They are not insurmountable. They are problems at this time largely because so little has been done to devise suitable investigative procedures.

Remote Sensing in Marsh Soil Studies

Because of the difficulty of movement through marshes, there has been much interest in the use of remote sensing techniques for marsh soil studies. Soil scientists were among the first to use remote sensing in their work. Aerial photographs were used for soil surveying as early as 1930. Black and white aerial photography is essential for any soils work in the marshes. It provides the best base map for this purpose. The use of remote sensing methods besides black and white aerial photography for soils work has been the subject of several investigations. Honea and Beatty report successful use of radar imagery for the detection of land-from textures. Simonette (1968) successfully used radar K imagery to separate soil associations. He was not successful in detecting individual series differences. Myers and his co-workers (1966 a, b, c) successfully determined soil salinity differences from black and white and color infra-red photography.

The most applicable work is that of Clark and Hannon (1968) in Australia. Their success in detecting marsh soil differences by aerial photography depended strongly on the association between soils and vegetation. Pestong (1969), in this country, seems to be the only one who has attempted to determine marsh soil properties by remote sensing. He indicated that the use of Ectachrome and infrared Ectachrome offer potential for this purpose.

While, no doubt, remote sensing techniques will be of value in marsh soil studies, there are certain inherent difficulties which seriously limit their value. The most important factor is the inability of remote sensors to penetrate below the surface. If a consistent relationship between marsh vegetation and marsh soils exists, then the probability of success is rather good. Unfortunately, the limited observations to date do not indicate that this association is consistent enough to serve as a reliable index. The basic problem of soil profile observation still remains. In the case of remote sensing, these observations serve as the essential ground truth.

Literature Cited

Baldwin, M., C. E. Kellog, and I. Thorp. 1938. "Soil Classification" in Soils and Men, USDA Yearbook of Agriculture.

Beatty, F. D. 1965. Geoscience Potentials of Side-Looking Radar. Rayatheon-Autometric Corp., Alexandria, Va.

Clark, L. D., and N. J. Hannon. 1968. The Mangrove Swamp and Salt Marsh Communities of the Sydney District. I. Vegetation, Soil and Climàte, J. Ecol. 55:753-771.

Harmon, Kathryn P., and J. C. F. Tedrow. 1969. A Phytopedologic Study of the Hackensack Meadowlands. Masters Thesis, Rutgers Univ.

Hill, David E., and Arthur E. Shearin. 1970. "Tidal Marshes of Connecticut and Rhode Island". Conn. Agr. Exp. Sta. Bull. 709, Feb. 1970.

Honea, R. B. 1970. Determination of Landform Textures from Radar Imagery. Tech. Dept. No. 16, ONR Contract N 00014-67-0102-0001.

Jackson, M. L. 1958. "Soil Chemical Analysis". Page 219, Prentice Hall, Inc., Englewood Cliffs, N. J.

Marbut, C. F. 1935. "Soils of the United States" in the USDA Atlas of American Agriculture. Pt. 3

Myers, V. I., D. L. Carter, and W. J. Rippert. 1966. a. Remote Sensing for Estimating Soil Salinity. J. Agr. Drain. Div. American Soc. (IV) Eng. 92(IR. 4): 59-68.

Myers, V. I., Weigand, M. D. Heilman, and J. R. Thomas. 1966. b. Remote Sensing in Soil and Water Conservation Research. Proc. 4th Symp. on Remote Sensing of the Environment. Inst. Sci. Tech. Univ. of Mich., Ann Arbor, Mich. 801-813.

Myers, V. I., B. R. Vessey, and W. J. Rippert. 1963. c. Photogrammetry for Detailed Detection of Drainage and Salinity Problems. Trans. Amer. Soc. Agr., Eng. G(4): 332-334.

Northeast Soil Survey Work Planning Conference Report. 1972.

Pederson, E. J. Personal Communications.

Pestong, R. 1969. The Evaluation of Multispectral Imagery for a Tidal Marsh Environment. ONR Contract 4430.
Multiband Photos for a Tidal Marsh. Photogrammetric Eng. May 1969.

Roberts, R. C., et al. 1939. Soil Survey of Washington and Kent Counties. USDA, Bur. Chem. and Soils, R. I.

Shearin, H. E., et al. 1942. Soil Survey of Newport and Bristol Counties, USDA, Bur. of Plant Industries, R. I.

Simonette, D. S. 1968. Thermatic Land Use Mapping with Spacecraft, Photography and Radar. NASA, FRAC, Status Review, Vol. 1, 8:1-20.

Soil Survey Staff, Division of Soil Surveys, USDA. 1960. "Soil Classification, a Comprehensive System".

Soil Survey Staff, Division of Soil Surveys, USDA. 1968. Supplement to Soil Classification System – Histosols.

Thorp, J., and G. D. Smith. 1949. "Higher Categories of Soil Classification". Order, Suborder and Great Soil Groups. Soil Science 67:117-126.

Van Lopik, J. R. 1969. Pedologic Aspects of Remote Sensing. Texas Instruments, Inc.

THE RELATIONSHIP OF MARINE MACROINVERTEBRATES TO SALT MARSH PLANTS*

John N. Kraeuter and Paul L. Wolf

University of Georgia Marine Institute
Sapelo Island, Georgia 31327

Tyrone Biological Laboratory, Lebanon Valley College
Annville, Pennsylvania 17003

Introduction

The picture of a salt marsh is not difficult to envision: a tidally inundated grassy area pockmarked with bare spots and pools, drained by small and increasingly larger creeks and bordered by shrubs and forests to the upland and by intertidal mudflats, sand bars and oyster bars or open bodies of water on the low side. This concept probably contains all major elements of the marsh, but in reality a salt marsh is a biological continuum between land and the estuary or ocean. This continuum includes the areas covered by halophytes, barren areas called pannes, pools surrounded by grasses and the creeks which drain the surface; to separate these areas will draw artificial boundaries, but for purposes of this paper the salt marsh is considered to begin where halophytes other than submerged aquatics appear. Similarly, the marsh ends when the grasses, rushes and salt worts end and shrubs begin. Pools surrounded by grasses and drainage creeks are not included in these zones. Invertebrate assemblages appearing beyond the marsh are those of the intertidal mudflat, sand bar, oyster bar, or similar subtital assemblages. Above the shrubs assemblages become increasingly terestrial in nature.

In spite of the numerous ecological works on salt marsh plants, most of which are concerned with zonation, primary production or nutrient requirements, the effects of the invertebrates on the vegetation have received little attention. Some workers have listed large numbers of marine invertebrate species inhabiting the marsh, but when the creeks, pools and similar areas are eliminated only a few species remain to dominate the invertebrates of the marsh (Table 1). Few of these genera or species of

TABLE I

Marine Invertebrates Living in Salt Marshes

Dexter, 1947 Cape Ann, Mass.	Wass, 1963 Wass & Wright, 1969 Virginia	Teal, 1962 Georgia	MacDonald, 1969 California	Green, 1968 Great Britain	Nicol, 1935 Scotland
		Annelida			Annelida
		Capitella capitata Laeonereis culveri Manayunkia aestuarina Neanthes succinea Streblospio benedicti Oligochaeta 3 spp.			Enchytraeus albidus
Mollusca	Mollusca	Mollusca	Mollusca	Mollusca	Mollusca
Littorina littorea Littorina obtusata Littorina saxatilis Melampus bidentatus Nassarius obsoletus Modiolus demissus Mytilus edulis	Littorina irrorata Melampus bidentatus Modiolus demissus Polymesoda caroliniana	Littoridina tenuipes Littorina irrorata Melampus bidentatus Modiolus demissus Polymesoda caroliniana	Assiminea californica Cerithidea californica Littorina newcombiana Littorina planaxis Littorina sulcata Melampus olivaceus Phytia myosotis	Hydrobia ulvae Limapontia depressa Phytia myosotis	Hydrobia ulvae Littorina saxatilis
Arthropoda	Arthropoda	Arthropoda		Arthropoda	Arthropoda
Balanus balanoides Gammarus locusta Orchestia platensis Philoscia vittata Carcinus maenas	Chthalamus fragilis Orchestia palustris Sesarma cinereum Sesarma reticulatum Uca minax Uca pugilator Uca pugnax	Orchestia grillus Orchestia platensis Orchestia uhleri Cyathura carinata Eurytium limosum Sesarma cinereum Sesarma reticulatum Uca minax Uca pugilator Uca pugnax		Orchestia gammarella Porcellio scaber Philoscia muscorum Sphaeroma rugicauda Carcinus maenas	Orchestia gammarella Porcellio scaber Philoscia muscorum Sphaeroma rugicauda Carcinus maenas

marsh-dwelling marine invertebrates have been shown to have direct effects on halophytes. Similar lists of species containing many of the same genera have been compiled for mangrove faunas (Macnae and Kalk, 1962; Macnae 1963; Warner, 1969).

Annelida

All annelids listed in Table 1 are burrowers and most are considered deposit feeders. The single exception to this is *Manayunkia aestuarina*, a filter feeder, listed by Teal (1962) as occurring in the Georgia salt marshes. The presence of so many annelids in the Georgia salt marshes and their notable absence in other areas may be due to greater study effort in the Sapelo Island marshes. Wass and Wright (1969) list *Laonereis culveri*, *Capitella capitata* and *Heteromastis filiformis* from Virginia marsh creeks, but they are not included on the marsh proper. McMahan (1972) has been able to show numerical differences of the oligochaete *Monopylephorus rubronivens* in an artificially polluted marsh and a natural control.

Gastropoda

All gastropods listed may be considered aufwuchs-detrital feeders. MacDonald (1969) believed all California marsh-inhabiting species rasped food of algae and detritus from the substrate. The littorinid snails of the New England coast are usually associated with the algae of the rocky shore. Their influence in the marsh has not been studied, but presumably their food is similar to that obtained from the rocks. Marples (1966) demonstrated that *Littorina irrorata* did not pick up radionuclide tracers from *Spartina*, but he was able to find significant quantities in *Littorina irrorata* when the sediments were labeled. Studies by Reimold and Durant (personal communication) when compared to those of Marples (1966) indicate that *Littorina* is primarily a detrital feeder. *Melampus*, and by association the other ellobiid snails such as *Pythia*, are detrital feeders. Hausmann (1932) reported *Oscillatoria*, diatoms, filamentous green algae and epidermal cell fragments from *Spartina* in *Melampus* stomachs. The cell fragments were presumably sloughed off the plant exterior. Kerwin (1972) reported *M. bidentatus* was most commonly associated with detritus accumulated at plant bases.

The only gastropod reported to directly eat halophytes is *Littorina scabra*. This snail is common in mangrove associations and Macnae (1963) reported it ate mangrove leaves. Other than this littorinid there is no evidence that gastropod grazing in any way harms the plants, and in fact this may keep the halophyte stems clean of algal debris.

Pelecypoda

All pelecypods listed are filter feeders, and as such, have no direct effect on the plants. The byssal threads of *Modiolus* attach to each other and the stems, but any effect of this on the plants is obscure.

Isopoda-Amphipoda

Orchestrid amphipods and the isopods *Philoscia*, *Porcellio* and *Sphaeroma* are all found under mats of dead, decaying plant debris. Smallwood (1905) found *O. palustris* would eat plant debris but was on omnivore feeding on flesh when it was available. Most amphipods and isopods have little demonstrable direct effect on the living marsh plants.

Decapoda

Crabs belonging to the genus *Eurytium* are secondary consumers and thus have no direct effects upon the halophytes. The genera *Uca* and *Sesarma* are generally considered to be detritus-algal feeders; however, Crichton (1960) and Daiber and Crichton (1967) have shown that at times *Sesarma reticulatum* feeds directly on living *Spartina*. During the summer months extensive areas of *Spartina* on mud creek banks and drainage ditches of Delaware marshes are reduced to stubble by the feeding of this species. Experiments in which *S. reticulatum* were fed *Spartina alterniflora* leaves indicate a crab will consume an average of 0.02 grams dry weight of *Spartina* per day. More recent work in Delaware indicates that this figure may be as high as 0.06 grams dry weight per day. Using the latter figure *Sesarma* consumes approximately 2.8×10^6 grams dry weight of *Spartina* from June to September in Delaware marshes.

Ropes (1968) reported that nearly 50% of the intertidal populations of the green crabs, *Carcinus maenas*, had *Spartina* in their stomachs. Nearly 25% of the diet of the green crabs sampled during his study was *Spartina*. Thus *Carcinus* may be important in marsh ecology of New England and northern Europe.

Insecta-Arachnida

Several notable studies have confirmed direct grazing on marsh grasses by insects. Smalley (1960) has shown the grasshopper *Orchelimum fidicinum* consumes less than 1% of the yearly production of *Spartina alterniflora* in Georgia marshes. The plant sucking bug *Prokelisia marginata* ingests 6% of this production (Smalley, 1959). Marples (1966) used tracer studies in these marshes and reported four species as confirmed grazers on *Spartina alterniflora*: *Orchelimum fidicinum*, *Prokelisia marginata*, the bugs *Trigonotylus* sp. and *Ischnodemus badius*. Marples (1966) also noted rapid uptake of labeled materials by the ant *Crematogaster*. Teal (1962) reported the ant *Crematogaster clara* bored into *Spartina* stems to make nests. Whether *C. clara* eats *Spartina* or whether the label uptake was due to nest building activities is not known. There is no doubt, however, that ants damage living *Spartina*. The Dolichopodidae were the only flies which became labeled in Marples' (1966) work. He concluded that both detrital and grazed foods were important in their diet. Marples' (1966) studies agreed with those of Odum and Kuenzler (1963) in finding that the marsh spiders are carnivores. Both these studies reported label uptake by spider populations only after considerable time.

The most complete taxonomic study of East coast marsh insects is that of Davis and Gray (1966). They examined the insect populations from different types of marsh halophytes in North Carolina and found that the dominant herbivores in all five marsh types (Table 2) were grasshoppers. The dominant grasshopper species were not the same in any two halophytic zones. Several species of ants were common in the *Spartina patens* zone and Davis and Gray (1966) believed they were eating the plant tissues. A second major group called plant sap feeders was noted to be present in large numbers in all halophytic zones but *Juncus*. A third group of insects was considered to live on plant secretions and a fourth was parasitic on the plant tissues and sap. Of these four groups three obviously have direct detrimental effects on the halophytes. The fourth, those living on plant secretions, may or may not be considered harmful. The insects of the North Carolina marshes seem to form definite assemblages associated with halophyte types, and thus parallel the study of spiders in North Carolina marshes by Barnes (1953). Unfortunately, Barnes (1953) did not attempt to define the food of the major species, but large spider populations must be beneficial to the halophytes by eliminating insects. The Acarina of English salt marshes have also been found in definite zones (Luxton, 1963, 1967); their effects on plants are not known.

Summarizing the effects of invertebrates on halophytes, we can divide both the marine invertebrates and the insects into two major groups: (1) those that live directly on plant tissues and (2) those living on detritus and algae. Teal (1962) has already inferred that the majority of the marine invertebrates are algal-detrital feeders and have no direct effect on the halophytes. In general, these may enter the terrestrial food chain by raccoons, clapper rails, ibis and waterfowl. They enter the marine food chain as larval forms or as food for crabs such as *Eurytium* and *Callinectes* or fish feeding in the marsh (Teal, 1962). The insects eating plant tissue and sap obviously do direct harm to the halophytes, and Teal (1959) estimated the insects, *Orchelimum* and *Prokelisia* utilized 4.6% of the *Spartina* production. This differs from the less than 1% and 6% estimates of Smalley (1959) for the same two species. The reasons for this difference are not known, but the change in percent utilization probably reflects changes in the estimates of marsh production.

Since Davis and Gray (1966) listed a great number of insect species in the *Spartina* marsh and Marples (1966) definitely established that insects other than *Orchelimum* and *Prokelisia* were feeding in Georgia marshes, it appears that insects consume approximately 10% of the yearly *Spartina* production. Cameron (1972) studied insect species diversity in the San Francisco Bay, California salt marshes. He reported herbivore diversity was highest in spring and lowest in winter, while saprovore diversity was highest in mid-winter and was positively correlated with litter accumulation. Cameron (1972) also noted increased diversity when plants were blooming suggesting greater predation at this time.

Williams and Murdoch (1972) found no appreciable insect or other invertebrate damage to *Juncus roemerianus* in North Carolina marshes. This seems to be in direct contrast to the list of insect species found by Davis and Gray (1966). Until the difference can be resolved it is difficult to assess the effect of insects on *Juncus*. In the Georgia marshes there is visible damage to

DISTRIBUTION OF INSECTS IN NORTH CAROLINA SALT MARSHES

Table 2. Trophic relationships of characteristic invertebrates from the herbaceous strata of five types of North Carolina salt marshes. (After Davis & Gray, 1966)

Feeding Habits	FOOD	DOMINANT PLANTS				
		Spartina alterniflora	*Spartina-Salicornia-Limonium*	*Juncus roemerianus*	*Distichlis spicata*	*Spartina patens*
Herbivorous	Plant Tissues	*Orchelimum fidicinium* *Conocephalus* spp. *Mordellistena* spp.	*Orphulella olivacea*	*Paroxya clavuliger* *Conocephalus* spp.	*Orphulella olivacea* *Conocephalus* spp. *Clinocephalus elegans* *Nemobius sparsalsus*	*Conocephalus* spp. *Mermiria intertexta* *Glyphonyx* sp. *Dorymyrmex pyramicus* *Pseudomyrmex pallida* *Iridomyrmex pruinosus*
	Plant Sap	*Prokelisia marginata* *Sanctanus aestuarium* *Draeculacephala portola* *Ischnodemus badius* *Trigonotylus uhleri*	*Prokelisia marginata* *Sanctanus sanctus*	*Keyflana hasta* *Rhynchomitra microrhina*	*Amphicephalus littoralis* *Spangbergiella vulnerata* *Delphacodes detecta* *Tumidagena terminalis* *Neomegamelanus dorealis* *Trigonotylus americanus* *Rhytidolomia saucia* *Cymus breviceps*	*Delphacodes detecta* *Tumidagena terminalis* *Neomegamelanus dorealis* *Haplarius enotatus* *Aphelonema simplex* *Trigonotylus uhleri* *Cymus breviceps*
	Plant Secretions	*Chaetopsis apicalis* *Chaetopsis fulvifrons* *Conioscinella infesta*	*Chaetopsis apicalis* *Chaetopsis fulvifrons* *Conioscinella infesta*		*Conioscinella infesta* *Oscinella ovalis*	*Conioscinella infesta* *Oscinella ovalis* *Hippelates particeps*
Carnivorous	Animal Tissues	*Isohydrocera tabida* *Collops nigriceps*	Spiders	*Erythrodiplax berenice* Spiders	*Naemia serrata* Spiders	*Isohydrocera aegra* Spiders
	Animal Body Fluids	*Dictya oxybeles* *Hoplodictya spinicornis* Spiders	*Dictya oxybeles* *Hoplodictya spinicornis* Spiders	Reduviids Asilids Spiders	*Tomosvaryella coquilletti* Reduviids Culicids Asilids Spiders	Reduviids Asilids Spiders
Omnivorous	Detritus	Ephydrids Dolichopodids *Littorina irrorata*	Ephydrids Dolichopodids		Ephydrids Dolichopodids	
Parasitic	Plant Tissues and Sap	Dipterous larvae	Dipterous larvae		Dipterous larvae	Dipterous larvae
	Animal Tissues and Body Fluids	Larvae of parasitic Hymenoptera	Larvae of parasitic Hymenoptera	Larvae of parasitic Hymenoptera	Larvae of parasitic Hymenoptera	Larvae of parasitic Hymenoptera

the tips of *Juncus roemerianus* by grasshoppers. Salt marsh insects may be eaten by fish and enter the marine food chain, but most seem to be consumed by spiders or birds and thus follow a more nearly terrestrial food chain.

Indirect effects on the plants

The majority of the effects of marine macroinvertebrates on marsh plants are indirect and few references exist on their interactions. A cursory view, with some general speculations, is all that can be offered.

Data on drainage through marsh sediments is scarce, but Redfield (1959) provided evidence suggesting stagnation of interstitial waters at least in the high marsh. Chapman (1959) showed appreciable drainage near the creeks at low water. Gallagher (1971) noted standing water in the tall *Spartina*, dwarf *Spartina* and *Distichlis* areas and this seems to support Redfield's contention that the marsh is saturated and stagnant. Pomeroy (1959) complained that during his bell jar experiments on algal productivity he often found it necessary to abandon the work because of seepage from the jar. He noted this was most often due to crab burrows and indicates a significant effect of burrowing organisms on marsh drainage. Observations such as this must be used with caution since tidal inundation does not create a uniform hydrostatic head.

Schwartz and Safir (1915) have classified the factors associated with the distribution of fiddler crab species as being moisture, substratum and vegetation, in order of their importance. *Uca pugilator* is found in areas of sparse vegetation or in areas entirely free of plants. *Uca pugnax* lives on a muddy substratum shaded by dense vegetation, and therefore continually moist. Schwartz and Safir (1915), however, state that the distribution of crabs may also be limited by the density of vegetation which, because of its profuse root system, makes burrowing difficult. Teal (1958) reported the absence of fiddler crabs in Georgia salt marshes along creek banks where *Spartina* is greater than 1.50 m in height and examined species substrate relations. Only *Uca pugnax* would be expected to be found here because of its preference for a vegetated muddy substrate. Wolf, Shanholtzer and Reimold (in press) found an average of 27 *Uca pugnax* per m^2 in the tall *Spartina* marsh as compared to $176/m^2$ and $196/m^2$ for the short *Spartina* and medium *Spartina* marshes, respectively. The factors which limit the numbers of fiddler crabs in the tall *Spartina* marsh have not been clearly established, but root system density has been mentioned as a factor. Preliminary data on root systems (Gallagher, personal communication) indicate that in tall *Spartina* roots would not interfere with burrowing, but the substrate may be too fluid to support long burrows. The work of Schwartz and Safir (1915) and Dembowski (1926) showed that fiddler crabs either abandon their burrows or change direction upon meeting an obstacle. These observations would support the hypothesis of root systems as limiting burrowing activities in dwarf *Spartina* areas and insufficient support for burrows may be responsible for Teal's (1958) data. *Sesarma reticulatum* primarily inhabits the tall *Spartina* marsh, and its burrows appear to be shallower and may not require as much support as those of *Uca*. The effects, either beneficial due to increased circulation or harmful due to root pruning and disruption, of burrowing on salt marsh halophytes remain essentially unassessed.

Kraeuter and Haven (1970) have shown that marsh invertebrates such as the snails *Littorina irrorata* and *Melampus bidentatus*, the crabs *Uca minax* and *Uca pugnax*, and the polychaete *Neanthes succinea* all produce solid feces. Haven and Morales-Alamo (1966) have shown the importance of bio-deposition by invertebrates and pointed out that filter feeding effectively increases grain size and aids in removal of particles from suspension. The nutritive value of invertebrate feces for animals has been demonstrated by Newell (1965), Johannes and Satomi (1966), and Frankenberg, Coles, and Johannes (1967). Kuenzler (1961) has shown that the ribbed mussel, *Modiolus demissus*, acts as a recycler of phosphorus in the Georgia salt marshes. He estimated that 4,700 ug P/m^2 day were deposited as pseudofeces. This amounted to a daily removal of about one-third of the particulate phosphorus from suspension (Kuenzler, 1961). Pomeroy, Shenton, Jones and Reimold (1972) have shown the marsh sediments contain relatively large quantities of phosphorus and it is doubtful that a net deposition of 1.7 g P/m^2 year on the marsh surface would be a significant contribution to the plants. The importance of pelletizing wastes by other members of the marsh community has not been assessed, but preliminary studies are now underway at Sapelo Island to determine the nutrients potentially available to the plants from invertebrate feces. The effect of these feces on the halophytes remains to be tested.

Grazing of snails such as *Littorina* and *Melampus* on the plant systems may clean halophytes of epiphytic algae. This may enhance growth by eliminating algal cover or have a detrimental effect by eliminating epidermal tissue. Other subtle effects are probably numerous and may be either harmful or beneficial, but they have yet to be tested or lucidated.

Indirect and direct effects of plants on animals

Direct effects of plants on animals are those which are intuitively obvious and may include protection from predation, a stabilizing platform or a food source. Crabs, snails, insects and other marsh invertebrates are less vulnerable to predation from fish or birds than if they existed on an open bank. Some birds and fish do penetrate the marsh to prey on these forms, but most predators remain on the marsh borders. The insects and *Sesarma* directly benefit through the use of the plants as food, and these effects have already been discussed.

The presence of halophytes in the marsh offers the marine invertebrates a stabilized surface in what would be an otherwise unstable horizontal platform. The mussel *Modiolus* can attach to the grass stems with its byssus and the barnacle *Chthamalus* is often found cemented directly to the stalks or leaves of *Spartina*. *Littorina* utilizes the halophytes as a vertical substrate to move over with the tides, and due to the root mat the burrowers such as *Uca* have a stabilized sediment layer. This stabilizing effect appears to be of primary importance to these marsh dwelling forms.

Indirect effects of plants such as amelioration of environmental fluctuations may be a simple reduction of wave action during high water or reductions of current velocities during flood and ebb. Effects of light, temperature and humidity will be discussed below as separate entities although they are clearly interrelated.

Light

The quantity and quality of light reaching the marsh surface may have profound effects on the food supply of marine invertebrates in the marsh. Studies have shown a reduction in incident light under marsh plants, but there are some differences in seasonal patterns of light. Blum (1968) reported that both *Distichlis* and *Spartina patens* exhibit little change in the reduction of sunlight throughout the year. Less than 5% of the incoming radiation reached the marsh surface beneath their dense mats. Creek bank *Spartina alterniflora* and dwarf *S. alterniflora* were shown to eliminate about 50% of the incident light in spring while similar areas eliminated about 90% upon reaching maximal growth in the fall (Blum, 1968).

Gallagher (1971) found similar patterns in the Canary Creek marsh of Delaware. Tall *Spartina* areas permitted less than 6 kilolux of illuminance to reach the marsh surface from June through December, but from January to April, 14 to 20 kilolux reached to marsh surface. This sudden change is primarily due to removal of dead *Spartina* by tides and ice scour. Gallagher (1971) also reported increases in illuminance on the dwarf *Spartina* marsh surface from February to March. This change was not as clear as in the tall *Spartina* and was apparently due to decay of leaves rather than loss of the entire plant. The light pattern in *Distichlis* was one of wide fluctuation from one month to the next, but in general, less light reached the marsh surface than in other plant zones.

Pomeroy (1959) did not find seasonal differences in the percent incident light reaching the surface under different height plants in the Georgia *Spartina* marshes. He was able to show approximately double the energy reaching the marsh surface in May when compared to December. Percent incident light reaching the marsh surface beneath *S. alterniflora* ranged from 5% at the creek banks to about 95% in sparse dwarf *Spartina* near the levee (Pomeroy 1959). At high water light radiation decreased to zero at the creek bank and to 70% on the levee.

Williams' (1962) work on Georgia marshes seems to indicate a slight seasonal pattern. While Pomeroy (1959) reported 5% and 95% incident light reaching the marsh surface in tall and dwarf *Spartina* respectively for both May and December Williams (1962) reported 23% (12-48%) and 47% (25-62%) respectively in January. Williams (1962) also provided data on percent incident light to the marsh surface for medium *Spartina*, 32% (28-40%) and *Juncus*, 16% (7-25%). The discrepancy between the dwarf *Spartina* figures of Williams (1962) and Pomeroy (1959) may be due to: (1)

definition of dwarf *Spartina*, (2) density of dwarf *Spartina*, (3) analytical differences, or (4) real differences with season. The latter possibility would indicate dwarf *Spartina* must be denser in winter than summer, but there is no evidence for this. Differences in measuration would mean none of the figures could be compared. The greatest possibility would seem to be in one of the first two choices. Data for tall *Spartina* may suffer from the same difficulties, but there is a suggestion that some seasonal differences may exist. More gradual removal of tall *Spartina* from the Georgia marshes lessens seasonality of light penetration when compared to northern marshes where export of creek bank materials is abrupt. A longer growing season in the more southern area also keeps a cover on the marsh surface.

Although his data indicated daily differences when algal production on the marsh surface was the greatest, Pomeroy (1959) found nearly constant daily production throughout the year. Gallagher (1971) found greatest production of edaphic algae in the tall *Spartina* and *Distichlis* areas during January to April, while the panne and bare bank did not change on a seasonal basis. Tall *Spartina* edaphic algal productions began to rise in March and remained at a high level until early September (Gallagher, 1971). The seasonality of light and its influence on edaphic algae is of primary importance to the aufwuchs-detrital feeders in the marsh. In southern marshes constant production sustains a year-round population and on warm winter days feeding could remain nearly normal. In northern marshes the greatest production of edaphic algae is at a time when most invertebrates are relatively inactive. Thus, the importance of light and the shading of plants bears a direct relation to the quantity and quality of food on the marsh surface. This in turn may affect the distribution patterns of the micro, meio and macrofaunas on the marsh surface.

Temperature

Temperature effects have long been known to affect the distributional pattern of benthic fauna. Blum (1968) reported that temperature fluctuations just under the vegetative mats of *Spartina patens* were nearly halved when compared to March, May, June and July air temperatures. He also discovered that one inch below the sediment surface temperature variations under creek bank *S. alterniflora*, dwarf *S. alterniflora* and *S. patens* were approximately 5.5^o, 3° and 2° C respectively in May; 2.5^o, 1.5^o and 0.5^o in June; but by July both dwarf *Spartina* and *S. patens* had approximately the same temperatures and variations. Creek bank *S. alterniflora* varied slightly more, but around the same temperature (Blum, 1968). Blum also pointed out that the marsh surface under the dense mat of *S. patens* did not warm as quickly as on other areas.

Gallagher (1971) found greatest temperature variations in all areas of Canary Creek marsh in Delaware in summer. Winter temperatures had the least variations. Gallagher (1971) indicated that except for one sampling period the panne surface was always warmer than the air while the bare bank was warmer in spring, summer and fall, but cooler in winter than daily air temperatures. The surface beneath the tall *Spartina* was cooler in spring and fall, equal or lower in winter and from March to May was warmer than the air temperature. Dwarf *Spartina* was similar, but was equal to or higher than air temperature from October to May and equal or lower from June to September. The relation between air temperature and surface temperature in the *Distichlis* zone was irregular.

Pomeroy (1959) explained daytime temperatures on Georgia *Spartina* marshes were influenced by insolation, shading by *Spartina*, the greenhouse effect and evaporation of water from the sediments. At night, surface temperatures approached that of the water, but daytime marsh surface temperatures are generally higher than those of the air or water. Sediment surface temperature under *Spartina* was cooler in winter and warmer in summer than on nearby bare areas. Pomeroy interpreted this as the result of the greenhouse effect and reduced evaporation. Gallagher (1971) found tall *Spartina* areas were cooler than the air throughout the year and both dwarf *Spartina* and *Distichlis* were generally cooler in summer and warmer in winter than the bare areas. Neither Pomeroy (1959) nor Gallagher (1971) found data on any of their marsh zones comparable with that of Blum (1968) for *S. patens*. These results indicate there is generally little chance that marsh grasses will ameliorate temperature fluctuations for marsh-dwelling invertebrates, but they may affect the rate of these changes.

Humidity

Blum (1968) found humidity was generally high around the bases of all plants. *Distichlis* and *S. patens* mats remained moist even when they were not inundated for several days. Pomeroy (1959) did not measure humidity, but he noted cracks on bare areas after several hours exposure. He also noted cracking is reduced or absent beneath *Spartina*. Gallagher (1971) noted that the panne and bare areas occasionally have extremely high salinities (135-185 o/oo). At the same time dwarf *Spartina* and *Distichlis* zones were affected to a lesser degree and no change was detected in tall *Spartina* (Gallagher, 1971). These data indicate that areas with halophytic cover have higher water retentive capacity than uncovered marsh surfaces.

High humidity has been shown to affect the distribution patterns of insects and intertidal rocky shore marine invertebrates (Kensler, 1967). The effect of temperature/humidity interactions on marsh organisms seems to be an area for more physiological studies. If the salt marsh could be thought of as an intertidal community with very wide zones, the patterns of distribution for marine macroinvertebrates could be compared with rocky intertidal areas.

Acknowledgements

The authors express their sincere appreciation for the criticisms of Drs. J. Gallagher, J. Hall and M. Wass. Their efforts have certainly improved the content and readability of this paper, but we, of course, are responsible for any errors which may appear.

Literature Cited

Barnes, R. D. 1953. The ecological distribution of spiders in non-forest maritime communities at Beaufort, North Carolina. Ecol. Monogr. 23:315-337.

Blum, J. L. 1968. Salt marsh *Spartinas* and associated algae. Ecol. Monogr. 38:199-221.

Cameron, G. N. 1972. Analysis of insect trophic diversity in two salt marsh communities. Ecology 53:58-73.

Chapman, V. J. 1959. Discussion of: Circulation of heat, salt and water in salt marsh soil by Redfield. Proc. Salt Marsh Conference held at Univ. Ga. Marine Institute, March 25-28, 1968, pp. 81-85.

Crichton, O. 1960. Marsh crab intertidal tunnel-maker and grass-eater. Estuarine Bull. 5:3-10.

Daiber, F. C. and O. Crichton. 1967. Caloric studies of *Spartina* and the marsh crab *Sesarma reticulatum* (Say). 1966-1967 Annual Pittman-Robertson Report to Delaware Board Game and Fish Comm. Project W-22-R-Z, Job No. 4, 20 pp.

Davis, L. V. and I. E. Gray. 1966. Zonal and seasonal distribution of insects in North Carolina salt marshes. Ecol. Monogr. 36:275-295.

Dembowski, J. B. 1926. Notes on the behavior of fiddler crabs. Biol. Bull. 50:179-201.

Dexter, R. W. 1947. The marine communities of a tidal inlet at Cape Ann, Massachusetts: A study in bio-ecology. Ecol. Monogr. 17:261-294.

Frankenberg, D., S. L. Coles and R. E. Johannes. 1967. The potential trophic significance of *Callianassa major* fecal pellets. Limnol. Oceanogr. 12:113-120.

Gallagher, J. L. 1971. Algal productivity and some aspects of the ecological physiology of the edaphic communities of Canary Creek tidal marsh. Ph.D. dissertation, University of Delaware, 120 pp.

Green, J. 1968. The biology of estuarine animals. Biology Series: University of Washington, Seattle, 401 pp.

Hausmann, S. 1932. A contribution to the ecology of the salt marsh snail, *Melampus bidentatus* Say. Amer. Natur. 66:541-545.

Haven, D. S. and R. Morales-Alamo. 1966. Aspects of biodeposition by oysters and invertebrate filter feeders. Limnol. Oceanogr. 11:487-498.

Johannes, R. E. and M. Satomi. 1966. Composition and nutritive value of fecal pellets of a marine crustacean. Limnol. Oceanogr. 11:191-197.

Kensler, C. B. 1967. Dessication resistance of intertidal crevice species as a factor in their zonation. J. Anim. Ecol. 36:391-406.

Kerwin, J. A. 1972. Distribution of the salt marsh snail (*Melampus bidentatus* Say) in relation to marsh plants in the Poropotank River Area, Virginia. Chesapeake Science 13:150-153.

Kraeuter, J. N. and D. S. Haven. 1970. Fecal pellets of common invertebrates of lower York River and lower Chesapeake Bay, Virginia. Chesapeake Science 11(3):159-173.

Kuenzler, E. J. 1961. Phosphorus budget of a mussel population. Limnol. Oceanogr. 6:400-415.

Luxton, M. 1963. Some aspects of the biology of salt-marsh Acarina. First International Congress of Acarology, Fort Collins, Colorado, pp. 172-182.

Luxton, M. 1967. The ecology of salt-marsh Acarina. J. Anim. Ecol. 36:257-277.

MacDonald, K. B. 1969. Quantitative studies on salt marsh faunas from the North American Pacific Coast. Ecol. Monogr. 39:33-60.

Macnae, W. 1963. Mangrove swamps in South Africa. J. Ecol. 51:1-25.

Macnae, W. and M. Kalk. 1962. The ecology of the mangrove swamps at Inhaca Island, Mocambique. J. Ecol. 50:19-34.

Marples, T. G. 1966. A radionuclide tracer study of arthropod food chains in a *Spartina* salt marsh ecosystem. Ecology 47:270-277.

McMahan, E. A. 1972. Relative abundance of three marsh-floor organisms in a sewage-affected marsh and in a sewage-free marsh. J. Elisha Mitchell Sci. Soc. 88(2):61-65.

Newell, R. 1965. The role of detritus in the nutrition of two marine deposit feeders, the prosobranch *Hydrobia ulvae*, and the bivalves *Macoma balthica*. Proc. Zool. Soc. London 144:25-45.

Nicol, E. A. T. 1935. The ecology of a salt-marsh. Mar. Biol. Assoc. U. K. J. 20:203-261.

Odum, E. P. and E. J. Kuenzler. 1963. Experimental isolation of food chains with the use of phosphorus - 32. *In* V. Schultz and A. W. Klement, Jr. (ed.) Radioecology. Reinhold, New York, pp. 113-120.

Pomeroy, L. R. 1959. Algal productivity in salt marshes of Georgia. Limnol. Oceanogr. 4:386-397.

Pomeroy, L. R., L. R. Shenton, R. D. H. Jones and R. J. Reimold. 1972. Nutrient flux in estuaries. Limnol. Oceanogr. Special Symposia I:274-291.

Redfield, A. C. 1959. Circulation of heat, salt and water in a marsh snail. Proc. Salt Marsh Conference held at Univ. Ga. Marine Institute, March 25-28, 1968, pp. 77-81.

Ropes, J. W. 1968. The feeding habits of the green crab, *Carcinus maenas* (L.). U. S. Fish and Wildlife Serv. Fish. Bull. 67:183-203.

Schwartz, B. and S. R. Safir. 1915. The natural history and behavior of the fiddler crab. Cold Spring Harbor Monogr. No. 8.

Smalley, A. E. 1959. The growth cycle of *Spartina* and its relation to the insect populations in the marsh. Proc. Salt Marsh Conference held at Univ. Ga. Marine Isntitute, March 25-28, 1968, pp. 96-97.

Smalley, A. E. 1960. Energy flow of a salt marsh grasshopper population. Ecology 41:672-677.

Smallwood, M. E. 1905. The salt marsh amphipod: *Orchestia palustris*. Cold Spring Harbor Monogr. 3:3-21, plus two plates and a map.

Teal, J. M. 1958. Distribution of fiddler crabs in Georgia salt marshes. Ecology 39:185-193.

Teal, J. M. 1959. Energy flow in the salt marsh ecosystem. Proc. Salt Marsh Conference held at the Univ. Ga. Marine Institute, March 25-28, 1968, pp. 101-107.

Teal, J. M. 1962. Energy flow in the salt marsh ecosystem of Georgia. Ecology 43:614-624.

Warner, G. F. 1969. The occurrence and distribution of crabs in a Jamaican mangrove swamp. J. Anim. Ecol. 38:379-390.

Wass, M. L. 1963. Checklist of the marine invertebrates of Virginia. Special Sci. Rept. No. 24 (revised), Virginia Institute of Marine Science, Gloucester Point, Virginia.

Wass, M. L. and T. D. Wright. 1969. Coastal wetlands of Virginia. Interim Report. Special Rept. in Applied Marine Science and Ocean Engineering No. 10, 154 pp. Virginia Institute of Marine Science, Gloucester Point, Va.

Williams, R. B. 1962. The ecology of diatom populations in a Georgia salt marsh. Ph.D. dissertation, Harvard University, 143 pp.

Williams, R. B. and M. B. Murdoch. 1972. Compartmental analysis of the production of *Juncus roemerianus* in a North Carolina salt marsh. Chesapeake Science 13:69-79.

Wolf, P. L., S. F. Shanholtzer and R. J. Reimold. (In press) Population estimates for *Uca pugnax* on the Duplin estuary marsh. Crustaceana.

RELATIONSHIP OF VERTEBRATES TO SALT MARSH PLANTS

G. Frederick Shanholtzer[1]
The University of Georgia
Marine Institute
Sapelo Island, Georgia 31327

Introduction

Significant salt marsh plant-vertebrate relationships exist in areas subject to tidal-saline waters. They are components of equilibrium communities (Hutchinson, 1953) in a pulse-stable ecosystem (Odum, 1971). Their associations assume both direct and indirect dimensions. Results of direct plant-vertebrate associations tend to be more immediate in their expression than indirect relationships. It is the intent of this paper to describe and evaluate these relationships.

Direct Relationships

Direct relationships involve spatial and physical utilization of salt marsh plants by vertebrates. Spatial utilization patterns are derived from the territory, home range, and behavior of vertebrate species using the halophyte dominated marsh environment, and the structure of the plants themselves.

Marsh vegetation provides a habitat volume and structural foundation for feeding (Fig. 1), reproductive and roosting activities of many vertebrates (Johnson et al., in press). Plant cover additionally provides a moderated thermal environment and refuge from predation. Thermal considerations are important during periods of avian incubation when excessive solar radiation can damage eggs and young.

Feeding

The feeding habitat as provided by halophytes varies with seasons, tides, plant species and height, and vertebrate species and numbers present. In Georgia salt marshes, seasonal dietary shifts occur in several species which result in different spatial orientation or activity patterns. For species such as the Seaside Sparrow, Red-winged Blackbird, and perhaps the rice rat (Sharp, 1967), a summer carnivorous diet shifts toward a more granivorous diet during autumn and winter months. This implies a vertical change in

---
[1] Present address: Dames and Moore, 14 Commerce Drive, Cranford, New Jersey 07016

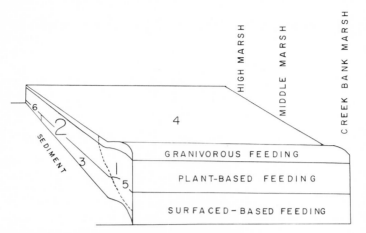

Figure 1. Structural foundations of the salt marsh.

utilization of available space. In contrast, the salt marsh harvest mouse (*Reithrodontomys raviventris*) in Pacific marshes maintains a primary diet of non-seed plant fiber (Fisler, 1965).

As plants grow during a year, the total volume of this feeding habitat increases until the peak of the growing season. Numbers and species of vertebrates using the salt marsh vary during the year. Peak periods of usage by birds occur during fall and spring migrations, with greater numbers of individuals and species in the autumn. Establishment of territory by nesting species during the breeding season reduces densities of individuals, and a rather clear delineation of feeding niches results (Fig. 1). Mink (*Mustela vison*) and raccoons (*Procyon lotor*) tend to use the marsh more in winter months than summer months (C. J. Durant, personal communication). Raccoons feed on fiddler crabs (*Uca* spp.) in the marsh and both species traverse the marsh surface enroute to feeding grounds along creek banks.

Tidal flooding of the marsh surface creates feeding habitat for fish. During warmer months red drum (*Sciaenops ocellata*) and mullet (*Mugil cephalus*), for example, actively feed above the marsh surface. The drum feed on small crabs and shrimp, and mullet on algae, plant detritus, and inorganics (Odum, 1966). Relatively little feeding by fish occurs over the marsh surface during spring tides in the coldest winter months (C. J. Durant, personal communication).

Reproduction

Nesting by marsh vertebrates often occurs within a relatively discrete area where feeding also occurs, and which is defended against other members of the same species. Other marsh vertebrates nest in colonies where feeding territories normally do not include nest territories. The Long-billed Marsh Wren (*Telmatodytes palustris*; Kale, 1965), and rice rat (*Oryzomys palustris*;

Sharp, 1967) are examples of the first group. Heron species assembledges exemplify the latter group (Shanholtzer, Hopkins, and Dopson, unpubl.; Bent, 1921). In this instance the area immediately around the nest is actively defended, while feeding is carried out at a variety of locations some distance from colonies. The Sharp-tailed Sparrow (*Ammospiza caudacuta*) displays an intermediate form of this behavior. Home ranges overlap and territories, if they exist, are not well defined.   In each case, however, nests are spatially located high enough in the grass to avoid inundation by all but highest spring or storm tides (Clapper Rail (*Rallus longirostris*) nests and eggs can tolerate some tidal inundation.

Johnston (1956) describes the spatial utilization of marsh grass by Pacific coast Song Sparrows (*Melospiza melodia*) along a tidal slough. Variation in breeding density over a six year period influenced only clutch size. *R. raviventris* may utilize old or abandoned *M. melodia* nests (Fisler, 1956).

Breeding and non-breeding vertebrates use marsh grasses as roosting (resting) sites. Surface roosting species include rails, mink (*Mustela vison*), and rabbits (*Sylvilagus* spp.). Plant roosting species are exemplified by swallows, marsh wrens, and Red-winged Blackbirds (*Agelaius phoeniceus*). Racks of dead marsh grass also act as resting sites for mink, river otter (*Lutra canadensis*), and various shorebirds.

Protection

Marsh plants, particularly in taller vegetation zones, provide a refuge for marsh vertebrates from predation. Hill (in Bent *et al.*, 1968) observed that "low windrows of dried and matted thatch...form an important shelter and hiding place for the Sharp-tailed Sparrows". It is notable that marsh rabbit and Clapper Rail activity in short grass marshes is greatest at dark and crepuscular hours, at a time when aerial predation is least effective. Fisler (1965) notes that *Salicornia* and other halophytes provide considerable cover for marsh harvest mice. Highest diurnal tides (December and January) in Pacific marshes expose these mice to avian predation.

No documentation is currently available to show that marsh vegetation shading provides a thermally more conducive environment for reproduction than nesting in an area of unrestricted solar radiation. I have noticed, however, that heron species which nest in exposed areas often spend a considerable length of time shading eggs and young from the sun during warm days.

Physical Utilization of Halophytes

Halophytes are used by vertebrates for nutritive and non-nutritive purposes. Consumption for nutritive purposes involves mainly granivorous and whole plant grazing. With the production of seeds (mainly by *Spartina alterniflora* in Georgia marshes), some birds and possibly the rice rat develop a granivorous mode of feeding. Seaside Sparrows, Sharp-tailed Sparrows, and Red-winged Blackbirds are some species which utilize this source of protein and energy.

Relatively little marsh grass is grazed while in the living state, and most of that is probably harvested by insects (Smalley, 1960). Deer, marsh rabbits, and rodents (*Microtus pennsylvanicus*, Shure, 1970; *Reithrodontomys megalotis*, Fisler, 1965) are a few vertebrate herbivores which enter the marsh for feeding. Most of these species restrict their activities to high marsh vegetation zones.

Several barrier islands along the Georgia coast have cattle and other large ungulates grazing adjacent salt marshes. These species locally consume large amounts of *S. alterniflora*. Though grazing conditions as this are not native to salt marshes in general, it is, however, appropriate to discuss the influence of these vertebrates on a salt marsh and the plants themselves. It is anticipated that sooner or later the question of the desirability or feasibility of cropping the grass for use in domestic vertebrate feeding or aquaculture will arise. *Spartina patens* marshes are already being mechanically harvested in New Jersey marshes (Hitchcock, 1972). A Sea Grant study at the University of Georgia Marine Institute of the effects of grazing by large herbivores on salt marshes has shown some interesting possible consequences.

For this study, two salt marsh sites, one ungrazed and one heavily grazed, were chosen along the Georgia coast (Fig. 2). Transect lines were established from creek bank to treeline at each marsh. Clip plots of grass samples (1 m^2) were taken along each transect. Plants collected were analyzed for species, living and dead, height, density, and dry weight. Collections were made at various intervals throughout the growing season with final samples being taken in November. One must be cautioned, however, against strictly comparing the two sites due to their different depositional environments and water regimes. By contrasting the two areas, we can see qualitative results which strongly correlate with grazing pressure.

Figure 3 leads us to a tentative conclusion that grazing affects vegetative zonation of halophytes. Characteristic high marsh plants (i.e., *Distichlis* and *Salicornia*) were distributed over a 240 m bank on the grazed marsh. In contrast, *Salicornia* assumed its typical high marsh pattern in the ungrazed marsh and was spread over a bank one half the distance found on the grazed marsh. Mean plant size (Fig. 4) was considerably smaller on the grazed marsh than the ungrazed marsh. This was due to harvesting by herbivores and possibly selection pressure favoring plant maturation at a smaller size (Table 1). *S. alterniflora* stem densities were also much greater on the grazed marsh (Table 2). Maintenance of short *Spartina* may reduce inter-and intra-specific competition for various resources such as light, minerals, etc., thus allowing *Distichlis* and *Salicornia* species to increase their distribution over the marsh.

Non-nutritive physical relationships between vertebrates and halophytes may be categorized as (1) utilization for nesting material or roosting, and (2) trampling due to animal movement. Species nesting in the marshes tend to

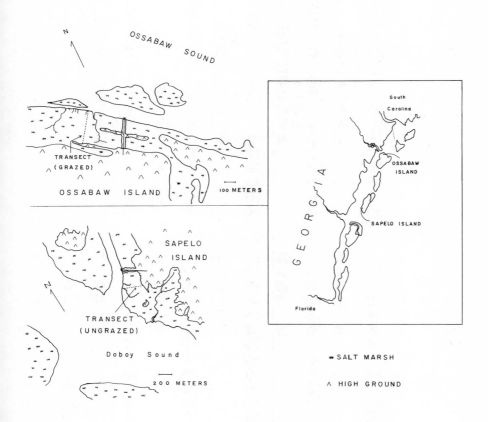

Figure 2. Salt marsh study sites along coastal Georgia.

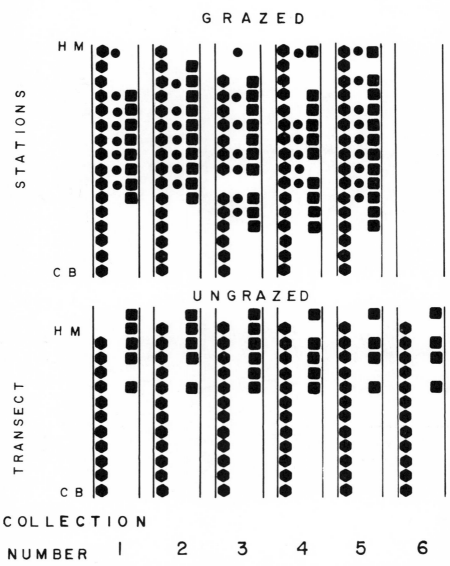

Figure 3. Distribution of marsh plants in six transects of a grazed and an ungrazed marsh. (HM = high marsh, CB - creek bank)

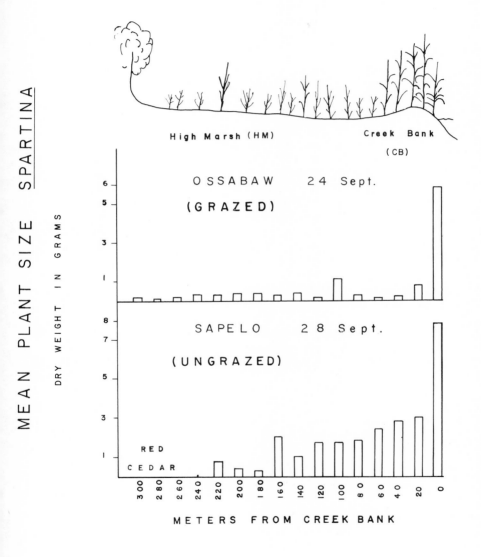

Figure 4. Mean plant size of marsh plants in a grazed an ungrazed salt marsh.

Table 1. Height of *Spartina alterniflora* at Inflorescense

Meters from Creek Bank	Mean Height in cm	Sample Size	S. E.
	Grazed Marsh		
300	23.4	2	6.1
260	29.7	9	2.6
100	19.4	9	2.3
80	27.0	12	2.1
60	25.2	7	2.8
	Ungrazed Marsh		
220	61.3	4	3.4
180	38.1	1	-----
160	68.2	11	4.1
120	60.1	11	4.4
100	63.6	17	2.9
80	81.5	4	6.5
60	86.4	3	3.7
40	59.5	15	3.3
20	65.2	25	2.4
Creek Bank	136.6	2	9.3

Table 2. Stem Density (Stems /m^2) of *Spartina alterniflora*

Station Location		Grazed Marsh	Ungrazed Marsh
High Marsh	−300m	782	-----------
	−280m	118	-----------
	−260m	672	-----------
	−240m	589	-----------
	−220m	980	378
	−200m	412	98
	−180m	375	208
Middle Marsh	−160m	442	321
	−140m	509	20
	−120m	510	201
	−100m	117	287
	− 80m	274	201
	− 60m	671	186
	− 40m	1256	159
	− 20m	506	158
Creek Bank		172	11

use available halophytes as the source of nesting material. For both birds and mammals, this consists of blades of grass interwoven to form a mat with a shallow central depression or enclosed structure with various openings, and a deep central depression. Locally, nesting can destroy a substantial amount of grass. Colonial avian nesting sites in the marsh are often hundreds of square meters in area, and are essentially a continuum of nest depressions formed from adjacent marsh grass. At the other extreme, relatively little marsh grass is destroyed in nest construction by highly territorial and non-colonial species. White Ibis (*Eudocimus albus*) on Georgia coastal marsh islands nest in dense colonies, destroying substantial areas of marsh vegetation (Shanholtzer et al., unpublished). In contrast, the rice rat, Long-billed Marsh Wren, and marsh rabbit (Caldwell, 1966) destroy minimal quantities of plant material.

The use of marsh grass for roosting purposes has already been discussed in part. Most plant based roosting by birds takes place in the medium to tall *Spartina* grass where presumably plants are better able to support the weight of the bird. Raccoons and mink construct beds of cord grass above high tide levels (S. A. Johnson et al., unpubl; Kale, 1965).

Trampling of marsh vegetation is pronounced only in those areas actively grazed. Mammal trails through the marsh are maintained in a trampled state, but do not amount to a significant area.

Indirect Relationships

It was previously stated that consequences of indirect vertebrate-salt marsh plant relationships are generally exhibited some time after an actual encounter between these biotic groups. This feature involves nutrient and material cycling in the marsh and dispersal of halophyte seeds.

Nutrient Cycling

Nutrient enrichment of the marsh via fecal input from colonial nesting birds may be considerable. A heron colony on a small marsh island in coastal Georgia (Satilla River, Camden Co., Georgia) consisted of 3000 to 4000 active nests in 1970 (Shanholtzer, Hopkins and Dopson, unpublished). An estimate was obtained of 3-4 metric tons (wet weight) of fecal material excreted on this island (Shanholtzer, unpublished). The significance of this input, however, has not been quantified. Preliminary studies of artifically fertilized vegetation zones have shown a two-fold increase in live standing crop of short *Spartina* at an application of 20.2 g/m^2 nitrogen applied as ammonium nitrate (J. L. Gallagher, personal communication). Since avian fecal and urinary matter is concentrated, this enrichment process may be locally significant.

Migratory and wintering birds add nutrient to the marsh surface. At times, tens of thousands of Tree (*Iridoprocne bicolor*) and Barn (*Hirundo rustica*) Swallows may be observed migrating and feeding above the marsh surface (GFS). These species and shorebirds and ducks (Reimold, 1968), for example, deposit fecal matter about the marsh surface. Quantification of this input, and its effects, would be difficult at best.

In essence, we can visualize both nutrient conversion within the salt marsh ecosystem as by rails and wrens, and an exchange from outside the system as from migrating birds and raccoons, mink, fish, and rabbits.

Material Export

Plant material is removed from the marsh by vertebrates for nest materials. The Boat-tailed Grackle and Red-winged Blackbird exemplify this behavior. These species remove material, some of which would otherwise be recycled within the salt marsh ecosystem. In relation to total plant biomass, however, plant removal as this is negligible.

Seed Dispersal and Removal

Seed dispersal of halophytes is assisted by vertebrate activities. Granivorous feeding by birds dislodges unconsumed seeds. Feeding by large herbivores additionally resulted in seed scattering. In the grazed marsh, masses of *S. alterniflora* seeds were suspended from the marsh during flooding tides, after apparently being dislodged by grazing animals. Suspension of these large amounts were never noticed on the ungrazed marsh (Shanholtzer, unpublished), indicating a more gradual loss of seeds and/or a lower density of seeding plants. It would also seem that the distribution and mixing of seeds (genomes) from different areas should be greater on the grazed marsh than the ungrazed marsh.

Conclusion

At the initiation of this project, I was impressed by how little experimental work had been done involving the role of vertebrates in the salt marsh. Isolated instances of thorough work have been done (i.e., Kale, 1965; Sharp, 1967), but certainly nothing comprehensive. What I have attempted to produce is a framework for future study of a functional nature. Species listings of vertebrates which occur in the marsh are numerous (Johnson *et al.*, in press; Roberts *et al.*, 1956; Bangs, 1898; and others). Now we should attempt to quantify these species' role in the ecosystem.

In many respects salt marsh grasslands resemble classic terrestrial grassland communitites. Principles of predator-prey interaction, utilization of grasses for material purposes, and consumption, for example, are likely the same. Consequently, the investigator can draw heavily from studies already undertaken in these habitats. The major distinctions between these areas involve differences in water and salinity regimes.

Finally, I must note the omission of one large vertebrate in this report. This organism's role in the salt marsh, however, has been documented countless times in a similar number of conservation publications. Thus, it is inappropriate to elaborate further on the role of this species in the salt marsh.

Acknowledgments

This research was funded in part by the U.S. Department of Commerce, Office of Sea Grant, to the University of Georgia. I wish to thank Drs. R. J. Reimold and J. L. Gallagher, and my wife Sheryl for critically evaluating this manuscript.

Literature Cited

Bangs, O. 1968. The land mammals of peninsular Florida and the coast region of Georgia. Proc. Boston Soc. Nat. Hist. 28(7):157-235.

Bent, A. C. 1921. Life histories of North American Gulls and Terns. Smithsonian Institution United States National Museum Bulletin 113. U. S. Government Printing Office. Washington, D. C. 337 pp.

Bent, A. C. and Collaborators. 1968. Life Histories of North American cardinals, grosbeaks, buntings, towhees, finches, sparrows, and allies. Dover Publications, Inc. New York. 602 pp.

Caldwell, L. D. 1966. Marsh rabbit development and ectoparasites. J. Mamm. 46:527-528.

Fisler, G. F. 1965. Behavior of salt marsh *Microtus* during winter high tides. J. Mamm. 42:37-43.

Hitchcock, S. W. 1972. Fragile nurseries of the sea. National Geographic. 141(6):729-765.

Hutchinson, G. E. 1953. The concept of pattern in ecology. Proc. Acad. Nat. Sci. (Phila.), 105:1-12.

Johnson, A. S., H. O. Hillestad, S. A. Fanning, and G. F. Shanholtzer. (In press) An ecological survey of the coastal region of Georgia. U. S. Government Printing Office. Washington, D. C.

Johnston, R. F. 1956. Population structure in salt marsh song sparrows. (Part II). Condor 58:254-272.

Kale, H. W. 1965. Ecology and bioenergetics of the long-billed marsh wren *Telmatodytes palustris griseus* (Brewster) in Georgia salt marshes. Publ. Nattall. Ornithol. Club, No. 5. Cambridge, Mass. 142 pp.

Odum, E. P. 1966. Regenerative systems: discussion (A. H. Brown, discussion leader). In: Human Ecology in Space Flight (D. H. Calloway, ed.). The N. Y. Acad. Sci. Interdisciplinary Communications Program, New York. pp. 82-119.

Odum, E. P. 1971. Fundamentals of Ecology. 3rd ed. W. B. Saunders Co. Phila. 574 pp.

Reimold, R. J. 1968. Evidence for dissolved phosphorus hypereutrophication in various types of manipulated salt marshes of Delaware. Doctoral Dissertation. Marine Laboratories, University of Delaware. Newark, Delaware. 169 pp.

Roberts, H. C., J. M. Teal, and E. P. Odum. 1965. Summer birds of Sapelo Island, Georgia: a preliminary list. The Oriole 21:37-45.

Shure, D. J. 1970. Ecological relationships of small mammals in a New Jersey barrier marsh habitat. J. Mamm. 51:267-278.

Smalley, A. E. 1960. Energy flow of a salt marsh grasshopper population. Ecology 41(4):672-677.

SALT MARSH PLANTS AND FUTURE COASTAL SALT MARSHES IN RELATION TO ANIMALS

Franklin C. Daiber

College of Marine Studies
And
Department of Biological Sciences
University of Delaware
Newark, Delaware 19711

Introduction

Warren (1911) reminded us that man has been manipulating tidal marshlands for agricultural purposes since antiquity. Shaw and Fredine (1956) considered the Swamp Land Acts of 1849, 1850, and 1860 to have paved the way for a century of exploitation of wetlands for agricultural purposes. Penfound and Schneidau (1945) described some of the ensuing detrimental results. Smith (1902) called attention to the role of tidal marshlands in mosquito breeding while Stearns, MacCreary, and Daigh (1940) pointed out the economic role these marshes have played in muskrat production. Shaw and Fredine (1956) and Barske (1961) (see Literature Cited for other references) stressed the value of tidal wetlands in providing food, shelter, and nesting sites for waterfowl and other wildlife. Niering (1961) talked about the contributions tidal marshes can make to education and scientific research. Their role as a source of food (Rankin 1961) and mineral resources (Sanders and Ellis 1961) has been identified. It is conceivable the fine marsh sediments may contribute to the separation of fresh ground water levels of the adjoining uplands and the salt waters of the bays and estuaries. Odum (1961) called attention to the very important role tidal marshes play in the great productivity of fish and shellfish in our estuarine and coastal waters. Daiber (1959) identified the uses to which these wetlands are put with the ensuing conflicts of interest.

Richard Pough (1961) reiterated the natural beauty of these coastal marshlands, of the near-sighted views of some people, and how, through the use of easements, this wild and muted beauty can be retained for the present and the future. Goodwin (1961) presented a five step course of action to maintain clean, productive coastal lands: (1) land acquisition, (2) protection of lands already in public ownership, (3) control of dredging and filling, (4) legal zoning, and (5) public education on a broad front. Odum (1961) expressed the need to consider: (1) a functional approach to both utilization and production, (2) the estuary as a whole, (3) the great diversity inherent in this ecosystem, and that one-crop development should be discouraged, and

(4) ways to cope with the man-made alterations which originate from other than biological motives.

Numerous authors have described the plant associations of tidal marshes and much of this has been brought together in Chapman's (1960) treatise. For the New England marshes, he identified the *Spartina alterniflora glabrae* (tall form of *alterniflora*) as the lowest most vegetation zone. This gives way with an increase in marsh level to *Spartina patens* which, in turn, is replaced by *Juncus gerardi*. *Distichlis spicata* is frequently associated with *S. patens*. Cottam et al. (1938) identified three vegetation zones as *Spartina alterniflora*, *Distichlis spicata*, and *S. patens*. Penfound and Hathaway (1938) described the vegetation zonation for the Gulf Coast marshes as did Purer (1942) for California marshes in San Diego County. Kerwin and Pedigo (1971) identified four marsh plant associations with their dominant species: edge marsh *S. alterniflora* (tallest forms), low meadow *S. alterniflora* (low form), salt-grass meadow *Distichlis spicata*, and high meadow *S. patens*.

Many authors attribute various factors to this zonation. DeVido (1936) identifies (1) likelihood for submergence, (2) type of substrate, (3) salinity of the soil and water, and (4) adaptation to mechanical action of wind and sun. Adams (1963) mentions soil pH, depth of the water table, and soil nutrients. Aeration of the soil as a factor is identified by DeVido (1936) and Chapman (1960).

However, tidal inundation and salinity are considered to be primary factors by many (Johnson and York 1951, Penfound and Hathaway 1938, Daigh, MacCreary, and Stearns 1938, MacNamara 1952, Cottam and Bourn 1952, Adams 1963) in determining marsh plant distributions. Allan (1950) identifies the presence of water as the great equalizer and salt as the great limiter. Cottam et al. (1938) said a 0.1 foot change in water level can effect distributions. Penfound and Hathaway (1938) report a less than 3-inch change in elevation can cause a transition from one community to the next. Adams (1963) identifies the mean elevation of occurrence above mean sea level divided by one-half of the mean tide range of the area as a characteristic constant for each salt marsh species. *Spartina alterniflora* flourishes in salt water, fresh water, and where frequent tidal flooding occurs but not where permanent flooding exists (Queen 1973). It fails to attain normal growth and flowering when the water table falls and is replaced by one of the hay species, *S. patens* or *Distichlis spicata*, which require less water. The cattail *Typha angustifolia* is associated with high water levels while *Scirpus olneyi* occupies an intermediate position (Daigh et al. 1938).

Salinity will effect plant distributions. In the Gulf Coast marshes, Penfound and Hathaway (1938) identify the dominants in the fresh water marsh as *Typha* spp., *Scirpus californicus*, and *Panicum hemitomon*, while the salt marshes are characterized by *Spartina patens*, *Distichlis spicata*, *Spartina alterniflora*, and *Juncus roemariannus*. Bourn and Cottam (1939) relate that wild rice *Zizania aquatica*, smartweeds *Polygonum* spp., and wild millet *Echinochloa* spp. are found in the fresher marshes, and that wild rice occupies the same water level position as the tall form of *Spartina alterniflora* in saline areas and *Spartina cynosuroides* in the intermediate brackish waters.

There are characteristic animal distributions associated with the various plant communities. Many workers have pointed out the relationship between vegetation and mosquito breeding (Smith 1902, 1907; DeVido 1936; Connell 1940; Florschutz 1959 (a); Catts et al. 1963), the general consensus being that where the marshes are characteristically inundated by regular tides, there is little or no mosquito reproduction. However, *Spartina patens* is identified as having a heavy mosquito population, usually *Aedes sollicitans*. Darsie and Springer (1957) point out breeding is highest

Average number of *Aedes* larvae and pupae per dip in salt marsh vegetation (Ferrigno 1958)

Plant Species	Mosquito Species		
	A. cantator	A. sollicitans	Total
S. patens	0.31	6.95	7.26
S. alterniflora	0.01	0.37	0.38
mixture	0.29	2.42	2.71

adjacent to the uplands associated with *Distichlis spicata*, *S. patens*, and *S. alterniflora*. Smith (1902, 1907) states mosquito breeding is lower where fiddler crabs are abundant and provide surface drainage by their burrowing. DeVido (1936) relates mosquito abundance to vegetation which is frequently flooded but where mosquito predators cannot get among the plants. Along this vein, Harrington and Harrington (1961) demonstrate that among fishes the area residents, such as the various Cyprinodont species, are the only important predators on mosquito larvae, that transients coming from the estuary onto the mash surface on a flooding tide generally ignore the larvae.

Surveys indicate larval green head flies (Tabanidae) are found in the wetter portions of the salt marsh (Hansen 1952). However, Gerry (1950) pointed out these flies develop primarily in the higher portions of the salt marshes reached only by the higher tides. Depressions where water tends to accumulate are not attractive as larvae, subjected to more than two days submergence, are apt to drown. He went on to say evidence indicates the larvae originate in the creeks from which they migrate to thatch piles at the head of the marsh. Jamnback and Wall (1959) found the tabanid larvae associated with several species of salt marsh vegetation but most abundant in *S. alterniflora* and *S. patens*. They always found them in conjunction with vegetation or thatch piles. Contrary to Gerry, they believe the larvae can survive in water for a long time. Olkowski (1966) found most larvae at the upper edge of the *S. alterniflora* with fewer individuals present as ground elevation increased toward S. patens. The mature larvae seem to be disseminated to higher ground by tidal action. MacCreary (1940) describes ditches filled with water and overhanging vegetation as being excellent sites for egg deposition and larval development of some tabanids.

More recent work, in contrast to the earlier reports (Rockell and Hansen 1970), demonstrates female tabanids oviposit primarily on vegetation of a certain height and this choice of site influences subsequent larval distributions. *Tabanus nigrovittatus* and *Chrysops* spp. were found in greatest abundance below the mean high water level on gently sloping banks where the cordgrass was about two feet tall. *Tabanus lineola* and *Tabanus* sp. were most numerous at higher soil elevations where the grasses were shorter. These investigations found few larvae along the ditches where the sparse tall *S. alterniflora* grows. Rockell and Hansen believe larval migration does not occur, based on an absence of seasonal changes in larval numbers and sizes at various elevations, the ability of larvae to emerge at low elevations where tidal inundation occurs, and the contiguous distribution of *Chrysops* spp. at low elevations suggesting the location of egg masses from which they hatched.

Davis and Gray (1966) identify the generally characteristic insect fauna for each North Carolina salt marsh plant community. They point out the highly dissimilar fauna for the two types of high marsh, *Distichlis* having a much larger number of insect species than *Spartina patens*.

MacDonald (1969) has done a detailed study of the molluscan fauna of the *Spartina-Salicornia* salt marshes and tidal creeks of the Pacific Coast. Each contains its distinctive fauna characterized by one or two widely distributed and abundant species. While other species are patchily distributed and low in numbers, the structure of each community is fairly uniform throughout the faunal province. Teal (1958) describes the distribution of fiddler crab species in a Georgia marsh being related to plant community and substrate. *Uca pugnax* shows a preference for a mud substrate and is found in all but the tall *Spartina* edge marsh. It is numerous only in the medium and short *Spartina* areas. The distribution of *U. minax* is determined by its preference for a muddy substrate and brackish water. It is found in certain parts of the short *Spartina* marsh. *U. pugilator* is limited by competition or interference in burrowing. It is found on a more sandy substrate, on the creek banks, and in the *Salicornia-Distichlis* marsh areas. Kerwin (1971) reports *Uca minax* widely distributed among vegetation types within the salinity range of 2-16 ppt. High population densities are associated with the tall and short *Spartina* zones and the edge of the *Distichlis-S. patens* community. The snail *Melampus bidentatus* is most abundant in the salt marsh with a frequency of 22.1 % and an average density of $7.24/m^2$. In the brackish water marsh, dominated by *Spartina alterniflora-Scirpus robustus* stage and *Spartina Cynosuroides*, there is a frequency of 2.9% and an average density of $0.23/m^2$ (Kerwin 1972).

There are various forms of bird life specialized to nest and feed in salt marshes (Urner 1935). Their distributions are related to the wetness of the marsh, creek, and pond depth, salinity, and accessibility of tides. Clapper rails are associated with a low wet marsh while birds like the bittern, pied billed grebe, and coot have a marked preference for a permanent pond.

The distribution of muskrats is related to salinity, as salinity increases the populations of muskrats decline. They prefer the less saline types of vegetation for food such as the brackish water 3-square sedges *Scirpus* spp. and the cattails *Typha* spp. (Stearns and Goodwin 1941). Muskrats are most abundant in the upper reaches of tidal streams and marshes where the water is fresher and tidal amplitude is reduced (Dozier 1947). Higher average weights are recorded for muskrats taken from the fresher, less brackish portions of a marsh (salinity 0-10 ppt. and vegetation-*Scirpus Typha*) as opposed to higher saline areas (salinity 15-43 ppt and vegetation-*Spartina* spp., *Scirpus*, *Typha*, *Juncus*) (Dozier, Markley, and Llewellyn 1948).

There has been some kind of tidal marshland manipulation going on for a long time. It has been motivated by the seeming need for more agricultural land (Warren 1911), the desire to reduce mosquito born diseases (Means 1903), insect pest species, and the desire to restore an ecological balance with particular reference to wildlife (Gabrielson 1936). Much of the present day tidal marsh changes have their bases in the understanding of the life cycle of the salt marsh mosquito *Aedes sollicitans* (Smith 1902). Since there is a well-established relationship between the natural distribution of salt marsh plant communities and their associated animal populations, how does tidal marshland management affect these floral and faunal associations?

a) Types

1. Water. The management of tidal marshlands can take several forms. Natural tidal inundation can be a factor limiting the distribution of mosquitoes. Connell (1940) pointed out that young *Aedes* larvae fail to appear in portions of a tidal marsh which is flooded by tides as frequently as 25 days per lunar month. He went on to say that abundant mosquito breeding can be expected in portions of *Spartina alterniflora* marsh where the frequency of tidal inundation is less than eight days per lunar month. This would imply that if all tidal wetlands were to remain as low marshes, as defined by Chapman (1960, p. 50), there would be few mosquitoes. However, it is well established that all tidal wetlands are not equally covered and flushed by the tides. Since the upper marshes characterized by *Spartina patens* retain pockets of water which facilitate mosquito breeding, some other form of water management is called for.

Stearns (1951) clearly stated that water management implies there can be various ways in which a water problem can be handled, and if used judiciously, such management might be the approach for those areas where there are conflicts of interest. Stearns considered drainage to be a form of water management in that it is a process of drawing off water from an area by degrees. To facilitate such drainage, ditches are dug by plows or entrenching machines or by blasting across the surface of the marsh in a regular pattern to move water from the marsh surface into tidal creeks. Any surface pools and low spots are connected to the streams by such ditching.

Another form of water management is to impound water behind dikes so an area is variously inundated. Christopher and Bowden (1957) considered water management, properly employed, as the single most potent mosquito control in fresh-water impoundments. They went on to say that for proper management, the purpose of the reservoir must be decided and this, in turn, will dictate the form of management that will be necessary which, in turn, will determine the kinds of mosquitoes such an impoundment will produce. Impoundments can take various forms: high level impoundments where water, other than rain, normally does not enter except when the manager wishes and is maintained above the tidal flow; low level impoundments where some water is retained behind a sluice gate as the tide recedes.

Permanent pools are often created on the marsh surface by blocking drainage from a low area, or by blasting (Provost 1948). Such basins are shallow and are often called "champagne" pools, a name derived from Clarke's (1938) likening the action of mosquito predators to bubbles in champagne. There are often ditches radiating outward from these pools across the marsh surface forming an inter-connecting network. Occasionally, such pools are associated with low level impoundments (Smith 1968).

Diking to exclude water from an area, as in the production of salt hay, is a form of water management (Ferrigno 1959). Levies constructed in parts of Gulf Coast marshes enhance the use of these areas by cattle, and when properly constructed with earth plugs in the borrow pits, water management is improved (Williams 1955).

2. Burning. Burning is another form of tidal marshland management. Lynch (1941) considered fire, before its use as a management tool, to have catastrophic effects. One purpose of controlled burning is to eliminate the holocaust created by wild conflagrations. Marsh fires fall into three classes. Cover burns are the most widely employed method for removal of accumulated dead vegetation. These burns do not affect the structure of the marsh since the fire does not reach the basal parts of the plants which are protected by a water cover. A clean cover burn provides large open areas whereas spotty cover burns provide dispersed open patches amongst the dense vegetation.

On the other hand, root burns or deep peat burns are carried out during a dry period. Root burns kill off the climax vegetation which, if left unburned, becomes a dense mass. Plants of lower succession groups are considered more desirable. Deep peat burns remove the accumulated peat often to the underlying mineral soil, thus creating open water areas when it rains.

Burning may serve one or more functions: improvement of habitat, promotion of food production, increasing availability of food, protection from fire, and facilitate trapping (Lynch 1941).

b) History

Embanking and drainage of tidal marshes for agricultural purposes dates back to antiquity. For nearly three centuries the coastal European countries bordering on the North Sea have engaged in this type of agriculture. There have been many failures in this country and few successes over the past two hundred years (Wright 1911, Penfound and Schneidau 1945). Successful ventures include rice growing in the South Atlantic coastal states and cranberries and truck farms along the North Atlantic coast, (Nesbit 1885, Means 1903, Wright 1907, Warren 1911).

At the turn of the century there was the recognition of the role of the mosquito in disease transmission and where the mosquito was controlled, diseases like malaria could be kept in check. It had long been recognized that tidal marshes breed mosquitoes, detracting from the peoples and their livestock's comfort and well being, as well as a hindrance to the development of the country in such localitites (Means 1903, Wright 1907).

Smith (1902), in describing the life cycle of *Culex sollicitans* Wlk (*Aedes sollicitans*), advocated filling those depressions on the marsh surface adjoining the uplands where the mosquito breeds, or drainage by ditching so tidal action flushes out such areas. Smith (1907) not only advocated ditching for mosquito control but advised such drainage would enhance the production of salt hay *Spartina patens* and *Juncus gerardi* by permitting water circulation and the reduction of water logged soils. This latter point was to become a focus of contention in subsequent years. Such ditches to be effective should be wide and deep, cutting through the sod into the mud because shallow ditches tend to fill. He advocated complete water removal from a completely diked marsh or its value would be reduced.

Diking as a means of water management for agricultural purposes reached elaborate levels (Warren 1911). Drainage by ditching for the rapid removal of water from the marsh surface became the primary means for mosquito control. With the passage of the years, many thousands of tidal marshland acres were ditched and progress was evaluated in miles of ditches dug each year. Such ditching probably reached its peak during the depression years of the 1930's when federal and state agencies with large appropriations for the relief of unemployment became involved in such activities (Stearns MacCreary, and Daigh 1940).

Up to this point, ditching was concerned only with the elimination of mosquito breeding. Little or no consideration was given to other consequences. While relief workers were making progress by digging miles of ditches, a few people with wildlife concerns began to question the wisdom of such activity, pointing out there were other organisms, both plant and animal, associated with these tidal marshes. Urner (1935) pointed out a number of game birds, waterfowl, and other bird species would be adversely effected by ditching. Bradbury (1938) described the adverse effects of mosquito control measures on waterfowl and shore bird populations in the Duxbury, Massachusetts, marshes. Cottam (1938) likened ditching for mosquito control to provide work for those on the relief roles to that of scalping an individual to cure a case of dandruff or burning the granary to get rid of the rats. Cottam and associates (Cottam et al. 1938, Bourn and Cottam 1939, 1950;

Cottam and Bourn 1952) noted ditching lowers the water table and has adverse effects on the waterfowl populations and their plant and invertebrate food supplies. Stearns, MacCreary, and Daigh (1939, 1940) called attention to similar adverse effects on muskrat populations.

While wildlife advocates were decrying the effects of ditches, mosquito control people adhered to the use of them for such control. Cockran (1935) did not believe ditching was harmful to muskrat production in tidal areas. He did say the water level should not drop more than two to three inches below the general marsh surface and that such levels could be maintained by flood gates. Later (1938), Cockran definitely stated ditching improved muskrat trapping in the State of Delaware. He went on to say that where the marshes were flooded in Sussex County, mosquito breeding increased with no increase in the growth of desirable vegetation. Headlee (1939) did not believe ditching lowered the water table or that it had any appreciable effect on marsh vegetation or adverse effect on wildlife. Travis et al. (1954) reported on the vegetation of 16 marshes in Volusia County, Florida, during the years 1939-1942 and again in 1953. During the first three years after ditching, there were no appreciable changes in the amount of cover for each plant species. During the 10-year period from 1942 to 1953, they reported a decrease in *Distichlis* and an increase in *Salicornia* and *Avicennia* (Black Mangrove). They went on to say none of the vegetational changes altered the marsh character 10 years after ditching was initiated. Any floral changes were related to tidal fluctuations because in northern Volusia County, where tidal ranges were small, vegetation changes were smaller than in the southern part of the county where tidal fluctuations were greater.

While arguments raged about the value of ditching as a water management technique for mosquito control, the concept of biological control began to be put forward. Gabrielson (1936) stated the need for information about plant and animal distributions as related to tidal inundation, the use of natural pools for food, shelter, and residence for the enemies of mosquitoes. He talked about artificial ponds associated with ditches for mosquito control. Glasgow (1939) pointed out biological control is not a simple thing. The effectiveness of mosquito control would be determined on a biological and ecological basis. Bradbury (1938) described their attempts in bringing waterfowl back to the Duxbury marshes. Prior to 1931, when mosquito control operations were completed, there were many different species of waterfowl on these marshes. During a 1936 survey evidence of this earlier bird, activity was still apparent with seven gunning stands evident in the marshes which had become dry and devoid of birds. The technique of restoration was based on the premise that mosquito larvae would be eaten by *Fundulus heteroclitus*, the mummichog minnow. The job was to create a habitat where fish could live at low tides and high temperatures. Former potholes were restored by damming outlets with sods. Care was taken to keep the water level about nine inches below the marsh surface, thus, keeping it free of water. Some potholes were deepened to assure sufficient water for

Fundulus to live in during dry periods. Controlled burning of salt hay made a variety of insects available for shore birds and it helped control mosquitoes by enhancing standing water evaporation. Ditches were partially blocked so water was retained but did not flow out over the marsh surface. Bird use was reported to immediately increase without any loss in mosquito control.

At the same time, Clarke (1938) erected three categories of fresh-water marshes: (1) permanent, holding water at all times; (2) intermittent, periodically wet and dry; and (3) temporary, holding water for only a few days. Clarke advocated permanent and intermittent marshes need not be drained, that it was better to control mosquitoes by encouraging their aquatic enemies in permanent pools on these marshes. Intermittent ponds can be a wildlife oasis by creating a central water hole with channels connecting shallow pools throughout the marsh. Protection and a resting site would be provided to wildlife by the central pool, while mosquito-eating fish, living in this pool, would radiate outward through the channels as the water level rose.

Price (1938) advocated a new approach to ditching without draining the marshland. The procedure was to dig a shallow ditch about 12 inches below the high tide mark but not to connect these ditches to tidal streams or guts, a blind ditch. Each high tide flushes the marsh potholes bringing in a new supply of fish flowing out over the entire marsh and on the ebb, collecting in the ponds, potholes, and ditches. The water level would be raised throughout the marsh, keeping water in the potholes and ponds that otherwise dried up. These depressions were freed of mosquitoes and maintained a good stand of widgeon grass, a prime habitat for ducks.

Cottam (1938) brought the concept of biological control into even sharper focus. Water control rather than drainage should be practiced in waterfowl areas in need of mosquito reduction. In such wildlife habitats, where permanent ponds are involved, mosquito control should be attempted by biological methods rather than by mechanical drainage because biological control methods usually induce less marsh modification. Before any ditching is done, a soil profile is needed, followed by ditching adjusted to the soil type to produce a minimal ecological impact. Cottam advocated the establishment of permanent pools serving as reservoirs for mosquito-eating fish with channels radiating outward, permitting these fish to get out over the marsh surface. Diking and water impoundment was advocated with the use of weir boards and sluice gates that would not restrict tidal flow yet would maintain a proper head of water in the marsh area.

Following the presentation of these several 1938 papers, numerous studies were carried out on the effects of water management through impounding and permanent pothole development on mosquitoes and on wildlife. MacNamara (1949) reported on the restoration of waterfowl and muskrats on a salt hay marsh that had earlier been important for waterfowl. In 1953, MacNamara reiterated the need for establishing permanent fresh-water ponds and recognition of the relationship between vegetation types and water levels to recognize mosquito breeding areas. Several workers (Chapman and

Ferrigno 1956; Catts 1957; Darsie and Springer 1957; Florschultz 1959a, b; Tindall 1961) noted the changes in mosquito species and numbers between natural, ditched, and impounded marshes. Other workers reported the general increases of various wildlife species in conjunction with impounded marsh areas (Darsie and Springer 1957; Florschutz 1959 a, b; Catts et al. 1963; Lesser 1965).

High level impoundments have their problems. They continue to breed mosquitoes although different ones flood private properties, enhance the encroachment of reed *Phragmites communis* and restrict nursery and food areas for marine fish (Ferrigno and Jobbins 1966). They also interrupt nutrient flow between the impounded marsh and the estuary (Reimold 1968). Following the principle established by Clarke (1938), some low level impoundment and/or champagne pool systems have been established (Shoemaker 1964, Bosik 1967, Smith 1968, Harrison 1970).

Another conflict of interests began to make itself evident. Ferrigno (1959, 1961) and Ferrigno and Jobbins (1966) point out that the production of salt marsh hay *Spartina patens* has a degrading effect on wildlife. By diking water, flow is excluded from these hay lands and the production of mosquiotes is greatly enhanced. (See section on effects of impoundments—birds.) Pesticides have been recommended as a control for mosquitoes within these diked areas but the use of these chemicals is so critical in terms of effects on waterfowl that Ferrigno and Jobbins (1966) strongly urged these diked areas to be restored to their former naturally flooded condition where periodic flushing occurs.

Lastly, Ferrigno, Jobbins, and Shinkle (1967) and Ferrigno and Jobbins (1968) advocate quality ditching of only heavy breeding areas. Deep, wide, and straight ditches that will not fill in as rapidly should be used in open marsh management, providing there is permanent flooding or good tidal circulation in breeding depressions. Sod should not be piled, but mashed into the surface, nor should it be placed on one side only forming a circulation barrier.

c) Effects of Ditching

1. Vegetation. Salt hay *Spartina patens* and *Juncus gerardi* production can be enhanced through drainage by ditching, facilitating water circulation. Smith (1907) was cognizant of the fact that diking would change salt marsh vegetation to that of a fresh-water marsh or an upland. Florschutz (1959a) noted no drastic vegetative changes on a ditched marsh over a short term, however, he did expect changes when he stated the ditches may eventually lower the water table so that *Hibiscus* and salt hay would replace *Spartina alterniflora*. The long term effect of ditching is generally considered to have a deleterious impact on tidal marsh vegetation by lowering the water table and replacing these plants with less desirable species from higher levels such as salt marsh flea bane *Pluchea camphorata* and salt marsh aster *Aster subulata* followed by *Iva* and *Baccharis* (Cottam et al. 1938). Prior to ditching, the mallows *Hibiscus oculiroseus* and *Kosteletzkya virginica*, the seaside

goldenrod *Solidago sempervirens, Aster novi-belgii, Bidens trichosperma, Pluchea camphorata,* ₁and the swamp milkweed *Asclepsias incarnata* were scattered generally throughout most of the marsh area adjoining Delaware City, Delaware, but they never dominated. Yet, in three years (1936-1938) following ditching, they dominated the marsh replacing *Scirpus olneyi* (Stearns, MacCreary, and Daigh 1940). The 3-square *S. olneyi* is found near the bordering high lands where fresh water enters. Ditching has the greatest effect on this species than on *S. patens* or *S. alterniflora* by lowering the water table and permitting intrusion of salt water in the Ocean City, Maryland, marshes (Cory and Crosthwait 1939). During a 10-year period, a pure stand of *Spartina alterniflora* covering 90% of the Mispillion marshes was replaced to a large degree by *Baccharis* (Bourn and Cottam 1950). Ditching drains the permanent pools, destroying the resident widgeon grass *Ruppia maritima* and dries out the marsh surface leaving it more vulnerable to destructive fires (Cottam et al. 1938). On the contrary, Cory and Crosthwait (1939) advocated ditching as a means for maintaining water levels in pools containing widgeon grass by permitting intrusion of water on the flooding tide. This would be particularly true in the summer when the pools would normally dry up. They maintained the ditches had to be shallow, otherwise the pools would be drained. Both Headlee (1939) and Travis et al. (1954) asserted ditching did not lower the water table nor was there any adverse effect on vegetation. *Spartina alterniflora* production can be enhanced in some places through ditching by increasing the edge effect (Ferrigno 1961).

2. Mosquitoes and Biting Flies. Smith (1902, 1907) advocated ditching as a means to control mosquitoes in those portions of the marsh considered to be good breeding areas. On the contrary, ditching *S. alterniflora* marsh is not considered an effective control of the salt marsh mosquito (Connell 1940). During an experimental ditching study, breeding was completely eliminated as no larvae or pupae of mosquitoes (*A. cantator, A. sollicitans, C. salinarius*) were observed within the study area of the Appoquinimink marshes (Stearns, MacCreary, and Daigh 1940). However, these same researchers point out that while ditching may control mosquitoes, the ecological side effects may be greater than the original objective (1939). Eight species of mosquitoes were collected from a drained marsh in the Assawoman Wildlife area in 1956 with *Aedes* spp. making up 92% of the collections. The populations varied with local tidal and rainfall fluctuations, and little breeding progressed beyond the first instar larval stage. This was attributed to the effectiveness of the drainage system as a means of controlling salt marsh mosquito production (Catts 1957). This work was continued in 1957-1958 by Florschutz (1959a) who reiterated the effectiveness of drainage in that he collected very few pupae in contrast to thousands of larvae. Drainage was more effective where the ditches were kept clear of encroaching vegetation. Florschutz recorded an increase in the numbers of *Culex salinarius* during a rainy season in contrast to low numbers during a dry year. He attributed this to the greater permanence of puddles on the marsh surface in the wet year.

MacCreary (1940) was of the opinion that ditching for mosquitoes may have helped control biting flies in some instances. However, to be effective, the ditches have to be free of overhanging vegetation as stated earlier. Gerry (1950) stated there was no relationship between the mosquito control ditches and the distribution of green head fly larvae. Rockell and Hansen (1970) found the greatest numbers of larvae six yards from the ditches in 1965-1966 and eight yards in 1967.

3. Other Invertebrates. Ditching has been deemed responsible for the marked declines in the invertebrate fauna of tidal marshes. Based on quadrat studies in ditched Delaware marshes in the course of one season, the invertebrates, consisting largely of molluscs and crustacea, were reduced in numbers in the following fashion:

Spartina alterniflora association — 43.5%
Scirpus robustus association — 92.6%
Distichlis spicata association — 71.9%
Spartina patens association — 84.2%

Table 1 depicts the differences attributed to ditching for these four plant associations. Not only is there a decline in numbers, but almost without exception there has been a reduction in the numbers of species. The years 1937-1938 were fairly wet years so the differences may have been even greater had the weather been drier. The greater differences for the *Scirpus robustus* association are particularly significant. The area occupied by this association is considerably smaller than for the others. The plants are more widely dispersed, thus the fauna is more accessible to wildlife and, lastly, the substrate is more porous and thus more easily drained (Cottam et al. 1938; Bourn and Cottam 1939, 1950).

The lowering of the water table enhances leaching and oxidation which produces more acid conditions, adversely affecting molluscs and crustacea dependent on alkaline conditions for shell building. Under anaerobic conditions, sulfates in the seawater are reduced to sulfides in the presence of organic matter. In this form these sulfides combine with the iron in the clay to form polysulfides. No further changes will occur if the soils remain wet. If they dry out, the sulfides oxidize to form sulfuric acid. This can reduce the pH to 2.5 or less (Neely 1962).

4. Birds. Most shore birds and waterfowl are adversely affected by ditching through a reduction in food supply, primarily molluscs and crustaceans as pointed out by Bourn and Cottam. Birds that need a low wet marsh, such as the clapper rail, will not be affected as long as the salinity and the water level are not changed. Birds that need a fairly constant water supply, like the bittern, piedbill grebe, and coot, will be seriously affected by such drainage. Black ducks, Willet, Virginia rail, and some of the herons will be adversely effected (Urner 1935). Bradbury (1938) reports drastic declines in waterfowl and shore birds following ditching of the Duxbury marshes. However, Florschutz (1959a, b) observed little difference between a natural and a ditched marsh. Wildlife species numbered 19 for 1957 and 29 for 1958, while

the ditched marsh had 20 for 1957 and 25 for 1958. The most common bird species observed on the ditched marsh during both years were gulls, green herons, crows, and red-wing black birds. Little evidence of waterfowl usage was noted.

Ditching can encourage *S. alterniflora*, which increases the edge effect, providing more shelter and nesting sites for clapper rail and black duck (Ferrigno 1961). Stewart (1951) found a high correlation between nest densities and the amount of edge between the tall (and dense) and the short (and sparse) growth form of *Spartina alterniflora*. The highest correlation (0.9747) existed when the edge consisted of 20 yards of short and 10 yards of tall *Spartina*. The lowest correlations pertained to pure stands of short (0.9180) and tall (0.9385) *Spartina*. Nest density in the best edge was 2.5 ± 0.3/acre with nests within 15 feet of the creeks. Stewart recommended ditches or creeks be constructed with sloping, rather than vertical sides, to produce the desired edge effect.

5. Muskrats. Cockran (1938) presents figures to support the contention that ditching has no adverse effects on muskrats in Delaware.

Sussex County — 95% ditched-trapping increased 15%
Kent County — 30% ditched-trapping increased 10%
New Castle County — 0% ditched-trapping increased 0%

Stearns, MacCreary, and Daigh (1939, 1940) strongly refute this contention. Their work in Kent and Sussex counties showed that muskrat marshes ditched for mosquito control have lower water tables which adversely affects the vegetation needed for food and house construction, thus, the animals migrate out of the area (see section on effects of ditching on vegetation). When the normal water level is at or very near the surface, the plants (*Scirpus olneyi* and *Spartina cynosuroides*), upon which the muskrat subsists, flourish best. Ground level in itself is not considered a factor in determining the location of a house, but there is a definite relationship between ground elevation and the height of the water table at the selected house site. The mean water level at 101 houses was 4.55 inches below the mean ground elevation (5.43 feet above local mean low water) with a range of 0.08-9.98 inches. This meant that the mean water level for 80% of the houses was above and 20% below the mean water level in the ditched area. (see following diagram derived from Stearns et al. 1940).

d) Effects of Impoundments

1. Vegetation. Diking has been done to encourage the production of salt marsh hay *S. patens* (Smith 1907, Ferrigno 1959). Smith remarked the diked area loses its value unless the water is completely removed. Ferrigno agrees water removal is necessary to enable machinery to cut and process the hay.

Impounding produces vegetation changes from *S. alterniflora–S. patens* to pond weed *Potomogeton berchtoldi* and *P. pectinatus*, widgeon grass, *Ruppia maritima*, and algal mats (during low water principally *Rhizoclonium*). Around the edges a variety of emergent species will appear; 3-square *Scirpus americanus*, rose mallow *Hibiscus moscheutos*, cattail *Thypha*, reed *Phragmites communis*, and switch grass *Panicum virgatum* (Springer and Darsie 1956).

This same pattern was noted by Florschutz (1959) and Tindall (1961) in the Assawoman and Little Creek Wildlife areas, respectively, in which salt marsh vegetation *Spartina, Distichlis, Scirpus, Hibiscus, Cladium, Baccharis,* and *Iva* were reduced and replaced by open water and emergent types *Potomogeton* and *Ruppia* beds, *Typha, Echinocloa, Cyperus,* and *Chara.* Florschutz noted *Typha* tended to expand in some places and decrease elsewhere. *Spartina patens* was greatly reduced on the inner flooded portions of the marsh, but flourished along the edge of the impoundment.

Mangold (1964), Shoemaker (1964), Smith (1968), and Harrison (1970) noted the replacement of *S. patens* by *S. alterniflora* during the flooding of low level impoundments and a champagne pool system. This was particularly true for early flooding. *Distichlis* associated with *S. patens* survived and flourished (Shoemaker) while alone and with *Juncus gerardi* it survived, but not well (Mangold). Flooding caused the disappearance of *Baccharis* and *Iva*, and the submergent horned pond weed *Zannichella palustris* increased and flourished. Smith (1968) noted the increase of *Baccharis*, common reed *Phragmites*, poke weed *Phytolacca americana*, and the fox tail grasses *Setaria faberii* and *S. magna* on the higher ground created by the embankments and spoil piles. The poke weed and fox tail grasses provide excellent food and shelter for wildlife. He also noted a decline of widgeon grass in the older impoundments when it had flourished in younger pools.

As Smith had noted in 1907, impounding a tidal marsh will change the salt marsh vegetation to that of a fresh-water swamp or that of an upland.

2. Mosquitoes. Impounding sharply alters the numbers and species composition of the mosquito population as compared to a natural tidal marsh (Chapman et al. 1954, 1955, 1956; Catts et al. 1963; Shoemaker 1964). *Aedes sollicitans* is invariably the most abundant mosquito in natural marsh conditions, making up as much as 96% by number of immatures dipped (Darsie and Springer 1957, Tindall 1961b). Chapman et al. (1954) identify *A. sollicitans* and *A. cantator* along with *Anopheles bradleyi* and *Culex salinarius*, the salt marsh group of mosquitoes, as being typical of a natural salt marsh. *Aedes* spp. can be essentially eliminated from impoundments while *Culex* and *Anopheles, Uranotaenia sapphirina,* and *Mansonia perturbans* breeding increases after impoundment (Chapman et al. 1954, 1956; Tindall 1961b; Franz 1963). *Mansonia* deposits its eggs in sedge tussocks and under mats of cattail debris under flooded conditions, and water level manipulation would be the most effective control (Hagmann 1953, Chapman and Ferrigno 1956). Springer and Darsie (1956) report *Anopheles* being eliminated along with the two *Aedes* species. Darsie and Springer (1957) note many of the permanent water mosquitoes are unimportant because of short flight patterns, biting habits, and other behavior.

Mosquito immatures dipped April to October, 1959 and 1960, Little Creek Wildlife Area, Little Creek, Delaware (Tindall 1961b)

	Impoundment				Natural Marsh			
	1959		1960		1959		1960	
	No.	%	No.	%	No.	%	No.	%
A. sollicitans	56137	96.1	76	0.4	1502	96.5	7203	99.9
Aedes sp.	62	0.1	10	-----	1	-----	-----	-----
TOTAL	56199	96.2	86	0.4	1503	96.5	7203	99.9
A. bradleyi	19	-----	677	3.5	1	-----	1	-----
Anopheles sp.	-----	-----	161	0.8	-----	-----	-----	-----
TOTAL	19	-----	838	4.3	1	-----	1	-----
C. salinarius	2143	3.7	17163	88.7	25	1.6	-----	-----
Culex sp.	69	0.1	743	3.8	27	1.7	-----	-----
TOTAL	2212	3.8	17906	92.5	52	3.3	-----	-----
Uranotaenia sapphirinia	-----	-----	512	2.7	-----	-----	-----	-----

Manipulation of water levels within impoundments seems to control the magnitude of breeding. MacNamara (1952) reports constant water levels produces mosquitoes while draw-down decreases breeding. Chapman and Ferrigno (1956) note heaviest breeding for *Aedes* at water depths 5-10 inches below meadow level, slightly below meadow level for *Culex*, and slightly above meadow level for *Anopheles*. Chapman et al. report summer draw-down controls *Mansonia perturbans* but they, along with Darsie and Springer (1957), report greatly enhanced *Aedes* broods following rains or reflooding. Catts et al. (1963) recommend moderate water levels of 9-12 inches compatible for reduced mosquito production and enhanced waterfowl usage. Tindall (1961b) suggests the higher water levels would reduce vegetation and expose the mosquito larvae to wave action and predators.

In contrast to flooding by impoundments, diking done to enhance the production of salt hay *Spartina patens* by preventing regular flooding of tidal marshland produces great broods of mosquitoes, primarily *Aedes sollicitans* and *Culex salinarius* (Ferrigno 1959).

Mosquito production from diked and undiked marsh land (modified from Table 2 of Ferrigno 1959).

Vegetation Type	C. salinarius	A. sollicitans	Dips	larvae-pupae
		Larvae-Pupae/Dip		
Undiked				
S. alterniflora	0	0.0001	8280	1
S. patens	0.11	2.74	1080	3293
P. virgatum	0.003	0	600	2
Woodland swamp	0.02	0.01	840	620
Diked				
S. alterniflora	0.26	4.22	360	1701
S. patens	0.72	3.54	2760	13376
S. cynosuroides	2.94	4.66	240	1988
P. virgatum	0.21	0.75	1320	2219
D. spicata	0	3.52	600	2761
J. gerardi	0	2.86	240	780
Typha	2.01	0.21	120	707

Certain *Tabanus* spp. appeared to be controlled by an accidental summer draw-down on high level impoundments. The species composition is altered in that *T. lineola* and *T. atratus* are more evident in the impoundment while *T. nigrovittatus* is more abundant in the natural tidal areas (Olkowski 1966). Harrison (1970) suggests low level impoundments have little influence on tabanid breeding, at least during the first year after impoundment. Anderson and Kneen (1969) report that *Chrysops fuliginosus* can be controlled by temporary impoundment of salt marshes. To be most effective, impoundment should occur when the species undergoes larval-pupal and pupal-imagnial molts, May through July in Connecticut.

3. Other Invertebrates. There is little specific information about the effects of impoundments on invertebrates. The micro-crustaceans have received some preliminary attention (Ruber and Jobbins 1963). Bosik (1967) lists some of the more obvious forms. Smith (1968) records fiddler crabs *Uca*, grass shrimp *P. pugio*, blue crab *Callinectes sapidis*, the snail *Melampus bidentatus*, and the mussel *Modiolus demissus* as being common, but they are common in natural marshes as well. Under salt marsh hay management, *Uca*, snails, and mussels decline in numbers. The first two are important in clapper rail and black duck

diets (Ferrigno 1961).

4. Fishes. Fish species typically associated with tidal creeks and marshes can be expected in impoundments. These include the mummichog *Fundulus heteroclitus*, striped killifish *Fundulus majalis*, eel *Anguilla rostrata*, and the sheepshead minnow *Cyprinodon variegatus*. Carp *Cyprinus carpio* and mosquitofish *Gambusia affinis* can be expected in less saline impoundments.

Part of the basic premise of water management and mosquito control on these marshes has been to provide a suitable habitat for mosquito-eating fish and the means for these fish to get at the mosquitoes. Both Mangold (1962) and Shoemaker (1964) believe the attraction of herons, bitterns, terns, etc. to low level impoundments is due to the increase in fish numbers. For reasons not fully understood, several species of fish-eating birds declined following impoundment in Florida (Provost 1969). *Fundulus* spp. can survive and reproduce in impoundments and provide an effective control of mosquito larvae provided water levels are high enough to permit the fish to forage amongst the vegetation (Alls 1969). The numbers of fish species increase and tend to shift toward the fresh-water forms with the production of young following impoundment including the bullhead *Ictalurus nebulosus*, the pickerel *Esox americanus*, and the sunfish *Lepomis gibbosus*. Bullfrogs *Rana catesbiana* and the snapping turtle *Chelydra serpentina* also appear in impoundments (Darsie and Springer 1957).

5. Birds. Impoundments have been established and developed for the restoration of wildlife, particularly waterfowl and shore birds. Bradbury (1938) reported on this for the Duxbury marshes while MacNamara (1949) demonstrated the fresh-water impoundments on the Tuckahoe, New Jersey, marshes were capable of producing large quantities of desirable waterfowl food by a complete draw-down. Eight inches of water tended to enhance muskrat food instead. MacNamara reported a kill of 1.91 ducks/hunter/day for the 1948 season compared to 0.79 ducks/hunter/day in 1947. Following restoration, several new species of ducks put in an appearance in the area; red head, ring-neck, surf-scoter, and the shoveller.

Other investigators have reported the increased usage of impoundments by birds (Catts 1957, Darsie and Springer 1957, Florschutz 1959, Tindall 1961, Mangold 1962, Shoemaker 1964, Lesser 1965, Smith 1968, Provost 1969). Darsie and Springer identified 86 bird species in contrast to 55 in the area prior to impoundment. Tindall reported a 3-fold increase. Smith sighted 62 species on the impoundments and 39 on the natural marsh areas. Several of these workers reported increased numbers of broods of young following impoundment, particularly black ducks. These impoundments offer emergent and submergent vegetation as food for ducks, fish and invertebrates as food for wading birds, open water for resting areas, and flooded cover for nesting sites.

Some bird species have declined in number with the advent of impoundments. The clapper rail has often disappeared and has been associated with the absence of fiddler crabs (Darsie and Springer 1957,

Mangold 1962, Shoemaker 1964). Both Mangold and Shoemaker also noted declines in the small birds, song sparrow, seaside sparrow, sharptail sparrow, and yellow throat warbler, primarily through loss of nesting sites and food. Smith (1968) reported marsh wrens and seaside sparrows relatively abundant, especially where the tide bush grew along ditch or pool margins; however, no comparative quantitative data was given. Provost (1969) indicated the decline in the dusky seaside sparrow following impoundment on Merritt Island, Florida. This bird is reported to prefer the *Distichlis* habitat. Fish-eating birds also declined on Merritt Island, especially a merganser. Provost could give no reason for such declines and went on to say that while six species of birds were reduced, there was not apparent effect on seven species and an increase in number was noted for 22 species.

While most workers have shown enhanced bird usage following impoundment, Ferrigno (1959, 1961) reported a decline in diked salt hay meadows.

Waterfowl and clapper rail usage and mosquito breeding for a New Jersey salt marsh. (Modified from Table 3, Ferrigno 1959)

Vegetation	Number of birds flushed. 10 censuses of 100 acres in each zone		
	Waterfowl	Clapper rail	Larvae & pupae
S. alterniflora tall form	1742	29	0
S. alterniflora short form	1239	33	0.0001
S. patens	285	3	3.05
diked hay meadow *S. patens*	111	1	4.13

6. Mammals. Evidence of mammals has increased with the creation of impoundments (MacNamara 1952, Catts 1957, Darsie and Springer 1957, Tindall 1961, Mangold 1962, Shoemaker 1964, Smith 1968). Darsie and Springer noted continued maintenance of high water levels within an impoundment tended to restrict muskrat usage but the recession of water from the vegetated margin during the summer enhanced plants attractive to muskrats and waterfowl. MacNamara and Tindall report increased numbers of muskrat houses while Mangold found muskrats to have preference for the fresher water impoundments.

Annual muskrat house count in New Jersey impoundments
(from MacNamara 1952)

Pond	County	1946	1947	1948	1949	1950	1951
1	Cape May	0	0	0	34	100	105
2	Cape May		12	37	8	22	56
3	Cape May		185	193	194	203	223
1	Atlantic	20	120	17	37	156	346
2	Atlantic			182	73	165	215
3	Atlantic				50	78	141

Whenever impoundments are flooded with high salinity water through storm action or by evaporation raising the salinity during a drought, muskrat populations decline through loss of food and drinking water (Dozier 1947). Smith (1968) noted no increase in small mammal populations, but increased evidence of predators about the impoundments suggested such small mammal populations had indeed expanded. Increased mammal activity was attributed to increased variety of habitat and increase in edge effect associated with embankments (Florschutz 1959).

e) Effects of Burning

The magnitude of burning will determine the effects on vegetation (Penfound and Hathaway 1938, Lynch 1941, Smith 1942). The character of the marsh vegetation is not altered during a cover burn when the marsh is flooded with water. Heavy accumulations of dead vegetation are removed in this fashion. The more severe root and peat burns kill off the climax vegetation and allow plants of lower successional stages to reappear. Intense and prolonged peat burns can bring about a reversion to the early hydric community with expanses of open water. Fire prevents the accumulation of organic matter and thus impedes the elevation of the marsh and succession to upland communities.

f) Effects of Filling

Filling a tidal marsh generally destroys it. This is a characteristic result where the fill rises above tidal height. The marsh plant and animals are buried under fill that is derived from a variety of sources, dredge spoils, construction wastes, sewage, and town dumps. Upland species will invade the fill and the giant reed grass *Phragmites communis* is a particularly characteristic and aggressive invader. It takes over almost to the complete exclusion of other plant species. Solid stands of this grass does not provide a suitable wildlife habitat.

Tidal marshes can be restored when filled by dredge spoil so long as the spoil does not exceed the intertidal height. New marshes have been created in North Carolina on spoil from navigational channel dredging operations. *Spartina alterniflora* has been particularly useful in establishing new marsh in the lower tidal reaches while *Spartina patens* has been successfully established at the higher tidal level. A secondary benefit could be the stabilization of the spoil so it would not drift back into the channels (Woodhouse, Seneca, and Broome 1972). *Spartina townsendii* has been used in Europe to stabilize mudflats for agricultural purposes, reduce channel filling and coastal erosion. It has a vigorous growth habit and at times has become a pest when introduced for the above purposes. It has been suggested that sound botanical advice should be sought before any introduction is made and that traditional reclamation techniques be modified rather than embark upon an expensive eradication procedure. Intensive grazing by sheep has been recommended as a management procedure (Ranwell 1967).

g) Effects of Feeding of Waterfowl and Mammals

Feeding activity of game birds and mammals can produce various ecological changes. Bourn and Cottam (1939) reported brant and snow geese turned to *Spartina alterniflora* with the disappearance of eel grass *Zostera*. A natural marsh can be markedly modified and mosquito breeding changed by large flocks of feeding snow geese. Small "eat-outs" produce the most mosquitoes while large "eat-outs" tend to produce ponds which support populations of mosquito-eating fish, sheepshead minnow *Cyprinodon*, and mummichog *Fundulus* (Ferrigno 1958).

Mosquito breeding on Egg Island marsh (Delaware Bay area) affected by feeding snow geese (Ferrigno 1958)

	Percent denuded marsh within 70 ft. radius of station marker	Vegetation Type	Water Depth Inches	Number Per Dip	Larvae-pupae Total	Number Dips
A	10-30%	S. patens	0-2	12.71	6607	520
B	19-40	S. alterniflora	0-6	1.56	2119	1355
C	50-90	mixed Spartina	1-8	0.05	25	460
D	50-90	S. alterniflora	1-6	0.03		490
E	100 (ponds)	mixed Spartina	5-24	0.002	3	1401

Geese (Snow, Blue, Canada) feed heavily on 3-square bullrush *Scirpus americanus*. They feed most easily on this plant when the flats are flooded and do so by "puddling out" the rhizomes. In so doing they make the seeds that have dropped into the mud available to ducks (Griffith 1940, Lynch et al. 1947).

Geese can lay waste to large acreage of muskrat marsh when they congregate in a local area in large numbers. They invade *Typha* marshes when water levels are low and damage *Panicum repens*, the dog tooth grass, when marshes are flooded. Such goose damage may initiate a muskrat "eat-out" which can have a greater impact by destroying more vegetation (Lynch et al. 1947). There is a decline in animal weight and in population size following the ruinous effect on marsh vegetation brought about by over population and under trapping (Dozier et al. 1948).

The ecological consequences resulting from goose and muskrat damage to a marsh are discussed in considerable detail by Lynch et al. (1947) and summarized in Figure 1. There seems to be no way to foretell which of these changes will result from a particular type of damage. In general, the sooner recovery can start following an "eat-out", the better the chance the marsh will become productive. Repair delayed one to three years usually results in unproductive climax marsh.

Muskrats have not damaged coastal cattle range in Louisiana and Texas in recent years. In earlier times, rat populations overran cattle range and rice crops but trapping now keeps their numbers down. During dry weather, cattle can destroy muskrat houses (Lynch et al. 1947).

Geese can severely damage cattle range, primarily in the fall, in that two-fifths of the wintering grounds of blue and snow geese are classed as cattle range. The extent of damage is determined by water levels. Low water causes geese to leave the cattle range and invade muskrat marshes. Normal water levels find the geese moving to new marsh areas before any one feeding ground is denuded. High water produces severe competition between cattle and geese (Lynch et al. 1947).

"Eat-outs" are beneficial to a great variety of shore birds, waders, and some ducks. They are attracted to the marsh plant seeds that become available and the minnows, crustaceans, insects, etc. that abound in new "crevys". "Eat-outs" are most attractive during the first year. After that their wildlife value declines. "Eat-outs" in 3-square marshes not only destroy muskrat and goose habitat but that of mallards as well (Lynch et al. 1947).

Burning can have an impact on such marshland feeding relationships. Burns during the summer will remove the coarser plant material cattle find less palatable. Cattle will be attracted to the succulent new vegetation and can cause muskrat damage by destroying the latter's houses. Summer burns will also drive muskrats out of a marsh for lack of house building materials. Fall burning of *S. patens* and other marsh grasses produces an attractive fodder for geese and cattle and, where the two co-exist in a marsh area, competition will be the outcome. Spring burning is more beneficial to muskrat populations

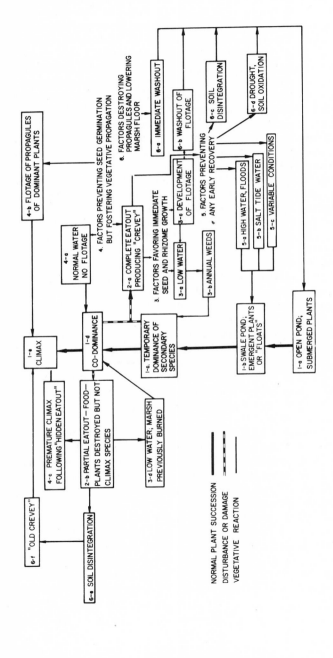

FIGURE 1
Ecological consequences of muskrat and goose damage to marshes (from Lynch et al. 1947). (Reproduced by permission of the senior author and the editor of the Jour. Wildlife Mgmt.).

(Griffith 1940, Smith 1942, Lay and O'Neil 1942, Lynch et al. 1947, Williams 1955, Neely 1962).

Another interesting feeding relationship has been reported from England for marsh vegetation and sheep grazing (Yapp et al. 1917, Ranwall 1961). Sheep seem to prefer the marsh to more upland pasturage. Practically all species on the permanent marsh are eaten except *Juncus maritimus* with a marked preference for *Armeria maritima* and *Festuca rubra* located in intermediate associations in a vertical zonation. Extention of pasturage was recommended by land reclamation, elimination of unproductive pans, and prevention of the spread of economically useless *Juncus* by diking.

Another advantage to this kind of pasturage was the freedom of the sheep from infestation of the liver fluke *Fasciola hepatica* by the absence of the intermediate host *Limnaea trunculata* from brackish and saline marshes while prevalent elsewhere.

An interesting interaction was the suggested use of sheep grazed salt marsh to provide turf for golf or bowling greens, particularly the fine grass *Festuca rubra* (Yapp et al., 1917).

Ranwell (1961) describes the vegetative changes in a tidal marsh brought about by sheep. Controlled sheep grazing can be carried out in the intertidal zone as long as it is confined to the upper reaches during the summer. Before the introduction of *Spartina "townsendii"*, *Puccinellia maritima* predominated near the mean high tide mark and was considered to be a fine sheep pasturage. *Spartina* is more aggressive, prefers a softer soil, and can survive grazing except where the ground is packed along sheep paths. Ungrazed *Spartina* marsh will be transformed into *Phragmites* or *Scirpus* marsh, taking about eight years. Grazed *Spartina* marsh will be changed to a predominantly *Puccinellia* association in about 10 years. Succession to *Puccinellia* is favored in more saline and sandy areas while *Phragmites* or *Scirpus* are enhanced in less saline muddy locations. Grazing will favor *Puccinellia* in muddy places.

VI. Discussion

Each tidal marsh community has its characteristic fauna. Numerous workers have pointed out that each vegetation type can be categorized according to the extent of certain mosquito breeding. Typically, as one proceeds in a *Spartina alterniflora* marsh from the frequently inundated tall form toward the less frequently flooded higher elevations, there is an increase in the number of salt marsh mosquito species, particularly *Aedes sollicitans*. Along with this increase in mosquitoes, there is the tendency to find greater numbers of biting flies. Whether the eggs of these flies are deposited on these higher elevations is unknown. The snail *Melampus* is found in the same locations with *Spartina patens, S. alterniflora* (short form), and *Distichlis*. Fiddler crabs (*Uca*) have an inverse relationship to *Melampus*. They are found near the creeks and among the tall form of *S. alterniflora*. With the presence of the snail and fiddler crab one can expect to see the black duck and clapper rail, respectively.

The life form and dispersion pattern of the tall form of *S. alterniflora* permits the mosquito-eating fish to move about among the plants on a flooding tide. The dense mats of *S. patens* and *Distichlis* prohibits such fish excursions, thus, mosquito larvae are freer of their fish predators.

The deeper surface pools, formed in a variety of ways (see Chapman's 1960 discussion of pannes and rotten spots), provide a residence for *Fundulus* and other mosquito-eating fish. These pools also provide submergent vegetation, *Ruppia* for example, an attraction to puddle ducks. By enhancing the edge effect, these pools also provide a greater variety of habitats, thus enabling more species to associate with the marsh.

Progression along the salinity gradient sees a change in vegetation from the saline *Spartina alterniflora* and *Juncus roemarianus*, through the brackish water species *S. cynosuroides* and *Scirpus olneyi*, to the fresher water *Zizania* and *Polygonum*. There is an associated reduction in tidal amplitude and increased numbers of muskrats along this gradient.

These relationships change with the natural sequence of events in ecological succession. There may be the orderly progression to a climax or, through some storm action or changes in sea level or wildfire, there will be a reversion to some earlier stage. In any case, the stage in the progression can be identified by its floral and faunal components. Allan (1950) called attention to the amazing uniformity of stages of marshland succession from Nova Scotia to Mexico. The same dominant plants occupy the same relative positions with respect to water table and salinity levels. Allan identified four stages in marshland succession. Stage One is dominated by *Spartina alterniflora*. *Distichlis spicata* dominates Stage Two, while Stage Three is dominated by *Spartina patens*. Allan considered this stage to be the most valuable in marshland development. Stage Three, along with Four which is dominated by *Spartina cynosuroides* and *Phragmites communis*, are viewed as edaphic grassland climaxes.

Man can and does bring about changes in this ecological succession by his application of water management techniques to tidal marshes. While ditching does control salt marsh mosquitoes and perhaps biting flies at certain times and locations, its overall effect is open to doubt. Water flow can be maintained in ditches if kept free of silt and vegetation by being dug straight and wide. In any case, marked ecological changes become evident in the marsh community. To begin with, the water table is lowered or, in effect, the elevation of the marsh is raised. Thus, there is a change from the low level, tall form of *S. alterniflora* marsh to the high level, short form of *S. alterniflora, S. patens-Distichlis spicata* marsh. Ditches, while intended to reduce mosquito breeding, can encourage it by creating more of the higher *S. patens* marsh through a lowered water table. Certainly, ecological succession to upland vegetation is hastened by enhancing the invasion of such vegetative types, and the decline of the more aquatic species such as *Ruppia*. The valuable muskrat foods *Scirpus* and *Typha* will give way in the process. Stands of *Scirpus olneyi* can also be adversely affected by the greater penetration of higher salinity water than would occur without ditching. Muskrat populations will decline because of changes in food, and the lowered water table under their houses. The average weights of animals also will decline with the change to less nutritive vegetation.

While this play is being acted out, another play of adveristy occurs involving water birds. Lowering the water table destroys the submergent food supply for the puddle ducks. Nesting sites are eliminated from the shores of these pools by a reduction of the edge effect. The food supply is also affected by the drastic reduction in invertebrates, especially molluscs and crustacea. While snails like *Melampus* are most frequently associated with the drier *S. patens-Distichlis* zone, numbers are reduced following ditching. Conversely, as was pointed out earlier, ditching can enhance the edge effect by encouraging the tall form of *S. alterniflora* with the associated fiddler crabs and clapper rails, provided water levels do not decline.

While ditches tend to hasten succession toward an upland community, burning and permanent impoundments, exclusive of diked salt hay marshes, have a retrogressive effect toward the open water community by flooding or lowering the marsh elevation through reduced accumulation of organic matter resulting from more rapid decomposition. The vegetational changes that occur are a reflection of the depth of flooding or intensity of burning. This regression away from the climax produces a vegetational community considered to be more valuable as food and shelter for many wildlife species. Increased numbers and species of waterfowl are attracted to these new conditions. Deeper water encourages submergent vegetation but changing water level promotes production of emergent plants along the boundaries of the impoundments which muskrats and some waterfowl species find more attractive. Such impoundments also tend to encourage fresher water types of vegetation, again favoring muskrats and waterfowl, and provide fish predators more opportunities to get at mosquito larvae.

Permanent impoundments eliminate salt marsh mosquitoes and possibly biting flies. Permanent water mosquito species neither travel as far nor have the feeding and behavioral traits that make the flooding water salt marsh species so obnoxious. This latter group are eliminated within the impoundment but the marsh acreage outside still produces *Aedes* spp. mosquitoes. Cost prohibits the construction and maintenance of dikes surrounding all tidal marshes. Furthermore, Reimold (1968) has shown that phosphorus exchange between a high level impoundment and the estuarine system is among the lowest for any kind of managed marsh. The reduced rate of exchange could have a retarding effect on the fertility of bays and coastal waters.

The wildlife principle of the edge effect in a tidal marsh is stimulated by burns and flooding. Spotty burns will produce new vegetation for feeding, surrounded by dense older plants for cover. Impoundments produce an edge in two ways. Dike construction provides a suitable habitat for upland vegetation such as pokeweed and foxtail grass which is valuable as wildlife food. The less desirable high tide bushes *Baccharis* and *Iva* are also encouraged although they do provide cover and nesting sites. These embankments and the adjoining emergent aquatic vegetation provide food, shelter, and nesting sites for a great variety of prey and predators (Smith 1968). They also serve as avenues of travel along with the adjacent borrow pits and ditches (Lay and O'Neil 1942, Williams 1955). Lesser (1965, p. 38) suggested that the extent of multiple edge effect of interspersed marsh vegetation can limit populations of resident waterfowl. When stands of emergent vegetation are in juxta-position with similar stands along the "borrow" pits, duck broods are observed more frequently and in greatest concentrations. In contrast, when only open water is available adjacent to the dike, there are fewer duck broods.

While high level impoundments were originally developed for wildlife usage, the maintenance of permanent water conditions does not prevent mosquito breeding. A diversity of habitat, including permanent pools, must be provided for wildlife. Mosquito-eating fish must have access to the larvae to provide effective control, yet have pools to retire to when the tide recedes. Nutrient exchanges between the marshes and the estuarine environment must be improved over that of high level impoundments. Potholing with radial ditching in heavy breeding areas can be a more effective control (Clarke 1938, Ferrigno 1958). The low level impoundment or champagne pool system appears to offer the best solution for these three needs. However, this system also has drawbacks. Spoil piles establish man-made high ground. While this spoil increases community diversity, it does create a disturbance. Muskrats can tunnel through any embankment and cause drainage. In order for any nutrient exchange to occur between such a marsh section and the estuary, there must be an ebb and flow of tide through a sluice gate. A very real engineering problem exists in trying to maintain these structures against the erosive power of the tidal flow.

Numerous workers emphatically state that great damage has been done to our tidal marshes by various drainage practices and other uses that are not natural to the habitat. Wildlife species and their habitats have suffered great and sometimes irreparable losses. With the exception of losses resulting from filling, much of this damage can be corrected. Kenny and McAtee (1938) stated that much marshland drainage and ditching had been carried out that should not have been done. Also, great abuse to animals and their rightful habitat have occurred through the persistence of man in the employment of trial and error methods. While mosquitoes seem to have been brought under general control, agricultural production has not been enhanced by the use of drained marshlands. The flow of energy and nutrients from the marshes to the estuarine ecosystem has been impeded.

Not only has damage been committed, but conflicts of interest have arisen. There have often been long standing conflicts between mosquito control people and wildlife interests, but this is not universally true. In recent years, conflicts between fisheries interests, real estate and industrial development, waste disposal concerns, and recreation have been added to the list. As Hawkes (1966) pointed out, with the exception of recreation most uses tend to degrade a marsh and mosquito control has upset ecological balances. Ferrigno (1961) identified a three-way conflict in the production of salt marsh hay in New Jersey. Development of hay production degrades a marsh for wildlife and encourages mosquitoes. As the climax vegetation *(Spartina patens)* is reached, mosquito breeding will be at a maximum and waterfowl usage will decline. Allan (1950) considered *S. patens* to be the most valuable stage in succession. It is often put to three uses which often conflict: intensive crop land management which has had limited success, grazing land for live stock, and valuable wildlife usage. According to Allan, when *S. patens* grows over a mineral soil, a decision must be made whether to maintain this vegetation at the climax stage or to encourage a secondary succession of annuals and seed bearing forbes. Livestock and fur bearers apparently do better on the climax condition while waterfowl prefer the secondary successional stage.

Any habitat management will alter the structure of the ecological community. Marsh management includes drainage and impoundment as a physical practice, while burning is identified as a cultural one (Smith 1942). Various investigators have advocated that biological control, with particular reference to mosquitoes, must be based on sound ecological principles (see Cottam and associates, Clarke, Gabrielson, Glasgow, MacNamara, etc.). One must recognize that each ecological community has its own unique floral components and that these develop in a normal sequence. Careful consideration must be given to the impact on this natural pattern by man's manipulations. Information is needed on species composition and distributions as related to the physical environment and associated species. There is need for a knowledge of the dynamics of plant ecology and the ability to recognize the various stages of primary and secondary succession. In addition, further data needs to be acquired on food and feeding habits, reproduction, population sizes, and age structure of the characteristic tidal marsh animals. Marshland management for wildlife and mosquitoes can be enhanced by inventorying the resource, determining relationships, and then facilitating the production and maintenance of food and cover.

Among others, MacNamara (1957) advocated the concept of multiple land use. He emphasizes that this does not mean an equality of land uses. He states these uses should be based on a broad interpretation of land classification and conservation, along with careful evaluation of existing potentials for the various uses under consideration. Good land use is based on good planning in relation to ecological parameters of the areas under consideration.

Unfortunately, tidal marshes have been subjected to short term goals with long term consequences. Some areas may never recover their natural state. Protective legislation is slow in coming. Riparian rights are still vague in most states (Hawkes 1966).

Conditions today have not advanced much beyond those reported by

Kenny and McAtee in 1938. However, signs are increasingly evident that we are becoming aware of our impact on the environment. By cooperating in management practices and by providing legal protection, we will continue to see and enjoy the diversity of marshland species in spite of the mosquitoes.

Acknowledgements

My thanks to those graduate students who spent many long hours working in Delaware's tidal wetlands assisting me in the development of a marsh program, to Norman Wilder and the old Game and Fish Commission for moral and financial support, and to Paul Catts for his critical review of the manuscript.

Literature Cited

Adams, D. A. 1963. Factors influencing vascular plant zonation in North Carolina salt marshes. Ecol. 44(3):445-456.

Allan, P. F. 1950. Ecological bases for land use planning in Gulf Coast marshlands. Jour. Soil and Water Conserv. 5(2):57-62, 85.

Alls, Ralph T. 1969. Killifish predation of mosquitoes in low level impounded Delaware salt marshes. Master's thesis, Univ. of Delaware.

Anderson, J. F., and F. R. Kneen. 1969. The temporary impoundment of salt marshes for the control of coastal deer flies. Mosq. News 29(2):239-243.

Barske, P. 1961. Wildlife of the coastal marshes. Connecticut Arboretum Bull. 12:13-15.

Bosik, J. J. 1967. The effectiveness of low-level impounded salt marshes in controlling the production of mosquitoes. Master's thesis, Univ. of Delaware.

Bradbury, H. M. 1938. Mosquito control operations on tide marshes in Massachusetts and their effect on shore birds and waterfowl. J. Wildlife Mgmt. 2(2):49-52.

Bourn, W. S., and C. Cottam. 1939. The effect of lowering water levels on marsh wildlife. Trans. N. Amer. Wildlife Conf. 4:343-350.

Bourn, W. S., and C. Cottam. 1950. Some biological effects of ditching tidewater marshes. Res. Rept. 19 Fish and Wildlife Serv., U. S. Dept. of Int. 17 pp.

Catts, E. P., Jr. 1957. Mosquito prevalence on impounded and ditched salt marshes, Assawoman Wildlife Area, Delaware, 1956. Master's thesis, Univ. of Delaware.

Catts, E. P., Jr., F. H. Lesser, R. F. Darsie, Jr., O. Florschutz, and E. E. Tindall. 1963. Wildlife usage and mosquito production on impounded tidal marshes in Delaware, 1956-1962. Trans. N. Amer. Wildlife Conf. 28:125-132.

Chapman, H. C., and F. Ferrigno. 1956. A three year study of mosquito breeding in natural and impounded salt marsh areas in New Jersey. Proc. N.

J. Mosq. Exterm. Assoc. 43:48-65.

Chapman, H. C., P. F. Springer, F. Ferrigno, and R. F. Darsie, Jr. 1955. Studies of mosquito breeding in natural and impounded salt marsh areas in New Jersey and Delaware in 1954. Proc. N. J. Mosq. Exterm. Assoc., 42:92-94.

Chapman, H. C., P. F. Springer, F. Ferrigno, and D. MacCreary. 1954. Studies of mosquito breeding in natural and impounded salt marsh areas in New Jersey and Delaware. Proc. N. J. Mosq. Exterm. Assoc. 41:225-226.

Chapman, V. J. 1960. Salt marshes and salt deserts of the world. Interscience Publ., N. Y. 392 pp.

Christopher, S., and N. W. Bowden. 1957. Mosquito control in Reservoirs by water level management. Mosq. News, 17(4):273-277.

Clarke, J. L. 1938. Mosquito control as related to marsh conservation. Proc. N. J. Mosq. Exterm. Assoc. 25:139-147.

Connell, W. A. 1940. Tidal inundation as a factor limiting distribution of *Aedes* spp. on a Delaware salt marsh. Proc. N. J. Mosq. Exterm. Assoc. 27:166-177.

Corkran, W. S. 1935. The relation of mosquito control to the muskrat industry on the salt marshes. Proc. N. J. Mosq. Exterm. Assoc. 22:19-23.

Corkran, W. S. 1938. New developments in mosquito control in Delaware. Proc. N. J. Mosq. Exterm. Assoc. 25:130-137.

Cory, E. N., and S. L. Crosthwait. 1939. Some conservation and ecological aspects of mosquito control. J. Econ. Entomol. 32(2):213-215.

Cottam, C. 1938. The coordination of mosquito control with wildlife conservation. Proc. N. J. Mosq. Exterm. Assoc. 25:217-227.

Cottam, C., and W. S. Bourn. 1952. Coastal marshes adversely affected by drainage and drought. Trans. N. Amer. Wildlife Conf. 17:414-421.

Cottam, C., W. S. Bourn, F. C. Bishopp, L. L. Williams, Jr., and W. Vogt. 1938. What's wrong with mosquito control? Trans. N. Am. Wildlife Conf. 3:81-107.

Daiber, F. C. 1959. Tidal Marsh: Conflicts and interactions. Estuarine Bull. 4(4):4-16.

Daigh, F. C., D. MacCreary, and L. A. Stearns. 1938. Factors affecting the vegetation cover of Delaware marshes. Proc. N. J. Mosq. Exterm. Assoc. 25:209-216.

Darsie, R. F., Jr., and P. F. Springer. 1957. Three-year investigation of mosquito breeding in natural and impounded tidal marshes in Delaware. Univ. of Del. Agr. Expt. Station Bull. 320. 65 pp.

Davis, L. V., and I. E. Gray. 1966. Zonal and seasonal distributions of insects in North Carolina salt marshes. Ecol. Monog. 36(3):275-295.

DeVido, L. A. 1936. Salt marsh vegetation and its relation to mosquito control. Proc. N. J. Mosq. Exterm. Assoc. 23:196-203.

Dozier, H. L. 1947. Salinity as a factor in Atlantic Coast tide water muskrat production. Trans. N. Amer. Wildlife Conf. 12:398-420.

Dozier, H. L., M. H. Markley, and L. M. Llewellyn. 1948. Muskrat investigations on the Blackwater National Wildlife Refuge, Maryland, 1941-1945. J. Wildlife Mgmt. 12(2):177-190.

Ferrigno, F. 1958. A two-year study of mosquito breeding in the natural and untouched salt marshes of Egg Island. Proc. N. J. Mosq. Exterm. Assoc. 45:132-139.

Ferrigno, F. 1959. Further study on mosquito production on the newly acquired Caldwalder Tract. Proc. N. J. Mosq. Exterm. Assoc. 46:95-102.

Ferrigno, F. 1961. Variations in mosquito-wildlife associations on coastal marshes. Proc. N. J. Mosq. Exterm. Assoc. 48:193-203.

Ferrigno, F., and D. M. Jobbins. 1966. A summary of nine years of applied mosquito-wildlife research on Cumberland County, N. J. salt marshes. Proc. N. J. Mosq. Exterm. Assoc. 53:97-112.

Ferrigno, F., and D. M. Jobbins. 1968. Open marsh water management. Proc. N. J. Mosq. Exterm. Assoc. 55:104-115.

Ferrigno, F., D. M. Jobbins, and M. P. Shinkle. 1967. Coordinated mosquito control and wildlife management for the Delaware Bay coastal marshes. Proc. N. J. Mosquito Exterm. Assoc. 54:80-94.

Florschutz, O. Jr. 1959a. Mosquito production and wildlife usage in natural, ditched, and impounded tidal marshes on Assawoman Wildlife Area, Delaware. Master's thesis, Univ. of Delaware.

Florschutz, O. Jr. 1959b. Mosquito production and wildlife usage in natural, ditched, and unditched tidal marshes at Assawoman Wildlife Area, Delaware. Proc. N. J. Mosq. Exterm. Assoc. 46:103-111.

Franz, D. R. 1963. Production and distribution of mosquito larvae on some New Jersey salt marsh impoundments. Proc. N. J. Mosq. Exterm. Assoc. 50:279-285.

Gabrielson, I. N. 1936. Information needed for a proper understanding of the effects of mosquito control work on the wildlife of tidal marshes. Proc. N. J. Mosq. Exterm. Assoc. 23:156-163.

Gerry, B. I. 1950. Salt marsh fly control as an adjunct to mosquito control in Massachusetts. Proc. N. J. Mosq. Exterm. Assoc. 37:189-193.

Glasgow, R. D. 1939. Biological control of mosquitoes; a special problem in applied ecology. Proc. N. J. Mosq. Exterm. Assoc. 26:20-25.

Goodwin, R. H. 1961. Connecticut's coastal marshes - a vanishing resource. Connecticut Arboretum Bull 12:1-36.

Griffith, R. E. 1940. Waterfowl management of Atlantic coast refuges. Trans. N. Amer. Wildlife Conf. 5:373-377.

Hagmann, L. E. 1953. Biology of *Mansonia perturbans* (Walker). Proc. N. J. Mosq. Exterm. Assoc. 40:141-147.

Hansens, E. J. 1952. Some observations on the abundance of salt marsh greenheads. Proc. N. J. Mosq. Exterm. Assoc. 39:93-98.

Harrington, R. W., Jr., and E. S. Harrington. 1961. Food selection among fishes invading a high subtropical salt marsh: from onset of flooding through the progress of a mosquito brood. Ecol. 42(4):646-666.

Harrison, F. J., Jr. 1970. The use of low level impoundments for the control of the salt-marsh mosquito, *Aedes sollicitans* (Walker). Master's thesis, Univ. of Delaware.

Hawkes, A. L. 1966. Coastal wetlands - problems and opportunities. Trans. N. Amer. Wildlife Conf. 31:59-78.

Headlee, T. J. 1939. Relation of mosquito control to wildlife. Proc. N. J. Mosq. Exterm. Assoc. 26:5-12.

Jamnback, H., and W. J. Wall. 1959. The common salt marsh Tabanidae of Long Island, New York. Bull N. Y. State Museum No. 375. 77 pp.

Johnson, D. S., and H. H. York. 1915. Relation of plants to tide levels. Carnegie Inst. Washington Publ. 206. 161 pp.

Kenny, F. R., and W. L. McAtee. 1938. The problem: drained areas and wildlife habitats. *In* Soils and Men. Yearbook of Agriculture, 1938.

Kerwin, J. A. 1971. Distribution of the fiddler crab (*Uca minax*) in relation to marsh plants within a Virginia estuary. Ches. Sci. 12(3):180-183.

Kerwin, J. A. 1972. Distribution of the salt marsh snail (*Melampus bidentatus* say) in relation to marsh plants in the Poropotank River area, Virginia. Ches. Sci. 13(2):150-153.

Kerwin, J. A., and R. A. Pedigo. 1971. Synecology of a Virginia salt marsh. Ches. Sci. 12(3):125-130.

Lay, D. W., and T. O'Neal. 1942. Muskrats on the Texas Coast. J. Wildlife Mgmt. 6(4):301-311.

Lesser, F. H. 1965. Some environmental considerations of impounded tidal marshes on mosquito and waterbird prevalence, Little Creek Wildlife Area, Delaware. Master's thesis, Univ. of Delaware.

Lynch, J. J. 1941. The place of burning in management of the Gulf Coast Wildlife Refuges. J. Wildlife Mgmt. 5(4):454-457.

Lynch, J. J., T. O'Neal, and D. W. Lang. 1947. Management significance of damage by geese and muskrats to Gulf Coast marshes. J. Wildlife Mgmt. 11:50-76.

MacDonald, K. B. 1959. Quantitative studies of salt marsh faunas from the North American Pacific Coast. Ecol. Monog. 39(1):33-60.

MacCreary, D. 1940. Report on the Tabanidae of Delaware. Univ. of Del. Agr. Exp. Sta. Bull. 226. 41 pp.

MacNamara, L. G. 1949. Salt marsh development at Tuckahoe, New Jersey. Trans. N. Amer. Wildlife Conf. 14:100-117.

MacNamara, L. G. 1952. Needs for additional research on mosquito control from the standpoint of fish and game management. Proc. N. J. Mosq. Exterm. Assoc. 39:111-116.

MacNamara, L. G. 1953. The production and conservation of wildlife in relation to mosquito control of state owned lands in New Jersey. Proc. N. J. Mosq. Exterm. Assoc. 40:74-79.

MacNamara, L. G. 1957. Multiple use of our lands, especially marsh land. Proc. N. J. Mosq. Exterm. Assoc. 44:103-106.

Mangold, R. E. 1962. The role of low-level dike salt impoundments in mosquito control and wildlife utilization. Proc. N. J. Mosq. Exterm. Assoc. 49:117-120.

Means, T. H. 1903. Reclamation of salt marsh lands, U. S. Dept. Agr. Bur. Soils Circ. 8 rev.

Neely, W. W. 1962. Saline soils and brackish waters in management of wildlife, fish and shrimp. Trans. N. Amer. Wildlife Conf. 27:321-334.

Nesbit, D. M. 1885. Tidal marshes of the United States. Misc. Spec. Rept. No. 7, U. S. Dept. Agr.

Niering, W. A. 1961. Tidal marshes: Their use in scientific research. Connecticut Arboretum Bull. 12:3-7.

Odum, E. P. 1961. The role of tidal marshes in estuarine production. The Conservationist. N. Y. Cons. Dept. 16:12-15, 35.

Olkowski, W. 1966. Biological studies of salt marsh tabanids in Delaware. Master's thesis, Univ. of Delaware.

Penfound, W. T., and E. S. Hathaway. 1938. Plant communities of the marsh lands of Southeastern Louisiana. Ecol. Monog. 8(1):1-56.

Penfound, W. T., and J. D. Schneidau. 1945. The relation of land reclamation to aquatic wildlife resources in southeastern Louisiana. Trans. N. Amer. Wildlife Conf. 10:308-318.

Pough, R. H. 1961. Valuable vistas: A way to protect them. Connecticut Arboretum Bull. 12:28-30.

Price, M. H. 1938. New developments in mosquito control in Rhode Island. Proc. N. J. Mosq. Exterm. Assoc. 25:111-15.

Provost, M. W. 1948. Marsh-blasting as a wildlife management technique. J. Wildlife Mgmt. 12(4):350-87.

Provost, M. W. 1969. Ecological control of salt marsh mosquitoes with side benefits to birds. Proc. Tall Timbers Conf. on Ecology. Animal Control by Habitat Mgmt. 1:193-206.

Purer, Edith A. 1942. Plant ecology of the coastal salt marsh lands of San Diego County, California. Ecol. Monog. 12(1):81-111.

Queen, W. 1973.

Rankin, J. S., Jr. 1961. Salt marshes as a source of food. Connecticut Arboretum Bull. 12:8-13.

Ranwell, D. S. 1961. *Spartina* salt marshes in southern England. I. The effects of sheep grazing at the upper limits of *Spartina* marsh in Bridgewater Bay. J. Ecol. 49(2):325-340.

Ranwell, D. S. 1967. World resources of *Spartina townsendii (Sensu Lato)* and economic use of Spartina marshland. J. Appl. Ecol. 4:239-256.

Reimold, R. J. 1968. Evidence for dissolved phosphorus hypereutrophication in various types of manipulated salt marshes of Delaware. Ph.D. diss., Univ. of Delaware.

Rockel, E. G., and E. J. Hansens. 1970. Distribution of larval horse flies and deer flies (Diptera:Tabanidae) of a New Jersey salt marsh. Ann. Ent. Soc. Amer. 63(3):681-684.

Ruber, E., and D. M. Jobbins. 1963. Dynamics of micro-crustacean populations on an impounded marsh. Proc. N. J. Mosq. Exterm. Assoc. 50:342-351.

Sanders, J. E., and C. W. Ellis. 1961. Geological aspects of Connecticut's coastal marshes. Connecticut Arboretum Bull. 12:16-20.

Shaw, S. P., and C. G. Fredine. 1956. Wetlands of the United States - their extent and their value to waterfowl and other wildlife. U. S. Dept. of Int. Fish and Wildlife Serv. Circ. 39. 67 pp.

Shoemaker, W. E. 1964. A biological control for *Aedes sollicitans* and the resulting effect upon wildlife. Proc. N. J. Mosq. Exterm. Assoc. 51:93-97.

Smith, D. H. 1968. Wildlife prevalence on low level impoundments used for mosquito control in Delaware, 1965-1967. Master's thesis, Univ. of Delaware.

Smith, J. B. 1902. The salt marsh mosquito, *Culex sollicitans*, Wlk. Spec. Bull. N. J. Agr. Exp. Sta. 10 pp.

Smith, J. B. 1907. The New Jersey salt marsh and its improvement. Bull. N. J. Agr. Exp. Sta. 207. 24 pp.

Smith, R. H. 1942. Management of salt marshes on the Atlantic Coast of the United States. Trans. N. Amer. Wildlife Conf. 7:272-277.

Springer, P. F., and R. F. Darsie, Jr. 1956. Studies on mosquito breeding in natural and impounded coastal salt marshes in Delaware during 1955. Proc. N. J. Mosq. Exterm. Assoc. 43:74-79.

Stearns, L. A. 1951. Introductory Statement: Symposium on water management and drainage for mosquito control. Proc. N. J. Mosq. Exterm. Assoc. 38:64-66.

Stearns, L. A., and M. W. Goodwin. 1941. Notes on the winter feeding of the muskrat in Delaware. J. Wildlife Mgmt. 5(1):1-12.

Stearns, L. A., D. MacCreary, and F. C. Daigh. 1939. Water and plant requirements of the muskrat on a Delaware tide water marsh. Proc. N. J. Mosq. Exterm. Assoc. 26:212-221.

Stearns, L. A., D. MacCreary, and F. C. Daigh. 1940. Effects of ditching on the muskrat population of a Delaware tidewater marsh. Univ. of Del. Agr. Expt. Sta. Bull. 225. 55 pp.

Stewart, R. E. 1951. Clapper rail populations of the Middle Atlantic States. Trans. N. Amer. Wildlife Conf. 16:421-430.

Teal, J. M. 1958. Distribution of fiddler crabs in Georgia salt marshes. Ecol. 39:185-193.

Tindall, E. E. 1961a. A two year study of mosquito breeding and wildlife usage in the Little Creek impounded salt marsh, Little Creek Wildlife area, Delaware, 1959-60. Master's thesis, Univ. of Delaware.

Tindall, E. E. 1961b. A two year study of mosquito breeding and wildlife usage in Little Creek impounded salt marsh, Little Creek Wildlife Area, Delaware, 1959-60. Proc. N. J. Mosq. Exterm. Assoc. 48:100-105.

Travis, B. V., G. H. Bradley, and W. C. McDuffie. 1954. The effect of ditching on salt-marsh vegetation in Florida. Proc. N. J. Mosq. Exterm. Assoc. 41:235-244.

Urner, A. 1935. Relation of mosquito control in New Jersey to bird life of the salt marshes. Proc. N. J. Mosq. Exterm. Assoc. 22:130-136.

Warren, G. M. 1911. Tidal marshes and their reclamation. Bull. Exp. Sta. U. S. Dept. Agr. 240:1-99.

Williams, R. E. 1955. Development and improvement of Coastal Marsh Ranges. *In* Water. Yearbook of Agriculture 1955, U. S. Dept. Agr. 444-450.

Woodhouse, W. W., Jr., E. D. Seneca, and S. W. Broome. 1972. Marsh building with dredge spoil in North Carolina. North Carolina State Univ. Agr. Exp. Sta. Bull. 445. 28 pp.

Wright, J. O. 1907. Reclamation of tide lands. U. S. Dept. Agr. Ofc. Expt. Sta. Rept. 1906:373-397.

Yapp, R. H., D. Johns, and O. T. Jones. 1917. The salt marshes of the Dovey estuary. J. Ecol. 5(2):65-103.

PART IV. APPLIED RESEARCH RELATED TO HALOPHYTES

REMOTE SENSING AS A TOOL FOR STUDYING THE ECOLOGY OF HALOPHYTES[1]

John L. Gallagher
The University of Georgia Marine Institute
Sapelo Island, Georgia 31327

Introduction

The basic ecological questions relevant to understanding halophyte communities are the same as those suggested by James Bonner (19) for the generalized ecosystem: First, who lives where and why? Second, who eats whom? Third, what is the physiology of togetherness? Modern sensors often coupled with airborne vehicles are capable of detecting much information relative to Bonner's three questions. Recent developments in instrumentation for quantifying and interpreting images produced by remote detectors have also given impetus to the current enthusiasm for remote sensing. These techniques used in conjunction with basic ground truth measurement can produce information over large areas quickly and at a low cost.

Types of Sensors

A large variety of sensors capable of examining various portions of the electromagnetic spectrum emitted or reflected from the plant community are available. Within the category of photographic sensors a number of film and filter combinations can be used to view numerous energy bands within the 0.36 u to 0.90 u range.

Single band photographic imaging (forming one image of a scene by sensitizing the emulsion with the energy from a band of radiation) is usually done with panchromatic black and white, color, black and white infrared or false color infrared films. Films of these types are available in a variety of sizes to fit a number of camera styles. For most studies aerial mapping cameras (9½" x 9½" format) are preferred. Stereo coverage of the study area is obtained by using 60% end and 30% side overlap on frames. The most useful format for viewing the film varies with the film type. Prints are the most useful product from black and white Panchromatic film. False color infrared film, such as Kodak Aerochrome type 2443, has a positive transparency as its initial and highest quality product. Color prints of somewhat lower quality than the initial transparencies may be produced using an internegative. Color film, such as Kodak Aerocolor negative 2445, produces a negative capable of being used to produce either positive transparencies or prints in either color or black and white.

In multiband photography several simultaneous pictures of a scene using several film types or a single film type filtered to produce images formed by different energy bands are recorded. Where more than one film type is to be

[1]University of Georgia Marine Institute Contribution Number 258.

used it is necessary to have several cameras attached together. If only one film type is to be used, an array of matched lenses may be focused on a single film. Usually four bands are exposed simultaneously, but the Itek Corporation has developed a camera which uses three lenses on each of three rolls of film, thus gathering information on nine bands (23). Aerial panchromatic or black and white infrared film is used in multiband photographic systems. One wave band may be utilized for interpreting one feature and another for a second feature. Several may be combined by projecting the images in register with white light thus simulating a number of black and white film filter combinations. Color simulations may be made by combining several black-and-white images each projected in a different primary color. With additive color viewing it is possible to simulate natural and false color infrared images (18).

Non-photographic imagery is useful for examining parts of the spectrum (reflected or emitted) which are beyond the range of photographic imaging. Holtzer (9) reviewed the uses and potential uses of these parts of the spectrum (ultraviolet, microwave, radar, and infrared). Selected and emitted infrared non-photographic imaging has received the widest application to date.

Scanners with infrared detectors can examine various bands from near the upper limit of photographic capability to 1000 u. In daytime imaging the energy in the 0.78 u to 3.0 u band is primarily reflected while that beyond 4.5 u is mostly self-emitted. Approximately equal quantities of reflected and emitted radiation occur in the 3.0 u to 4.5 u band during daytime imaging (9). Nighttime imaging with detectors made of indium antimonide, mercury cadmium telluride or other combinations may be used to eliminate the reflected energy. Each detector has its particular range of greatest sensitivity; therefore, different filter-detector combinations may be used depending on the wave band selected for imaging.

As an alternative to accumulating information on film, it may be stored on magnetic tape. Visible, infrared and ultraviolet spectral signals stored in this manner may be used to print computer produced maps which enhance certain spectral qualities (23).

An addition to the above remote sensing procedures, radio-location telemetry may prove to be a useful tool for following the movement of animals in halophyte communities. Brander and Cochran (2) reviewed the current state of the art and discussed applications to date.

Uses of Remote Sensing

The greatest use of remote sensing in answering the questions of halophyte ecology has come in determining "who lives where and why?", with the emphasis to date on the first part. Olson (15) was able to recognize general types of salt marsh communities from panchromatic black-and-white or color aerial photographs. Stroud and Cooper (20) used Kodak Ektachrome Infrared Aero film type 8443 to differentiate vegetative types in North Carolina salt marshes. Color infrared photographs were more useful for the differentiation of vegetative types in California marshes than color, panchromatic or nine-lens multiband imagery in the 0.4 u to 0.9 u range (16). Pure stands of *Spartina alterniflora*, *Juncus roemerianus* and mixed stands of *Salicornia virginica*, *Distichlis spicata* and *Limonium nashii* are clearly discriminable on color (Kodak Aerocolor negative 2445) and false color infrared (Kodak Aerochrome type 2443) in studies by Reimold, Gallagher and Thompson (17) in Georgia. The best separation of the species in Georgia was achieved with late summer photography while patterns were least distinct in February when dead plant parts dominated the scene. The necessity of combining seasonal preference, sun angle restriction (reflections of the sun on the wet mud are a serious problem when the sun angle is above 70º), low tide designation and cloud-free, haze-free atmospheric specifications place severe limitations on the number of acceptable days per month in Georgia. These specifications are, however, necessary for satisfactory results.

Grimes and Hubbard (8) considered October to be the preferred season for photography (considering atmospheric as well as botanical factors) in Great Britain. False color infrared imagery produced the most useful information for species separation. *Puccinellia maritima*, *Spartina anglica*, *Scirpus maritimus* and *Juncus* species could be separated on the basis of color and texture. *Phragmites communis* could be distinguished stereoscopically because of its great height.

The best time of year to get maximum discrimination between species in a particular location can be determined by taking a series of photographs at selected times during the year or by comparing seasonal spectral signatures obtained with a spectoradiometer. Figure 1 compares a spectral reflectance curve of stands of *Spartina alterniflora* and *Juncus roemerianus* in Georgia in August.

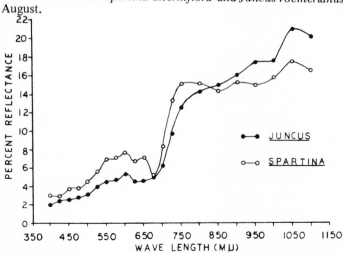

Figure 1. Spectral reflectance of **Juncus roemerianus** and **Spartina alterniflora** stands as viewed from above in August.

Pestrong (16) found nine band imagery to be excessive in his study of California marshes. He concluded a synchronized four camera system utilizing Ektachrome, Ektachrome false color infrared, black and white infrared and panchromatic (filtered for the 550 u to 630 u band) films would give an equivalent amount of information.

The interpretation of the photography may be done with single frames and the unaided eye. Stereo viewing in a plotter such as the Kern PG 2 has certain advantages: magnification, depth perception and color balance. The color balance problem comes from the different appearance of the sun and shade sides of the scene as depicted in Figure 2. By flying east-west and using adjacent frames in the flight line in the stereo plotter the tones will be integrated. The effect is also reduced by flying higher thus reducing the relative amount of the side of the plant exposed.

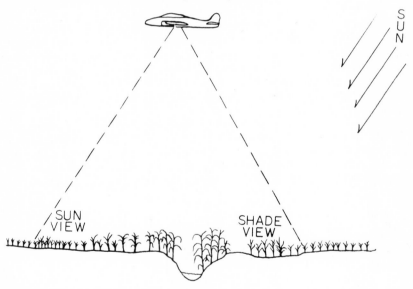

Figure 2. Sun and shade problem resulting from oblique sun angle.

Additive color viewing may be useful in interpreting multi-image photography. By combining various narrow band images projected with colored lamps it is possible to emphasize different features (18).

Non-photographic imaging can also be used to prepare vegetative maps. Gallagher, Reimold and Thompson (6) were able to separate *Juncus roemerianus* and *Spartina alterniflora* by using a 2.0 u to 3.5 u window during the daylight hours at low tide (Figure 3). Because of distortion, production of high quality maps from this imagery is more difficult than from photographs. This scanning type of imagery is useful if it isn't practical to wait for good

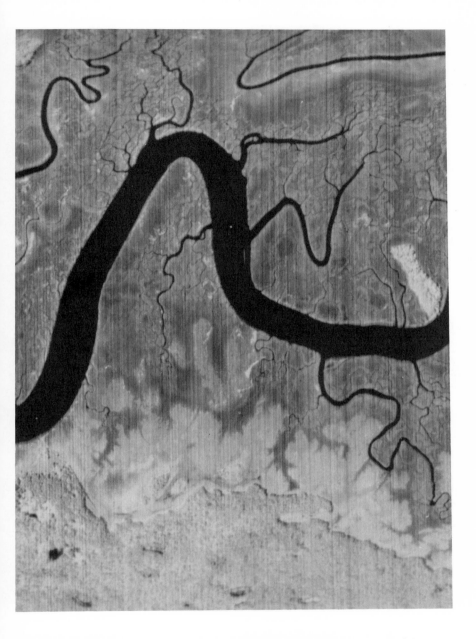

Figure 3. Non-photographic scanning infrared imagery of a Duplin River marsh (2.0 μ to 3.5 μ, low tide, February).

weather since this type of imagery can be done at night. A window other than the 2.0 u to 3.5 u would probably be used for the nocturnal imaging. Only emitted radiation would be detected and the window giving maximum differences would have to be determined experimentally.

Remote sensing of "who lives where?" is thus possible using a number of photographic imaging techniques, as well as infrared scanning imagery and telemetry. The question is straightforward and most remote sensing work with halophyte ecology has been directed toward answering it. "Why" various species live where they do is a more difficult question. Although remote sensing techniques have not been widely applied to this question some work is being done and more will likely develop in the future.

The relationship of tidal flooding to the zonation exhibited by intertidal halophytes is being examined in the coastal marsh herbaceous species. Infrared scanning imagery in the wave band 8.0 u to 12.5 u clearly shows the advancing front of water as the tide rises. Figure 4 shows four states of tidal rise. These patterns can be checked against plant zonation patterns (species distribution and stand morphology) associated with the tidal flooding. Preliminary examination of this imagery, associated photography and site visitation indicates *Juncus roemerianus* zones are flooded later than *Spartina alterniflora* areas. Within *S. alterniflora* stands the tall robust plants are flooded first, the medium density, medium height plants second, small high density plants third, medium height robust plants fourth and the nearly bare levees last. A description of these stands is shown in Table 1. Traditional elevation measurements used to estimate flooding assume the water will be at the same height in the marshes as it is at a tidal station some distance away. The infrared scanning imagery technique follows the water front and avoids problems caused by the resistance to flow created by the vegetation.

The nutrient status of *Spartina alterniflora* was assessed by false color infrared photography at the University of Georgia Marine Institute. Levels of nitrogen enrichment were easily identified two months after addition of the nutrient.

Other plant characteristics associated with environmental conditions have been examined in agricultural situations. Nitrogen deficient sweet pepper leaves have higher reflectance in the 0.38 u to 0.70 u region of the spectrum than normal leaves because of lower chlorophyll concentrations (22). In the near infrared (0.70 u to 1.00 u), deficient leaves also reflected more than normal leaves. This difference may be due to smaller cells since near infrared reflectance has been associated with air cavity-cell wall interfaces in the leaf (10). Changes in leaf structure therefore result in changes in reflectance.

Salinity (14) and moisture stress (21, 11) have been studied in agricultural crops. Care must be exercised in interpreting spectral reflectance changes since similar "symptoms" may be caused by several stresses. Reflectivity of plant communities changes with many factors: surface dirt (4), age (4), nutrient level (21), salinity (14), moisture (11); the proper interpretation

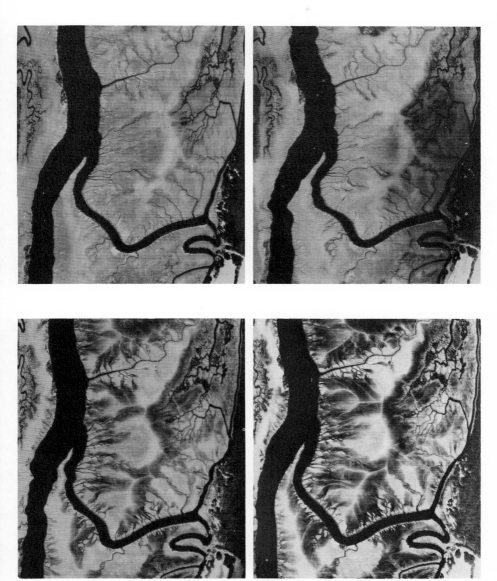

Figure 4. Non-photographic scanning infrared imagery of a Duplin River marsh showing four stages in tidal rise (8.0 μ to 12.5 μ, October). Low tide - upper left, high tide - lower right.

Table 1. Description of _Spartina alterniflora_ stands arranged in order of flooding sequence as interpreted from scanning infrared imagery. $\bar{Y}$ = arithmetic mean, CV = coefficient of variation.

	Flooding sequence							
	1		2		3		4	
	$\bar{Y}$	CV	$\bar{Y}$	CV	$\bar{Y}$	CV	$\bar{Y}$	CV
DENSITY								
culms m^{-2}	98	(19)	142	(26)	206	(39)	167	(26)
HEIGHT CLASS								
(percent abundance)								
0.01 - 0.50 m	28		51		79		36	
0.51 - 1.00 m	16		47		17		64	
1.01 - 1.50 m	24		2		3		0	
> 1.51 m	32		0		1		0	
LIVING MATTER								
dry weight (g m^{-2})	1665	(41)	499	(25)	331	(34)	630	(42)
dry weight (percent)	27		34		36		36	
culm size (g)	17.0		3.5		1.6		3.8	
leaf width (cm)	16.4		11.8		8.8		11.2	
CHLOROPHYLL "A"								
(mg g^{-1} of fresh weight of leaf)	1.22		0.64		0.57		0.71	

must be carefully verified by ground truth.

Answering the second of Bonner's questions, "Who eats whom?", is primarily a question of understanding the detritus food web for halophyte communities. Microscopes and liquid scintillation counters are often the best instruments for examining these Lilliputian interactions. The most widespread use of remote sensing (in what is the normal scale for remote sensing) in answering this question has been in assessing the magnitude of the base of the food web.

Spatial and temporal variations in the standing crop of plants in a Georgia salt marsh have been evaluated by Reimold, Gallagher and Thompson (17). Kodak Aerochrome infrared 2443, Aerocolor negative 2445, Infrared Aerographic 2424, and scanning imagery from a Bendix Thermal Mapper were compared. Within the *S. alterniflora* zone various levels of production were seen from ground surveys. Patterns related to this production were also seen on photographic and non-photographic imagery of the marshlands.

Color and to some degree texture was used to delineate mapping units with false color infrared photographs. Ground truth consisting of biomass data collected by the harvest method as outlined by Milner and Hughes (13) indicated "bright red" areas along the stream bank have a mean maximum standing crop of 1665 g m^{-2} dry weight of living matter. "Light red" areas averaged 630 g m^{-2}, "blue-red" 499 g m^{-2}, "blue" 331 g m^{-2}. The high levees have little if any plant growth and also appear blue. They can be separated from the "blue" vegetated area by their finer texture.

Color photographs were also examined and ground truth data applied to the production categories but they are more difficult to interpret because the color patterns are less discrete. Black and white infrared has little to recommend it for these purposes. Figure 5 shows 8.0 u to 12.5 u scanning infrared imagery of a pure *S. alterniflora* stand where the mapping units were three shades of gray. All factors considered (cost, ease of interpretation, information gained), false color infrared photography appears best for Georgia (7).

Remote sensing techniques have not been applied to answering questions related to "the physiology of togetherness" in halophyte communities. In agricultural and forest settings multiband photography has been used to study insect and disease interactions. Corn leaf blight infections were detected by Bauer *et al.* (1) with several different films and infrared scanning. At least three levels of severity of infection could be detected.

The earliest symptoms of insect attack or disease infection in timber stands often is a loss of near-infrared reflectance (4). Since these changes are not visible to the human eye they may be detected by infrared imagery before they can be detected visually. Weber and Polcyn (23) have examined numerous stress parameters in forest trees using a multispectral system which has twelve channels in the 0.4 u to 1.0 u range, three in the 1.0 u to 2.6 u range and either a UV detector or a IR detector (8.0 u to 14.0 u). French and Meyer (5) located oak wilt infected trees on color infrared photography.

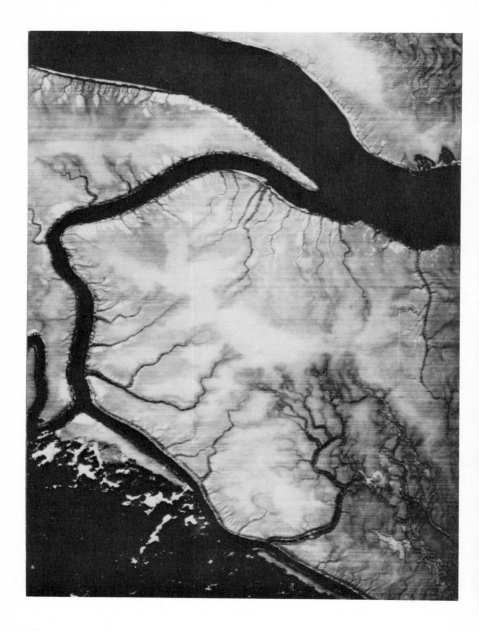

Figure 5. Non-photographic scanning infrared imagery of a Duplin River marsh **Spartina alterniflora** stand (8.0 u to 12.5 u, August).

Competitive pressure between forest trees has been estimated from aerial photographs (12).

The detection of inter and intra specific stress in agricultural and forest systems is still in the developmental stages. Because of economic considerations remote sensing techniques for these systems are more highly developed than those for halophyte communities.

As a result of pressure on these communities by man's development of the coastal zone and his new-found appreciation of ecological principles more effort will be directed toward understanding processes in these natural systems and toward managing them.

The most widespread use of remote sensing techniques to date has been in mapping the distribution of various halophyte stands. Recently ecologists are beginning to seek answers to questions of production, the detection of nutrient changes and the sequence of inundation of the various stands of intertidal halophytes. Detection and analyzing techniques, as well as interest, are developing rapidly and new applications of remote sensing to answering questions of halophyte ecology will soon be realized.

REFERENCES

1. Bauer, M. E., P. H. Swain, R. P. Mrocezynski, P. E. Anuta and R. B. MacDonald. 1971. Detection of southern corn leaf blight by remote sensing techniques, p. 53-54. Proc. of the 7th Int. Symposium – Remote Sensing of Environment. Univ. of Mich., Ann Arbor. 218 p.
2. Brander, R. L. B., and W. H. Cochran. 1969. Radio-location telemetry. p. 95-103. In R. H. Gilles Wildlife Management Techniques. 623 p. The Wildlife Society, Washington, D. C.
3. Carter, V. P. and R. R. Anderson. 1972. Interpretation of wetlands imagery based on spectral reflectance characteristics of selected plant species. p. 580-598. In W. J. Kosco, Proceedings of the 38th Annual Meeting, American Society of Photogrammetry, Washington, D. C. 636 p.
4. Cowell, R. N. 1970. Applications of remote sensing in agriculture and forestry. p. 164-223. In National Research Council. Remote Sensing with Special Reference to Agriculture and Forestry. National Academy of Sciences, Washington, D. C. 424 p.
5. French, D. W. and M. P. Meyer. 1969. Aerial survey for oak wilt and Dutch Elm disease in Bloomington, Minnesota. p. 77-80. In E. B. Knipling (Ed.), Aerial color photography in the plant sciences. Univ. of Florida, Gainesville. 173 p.
6. Gallagher, J. L., R. J. Reimold and D. E. Thompson. 1972. Remote sensing and salt marsh productivity. p. 338-348. In W. J. Kosco, Proceedings of the 38th Annual Meeting, American Society of Photogrammetry, Washington, D. C. 636 p.
7. Gallagher, J. L., R. J. Reimold and D. E. Thompson. (In press) A

comparison of four remote sensing media for assessing salt marsh primary production. Proceedings of the VIII International Symposium on Remote Sensing of Environment. The University of Michigan, Ann Arbor.
8. Grimes, B. H. and J. C. E. Hubbard. 1971. A comparison of film type and the importance of season for interpretation of coastal marshland vegetation. Photogrammetric Record 7:213-222.
9. Holzer, M. R. 1970. Imaging with nonphotographic sensors. p. 73 to 163. *In* National Research Council. Remote Sensing with Special Reference to Agriculture and Forestry. National Academy of Sciences, Washington, D. C. 424 p.
10. Knipling, E. B. 1969. Leaf reflectance and image formation on color infrared film. p. 17-29. *In* P. I. Johnson (Ed.), Remote Sensing in Ecology. Univ. of Georgia Press. 244 p.
11. Kramer, P. J. 1959. Transpiration and the water economy of plants. Plant Physiology. Academic Press. 701 p.
12. Latham, R. P. 1972. Competition estimator for forest trees. Photogrammetric Engineering 38:48-53.
13. Milner, C. and R. E. Hughes. 1968. Methods for the measurement of the primary production of grassland. Burgess and Son, Ltd. Abingdon. 70 p.
14. Myer, V. I., D. L. Carter and W. J. Rypert. 1966. Remote sensing for estimating soil salinity. J. Irrig. Drain. Div., Amer. Soc. Civ. Eng. 92:59-68.
15. Olson, D. P. 1964. The use of aerial photographs in studies of marsh vegetation. Maine Agr. Exp. Sta. Bull. 13. 62 p.
16. Pestrong, R. 1969. Multiband photos for a tidal marsh. Photogrammetric Engineering 35:453-470.
17. Reimold, R. J., J. L. Gallagher and D. E. Thompson. 1972. Coastal mapping with remote sensors. p. 99-112. *In* Proceedings of a Symposium on Coastal Mapping. American Society of Photogrammetry. Washington, D. C. 319 p.
18. Ross, D. S. 1972. Simple multispectral photography and additive color viewing. Presented before American Society of Photogrammetry. Washington, D. C. March.
19. Salisbury, F. B. and C. Ross. 1969. Plant physiology. Wadsworth Publishing Company, Inc. Belmont. 747 p.
20. Stroud, Linda M. and A. W. Cooper. 1969. Color-infrared aerial photographic interpretation and net productivity of a regularly flooded North Carolina salt marsh. Water Resources Research Institute of the Univ. of N. C. Report No. 14. 86 p.
21. Thomas, J. R., V. I. Myers, M. D. Heilmond and C. L. Wiegand. 1966. Factors affecting light reflectance of cotton. p. 305-312. *In* Proc. 4th Symp. Remote Sensing of Environment. Univ. of Mich., Ann Arbor.
22. Thomas, J. R. and G. F. Oerther. 1972. Estimating nitrogen content of

sweet pepper leaves by reflectance measurements. Agronomy Journal 64:11-12.
23. Weber, F. P. and F. C. Polcyn. 1972. Remote sensing to detect stress in forests. Photogrammetric Engineering 38:163-175.

STABILIZATION OF COASTAL DREDGE SPOIL WITH SPARTINA ALTERNIFLORA[1]

Ernest D. Seneca

Departments of Botany and Soil Science, North Carolina State University, Raleigh, North Carolina 27607

Abstract

Stabilization of dredge spoil along the North Carolina coast with *Spartina alterniflora* Loisel. can be obtained within two years. Experimental plantings made with transplants from nature were designed to study source of transplants, spacing, time of year to plant, depth to plant, and the correlation between plant response over an elevational gradient and tidal amplitude. Plant response was measured over substrate material with various physical and chemical differences. On coarse substrate material, planting operations can be mechanized by using a tractor and a modified tobacco planter. Seeding experiments revealed that seedling establishment is restricted to a much narrower elevational range than that over which transplants are successful. Laboratory studies indicated that seed germinate well in a 35-20°C alternating diurnal thermoperiod but that they must be kept moist during storage prior to germination. Controlled seedling growth experiments indicated best growth under long-day conditions in a 26-22°C or 30-26°C alternating diurnal thermoperiod. Planting of the intertidal zone of dredge spoil with *S. alterniflora* not only can result in stabilization of the spoil material, but also initiates the development of estuarine salt marsh.

Introduction

The Corps of Engineers maintains about 1,500 miles of navigation channels in the sounds and estuaries of North Carolina. Channel maintenance is accomplished by dredging material from areas where the channels have become filled in and depositing the spoil several hundred feet away. Annual maintenance costs range from two to three million dollars. Heretofore, man has taken no steps to stabilize this spoil material and much of it finds its way back into the channels through wind and wave activity.

The ecological value of low, regularly-flooded salt marshes has been well established (Teal 1962, de la Cruz and Odum 1967, Cooper 1969, Williams

---

[1] The research reported on in this paper is supported by the Coastal Engineering Research Center, U. S. Army Corps of Engineers; the NOAA Office of Sea Grant, Department of Commerce, Grant No. GH-78; and the North Carolina Coastal Research Program.

and Murdock 1969), but their acreage in North Carolina has been diminished through man's activities. They have been considered waste lands by many people and consequently have been the sites of waste disposal and filling operations. Until legislation prohibited the unrestricted dredging and filling of these valuable ecological systems, man filled them in with spoil material to make them suitable for building sites.

With these facts in mind, and the encouragement of several ecologists, we decided to undertake an experimental project to establish *Spartina alterniflora* Loisel. (a highly productive grass of the low regularly-flooded salt marsh) in the intertidal zone on certain of these spoil deposits. Our objectives were twofold: (1) to stabilize the intertidal zone of dredge spoil and (2) to construct salt marsh acreage for the State. The planting of marsh plants on tide lands has been accomplished in Europe for the purpose of land reclamation, and channel and harbor protection (Ranwell 1967), but we found no published accounts of similar activity in this country. Consequently, our initial work involved simple studies to determine: type and source of transplants, spacing, depth for transplant, site requirements, time of year to transplant, and the correlation between plant response over an elevational gradient and tidal amplitude. Many of these same factors were also studied with regard to seeding. Plant response to substrate material with various physical and chemical differences was gauged by productivity determinations. This paper summarizes results of a current research program which began in the fall of 1969. More detailed results, together with illustrations, have been presented in a recent publication (Woodhouse, Seneca, and Broome 1972).

Seeding Studies

Field observations indicated that seedlings of *S. alterniflora* were abundant in certain areas, important in filling in open areas of marsh, and colonized new areas. An earlier study (Mooring, Cooper and Seneca 1971) established the basic requirements for storage of seed and their germination. Newly established marshes were found to produce higher percentages of viable seeds. Seedheads can be harvested by hand, but the operation can be speeded up several fold by using a reel, cutting blade, and canvas catchment bag mounted on the front of a two-wheeled garden tractor operated by one man. Following collection, seedheads can be threshed in a small grain thresher and stored in seawater at $4^{o}C$.

Several methods of seeding dredge spoil have been tried, but any method that will get seed covered and hold them in place until germination is satisfactory. Raking seed in by hand with a hoe fork has yielded satisfactory results. In scaling up seeding operations, we have obtained good seedling stands by dragging a harrow over the area with a tractor, spreading the seed by hand, and then dragging the harrow over the area again.

Although seed will germinate in March in the field, best stands have resulted from seedings occurring from late April through May. Fewer seed are washed away or otherwise disturbed prior to germination in these late spring seedings. The likelihood of damaging storms is much less as the growing season progresses.

Seed germinate and young seedlings begin to grow over most of the intertidal zone, but seedling survival is limited to the upper part of this zone near mean high water. Consequently, if a wider part of the intertidal zone is to be stabilized or the area is more exposed initially, transplants must be used. Where feasible, direct seeding offers a substantial savings in labor and cost as compared to transplanting.

Transplanting Studies

Initially, transplanting was done by hand and the most efficient method was for men to work in pairs. One man would open a hole with a dibble or narrow-pointed shovel and the other man would place the single-culm transplant into the hole. In areas with saturated substrate this teamwork is essential because the holes will close quickly. We have found that plants should be planted about the same depth as they were growing naturally. Contrary to transplanting in other areas, moisture is not a problem and planting depth is not critical from that standpoint. Transplanting of rhizomes alone was not found to be a satisfactory method of establishing a stand. Hand-planting is highly successful and must be used in areas too unstable for vehicles, however, it is very time-consuming. Therefore, we mechanized the planting operation on areas with coarse substrate by using a tractor and a modified tobacco planter. Regardless of the planting method employed, the most time-consuming task of the transplanting operation is digging transplants. Areas of sand accretion wi hin the marsh, and newly established areas of salt marsh on fresh substrate are the most desirable areas to obtain transplants. Plants growing in these sites are usually very healthy with well-developed root systems and are more easily dug and separated than those from areas of old marsh. Transplants are dug by hand with shovels, separated into single culms, washed, and kept moist until transplanting. We have dug plants with a tractor-mounted back hoe when large numbers of transplants were needed. Observations indicate that the natural marsh recovers from these digging operations primarily by lateral rhizome development the following growing season.

A 3 x 3 ft. spacing of transplants has yielded satisfactory results with complete cover attained in two growing seasons. The 3 x 3 ft. spacing is a reasonable compromise considering the cost involved to plant a closer spacing and the stand obtained. Although a closer spacing might yield more above ground growth the first season, two growing seasons are required to develop a network of roots and rhizomes which binds the substrate together regardless of spacing.

Productivity is high for November through early March transplants, but survival percentages are often low. Conversely, later transplants produce less biomass but have much higher survival percentages. Thus, we have found that the best season to transplant is from late March through June when there is a good balance between productivity and survival. As with seeding operations, the likelihood of damaging storms is less as the growing season progresses until late summer when hurricanes become more probable. We have obtained excellent survial from midsummer transplants, but production by peak standing crop in early fall is very low compared with late spring plantings.

Although transplants survived over most of the elevational gradient from mean low water to the spring high tide line, there were zones of differential growth response. A zone of optimum growth occurred slightly below mean high water. Above and below this zone growth was not as good. Elevation becomes more critical in areas of low tidal amplitude where the intertidal zone may be restricted to a very narrow band. More extensive areas can be transplanted where tidal amplitude is greater.

Controlled Growth Studies

Laboratory, greenhouse, and controlled growth chamber studies were designed to complement our field studies and to help us better interpret plant response in the field. Germination tests in the laboratory confirmed results of earlier work by Mooring, Cooper, and Seneca (1971) that seed must be stored moist and at low temperatures (2-4°C) prior to germination in a 35-30°C alternating diurnal thermoperiod. Broome (1972) concluded that seed collections from diverse populations along the North Carolina coast did not respond identically to various storage treatments because of differential stages of seed maturity. Although seed should be harvested as near maturity as possible and stored in seawater at 2-4°C, he feels that it is necessary to compromise on complete maturity due to the fact that many seed will be lost due to shattering if harvesting is delayed too long.

Seedling growth studies in the Southeastern Plant Environmental Laboratory, North Carolina State Unit (phytotron) indicated best seedling growth under long-day conditions in 26-22 and 30-26°C thermoperiods (Seneca and Broome, 1972). Seedlings grown under short-day conditions were shorter, contained less biomass, produced more culms and rhizomes, had less shoot moisture, and contained higher chlorophyll concentrations than those under long-day conditions at the same temperatures. Phytotron results correlated well with seedling growth response in the field where the greatest rate of growth occurs in June under long daylengths with flowering occurring in August. These controlled growth studies have established the necessary photoperiod and thermoperiod requirements so that we can now investigate various nutrient relationships.

Nursery Propagation

Because obtaining transplants from nature is the most time-consuming phase of the transplanting operation, nursery production is well worth consideration. Also, when other groups begin constructing marsh on dredge spoil, demands on natural marsh for transplants may become too great and nursery production will become a necessity. Our initial trials indicate that nursery production on an upland site with proper irrigation methods is feasible. However, techniques and procedures to make the operation more practical need to be refined. We have established several nursery plots along the coast on dredge spoil both from seed and transplants to serve as a source of transplants for future work.

Summary

Planting of the intertidal zone of coastal dredge spoil with *S. alterniflora* can result in stabilization after two growing seasons with the concomitant establishment of estuarine salt marsh. Direct seeding requires less labor and less over all cost but can produce a stand over only the upper portion of the intertidal zone. Transplanting is more costly, but can be mechanized on coarse substrate and can produce a complete cover over the entire intertidal zone in two growing seasons under favorable growing conditions. Plant response to fertilizer, controlled environment conditions, and nursery production of *S. alterniflora*, as well as use of other plant species on dredge spoil at higher elevations, are problem areas currently being evaluated and will be reported on later.

Literature Cited

Broome, S. W. 1972. Stabilizing dredge spoil by creating new salt marshes with *Spartina alterniflora*. Proc. N. Carol. Soil Sci. Soc. 15:136-149.

Cooper, A. W. 1969. Salt marshes. Pages 563-611 *in* Coastal Ecological Systems of the United States. Edited by H. T. Odum, B. J. Copeland, and E. A. McMahon (Ed.). (Unpubl. Man.)

de la Cruz, A. A., and E. P. Odum. 1967. Particulate organic detritus in a Georgia salt marsh-estuarine ecosystem. Pages 383-388. *in* Estuaries. Edited by G. H. Lauff, Amer. Assoc. Advanc. Sci.

Mooring, M. T., A. W. Cooper, and E. D. Seneca. 1971. Seed germination response and evidence for height ecophenes in *Spartina alterniflora* from North Carolina. Amer. J. Bot. 58:48-55.

Ranwell, D. S. 1967. World resources of *Spartina townsendii* (Senu Lato) and economic use of *Spartina* marshland. J. Ecol. 55:239-256.

Seneca, E. D., and S. W. Broome. 1972. Seedling response to photoperiod and temperature by smooth cordgrass, *Spartina alterniflora*, from Oregon Inlet, North Carolina. Chesapeake Sci. 13: 212-215.

Teal, J. M. 1962. Energy flow in the salt marsh ecosystem of Georgia. Ecology 43:614-624.

Williams, R. B., and M. B. Murdock. 1969. The potential importance of *Spartina alterniflora* in conveying zinc, manganese, and iron into estuarine food chains. *in* Proc. 2nd National Symposium on Radioecology. Edited by D. J. Nelson and F. C. Evans (Ed.). USAEC, CONF-670503.

Woodhouse, W. W., Jr., E. D. Seneca, and S. W. Broome. 1972. Marsh building with dredge spoil in North Carolina. N. Carol. State Univ. Agric. Exp. Sta. Bull. 445. P. 28.

EFFECTS OF HERBICIDES ON THE SPARTINA SALT MARSH

Andrew C. Edwards and Donald E. Davis*

Department of Botany and Microbiology
Auburn University Agricultural Experiment Station
Auburn, Alabama 36830

Abstract

During recent years the use of herbicides has increased steadily and the use of organic arsenical herbicides such as MSMA (monosodium methanearsonate) has increased even more rapidly than the others. Some of these herbicides end up dissolved in run-off water or absorbed to water-borne soil colloids in estuaries. The tidal action that maintains the fertility of the salt marsh may also transport herbicides at sufficient concentrations to be injurious to the productivity of the salt marsh ecosystem. In this paper the literature dealing with the use of herbicides to control *Spartina* is reviewed. The mode of action, effects on non-target organisms, persistence, and possible effects of these herbicides on the salt marsh are discussed. Preliminary results from experiments conducted on Sapelo Island, Georgia, on the effects of MSMA on *S. alterniflora* and *Littorina irrorata* are reported. MSMA effects on ^{32}P uptake by *S. alterniflora* are also given.

Aquatic pollution by domestic, industrial, and agricultural activities has become a serious problem to salt marsh and adjacent estuarine systems. Pollutants discharged throughout the entire watershed may ultimately end up in coastal areas. The same tidal action that maintains high fertility in salt marshes (12) may cause these pollutants to be swept repeatedly back and forth through salt marshes. (Figure 1).

One of the major sources of agricultural pollutants in the southeastern U. S. is pesticides used in cotton production. These pesticides are particularly important in salt marshes because much of the cotton is grown in delta areas near large rivers and because more pounds of pesticides are used per acre of cotton than for any other crop grown in this area. Organic arsenicals constitute nearly one-tenth of the total amount of pesticides used on cotton.

*Graduate Teaching Asst. and Alumni Prof., respectively.

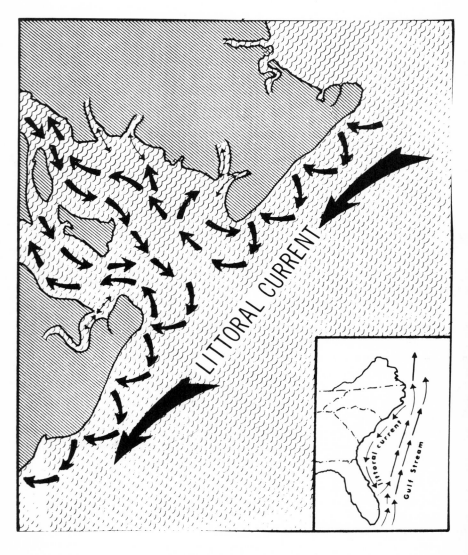

Figure 1. Effect of various concentrations of MSMA on the standing crop of living **Spartina alterniflora.**

Based on the 1969 and 1970 cotton acreages for Alabama, Georgia, Mississippi, South Carolina, and Texas, and assuming two applications per year of two pounds per acre (1b/A) or about 2.2 kg per hectare (kg/ha), 8.25 million pounds (3.7 million kg) of organic arsenicals are used per year in the southeastern United States.[1] A 1967 survey estimated that 284 thousand pounds or 130 thousand kg of organic arsenicals were used in Sunflower, Washington, and Bolivar Counties of Mississippi alone.[2]

This paper will deal with the possible effects of various herbicides on the salt marsh ecosystem and with a pilot study carried out in 1971 on the effect of MSMA on the growth of smooth cordgrass (*Spartina alterniflora*) and the closely associated salt marsh periwinkle (*Littorina irrorata*) and on the uptake of ^{32}P by *Spartina alterniflora*. The chemical names of all herbicides discussed are listed in Table 1.

Although our concern is with the possible deleterious effects of small concentrations of herbicides on the productivity of the salt marsh ecosystem, the best source of information on potentially damaging herbicides is literature dealing with herbicides used to control *Spartina*. The primary reasons for attempting to control *Spartina* are to improve navigation, to intersperse open water areas with marsh and thus increase carrying capacity for waterfowl and fur bearers, to facilitate drainage, and to keep recreational beaches clear of colonizing species.

Among the herbicides that have been used to control cordgrass are: dalapon, fenuron, paraquat, bromacil, amitrole-T, and diuron. For each of these herbicides we will discuss the effects on *Spartina*, mode of action, persistence, and possible effects on other components of the salt marsh ecosystem. *The Herbicide Handbook of the Weed Society of America (Weed Sci. Soc. 1970)* and Pimmentels'(1971) book on the effects of pesticides on non-target organisms are the primary sources of the information on herbicides reported here. References to the original investigations may be obtained from these books.

The earliest attempt to control *Spartina* with herbicides is a report by Ranwell and Downing (1960) on the use of dalapon, fenuron, and a pelleted combination of fenuron, 2, 4-D, and sodium borate. Dalapon was the most effective herbicide tested and 56 kg/ha gave 80% to 100% kill. However, reinvasion by seedlings the following year suggested that repeat applications would be necessary. Taylor and Burrows (1968) also found that, although 56 kg/ha gave good initial kill, regrowth occurred the next year. However, 112 kg/ha gave 98% to 100% kill with little regrowth the second year. Bascand (1968) reported good control with 45 kg/ha dalapon whether applied in one or split plot applications.

[1]**Report to President's Office of Science and Technology. Subpanel on Arsenic Sources, Uses, and Effects.**

[2]**Pesticide Study by Delta Branch Exp. Sta. of the Miss. Agr. Exp. Sta.**

Table 1. List of chemical names and formulas for herbicides mentioned.

Common name	Chemical name
amitrole-T	3-amino-s-triazine plus ammonium thiocyanate
bromacil	5-bromo-3-sec-butyl-6-methyluracil
cacodylic acid	hydroxydimethylarsine oxide
2,4-D	(2,4-dichlorophenoxy) acetic acid
dalapon	2,2-dichloropropionic acid
dicamba	3,6-dichloro-o-anisic acid
diuron	3(3,4-dichlorophenyl)-1,1-dimethylurea
dichlorobenil	2,6-dichlorobenzonitrile
DSMA	disodium methanearsonate
fenac	(2,3,6-trichlorophenyl) acetic acid
fenuron	1,1-dimethyl-3-phenylurea
MSMA	monosodium methanearsonate
paraquat	1,1-dimethyl-4,4'-bipridinium ion

Dalapon is ordinarily applied as foliar spray at 1 to 22 kg/ha and is primarily used for control of grass (Weed Sci. Soc. 1970). It is easily absorbed by either the roots or leaves and is translocated readily throughout the plant. It breaks down rapidly and completely in the soil and several microorganisms are able to bring about this degradation (Foy 1969). The mode of action of dalapon is quite complex and it seems probable that some of the effects observed for microorganisms are not applicable to higher plants. Crafts (1961) in an earlier, simplified statement reported three possible modes of action: (1) competitive and non-competitive inhibition of pyruvate metabolism, (2) inhibition of pantothenic acid synthesis, a precursor of Co-A, and (3) precipitation or denaturation of a pyruvate-protein complex.

Butler (1963) reported that 48- and 96-hour exposures to 1.0 ppm dalapon had no noticeable effect on the longnose killifish (*Fundulus similis*) or the growth of American oyster (*Crassostrea virginica*), respectively. However, in the same investigation, a 48% increase in paralysis or death for the brown shrimp (*Penaeus aztecus*) was observed. The LC 50 (lethal concentration for 50%) for the fathead minnow (*Pimephales promelas*) was 440 ppm for a 48-hour exposure (1962) and goldfish (*Carassius auratus*) mortality was 0 after a 24-hour exposure to 100 ppm and 100% for a 24-hour exposure to 500 ppm and above (Weed Sci. Soc. 1970). The oral LD 50 (lethal dose for 50%) for rats is 7500 mg/kg.

Because dalapon is readily degraded by microorganisms and is relatively nontoxic to *Spartina* it seems unlikely that damaging concentrations would occur as pollutants in salt marshes. The 45 to 112 kg/ha recommended for control of *Spartina* could give concentrations in the water sufficient to cause damage to other organisms, particularly the brown shrimp.

Fenuron, a substituted urea, controls *Spartina* to some degree although it is less consistently effective than dalapon (Bascand 1968, Taylor and Burrows 1968). Taylor and Burrows reported that a spring application of at least 67 kg/ha was necessary to give complete kill for one year and at least two 67 kg/ha treatments in one year were necessary to maintain adequate control into the second year. They considered this rate uneconomical. It is interesting to note that 67 kg/ha applied in October had no observable toxicity to *Spartina* either in the year of application or during the following year.

Fenuron is most often used to control woody plants. It is applied to the soil where it is absorbed by the roots and moves up in the transpiration stream (Weed Sci. Soc. 1970). It is a potent inhibitor of the Hill reaction component of photosynthesis (Anders 1964). Fenuron is fairly persistent. High rates of application may last for over one year but this is unusual (Weed Sci. Soc. 1970). Adsorption is least in sandy soils and highest in soils with a high humus or clay content, particularly clays with a high cation exchange capacity.

Fenuron has a low mammalian eoxicity. The oral LD 50 is 6,400 mg/kg for rats. Butler (1963) reported that spot (*Leiostomus xanthurus*) were unaffected by a 48-hour exposure to 1 ppm. There are no reports in the literature on effects on crustaceans, molluscs, or other marine invertebrates. It seems unlikely that pollution concentrations of fenuron are apt to be a problem in salt marshes since it is not widely used.

Taylor and Burrows (1968) reported that 7 to 10 kg/ha of paraquat gave 75% to 100% kill of *Spartina*. They also showed that foliarly applied paraquat was translocated to the underground parts and from there to untreated aerial shoots. They demonstrated that susceptibility of *Spartina* to paraquat decreased with age and stated that its use to control mature *Spartina* is inadvisable until more information is available.

Paraquat is applied as a foliar spray to bring about contact kill of weeds and most often is used in non-crop situations. Paraquat is more active in green plants carrying on photosynthesis (Davis 1972). A unique characteristic of paraquat is its very rapid and complete detoxication upon contact with soil because it is very tenaciously adsorbed to the negatively charged clay surfaces including those between the layers in montmorillonite clays (Funderburk 1969). The chemical, as such, is very persistent in soils and is subject to only slow microbial degradation, however, it is not phytotoxic because it is so tenaciously bound.

While paraquat is relatively toxic to mammals, the oral LD 50 for rats is 150 mg/kg, it is generally considered safe to fish and wildlife (Weed Sci. Soc. 1970). Butler (1963) reported no effect on shell growth of oysters exposed to 1.0 ppm paraquat for 96 hours; however, a 53% reduction in phytoplankton productivity was observed following a 4-hour exposure to 1.0 ppm. No noticeable effects were observed when brown shrimp were exposed to 1.0 ppm for 48 hours. Accumulation of paraquat has been demonstrated in bluegills (*Lepomis macrochirus*) and goldfish (Cope 1963, Funderback 1969). Treatment for 48 hours with 1.0 ppm had no effect on longnose killifish, rainbow trout (*Salmo gairdneri*), green sunfish (*Lepomis cyaniellos*), bluegills, or channel catfish (*Ictalurus punctatus*) (Pimentel 1971). The LC 50 for bluegills was 400 ppm. In view of the very rapid adsorption on clays and relatively low toxicity to fish species, it seems doubtful that pollution concentrations of paraquat are apt to be of concern. However, it should be borne in mind that applications of paraquat in the salt marsh itself might do serious damage to the sensitive phytoplankton and to species dependent on them.

Taylor and Burrows (1968) obtained complete kill of *Spartina townsendii* with 22 kg/ha of bromacil or 5.6 kg/ha of bromacil in combination with 4.5 kg/ha of amitrole-T. More than 99% kill was present two years after the 22 kg/ha application. They also achieved nearly 100% kill with 5.6 kg/ha of bromacil alone but 4.5 kg/ha of amitrole-T was much less effective. Bascand (1968) obtained good temporary control of *S. alterniflora* with 6.7 kg/ha of amitrole-T but found bromacil to be ineffective at rates as high as 36 kg/ha.

Bromacil is a specific and potent inhibitor of photosynthesis and usually is applied as a foliar spray or as granules to the soil surface on non-crop lands for the control of perennial grasses, broadleaved weeds, and certain woody species (Crafts 1961). It is absorbed more readily by roots than leaves (Weed Sci. Soc. 1970). It persists in soils for more than one year only if applied at soil sterilant rates (25-30 kg/ha). It is subject to microbial degradation and is not adsorbed to colloidal soil particles as tenaciously as many herbicides. Bromacil has relatively low to medium toxicity; LC 50 for rainbow trout and bluegills is 100 ppm for 24 hours or 70 to 75 ppm for 48 hours. Pollution concentrations are not expected to constitute an appreciable hazard to salt marshes.

Amitrole, the primary active ingredient in amitrole-T, is used as a foliar spray to control certain perennial broadleaved weeds and quackgrass either in non-crop land or before the crops are planted (Weed Sci. Soc. 1970). It is absorbed by both the roots and leaves and readily translocates throughout the plant where it inhibits chlorophyll formation and prevents regrowth from buds. It is readily broken down by microbes in warm moist soil and thus persists for only two or three weeks. It is essentially nontoxic to fish and wildlife. Because of its very short persistence in soil and low toxicity, it seems unlikely to be a problem in salt marshes. However, Grzenda, Nicholson, and Cox (1966) reported that amitrole applied at the rate of 1.0 ppm persisted in water for 201 days and significant quantities were detected in the hydrosoil. Apparently the microbes involved in amitrole degradation in agricultural soils are less active in water or hydrosoil. Ammonium thiocyanate, the other constituent in amitrole-T, has been found lethal to fish in concentrations of 200 ppm (1957).

Diuron, another substituted urea, with many similarities to fenuron was tried by Bascand (1968) but even 67 kg/ha did not significantly affect the growth of *Spartina alterniflora*. It persists longer in the soil than fenuron and for that reason it is of somewhat greater concern. It is somewhat toxic to fish. The 48-hour LC 50 for white crappies (*Pomoxis annularis*) is 6 ppm whereas it is 48 ppm for largemouth bass (*Micropterus salmoides*) and 16 ppm for coho salmon (*Oncorhynchus kisutch*) (Weed Sci. Soc. 1970). Exposure of American oysters to 1.8 ppm for 48 hours decreased shell growth 50% (Pimentel 1971). Phytoplankton exposed to 1.0 ppm for 48 hours showed an 87% decrease in production (Butler 1963). The sensitivity of phytoplankton and oysters suggest that diuron is one of the chemicals of possible concern in the salt marsh.

MSMA is applied as a foliar spray for the control of weeds in the grass family in broadleaved crops and on non-crop land. Not as much is known about the mode of action of this herbicide as for those previously mentioned. It is moderately toxic to mammals. It has not been tried for *Spartina* control. Because MSMA and the other organic arsenicals such as DSMA and cacodylic acid are so widely used in delta regions and because so little is known about them, we felt that an investigation of their effects on the salt marsh was needed. Our preliminary results indicate that *Spartina alterniflora* is quite resistant to MSMA as it also is to many other herbicides.

The relatively high tolerance of *Spartina* to herbicides is of considerable interest. Several environmental and morphological factors may be involved in its tolerance. Foliarly applied herbicides that are not absorbed rapidly may be washed off by tides in areas or times where inundation occurs once or twice a day. The thin layer of mud that covers the aerial portions after the tide has receded may decrease herbicide uptake by adsorption of the herbicide on the clay colloids or by preventing direct contact between the herbidice and leaf surface. The high clay content, including the high exchange capacity montmorillonite, and humus in the marsh soils may adsorb and thus inactive soil-applied herbicides.

The adaptation of these halophytes to alternate aerial and aquatic environments involves specialized epidermal features (Sutherland and

Eastwood 1916). A heavy waxy cuticle covers all exposed aerial portions and it is extremely resistant to wetting. The upright growth and linear grooves on the adaxial surface of leaves facilitates rapid run-off of a foliar spray. Hydathodes secrete salt solution thereby decreasing salt accumulation and enabling the plant to overcome the osmotically difficult environment in which it thrives. Evaporation of this secretion results in a crystalline deposit over much of the leaf surface which may, as in the case of the silt deposit on the leaf, effectively decrease phytotoxic action. The complex network of rhizomes and the deep rooted nature of *Spartina* make it difficult to control with soil-applied herbicides. Complete coverage of all aerial parts with herbicides is difficult to achieve and unless phytotoxic concentrations are translocated to all parts of the plant, those parts escaping lethal concentrations can furnish a good start for regrowth the next season.

Experimental Procedures

Field Studies. This study was conducted in 1971 in a diked area, still open to tidal flux, near the University of Georgia Marine Institute at Sapelo Island, Georgia. Plots were located in a monospecific stand of *Spartina alterniflora* just interior to a natural levee paralleling a creek which furnishes the primary route of tidal flux into and out of the area. At this location, the average height of mature shoots in June was nearly one meter.

Twenty 3 m by 6 m plots were established in four groups of five each. Each plot was isolated from adjacent plots by mowed strips 1 m wide. Groups of plots were isolated from each other and the rest of the marsh by a 2 m-wide strip. Treatments were randomized within blocks and replicated four times. One liter of MSMA solution was applied with a hand sprayer in the following concentrations and frequencies: 9,000 ppm once weekly for three weeks; 100 ppm, 10 ppm, 1 ppm once weekly for five weeks; and a control (see Table 2 for dates of spraying). One application of 9,000 ppm equals 4.5 kg/ha.

On August 29-30, 1971, *Spartina alterniflora* and the salt marsh periwinkle (*Littorina irrorata*) were harvested from two randomly selected 1 m^2 quadrats located within the central 2 m by 5 m area of each plot. The harvested plant material was initially divided into living and dead shoots. Living *Spartina* shoots were subdivided into 3 size classes; tall (greater than 1.0 m), medium (0.5-1.0 m), and short (less than 0.5 m). Total number of shoots and their fresh weight were determined for each class. Dry weights were determined for subsamples by drying to a constant weight at 80°C. These values were used to calculate the total dry weight for each size class. Fresh and dry weights were determined for the dead *S. alterniflora*.

All periwinkles found within the sample areas were counted, dried to a constant weight at 105°C, and ashed for six hours at 500°C.

On November 12, the number of flowering shoots were determined for two randomly chosen 1 m^2 areas in each plot.

Table 2. Dates and concentrations of MSMA applications to a Georgia salt marsh during the summer of 1971.

Concentration (mg/L)	No. of Applications	Dates
9000	3	July 27, Aug. 10, Aug. 18
100	5	July 27, Aug. 2, Aug. 10, Aug. 18, Aug. 21
10	5	July 27, Aug. 2, Aug. 10, Aug. 18, Aug. 21
1	5	July 27, Aug. 2, Aug. 10, Aug. 18, Aug. 21

Laboratory Studies on Phosphorus Uptake. A stock of *Spartina alterniflora* was maintained under natural lighting in the glasshouse at Auburn University. Plants were grown in vermiculite and watered weekly with Hoagland and Arnon's nutrient solution and, as needed, with tap water. No attempt was made to duplicate the salinity of seawater. Twenty-five young plants were separated from parent culms and the roots carefully removed from the vermiculite. These individuals were transferred to 500-ml opaque plastic containers filled with unaerated Hoagland and Arnon's nutrient solution. The roots were submerged in the nutrient medium and the tops extended through holes in the center of the lids of the containers. The plants were then transferred to an environmental growth chamber (25°C, 14-hour day, 20°C, 10-hour night, and 60% relative humidity). After the plants were well established, usually 10 to 14 days, unhealthy plants were discarded. The remaining plants were divided into groups according to size and then within sizes randomly divided into three experimental units which were treated with 0, 5, and 50 ppm of MSMA.

To simulate arsenical exposure via tidal inundation, each plant was removed from its container and the root system gently wrapped in saturated paper toweling and enclosed in a plastic bag. The plants were then suspended in an inverted position such that the aerial portions were submerged in the herbicidal solution contained in large battery jars. The shoots were submerged in the solutions in this manner for two hours twice each day for five consecutive days. Between treatments, plants were returned to the nutrient solutions. These exposure times would approximate the inundations that occur in salt marshes on the Georgia coast during spring tides.

After the herbicidal treatment was completed, the plants were allowed to continue growing for three days to allow MSMA toxicity symptoms to develop. On the fourth day the nutrient medium was replaced with one-half strength Hoagland and Arnon's solution containing 88 u Ci/L of ^{32}P. Uptake was allowed to continue through a full light period of 14 hours.

The plants were dried for 48 hours at 105°C and weighed, and the aerial shoots were excised and ground in a Wiley Mill. Weighed increments of shoot material were added to an aqueous scintillation media and radio-assayed in a liquid scintillation spectrometer. Observed counts were corrected for quenching.

Results and Discussion

The primary objective of this research program is to determine whether organic arsenical herbicides may be expected to cause damage either by a direct effect on various components of the ecosystem or by interference with cycling of phosphorus in the marsh. The data presented here are the preliminary results of the 1971 pilot studies. The research will be continued for an additional three years. Definitive conclusions will not be possible until the completion of the study.

It is not known at this time what levels of organic arsenicals may be reaching coastal areas. This is currently being investigated by another research group. The concentrations used in this study are well above those expected to occur. Arsenicals washed from the land are greatly diluted by run-off water and large quantities are adsorbed by clay particles in the water. The highest concentration, 9,000 ppm, was chosen in order to assess the effects of an accidental spraying of rates often used for weed control. The three lower concentrations were chosen to establish and evaluate the detrimental effects of logarithmically decreasing concentrations.

Field Studies. Observations made at the time of harvesting revealed only very limited morphological effects of MSMA on *Spartina alterniflora* and only at the two higher rates, 9,000 and 100 ppm. Symptoms of MSMA damage were increased necrosis of the leaf tips and curling and necrosis of leaf margins. Injury symptoms were not observed on leaves formed subsequent to treatment which suggests that either the herbicide is not readily absorbed and translocated by *Spartina* or else it is not very toxic to the plant. There was no evidence of injury to plots treated with either 1 ppm or 10 ppm.

Fig. 2 presents the effects of the various treatments of MSMA on the standing crop of *Spartina alterniflora*. It appears that 100 ppm applied five times decreased the standing crop of living *Spartina*. However, the decrease was only significant at the 10% level.

The highest rate of MSMA (9,000 ppm) reduced the number of flowering shoots to only 11% of the control (Fig. 3). This decrease was also only significant at the 10% level. A smaller decrease was observed on those plots receiving 100 ppm.

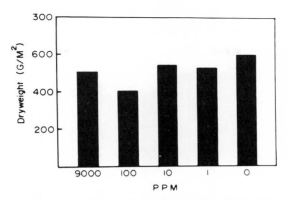

Figure 2. Effects of various treatments of MSMA on the standing crop of Spartina alterniflora.

The salt marsh periwinkle spends most of its adult life on the shoots of *Spartina* where it probably feeds on epiphytic microorganisms and/or detritus. Therefore, this animal would receive direct contact with any foliarly applied herbicides and would ingest the material as it grazed on the stems. The dry wt/m^2 of standing crop of *Littorina* was reduced 50% on those plots receiving three applications of 9,000 ppm MSMA (Fig. 4). This decrease primarily reflects a decrease in numbers rather than growth and was not statistically significant because of the large variation in the spatial distribution of the population within the area. It is not known whether MSMA at the lower concentrations had any adverse effect on the physiology of the organism.

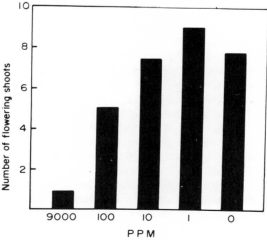

Figure 3. Effect of various concentrations of MSMA on number of flowering shoots of **Spartina alterniflora** per m^2 on November 12, 1971.

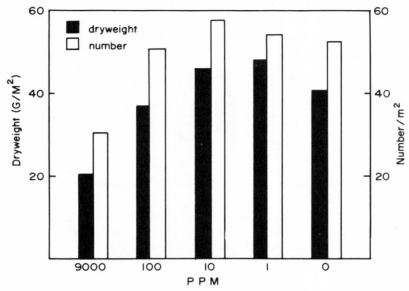

Figure 4. Effect of various concentrations of MSMA on dry weight per square meter of Littorina.

Laboratory Studies on Phosphorus Uptake. Pomeroy et al. 1967 indicated that *Spartina alterniflora* acts as a nutrient pump in extracting phosphorus from the sediments. Upon death and subsequent decomposition, phosphorus is released to other organisms in the detritus food chain. Consequently, we investigated the effects of 50, 5, and 0 ppm MSMA on phosphorus absorption by *Spartina* as indicated by their uptake of ^{32}P (Table 3). Analysis of variance did not show any significant variation between treatments. During this laboratory investigation it was observed that MSMA was much more toxic than when applied as a foliar spray in the salt marsh. When 100 ppm was applied 10 times, desiccation and burning occurred on over 50% of the aerial portions of the plant. This amount of injury was not observed in the field tests. It is not known whether this increased toxicity was because the plants were younger, grown under nonsaline conditions, or because of the method of application.

We feel that the most probable explanation for the higher toxicity in the laboratory is that submergence for two hours results in significantly more absorption and translocation than a foliar spray with the same concentration. The major cite of entry on foliar-applied herbicides is through the cuticle with minor entry through the stomates. Crafts (1961) attributed differences in absorption of herbicides under varying conditions of humidity to changes

Table 3. Effect of submerging the tops of *Spartina alterniflora* for two hours twice each day for five consecutive days in 0, 5, and 50 ppm MSMA on the subsequent accumulation of ^{32}P in shoots from ^{32}P supplied in nutrient solution to the roots.

Concn MSMA (mg/L)	\multicolumn{5}{c}{Replicates}	Av.
	1	2	3	4	5	
	\multicolumn{5}{c}{Specific activity[a] (cmp/mg)}	
20	1,200	2,000	3,700	4,000	8,400	4,000
5	2,400	3,700	8,200	7,000	7,000	5,700
0	1,200	3,200	9,900	7,600	8,600	6,100

[a]Individual values are average from three subsamples.

in moisture levels in the cuticle. It is thought that the cuticle consists of a somewhat open sponge-like material consisting of lipoid framework interspereed with pectin strands and pores. When in the saturated condition, these pores are filled with water and the pectin is highly hydrated. A water soluble herbicide molecule would, therefore, make instantaneous contact with the water continuum of the leaf and a pathway of entry would be established.

Conclusion

Based on the limited information collected in 1971, organic arsenical pollution at the levels anticipated in the tidal flux would not have an appreciable effect on *Spartina alterniflora*, the dominant halophyte of southeastern salt marshes. Concentrations of 100 ppm applied several times during the growing season apparently reduced the standing crop. It is highly unlikely, however, that concentrations would approach even 1 ppm in tidal waters as it is tenaciously adsorbed to clay particles. In the salt marsh environment, decomposition by microbial and chemical means would probably result in the organic arsenic being reduced to volatile alkyl arsines. Some methyl arsine is generated from cacodylic acid in soil under flooded conditions but it is trapped by the soil as an insoluble salt if the soil is high in aluminum or iron.

During the summer of 1972 the experiments were repeated with a modified approach. Except for the addition of a concentration of 0.1 ppm, the same rates were applied but this time twice a day for five consecutive days, once a month for three consecutive months. It is thought that this regime is more analogous with exposure to arsenicals in the tidal water. In addition, four metal cylinders 10 m in diameter have been sunk in the *Spartina* marsh. These are filled 0.5 m deep with seawater, treated with sufficient MSMA to achieve a concentration of 1.0 ppm, held for two hours, and emptied. The same number and frequencies of applications are being used as for the sprayed plots.

Arsenic residue determinations will be made on *Spartina, Littorina*, the Atlantic ribbed mussel (*Modiolus demissus*), mud, and seawater samples taken immediately after and one month after the final treatment.

Acknowledgements

The authors would like to acknowledge the continuing cooperation of the Sapelo Island Research Foundation and the staff of the University of Georgia Marine Institute with special thanks to Drs. R. J. Reimold, J. Gallagher, and A. Greene. The U. S. Dept. of Agr. supplied the funds to travel to the 25th AIBS meeting and make this presentation.

Literature Cited

Audus, L. J. 1964. The physiology and biochemistry of herbicides. Academic Press, New York and London.

Bascand, L. D. 1968. The control of *Spartina* species. Proc. New Zeland Weed Conf. 108-113.

Butler, P. A. 1963. Commercial fisheries investigations. Pages 11-25 in Pesticide and Wildlife Studies. U. S. Fish and Wildl. Serv. Circ. 167.

Carter, Mason. 1969. Amitrole. Pages 187-206 in Degradation of Herbicides. Edited by P. C. Kearney and D. D. Kaufman. Marcel Dekker, Inc., N. Y.

Cope, O. B. 1963. Sport fishery investigations. Pages 26-42 in Pesticide and Wildlife Studies. U. S. Fish and Wildl. Serv. Circ. 167.

Crafts, A. S. 1961. Chemistry and Mode of Action of Herbicides. Interscience Publishers, New York and London. 269 pp.

Davis, D. E. 1972. Effects of herbicides on plant physiological processes. Pages 49-65 in Proc. Scotts Turfgrass Res. Conf. Volume 3 – Weed Controls. O. M. Scott & Sons Co., Marysville, Ohio.

Foy, C. L. 1969. The chlorinated aliphatic acids. Pages 207-253 in Degradation of Herbicides. Edited by P. C. Kearney and D. D. Kaufman. Marcel Dekker, Inc., N. Y.

Funderburk, H. H., Jr. 1969. Diquat and paraquat. Pages 283-298 in Degradation of Herbicides. Edited by P. C. Kearney and D. D. Kaufman. Marcel Dekker, Inc., N. Y.

Grzenda, A. R., H. P. Nicholson, and W. S. Cox. 1966. Persistence of four herbicides in pond water. J. Amer. Water Works Assoc. 58:326-332.

Johnson, A. S., H. O. Hillestad, S. A. Fanning, and G. F. Shanholtzer. 1971. An Ecological Survey of the Coastal Region of Georgia. A Report to the National Park Service. 254 pp. (Unpublished).

Odum, E. P. 1961. The role of tidal marshes in estuarine production. N. Y. State Conservationist 70-79.

Odum, E. P., and A. A. de la Cruz. 1967. Particulate organic detritus in a Georgia salt-marsh estuarine ecosystem. Estuaries 83:383-388.

Pimentel, David. 1971. Ecological Effects of Pesticides on Non-Target Species. U. S. Govt. Printing Office, Washington, D. C. 220 pp.

Pomeroy, L. R., and R. E. Johannes, E. P. Odum, and B. Roffman. 1967. The phosphate and zinc cycles and productivity of a salt marsh. Pages 412-419 in Proc. 2nd Radioecology Symposium. Edited by D. Nelson and F. C. Evans. Clearinghouse for Fed. Sci. and Tech. Inf., Nat. Bureau of Stand., U. S. Dept. of Commerce. Springfield, Virginia.

Pomeroy, L. R., L. R. Shenton, R. D. H. Jones, and R. Reimold. 1972. Nutrient flux in estuaries. Pages 274-291 in Vol. 1, Proc. Symp. on Nutrients and Eutrophication: The Limiting Nutrient Controversy. Edited by D. E. Lechens. Amer. Soc. of Limnol. and Oceanog., Inc. Allen Press, Inc.

Ranwell, D. S., and B. M. Downing. 1960. The use of dalapon and substituted urea herbicides for control of seed-bearing *Spartina* (cord grass) in inter-tidal zones of estuarine marsh. Weeds 8:78-88.

Smalley, A. E. 1959. The growth cycle of *Spartina* and its relation to insect populations in the marsh. Proc. Salt Marsh Conf. Univ. of Ga. Press. Athens, Ga.

Springer, P. F. 1957. Effects of herbicides and fungicides on wildlife. Pages 87-106 in North Carolina Pesticide Manual.

Surber, E. W., and Q. H. Pickering. 1962. Acute toxicity of endothal, diquat, hyamine, dalapon, and silvex to fish. U. S. Fish and Wildlif. Serv., Progr. Fish Culture 24:161-171.

Sutherland, G. R., and A. Eastwood. 1916. The physiology anatomy of *Spartina townsendi*. Ann. Bot. 30:333-351.

Taylor, M. C., and E. M. Burrows. 1968. Chemical control of fertile *Spartina townsendi* (S.L.) on the Cheshire shore of the Dee Estuary. I. Field trials on *Spartina* sward. Weed Res. 8:170-184.

Taylor, M. C., and E. M. Burrows. 1968. Chemical control of fertile *Spartina townsendi* (S.L.) on the Cheshire shore of the Dee estuary. II. Response of *Spartina* to treatment with paraquat. Weed Res. 8:185-195.

Teal, J. M. 1962. Energy flow in the salt marsh ecosystem of Georgia. Ecology 43:614-624.

Weed Science Society of America. 1970. Herbicide Handbook of the Weed Society of America. 2nd Ed. W. F. Humphrey Press, Inc. Geneva, N. Y. 368 pp.

NUTRIENT LIMITATION IN SALT MARSH VEGETATION[1]

Ivan Valiela
Boston University Marine Program
Marine Biological Laboratory
Woods Hole, Massachusetts 02543

and

John M. Teal
Woods Hole Oceanographic Institution
Woods Hole, Massachusetts 02543

Introduction

Salt marsh plants, particularly *Spartina alterniflora*, commonly show great variation in standing crop and production from site to site. Marsh plants are usually taller toward the southern end of their geographical range (Wass and Wright 1969, Valiela, Teal and Van Raalte, in press). However, the most impressive variation in production occurs within individual marshes. For example, in Barnstable Marsh, Mass., *S. alterniflora* is over 2 m tall in some creek-side sites, but only reaches 20 cm in height in some areas at higher elevations. Vegetation also tends to be greener near the creeks compared to elsewhere in the marsh.

A number of studies attempt to correlate various environmental factors, such as salinity and tidal height, to standing crops and marsh production (Adams 1963, Stalter and Batson 1969, Good 1972). The results have been often inconclusive and, due to the nature of the correlational approach, fail to directly identify the causes of the large differences in primary production. Similarly, laboratory experiments (Mooring et al. 1971, Phleger 1971), while quite specific as to the variables manipulated, are perforce carried out in an artificial environment. The needed compromise is to perform field experiments, altering specific variables in the real world, with an opportunity to evaluate responses from a variety of components simultaneously.

We have claimed in a previous paper (Teal and Valiela, in press) that nutrients were the most likely factor determining how much plant growth

[1]Contribution No. 2955 from the Woods Hole Oceanographic Institution. This work was supported by N. S. F. Grants GA28365 and GA28272. We thank Nell Backus, Nat Corwin, Nancy McNelly, Helen Ortins, Warren Sass and Eric Teal for help in conducting this study.

occurs in a salt marsh. Since salt marshes are repeatedly washed by seawater, it seems likely that micronutrients are not limiting to plant growth. The anoxic marsh sediments store large amounts of a great variety of elements (Tyler 1971), most of which are probably sequestered from seawater. Although *S. alterniflora* is known to have a high requirement for iron (Adams 1963, Mooring et al. 1971), marsh sediments hold large concentrations of iron as evidenced by the black color of marsh soils, largely due to FeS. Among major nutrients, potassium, for example, averages about 380 mg/l (Goldberg 1963) in full seawater. Tyler (1971) reports an abundance of available K, as well as Mg and Ca, throughout the year in marsh sediments.

Nitrogen seems to be the most likely macroelement limiting growth since the phosphorus content of the top 10 m of Georgia marsh sediments are high enough to provide for growth of plants for 500 years (Pomeroy et al. 1969). However, it is not clear if P may be limiting within one growing season since storms and meandering creeks are unlikely to have enough time to expose sediments as new sources of phosphorus. Most grass roots are found relatively near the surface and the amount of P available there may become quite low (Tyler 1967). Further evaluation of the importance of both N and P is therefore needed. The experiments described below measure the effects of phosphorus and nitrogen enrichments on primary production by higher plants on Great Sippewissett salt marsh, Buzzards Bay, Massachusetts.

A number of compounds can be used in enrichment experiments as nitrogen sources, each with different properties. Tyler (1967), in a *Juncus* marsh in the Baltic, conducted a series of small, short-term fertilization experiments in which he observed a significant growth response to ammonium enrichment and no response to nitrate fertilization. Although not mentioned by the author, denitrifying bacteria probably converted the nitrates into free nitrogen so that little, if any, of the added nitrate may have been available to *Juncus*.

We chose to use urea as the source of nitrogen because it was easily available as commercial fertilizer, was not subject directly to denitrification, and was easily hydrolyzable to ammonium, a form readily available to plants. Some volatilization of NH_3 may have occurred after hydrolysis but such losses may be low (Raps 1971, P. Mangelsdorf, personal communication).

Methods

The urea fertilizer (46% N by weight) was broadcast every two weeks onto circular plots 10 m in radius from May to mid-November 1971 and 1972. The urea was added at the rate of 11.2 g/two weeks.

The phosphate fertilizer (20% phosphate by weight) was applied at 64.8 g/two weeks during 1971 and at 13 g/two weeks during the 1972 growing season.

Each enrichment treatment was applied to two replicate plots. Two similar control plots were also established. Samples of water from the marsh sediments were obtained by removing a core 10 cm deep and collecting the water which percolated into the core hole. The samples were then preserved by adding a few drops of mercuric chloride and were subsequently filtered using HAO.45 u filters.

The concentrations of NH_4-N in the sediment water were determined using the method of Solorzano (1969). NO_2 plus NO_3 nitrogen was measured by the procedure of Wood, Armstrong and Richards (1967), and phosphate phosphorus was measured as recommended by Strickland and Parsons (1968).

The experimental plots were desinged to include two types of marsh vegetation. Low marsh refers to areas covered by most tides and is dominated by *Spartina alterniflora*. High marsh, dominated by *Spartina patens* and *Distichlis spicata*, lies above mean high tide. Other species of higher plants were also present (Table 1) but their occurrence was irregular and their contribution to standing crops generally small.

Standing crops of marsh vegetation were measured in 1971 by harvesting two rectangular quadrats 0.05 m^2 in area in each of the two marsh types at intervals during the growing season. One circular quadrat of 0.1 m^2 was collected from the plots at intervals during 1972. The two 0.05 m^2 samples collected in 1971 were pooled to obtain one 0.1 m^2 sample from each plot. Samples of 0.1 m^2 provided fairly stable estimates of means and standard errors (cf. Fig. 3) and were also small enough to allow repeated sampling within our 314 m^2 treated plots without drastic damage to the marsh. Sample quadrats were also collected in high and low marsh about one or two meters outside each of the treated plots. Each sample taken inside the plots could then be paired with one taken outside the plot but in similar vegetation. This allowed the separation of variation due to site variability from variation due to treatments. After sampling, the plants were separated according to species and dried overnight at 60°C and weighed. In 1971, subsamples of the harvested plants were then taken for semimicro-Kjeldahl analysis for total nitrogen content and combusted to determine ash-free organic matter. In 1972, total nitrogen and carbon were measured with an elemental analyzer.

Cores 6.5 cm in diameter were taken in low marsh sites within the experimental plots to depths varying from 20 to 45 cm. Each core was then vertically divided in half. One-half was thoroughly washed to separate roots and rhizomes from the mineral particles. Roots and rhizomes were then weighed and analyzed for total nitrogen and combusted to obtain ash-free dry weights.

Results and Discussion

Fig. 1 shows the NH_4-N concentrations in sediment water in each of the three treatments. During 1971 the ammonium nitrogen contents of the sediment water in the phosphate-treated and control plots were steadily depleted as the growing season progressed. Where urea was added there was a marked increase in the concentration of dissolved NH_4-N. These differences

Table 1. Principal species of higher plants found in the two main marsh types within the experimental plots.

Low Marsh

Spartina alterniflora (salt marsh cord grass)

Salicornia virginica, S. bigelovii, (glass worts)

Suaeda maritima (sea blite)

Spergularia marina (sand spurrey)

High Marsh

Spartina patens (salt meadow grass)

Distichlis spicata (spike grass)

Juncus gerardi (black rush)

Limonium carolinianum (sea lavender)

Aster tenuifolia (salt marsh aster)

were maintained through the winter and the next growing season. There is only a suggestion of depletion of NH4-N in the phosphate and control plots during 1972. However, regardless of the fluctuations occurring over a growing season, urea-enriched plots showed substantially greater concentrations of NH4-N than either the phosphate or the control plots. Most of the sediment water values are for low marsh sites. However, the few high marsh measurements taken in 1971 are similar to the low marsh values.

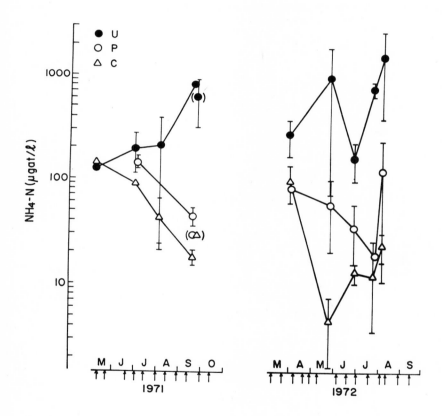

Fig. 1. Concentration (mean + std. error) of ammonium nitrogen (NH4-n) in the sediment water in plots subjected to urea (U), phosphate (P) and control (C) treatments. All points are for low marsh sites, except for points in parenthesis which indicate values for high marsh. The arrows along the horizontal axis indicate the dates of fertilization.

Fig. 2 contains the results of PO_4-P determinations in sediment water in each treatment. During the first season there was a rapid build-up of dissolved phosphate in the sediment water in the PO_4 fertilized plots. There is no indication, as in the case of NH_4-N, of depletion of PO_4-P in the sediment water in any of the treatments. The differences in PO_4-P were maintained very well through the winter. After the phosphate dosage was reduced to one-fifth during the 1972 growing season, the concentrations of dissolved phosphate decreased as time went on. However, at all times the phosphate-enriched plots showed phosphate concentrations about one order of magnitude higher than the urea-fertilized plot and the controls.

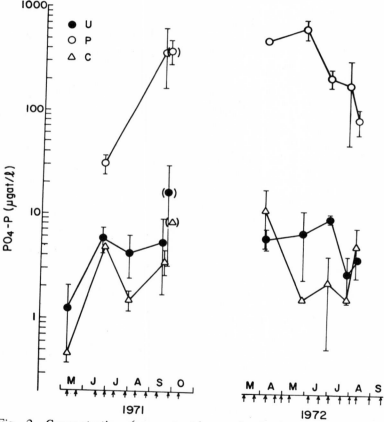

Fig. 2. Concentration (mean + std. error) of phosphate phosphorus (PO_4-P) in sediment water in plots subjected to the control (C), urea (U), and phosphate (P) treatments. All points are for low marsh sites, except for points in parenthesis which indicate values for high marsh. The arrows along the horizontal axis indicate the dates of fertilization.

The measurements of NO_2 plus NO_3 nitrogen all fell within a range of 0.5 to about 30 ugat/l with no differences ascribable to treatments and are not included here. The small concentration of oxidized nitrogenous forms relative to NH_4-N probably merely reflect the reduced condition of the sediments.

Fig. 3 shows the dry weights of live above-ground vegetation for 1971 and 1972 for low and high marsh for the three treatments.

The standing crops of above-ground vegetation are consistently greater where urea was supplied than in either the phosphate enrichment or the control in both 1971 and 1972. The increase in phosphate plots relative to controls in 1972 is only apparent. When the standing crops inside the plots are subtracted from the corresponding standing crops outside the plots, thereby removing site effects (Fig. 4), it becomes clear that the phosphate differences were only due to site variation and not to the enrichment with phosphorus. Similar subtractions did not alter any of the other results and are not shown.

The standing crops of the dominant plant species follow the pattern of the total vegetation. Fig. 5 shows the differences of samples obtained inside and outside the experimental plots for *S. alterniflora* in low marsh, *S. patens* and *D. spicata* in high marsh. All three species, and particularly *D. spicata*, show increased growth under urea enrichments and no response to phosphate enrichments. These patterns in the individual species were found in both 1971 and 1972 but for brevity only the 1972 data are shown.

Fig. 5 also shows the standing crop of *S. alterniflora* in high marsh sites. In general, *S. alterniflora* grows poorly and shows no response to enrichments in this habitat. Either some other factor is limiting or competition from other plant species prevents increased growth of *S. alterniflora* in high marsh habitats.

Table 2 includes values of percent nitrogen and ash-free organic matter for above-ground samples of the three main grass species for the urea and control treatments. This data, collected at the end of the first growing season, shows that for all three species the urea treatment increased the nitrogen content. This is emphasized when the ratio organic matter to nitrogen is used as a standardization.

The percent nitrogen values obtained for *S. alterniflora* toward the end of the second season of the experiments did not show a substantial increase in the urea plots (Table 3). However, the C/N ratio in the urea plots are substantially lower than either the control or phosphate C/N ratios, again showing nitrogen enrichment of the plant tissues.

The increased nitrogen content of grasses in urea-fertilized plots was to be expected in view of the results reported in the extensive agricultural literature on the subject. However, it is of interest here since Teal (1962) calculated that about one-half of the yearly above-ground production of Georgia salt marsh plants was exported to deeper waters. In situations where waters

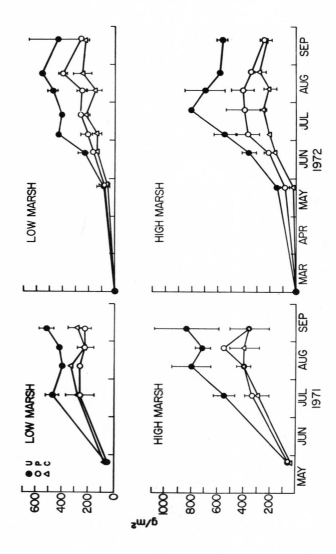

Fig. 3. Dry-weight standing crops of salt marsh vegetation under the control (C), phosphate (P) and urea (U) fertilizations during 1971 and 1972.

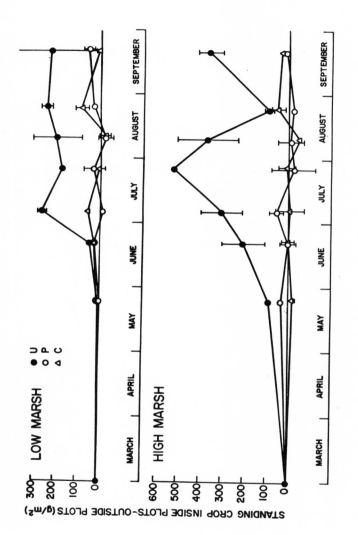

Fig. 4. Dry weight (mean + std. error) of plants inside minus dry weight of plants outside the experimental plots for 1972.

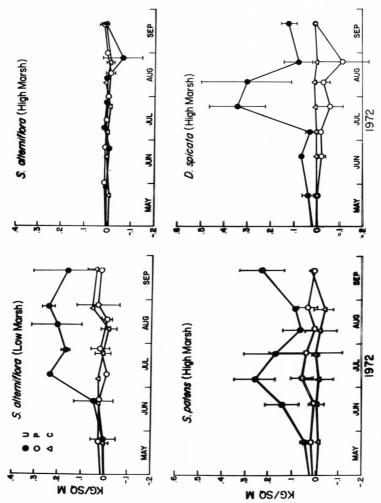

Fig. 5. Dry-weight of total live standing crops of individual plant species inside minus their standing crops outside the experimental plots for 1972. The means + std. errors for the differences are plotted. The values shown are for *Spartina alterniflora* in low and high marsh, and *S. patens* and *Distichlis spicata* in high marsh.

Table 2. Percent nitrogen and ash-free organic matter in above-ground vegetation for each of the three dominant salt marsh grasses for the urea and control treatments. Samples collected August 1971.

Spartina alterniflora	%N	%OM	OM/N
CONTROL	1.29	87.7	68.0
UREA	1.69	86.0	50.9
Spartina patens			
CONTROL	0.80	94.8	118.5
UREA	1.22	93.1	76.3
Distichlis spicata			
CONTROL	0.86	73.0	84.9
UREA	0.90	68.8	76.4

within salt marshes become enriched in nitrogen, the resulting export of dead and decaying plants will carry organic matter enriched in nitrogen out to the estuaries. This may be of importance in the productivity of estuaries since nitrogen is suggested to be a primary limiting factor to algal production in coastal waters (Ryther and Dunstan 1971).

Fig. 6 presents preliminary data on below-ground biomass in two cores from the urea and control treatments. The most notable of this New England salt marsh is the relatively shallow distribution of roots and rhizomes, virtually all the living material being located in the top 10 cm. There is no apparent growth response of below-ground biomass to fertilization with nitrogen. The living biomass in cores from urea-enriched plots (1.10 and 1.06 gm), if anything, is less than in the controls (1.46 and 1.80 gm). However, analysis of more cores are needed to establish whether this is a significant difference. At least there seem to be no large changes in underground plant parts comparable to the above-ground response to nitrogen additions. However, the belwo-ground dry-weight standing crops calculated from the cores (331.5, 319.5, 440.0, and 542.5g/m^2) are roughly comparable to the above-ground standing crops.

The nitrogen and organic matter contents of underground plant parts are shown in Table 4. Since these cores were taken in low marsh, *S. alterniflora* contributes almost all the biomass. Roots and rhizomes growing in urea-enriched plots contain more nitrogen than those from phosphate or

Table 3. Percent nitrogen, carbon, and C/N ratios determined with an elemental analyzer on shoots of *S. alterniflora*. Three plants are included in each measurement. Two samples of three plants were collected from each of the two replicate plots under the urea, phosphate, and control treatments, and two subsamples were analyzed in each sample. The means and standard errors of the subsamples were weighed and pooled using the procedures of Yates (1949), who uses the reciprocal of the variance as the weighing factor. The samples were collected August 1972.

	%N	%C	C/N
CONTROL	0.62	19.40	31.3
	0.66	18.93	28.6
	0.80	22.0	27.5
	0.64	19.88	31.1
x̄±s.e.:	0.64 ± 0.02	19.25 ± 0.23	29.7 ± 1.08
UREA	0.58	10.61	18.4
	0.49	8.85	18.4
	0.65	11.93	18.3
	0.43	7.64	17.8
x̄±s.e.:	0.54 ± 0.04	9.74 ± 0.81	18.4 ± 0.01
PHOSPHATE	0.46	14.37	31.2
	0.52	15.30	29.4
	0.42	9.50	22.6
	0.37	10.07	27.3
x̄±s.e.:	0.43 ± 0.02	11.16 ± 0.24	29.6 ± 0.84

control plots. The organic matter to nitrogen ratios in the urea treatments can be shown to differ significantly from the other treatments using Duncan's multiple range test (Steel and Torrie 1960). There are no significant differences in the O.M./n ratio between the control and phosphate treatments for roots or rhizomes. These accumulations of nitrogen in underground plant parts were the effects of fertilization and uptake during the 1971 growing season. The enrichment is particularly noticeable in the rhizomes, which were grown in 1971 and provided the starting growth for the 1972 growing season.

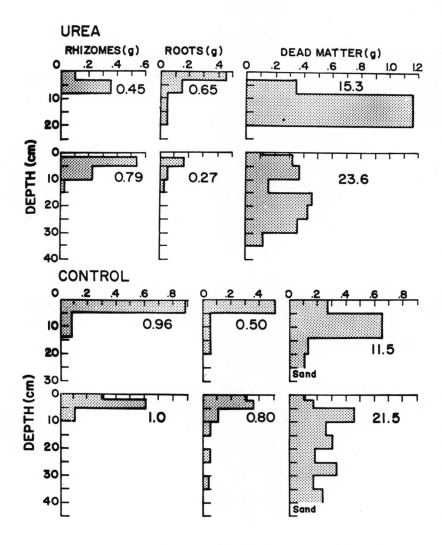

Fig. 6. Dry-weights of rhizomes, roots, and dead roots and rhizomes in cores taken in urea and coastal plots. The histograms show the vertical distribution of underground plant parts and the numbers in each graph are the total weight of living plant material over the whole core profile. The surface area of each vertical profile is 33.2 cm^2.

The storage of such supplies may be at least one way in which production in one year could influence production for the next growing season. In addition, the supply of nitrogen in the over-wintering roots and rhizomes may also be involved in the one-season lag in production observed in a number of fertilization and transplantation experiments.

Table 4. Percent nitrogen and ash-free organic matter for underground parts in each treatment. Values entered are for two replicate cores taken in low marsh sites. Samples collected in March 1972.

RHIZOMES		%N	%OM	OM/N
	CONTROL	0.247	90.15	364.98
		0.320	89.94	281.06
	$\bar{x}$±s.e.:	0.284 ± 0.04	90.05 ± 0.10	323.02 ± 42.0
	UREA	0.953	85.81	90.04
		0.520	80.14	169.50
	$\bar{x}$±s.e.:	0.737 ± 0.22	86.98 ± 1.16	129.77 ± 39.7
	PHOSPHATE	0.405	82.07	202.64
		0.322	87.78	272.61
	$\bar{x}$±s.e.:	0.364 ± 0.04	84.93 ± 2.85	237.63 ± 38.0
ROOTS				
	CONTROL	0.420	84.02	200.05
		0.409	84.61	206.87
		0.415 ± 0.01	84.32 ± 0.30	203.46 ± 3.4
	UREA	0.525	79.74	151.89
		0.453	78.76	173.86
		0.489 ± 0.04	79.25	162.88 ± 11.0
	PHOSPHATE	0.259	80.32	310.12
		0.349	83.44	239.08
		0.304 ± 0.05	81.88 ± 1.56	274.60 ± 35.5

As mentioned above, salt marsh vegetation tends to be taller, heavier, and greener near the mean low tide marks and particularly so near the margins of tidal creeks. Mooring et al. (1971) collected seedlings of the tall and short forms of *S. alterniflora* and found that the differences disappeared when the plants were grown in the laboratory under identical conditions. Shea et al. (1972) showed by electrophoretic and transplantation studies that the tall and short forms of *S. alterniflora* were only ecophenes with no genetic differences. Other transplant studies involving exchanges of plants from high and low marsh sites (Stalter and Batson 1969) seemed to show that plant height in *S. alterniflora* was fixed genetically. However, these latter experiments were conducted with small pieces of marsh sod and only over one season, and Shea et al. (1972) and some of our unpublished work in fertilized plots suggest that marsh plants may delay responses to transplantation or fertilization for as long as one growing season. It seems, therefore, that the marked differences in standing crops, height, and color are due to some feature of the habitat rather than genetic differences.

Our fertilization experiments, some of which are recorded above, have succeeded in converting, within one site, short, sparse grasses into a sward which looks very much like the vegetation found at a lower tide level (Teal and Valiela, in press). For all three dominant species of salt marsh grasses, plants anywhere in our plots fertilized with nitrogen are heavier, greener (due to increased chlorophyll content), taller, and show thicker stems and wider leaves, much like unfertilized creek-side plants. We infer from this that the feature of the habitat which is of importance in determining grass standing crop and the other observed differences is nutrient supply, and in particular, nitrogen, as shown by the results reported here.

Seneca (personal communication) in North Carolina has obtained increases in growth of transplanted *S. alterniflora* in plots fertilized with either phosphorus or nitrogen. However, the transplants were placed on dredge spoils which consisted of sand and some silt. Such sediments may have likely been rather poor in both phosphorus and nitrogen since the coarse grain would not have offered the large adsorptive area typical of the clayey organic muds found in salt marshes.

If tidal water is the main source of nutrients for a salt marsh (Blum 1968), it seems reasonable that plants growing in lower sites have access to more nutrients than those growing at higher elevations. Sediments at creek bank sites would be covered by more nutrient-laden water and for longer periods than sediments at higher sites. Since marsh sediments have a marked ability to retain and to remove nutrients from the water column, (Pomeroy et al. 1969; Valiela, Teal and Van Raalte, in press) creek bank sediments would be supplied with more nutrients than high marsh sediments. This, in turn, could produce the differences in standing crops, color and height observed in salt marsh vegetation.

Summary

Fertilization with urea and phosphate produced significant increases in dissolved NH_4-N and PO_4-P, respectively, in the sediment water of treated salt marsh plots.

Standing crops of salt marsh plants increased in the urea-fertilized plots while the standing crop in swards undergoing phosphate enrichments resembled control standing crops. This response pattern held for total above-ground vegetation and for individual standing crops of *Spartina alterniflora*, *S. patens*, and *Distichlis spicata*, the dominant grasses. Nitrogen supply therefore is one of the most important limiting factors for salt marsh vegetation. Preliminary results suggest, however, that roots and rhizomes will not show an increase in standing crops in response to fertilization with either phosphorus or nitrogen.

The urea enrichments also resulted in a general increase in nitrogen content of above-ground plant tissues, roots and rhizomes. The increased standing crops and nitrogen contents of above-ground plant parts may be an important source of nitrogenous organic materials exported from salt marshes to nitrogen-poorer estuarine waters, particularly where the salt marsh is subject to nitrogen enrichment.

The plots fertilized with urea resemble swards found at lower elevations in salt marshes, suggesting that the great variations in height, color, and standing crop seen in marshes are related to nitrogen supplies. Since nutrients are brought into salt marshes largely by tidal water, it seems that marsh sediments and microflora are able to remove substantial amounts of dissolved nutrients from tidal water columns and, therefore, marshlands are of importance as removers of dissolved nutrients, particularly nitrogen, from estuarine waters. Dissolved nutrients may then be incorporated into vegetation and exported as organic nitrogen in dead and decaying vegetation.

References

Adams, D. A. 1963. Factors influencing vascular plant zonation in North Carolina salt marshes. Ecology 44:445-456.

Blum, J. L. 1968. Salt marsh spartinas and associated algae. Ecol. Mono. 38:199-221.

Goldberg, E. D. 1963. The oceans as a chemical system. Pages 3-25 *in* The Sea. Vol. 2. Edited by M. N. Hill. John Wiley & Sons, New York. 554 pp.

Good, R. E. 1972. Salt marsh production and salinity. Bull. Ecol. Soc. Amer. 53:22 (abstract).

Mooring, M. T., A. W. Cooper, and E. D. Seneca. 1971. Seed germination response and evidence for height ecophenes in *Spartina alterniflora* from North Carolina. Amer. J. Bot. 58:48-55.

Phleger, C. F. 1971. Effect of salinity on growth of salt marsh grass. Ecology 52:908-911.

Pomeroy, L. R., R. E. Johannes, E. P. Odum, and B. Roffman. 1969. The phosphorus and zinc cycles and productivity of a salt marsh. Pages 412-419 *in* Proc. 2nd Nat. Symp. on Radioecology. Edited by D. J. Nelson and F. C. Evans.

Raps, M. E. 1971. Study of ammonia diffusion across the air/water surface of the Sea Grant ponds. Pages 103-113 *in* Ann. Rept. 1970-1971 to

Office of Sea Grant Programs, Grant No. GH103. Edited by E. J. Kuenzler and A. F. Chestnut. 345 pp..

Ryther, J. H., and W. M. Dunstan. 1971. Nitrogen, phosphorus, and entrophication in the coastal marine environment. Science 171:1008-1013.

Shea, M. L., R. S. Warren, and W. A. Niering. 1972. The ecotype/ecophene status of the varying height forms of *Spartina alterniflora*. Bull. Ecol. Soc. Amer. 53:17 (abstract).

Solorzano, L. 1969. Determination of ammonia in natural waters by the phenolhypochlorite method. Limnol. Oceanogr. 14:799-801.

Stalter, R., and W. T. Batson. 1969. Transplantation of salt marsh vegetation, Georgetown, South Carolina. Ecology 50:1087-1089.

Steel, R. G. D., and J. H. Torrie. 1960. Principles and Procedures of Statistics. McGraw-Hill Book Co. Inc., New York. 481 pp.

Teal, J. M., and I. Valiela. 1972. Production and physiology of *Spartina alterniflora* Loisel, a plant of extreme environments. INTECOL Symp. on Physiological Ecology of Plants and Animals in Extreme Environments. Dubrovnik, March (in press).

Tyler, G. 1967. On the effect of phosphorus and nitrogen, supplied to Baltic shore-meadow vegetation. Botaniska Notiser 120:433-447.

Tyler, G. 1971. Distribution and turnover of organic matter and minerals in a shore meadow ecosystem. Oikos 22:265-291.

Valiela, I., J. M. Teal, and C. Van Raalte. 1972. Nutrient and sewage enrichment experiments in a salt marsh ecosystem. INTECOL Symp. on Physiological Ecology of Plants and Animals in Extreme Environments. Dubrovnik, March (in press).

Wass, M. L., and T. D. Wright. 1969. Coastal Wetlands of Virginia. Special Report in Applied Marine Science No. 10. Virginia Inst. of Mar. Sci. 154 pp.

Wood, E. D., F. A. J. Armstrong, and F. A. Richards. 1967. Determination of nitrate in sea water by cadmium-copper reduction to nitrite. J. Mar. Biol. An. U. K. 47:23-31.

Yates, F. 1949. Sampling Methods for Censuses and Surveys. Charles Griffin & Co., Ltd., London. 440 pp.

THE POTENTIAL ECONOMIC USES OF HALOPHYTES

Peta J. Mudie

Foundation for Ocean Research
Scripps Institute of Oceanography
La Jolla, California 92037

Introduction

The presence of an excessive amount of salts in irrigation or soil water has been considered a major threat to the permanence of irrigation agriculture throughout the world (Allison 1964). A variety of factors determine the magnitude of salinity or alkalinity at which crop growth is adversely affected. As a broad generalization, however, when conventional agricultural soils and practices are involved, the yields of most major crops are likely to be seriously inhibited by irrigation water containing about 1,500 to 2,000 mg/l TDS (total dissolved solids) or about 15 me/l Na^+, and by a soil water salinity of about 5,000 mg/l (U. S. Salinity Lab 1954).

Despite the historic record of salinity problems subsequent to intensive irrigation (Bernstein 1962, Raheja 1966), the expansion of irrigation agriculture in the arid and semi-arid regions of the world is considered by some as one of the most important possibilities for increasing world food production in the near future (Rains and Epstein 1967). Furthermore, the feasibility of using slightly saline urban and agricultural waste water, large underground reservoirs of brackish water, or even seawater, for the cultivation of economic plants presents challenging and exciting opportunities for conserving dwindling fresh water supplies. The ability of halophytes to tolerate highly saline soils and Na^+ concentrations of 66 me/Kg (0.5% NaCl) or more, suggests that they may be useful in supplying clues for combating salinity problems and for expanding agriculture in regions of limited fresh water.

During the past two years, the Halophyte Research Group at the Scripps Institution of Oceanography has been conducting exploratory studies on the use of highly saline water for agriculture. Part of this research has been aimed at investigating the potential for exploiting halophytes through direct use and by crossbreeding or graftage with glycophytes of economic importance. The preliminary findings of these studies are reviewed by Mudie et al. (1962). Comprehensive reviews on the economic uses of halophytes have also been made by Chapman (1960, 1972) who primarily emphasized the reclamation of naturally saline soils and the exploitation of coastal marshlands, rather than the cultivation of salt-tolerant plants. Other information on the existing economic uses of halophytes is fragmentary. One of the main obstacles to predicting the usefulness of halophytes at this time is the relatively scanty data on the precise limits of salt tolerance of individual species. This problem is compounded by the variety of methods that have been used to quantify the salinity of salt marsh soils and by the difficulty of relating ecological salinity values to the salinity criteria commonly used by agriculturalists. Questions regarding the possible exploitation of the genetic salt tolerance of halophytes through crossbreeding with glycophytic crops are clouded with virtual ignorance of the genetic basis for, and the heredity of, salt tolerance. Furthermore, the prospects for halophyte genetic manipulation must be weighed against the possibility of increasing the salt tolerance of conventional glycophytic crops through genetic selection or by use of special cultivation conditions.

Over and above these pragmatic problems of inadequate scientific data lies the complexity of predicting the future socio-economic and political attitudes that will ultimately dictate both the acceptance of unconventional halophytic products and the need for expanding agriculture through the use of salt-tolerant plants. Views on the prospects of new plant products range from the enthusiasm of Pirie (1969) to highly restrained attitudes based on the failure of relatively recent attempts in the U. S. A. to exploit novel economic plants, such as canaigre *Rumex hymenosepalus* (Bender 1963), guayule *Parthenium argentatum*, and Russian dandelion *Taraxicum kok-saghyz* (Baker 1965) and on the fact that the variety of economic plant species seems to be dwindling rather than expanding, at least in the tropics (Raven and Curtis 1970).

It is not my intent, however, to debate future trends in world agriculture. In this review on halophyte economic potentials, I have chosen to make the implicit assumption that the need for both large-scale expansion of economic plant resources and the utilization of saline water or soils for agricultural purposes will continue. I shall aim at outlining the possible roles that halophytes might play in meeting these needs and at directing attention towards the kind of information that is required to improve our evaluation of their potentials. In short, my argument is that much might be gained from greater emphasis on increasing the salt tolerance of economic plants in contrast to the past trend of research that has been primarily directed towards the reduction of agricultural soil and irrigation water salinity.

Agriculture and Salinity: Dimensions of the Problem

The literature describing various aspects of the agricultural salinity problem, including historical trends, hydrological and pedological characteristics, crop responses, and reclamation procedures, is voluminous and cannot be considered in this paper. However, the interested reader may refer to a number of comprehensive reviews for detailed background information (Magistad 1945, Hayward and Wadleigh 1949, Hayward 1954, Kelley 1951, Hayward and Bernstein 1958, Bernstein and Hayward 1958, Bernstein 1962, Allison 1964, Raheja 1966). Here, only the global extent of the agricultural salinity problem will be delimited in order to provide a broad framework for consideration of the need for new approaches to the problem.

The geographical magnitude of the salinity problem may be evaluated from three viewpoints: 1) by the extent of naturally saline soils that are not in production because of the limited salt tolerance of economic crops; 2) by the extent of arable soils that have become unproductive as a result of salt accumulation under agricultural practices, and 3) by the area of nonsaline soils that is not farmed because the only available water is saline. Naturally saline "problem" soils include both the tidally-influenced solonchak soils of low-lying coastal regions, and the pedocal-type soils and solods that characterize arid and semi-arid regions. The former are relatively limited in extent since the entire land-sea interface comprises considerably less than 2% of the land surface. In contrast, semi-arid and arid regions constitute roughly one-third (4.8×10^9 ha or 18.86×10^6 sq. miles) of the earth's land area (Raheja 1966), and highly saline soils are prevalent in about half this area.

Some estimates of the extent of naturally saline soils in potentially arable regions are listed in Table 1. This table also includes a few examples of the magnitude of soils that have been rendered unproductive as a result of salt accumulation following intensive irrigation. The overall scale of the problem of salinization under irrigation agriculture can be gauged from the fact that more than 160×10^6 ha of the world's arable lands are under irrigation (Israelson and Hansen 1962) and one-third of this area (53×10^6 ha) is estimated to be plagued with salinity problems (Van Aart 1971).

It is beyond the scope of this report to calculate the extent of land that could be farmed if saline water supplies, such as ground water, sewage effluent, or seawater, were to be employed for irrigation. Coastal deserts border the oceans for a total of almost 30,000 km and about half of their area lies below 33 m, which is considered to be the practical upper limit to which irrigation water may be elevated (Meigs 1966). Sand dunes, which perhaps offer the greatest promise for irrigation with extremely saline water, cover about 1.3×10^9 ha (5×10^6 sq. miles), or about 9% of the earth's land surface (Boyko 1966), but the area of dunes within an economically feasible pumping distance from the coast or saline ground water reservoirs has not yet been determined. Vast areas of brackish ground water (4,000-10,000 mg/l

Table 1.   Estimates of the extent of salt-affected agricultural or potentially arable soils in various contries.

Country	Extent (ha x 10^6)	References*
USSR	Salt-affected: 75	[1]
	Solonetz: 50-60	[3]
	Saline/sodic: 89% irrigated area in central Asia	[3]
China	Salt-affected: 20	[1]
	Saline: 0.02 in Sinkiang region	[2]
Iraq	Saline: 3.68 (=80% irrigable land)	[4]
India	Saline/sodic: 8	[4]
	Sodic ("usar"): >0.8	[5]
	Saline: >25% irrigated area of Punjat	[2]
	Saline/sodic: ~0.21 in Little Rann of Cutch	[2]
W. Pakistan	Saline: 1.2/5.06 (=24%) irrigated land; ~0.53 out of cultivation	[2]
Spain	Saline/sodic: >0.1 in lower valley of Guadalquiver	[2]
Yugoslavia	Salt-affected: >0.2 mostly in S. Pannonian Plain	[1]
Hungary	Salt-affected: >0.5	[1]
Egypt	Saline (in 1960): 0.12; reclamation proceeding at rate of 8 x 10^3 ha/yr	
Canada	Alkali: ~0.36 in Saskatchewan	[5]
	Solonetz: ~2.84 south and east of Edmonton	[5]
USA	Saline: 0.12 in SW, including San Joaquin Valley	[2]
	0.162 agricultural land in N. Dakota	[6]
	Salt-affected: 25% total irrigated acreage	[5]
Puerto Rico	Salt-affected: 36% soils in Lajas Valley	[1]
Peru	Saline: 0.004 irrigated coastal valley soils	[7]
	Saline/sodic: ~0.051	[7]
	Sodic: ~0.016	[7]
N.E.Brazil	Saline/sodic: 0.031-0.038 irrigated land	[5]
Australia	Saline: 20% area irrigated by Murrumbidgee	[2]
	Salt-affected: 0.405 in W. Australia wheat belt	[1]
	Solonetz: 4.925 in W. Australia	[5]
	Solonized brown soils ("mallee"): 42.5 in South	[5]

* [1] Bernstein 1962; [2] Raheja 1966; [3] Strogonov 1964; [4]; van Aart 1971; [5] Hayward 1954; [6] Doering & Benz 1972; [7] Anon 1966.

TDS) underlie many regions of the world (Dixey 1966, Meigs 1966). However, this type of water is extremely variable in ionic composition and each source should be evaluated independently to determine its potential agricultural use.

The Economic Uses of Halophytes

The ability of halophytes to tolerate relatively high soil and water salinities invites consideration of their use in agriculture, horticulture, and industry in water-short and saline soil regions. In the broad sense, halophytes include all salt-tolerant plant micro-organisms (marine phytoplankton, fungi, bacteria) and seaweeds, in addition to salt-tolerant vascular plants. However, with the exception of *Spirulina platensis*, a blue-green alga, only the uses of vascular plants will be considered at this time.

The compilation of a list of economic uses of halophytes denotes an *a priori* ability to distinguish between halophytic and glycophytic taxa. In reality, however, the boundary between salt-tolerant glycophytes and miohalophytes appears to be highly tenuous. Furthermore, use of the commonly accepted criterion of halophytism, namely tolerance of 0.5% NaCl in the soil water (Chapman 1960), demands information on the soil salinity with which a species is associated. Unfortunately, for the most part, this information is lacking for individual taxa. However, species characteristic of the lower littoral zone of coastal marshes and inland saline sinks are usually continually subjected to a soil water salinity of at least 0.5% NaCl and thus may safely be assumed to be halophytes. In contrast, upper marsh soils often undergo wide seasonal fluctuations in moisture content and salt concentration. Without precise measurements of seasonal and vertical changes in NaCl and correlated observations of plant growth characteristics (e.g. dormancy period, depth of active roots), it is impossible to determine whether or not a high marsh species conforms with the 0.5% NaCl criterion. The general importance of this problem will be discussed later.

As a first approach to cataloguing the economic uses of halophytes, I chose to consider all species recorded from, or in the immediate vicinity of, naturally saline habitats. A taxonomic list of putative halophytic species was compiled from the literature on salt marshes, salt deserts, saline sinks, and mangrove swamps. Each species was assigned to a salinity group on the basis of its recorded salt tolerance, or in the absence of this data, on the basis of its position in the marsh (high, middle, or low) and/or its association with halophytes of known salt tolerance. Three salinity groups were used, viz. euhalophytes, mesohalophytes, and miohalophytes, which broadly correspond with the meso-and eu-euhalophytes, mesohalophytes, and miohalophytes of Chapman (1960). This taxonomic list, which includes approximately 550 species belonging to 220 genera in 75 families, may be obtained from the author upon request.

Two methods of exploiting halophytes were considered:

1. Direct uses — halophytes of existing economic significance, such as species of *Spartina* and *Rhizophora*, could be put to more widespread use, or species with potential economic value, e.g. *Juncus maritimus* (high cellulose content) and *Atriplex hortensis* (occasionally used as a vegetable), could be introduced as novel economic products.
2. Indirect uses — the genetic salt tolerance of halophytes might be used to increase the salt tolerance of conventional economic plants through hybridization or graftage. Discussion of the indirect uses of halophytes will be deferred until a later section on genetic improvement of salt tolerance.

Halophytes that have some present direct economic value (other than local use for sand-binding or marsh reclamation) are listed in Table 2, along with their reported ranges of salt tolerance and/or their estimated tolerance (salinity group). Halophytes with minor, subsistence-level uses in restricted areas and those with historical food or medicinal uses are listed in Table 3. This table also includes halophytes with characteristics that appear to make them suitable for some economic purpose. Most of the species in Table 3 represent incompletely evaluated resources and only tenuous judgements can presently be made concerning their potential usefulness.

It will be noted that Table 2 includes a number of crops that are widely cultivated in nonsaline soils (e.g. *Cocos nucifera*, *Nypa fruticans*, *Phoenix dactylifera*, *Phormium tenax*, *Cynodon dactylon*, *Beta vulgaris*, *Trifolium* spp., *Lotus corniculatus* var. *tenuifolius*). The cultivars of these species are not usually considered to be halophytes, and intensive selection for properties other than salt tolerance has undoubtedly reduced their halophytic tendencies. It is interesting, however, that most of the widely-grown crops in Table 2 are noted for their relatively high salt tolerance in agricultural soils (see Table 6) and some (e.g. beets, New Zealand spinach) are highly resistant to extreme salt stress under special cultivation conditions (Mudie et al. 1972). It is thus suggested that more attention should be paid to determining the salt tolerance of wild populations of these species since these wild gene pools may well contain useful germ plasm for increasing the salt-tolerance of crops.

The prospects appear to be rather low for large-scale use of halophytes as novel foods for human consumption even though several species have edible leaves, fruits or seeds and archaeological evidence shows that the seeds of *Atriplex*, *Suaeda* and *Distichlis spicata* were eaten by prehistoric New World Indians (Heizer and Napton 1969). The relatively small, chaffy nature of the fruits of most edible halophytes would make harvesting tedious, and questionable palatability and nutritional value of the seeds make it unlikely that the intensive effort needed to develop an acceptable food product would be warranted. Furthermore, the general outlook for the development of important new food plants is considered to hold low promise (Pirie 1969).

Perhaps the most promising novel halophyte used as food is *Spirulina platensis*, a blue-green alga eaten by the Kanembu tribe of the Lake Chad region (Leonard 1969). This alga inhabits lake waters with a high Na_2CO_3

LITERATURE REFERENCES AND ABBREVIATIONS FOR TABLES 2 AND 3

(See "Literature Cited" for full citation.)

EU = euhalophyte; MES = mesohalophyte; MIO = miohalophyte (see text for full explanation)

(1) Hayward and Bernstein 1958; (2) Udell et al. 1969; (3) Mudie, this paper; (3a) Mudie, unpublished; (4) Wheeler 1950; (5) Chapman 1960; (6) Hannon and Bradshaw 1968; (7) Ranwell 1967; (8) Mobberley 1956; (9) Ungar 1965; (10) Macnae 1968; (11) Boyko 1966; (12) Headdon 1929; (13) Chapman 1972; (13a) Chapman pers. comm.; (14) Ungar 1967; (15) Ungar 1966; (16) Tewari 1967; (17) Lunt, Youngner and Oertli 1961; (18) Penfound and Hathaway 1938; (19) Seneca 1972; (20) Adriani 1945; (21) Shreve 1942; (22) Morton 1965; (23) Youngner, Lunt and Nudge 1967; (24) Gausman et al. 1954; (25) Aston 1933; (26) Davies 1931; (27) Greenway 1968; (28) Turner and Bell 1971; (29) Clawson 1937; (30) Fryxell, Pers. comm.; (31) Atchley 1938; (32) Darlington and Wylie 1955; (33) Gledhill 1963; (34) Bloch et al. 1954; (35) Singh 1970a; (36) Elzam and Epstein 1969; (37) Hayward 1954; (38) Gupta and Gulati 1967; (39) Furr and Ream 1968; (40) Firmin 1968; (41) Chatterton et al. 1970; (42) Kearney and Peebles 1960; (43) Gunn 1972.

Table 2. HALOPHYTIC SPECIES OF ECONOMIC SIGNIFICANCE

Numbers in parentheses refer to sources of information; for key to these and the salinity group symbols, see p.

(= values in % soil solution, unless specified; * = minor use)

FAMILY	Species and Common Name	Use	Salinity Group	Salinity Range+ TDS		NaCl/Cl	
AGAVACEAE	*Phormium tenax*, J. R. & C. Ford (NEW ZEALAND FLAX)	ornamental, fiber	MIO			0 - ~1.5 NaCl	[3a]
AIZOACEAE	*Tetragonia expansa* Murr. (NEW ZEALAND SPINACH)	vegetable	MIO			) - 3.0 NaCl	[3]
CHENOPODIACEAE	*Atriplex argentea* Nutt. (SALT BUSH) *A. confertifolia* (Torr. & Frem.) Wats.	forage [12]	MIO ?	0.22-1.1	[9]		
	A. nummularia Lindl. (SALT BUSH) *A. semibaccata* R. Br. (AUSTRALIAN SALT BUSH)	forage [12] forage [27] forage [12]	MIO ? EU MIO	0.5 -0.8	[21]	1 -300 me/1 NaCl ? - 1.2 NaCl	[27] [37]
	Beta vulgaris L. (BEET)	vegetable, sugar, fodder	MIO			0 3.0 NaCl	
	Kochia scoparia Schrad. (SUMMER CYPRESS)	ornamental, fodder	MES ?	0.34-1.3 soil d.w.	[15]		
COMBRETACEAE	*Lumnitzera littorea* Voigt *L. racemosa* Willd.	timber [10] timber [13]	MES ? EU	? - 9.0	[10]		
COMPOSITAE	*Baccharis halimifolia* L. (GROUNDSEL BUSH)	ornamental [32]	MIO ?	0 - 1.98	[18]		
	Matricaria chamomilla L. (FALSE CAMOMILE)	medicine, flavoring, perfume [35]	MIO				
EUPHORBIACEAE	*Excoecaria agallocha* L.	paper [13]	MIO				
GOODENIACEAE	*Selliera radicans* Cav.	ornamental [32]	MIO ?			0.3-1.0 % NaCl d.w.	[26]
GRAMINEAE	*Agropyron elongatum* (Hort.) Beauv. (TALL WHEATGRASS)	forage	MIO			0.1-0.5 M NaCl	[36]

Table 2. Continued HALOPHYTIC SPECIES OF ECONOMIC SIGNIFICANCE

Numbers in parentheses refer to sources of information; for key to these and the salinity group symbols, see p.

( ) = values in % soil solution, unless specified; * = minor use

FAMILY	Species and Common Name	Use	Salinity Group	TDS	Salinity	Range+ NaCl/Cl−	
	A. junceum (L.) Beauv.	forage, sand-binding	MES ?	? -3.2	[11]		
	Agrostis alba L. var. *palustris*	forage, hay, turf	MIO	0.03-0.68	[9]		
	=*A. stolonifera* L.(=*A. palustris* Huds.) (RED TOP)	[4]					
	A. tenuis Sibth. (COLONIAL BENT)	turf [4]	MIO	0-200 me/l	[23]	0 - 2.3 NaCl	[6]
	Chloris gayana Kunth (RHODES GRASS)	forage	MIO	0-180 me/l	[24]	90 me/l NaCl	[17]
	Cymbopogon nardus (L.) Rendle (CITRONELLA GRASS)	oil, perfume [38]	MIO	0-0.45			
	Cynodon dactylon Pers.(BERMUDA GRASS)	lawn, forage [4]	MIO	0-0.45	[24]	? - 0.6 NaCl	[37]
	Distichlis spicata (L.) Greene (SALT GRASS)	forage [1,2]	EU	0.04-2.52	[9]	0.75 3.0 NaCl	[37]
	Festuca arundinacea Schreb. (TALL FESCUE)	forage, turf [4,17]	MES/MIO	0-180 me/l		90 me/l NaCl	[17]
	F. rubra L. (RED FESCUE)	forage, turf [4]	MES			0 - 2.3 NaCl	[6]
	Hordeum maritimum Huds. (SEA BARLEY)	forage [37]	MIO ?			? -1.2 NaCl	[37]
	H. marinum L. (STERILE BARLEY)	forage [37]	MIO			? -1.2 Na Cl	[37]
	Panicum antidotale Retz. (BLUE PANICUM)	forage, sand-binding	MIO	0-0.45	[24]		
	P. virgatum L. (SWITCH GRASS)	forage	MIO ?	0.003-0.68	[9]		
	Paspalum distichum L.(KNOT GRASS)	forage	MIO ?				
	P. gayanum E. Desv.	forage	MIO ?				
	P. vaginatum Sw.	forage [37]	MIO ? [37]				
	Pholiurus incurvus (CURLY RYE)	forage [37]	MIO ? [37]				
	Puccinellia distans (L.) Parl. SALT GRASS)	turf	EU	0-330 me/l	[17]		
	P. maritima (Huds.) Parl. (SEA POA)	forage [5]	EU	0.55-4.97	[18]	0.56-2.08 Cl−	[20]
	Spartina alterniflora Loisl. (SMOOTH CORDGRASS)	forage, silage[7]	EU	0.55-2.0	[18]		
	S. cynosuroides (L.) Roth. (BIG CORDGRASS)	hay [7]	MIO	0.12-3.9	[18]	0 4 4 NaCl	[19]
	S. patens (Ait.) Muhl. (HAY CORDGRASS)	hay, forage [7]	EU	0.18-1.00	[9]	0.03-0.36 Cl−	[9]
	S. pectinata Link (PRAIRIE CORDGRASS)	hay, brushes [7]	EU				
	S. spartinae (Trin.) Merrill	hay*, brushes	MIO ?[8]				
	S. townsendii H. & J. Groves (CORDGRASS)	hay*, silage, forage [7]	EU			0.7-1.41 Cl−	[20]
	Sporobolus airoides (Torr.) Torr. (ALKALI SACATON)	hay, forage [4]	MIO	0.003-1.60	[9]		
	S. pyramidatus (Lam.) Hitchc. (RAT'S TAIL GRASS)	forage, grain*	MIO	? - 0.85	[9]		
	Stenotaphrum subsecundum (Walt.) O. Kuntze (ST AUGUSTINE GRASS)	lawn, forage	MIO				
JUNCACEAE	*Juncus maritimus* (Lam.) (SEA RUSH)	fiber, paper [34]	MIO			0.35-0.95 NaCl	[25]

573

Continue_

Table 2. HALOPHYTIC SPECIES OF ECONOMIC SIGNIFICANCE

Numbers in parentheses refer to sources of information; for key to these and the salinity group symbols, see p. 9

( ) = values in % soil solution, unless specified; * - minor use)

FAMILY	Species and Common Name	Use	Salinity Group	Salinity TDS	Range$^+$ NaCl/Cl$^-$
	Derris heterophylla Willd. (DERRIS)	insecticide (India) [13a]	MIO		
	Lotus corniculatus L. var. *tenuifolius* (NARROW-LEAF BIRD'S FOOT TREFOIL)	forage	MIO	0.5-0.75 [37]	
	Medicago hispida Gaertn. (BUR CLOVER)	forage [37]	MIO		? - 0.8 NaCl [37]
	Melilotus indicus (L.) ALL. (SOUR CLOVER)	forage	MIO		
	Trifolium fragiferum L. (STRAWBERRY CLOVER)	forage	MIO		
	T. repens L. (WHITE CLOVER)	forage	MIO		
	T. resupinatum L. (PERSIAN CLOVER)	forage	MIO		
	T. tomentosum L.	forage [37]	MIO		? - 0.6 NaCl [37]
MALVACEAE	*Hibiscus tiliaceus* L.	ornamental, fiber* [32]	MIO		
	Sida hederacea (Doug.) Torr. (ALKALI MALLOW)	ornamental	MIO		
	Thespesia populnea Soland. ex Correa (PORTIA TREE)	ornamental, dye* [32]	MIO		
MELIACEAE	*Xylocarpus moluccensis* Roem.	cabinet wood [10]	MIO ?		
NYPACEAE	*Nypa fruticans* Wurmb. (NYPA PALM)	alcohol, sugar, fiber	MIO		
PALMACEAE	*Cocos nucifera* L. (COCONUT PALM)	alcoholic liquors, beverages, nuts, oil, sugar, textile, timber	MIO		? - 0.5 NaCl
	Oncosperma filamentosa Bl. (NIBUNG PALM)	fiber, vegetable, wood [32]	MIO [10]		
	O. tigillaria (Jack.) Ridl.	fiber, wood [32]	MIO [10]		
	Phoenix dactylifera L. (DATE PALM)	fiber, fruit, sugar	MES ?	0.3-2.4 [39]	
	P. reclinata Jacq.	fiber, ornamental	MIO [10]		
	Armeria maritima (Mill.) Willd. (SEA PINK/THRIFT)	ornamental	MIO		0 - 1.5 NaCl [3]
PLUMBAGINACEAE	*Limonium bellidifolium* (Gouan) Dum. (SEA LAVENDER)	ornamental	EU/MES		0.02-0.32 Cl [20]
	L. gmelini Kuntze	ornamental	MIO ?		
	L. humile Mill.	ornamental	MES		
	L. sinuatum (L.) Mill. (STATICE)	ornamental	MES ?		
	L. spicatum Kuntze	ornamental	MIO ?		
	L. vulgare Mill.	ornamental	EU/MES		

Continued

Table 2. HALOPHYTIC SPECIES OF ECONOMIC SIGNIFICANCE

Numbers in parentheses refer to sources of information; for key to these and the salinity group symbols, see p.

(= values in % soil solution, unless specified; * - minor use)

FAMILY	Species and Common Name	Use	Salinity Group	Salinity Range[+]	
				TDS	NaCl/Cl[-]
POLYGONACEAE	*Coccoloba uvifera* (L.) L. (SEA GRAPE)	ornamental, fruit* [13]	MIO		
RHIZOPHORACEAE	*Bruguiera gymnorhiza* (L.) Lam.	timber, fuel [13]	EU	1.0-2.5 [10]	
	Ceriops tagal (Per.) C.B. Robinson	tannin, fuel, adhesive* [13]	EU ?	? -6.0 [10]	
	Rhizophora apiculata Blume	timber, tannin* [13]	MES ?	? < 1.5 [10]	
	R. harrisoni (=*R. brevistyla*)	tannin, charcoal* [13]	MES ?		
	R. mangle Roxb. (=*R. racemosa*) (RED MANGROVE)	tannin, medicine*[13,22]	EU		
	R. mucronata Lam.	timber, tannin* [13]	EU	1.5-5.5 [10]	
SONNERATIACEAE	*Sonneratia acida* L.	paper [13]	MES ?		
STERCULIACEAE	*Heritiera littoralis* (L.) Dyrand.	timber, fuel [13]	MIO ?		
	H. monor (SUNDRI)	timber, rayon	MIO ?		

575

Table 3: MINOR USES AND POTENTIAL USES OF HALOPHYTIC SPECIES

(* refers to potentially useful species)

FAMILY	Species and Common Names	Use	Salinity Group
AIZOACEAE	* *Mesembryanthemum australe* Soland.	horticulture	MES ?
	M. crystallinum L.	horticulture, vegetable [32]	MIO
	* *M. nodiflorum* L.	horticulture	MIO
AMARYLLIDACEAE	*Pancratium maritimum* L. (SEA LILLY)	ornamental, vegetable, fiber [31]	MIO
APOCYNACEAE	* *Cerbera odallum* Gaertn. (RUBBER VINE)	ornamental	MIO
	* *Rhabdadenia biflora* (Jacq.) Muell. Arg.	ornamental	MIO
	* *Urechites lutea* (L.) Britt. var. *lutea* (WILD ALLEMANDER)	ornamental	MIO
AVICENNIACEAE	*Avicennia alba* Bl.	buffalo fodder [21], timber, medicine [13]	EU
	A. germinans (L.) L. (=*A. nitida* L.) (BLACK MANGROVE)	cattle fodder [3a], fuel, medicine [13]	EU
	A. marina (Forsk.) Vierh. (BLACK MANGROVE)	camel fodder, timber, fuel [13], sheep & cattle fodder, tannin [40]	EU
	A. officinalis L.	edible seedlings [13]	MES ?
BOMBACACEAE	*Campostemon schultzii* Mast.	paper [13]	MIO ?
BORAGINACEAE	*Heliotropium curassavicum* L. (CHINESE PUSLEY)	ornamental, medicine [32]	MES
CELASTRACEAE	*Tricerma phyllanthoides* (Benth.) Lund. (*Maytenus phyllanthoides*)	edible fruit (baja Calif.), * ornamental	MES/MIO
CHENOPODIACEAE	*Atriplex canescens* (Pursh.) Nutt.	forage [42]	MIO ?
	A. elegans (Moq.)D. Dietr.	forage, vegetable (N. Am. Indian) [42]	MIO ?
	A. hortensis L. (ORACHE)	potherb	MES
	A. patula L. ssp. *hastata* (L.) Hall & Clem.	potherb	MES
	A. polycarpa (Torr.) Wats. (ALL-SCALE)	forage [41, 42]	MES
	A. lentiformis (Torr.) Wats. (LENS-SCALE)	forage [42]	MIO ?
	A. wrightii S. Wats.	grain (N. Am. Indian) [42]	MIO ?
	Nitraria retusa (F.) Aschers.	forage [40]	EU
	Salicornia brachiata Roxb.	forage [16]	EU
	S. europaea L. (GLASSWORT, SAMPHIRE)	potherb [5]	EU
	S. herbacea L. (=*S. stricta*)	potherb [16]	EU

Table 3: MINOR USES AND POTENTIAL USES OF HALOPHYTIC SPECIES Continued
(* refers to potentially useful species)

FAMILY	Species and Common Name	Use	Salinity Group
	S. virginica L. (PICKLEWEED)	N. Am. Indian potherb [28]	EU
	Sarcobatus vermiculoides (Hook.) Torr.	forage [42]	MES ?
	Suaeda maritima (L.) Dum. (SEA BLITE)	potherb	EU
	S. nudiflora Moq.	potherb [16]	EU ?
COMPOSITAE	* *Aster tripolium* L. (SEA ASTER)	ornamental	MES
	Artemisia maritima L. (SANTONIN)	insecticide, medicine	MIO
	* *Cotula coronopifolia* L. (BRASS BUTTONS)	ornamental	MES
	Lasthenia glabrata Lindl.	ornamental	MES ?
	Grindelia humilis H. & A.	ornamental	MIO
	* *G. latifolia* Kell.	ornamental	MIO
	* *G. stricta* D. C.	ornamental	MIO
	Jaumea carnosa (Less.) Gray (JAUMEA)	ornamental	EU ?
	Pluchea purpurascens (S.W.) D. C. (SALT MARSH FLEABANE)	ornamental	MIO
CONVULVLACEAE	*Convolvulus sepium* L. (BINDWEED)	medicine, vegetable [32], * ornamental	MIO
	C. soldanella L. (BEACH MORNING GLORY)	medicine, vegetable [32], * ornamental	MES/MIO
	Cressa cretica L.	cattle fodder [16]	MES/MIO ?
	* *Ipomoea alba* L. (MOON VINE)	ornamental	MIO
	* *I. pes-caprae* (L.) R. Br.	ornamental, sand-binder	MIO
CRUCIFERAE	*Cakile maritima* Scop. (SEA ROCKET)	vegetable [32]	MES
	Cochlearia anglica L. (LONG-LEAVED SCURVY GRASS)	medicine, spice [32]	MES/MIO
	C. officinalis L. ssp. *officinalis* (SCURVY GRASS)	vitamin C [3], medicine, spice [32]	MES ?
	Crambe maritima L. (SEA KALE)	vegetable [32]	MIO ?
	Rhaphanus maritimus Sm. (SEA RADISH)	vegetable [32]	MIO ?
FRANKENIACEAE	* *Frankenia grandifolia* Cham. and Schlecht.	ornamental	EU
GRAMINEAE	*Distichlis palmeri* (Vasey) Fassett ex L. M. JOHNSTON (DESERT SALT GRASS)	grain (N. Am. Indian) [30]	MES ?
	Panicum amarulum Hitch. & Chase	dune binder [19], * forage	MES
JUNCACEAE	*Juncus maritimus* Lam. (SEA RUSH)	paper [5]	MES
JUNCAGINACEAE	*Triglochin palustris* L. (MARSH/ARROW GRASS)	fodder [8]	MES/MIO?
LEGUMINOSAE	*Anthyllis vulneraria* L.	fodder, ornamental [32]	MIO
	* *Derris eoxastophyllum* (L.) Benth. (DERRIS)	insecticide	MIO
	Dichrostachys glomerata Hutch. & Dalz.	fiber, timber, medicine [32], * ornamental	MIO
	Prosopis juliflora (Sw.) D.C. (MESQUITE)	alcohol, forage, resin, legume [32, 40]	MIO
	P. pubescens Benth. (SCREW BEAN)	legume (N. Am. Indian)	MIO
MELIACEAE	*Xylocarpus granatum* Koenig.	tannin, oil [13]	MES [34]

577

Continued

Table 3: MINOR USES AND POTENTIAL USES OF HALOPHYTIC SPECIES

(* refers to potentially useful species)

FAMILY	Species and Common Name	Use	Salinity Group
PLANTAGINACEAE	*Plantago coronopus* L. (BUCK'S HORN)	vegetable [32]	MIO ?
	* *P. maritima* L. (SEA PLANTAIN)	vegetable [32]	MES ?
PORTULACACEAE	*Sesuvium portulacastrum* L.	potherb, ornamental	EU/MES
	S. verrucosum Raf.	potherb, ornamental	MES ?
	Trianthema portulacastrum L.	potherb, ornamental	MIO ?
PTERIDIACEAE	* *Acrostichum aureum* L. (LEATHER FERN)	ornamental	MES/MIO
	* *A. speciosum* Willd.	ornamental	MES/MIO
	* *Belvinum serratulum* Richard	ornamental	MIO
RHIZOPHORACEAE	*Bruguiera gymnorhiza* (L.) Lam.	tannin, timber, edible seedlings [13]	
	B. sexangula Poir (=*B. eriopetala* W. & A.)	tannin, timber, edible seedlings [13]	
Rosaceae	*Potentilla anserina* L. (SILVERWEED)	ornamental, medicine, vegetable [32]	MIO ?
SONNERATIACEAE	*Sonneratia alba* G. Smith	forage, timber [13]	MES
TAMARICACEAE	*Tamarix africana* Poir. (TAMARISK)	ornamental, windbreak [32]	MIO
	T. gallica L. (TAMARISK)	ornamental, windbreak [32]	MIO
	T. pentandra Pall. (TAMARISK)	ornamental, windbreak [32]	MIO
UMBELLIFERAE	*Crithmum maritimum* L. (ROCK SAMPHIRE)	spice, vegetable [32]	MES/MIO
ZOSTERACEAE	*Zostera marina* L. (EEL/SURF GRASS)	vegetable and food flavoring (N. Am. Indian) [28]	EU

and Na$_2$SO$_4$ content. It is easily raised in laboratory cultures and can be harvested as a solid fibrous mat (Chapman, pers. comm.). The protein content of *Spirulina* is extremely high (45-68% of the dry weight) and most essential amino acids are present in adequate quantities.

The majority of economically important halophytes are presently used for forage purposes such as grazing, fodder, hay, and silage. Of the euhalophytic forage species, most attention has been given to the use of *Spartina townsendii* s.l. This vigorous allopolyploid, which is thought to have evolved only 100 years ago, now covers 21-27.7 x 10^3 ha. of marshland in various temperate parts of the world and is widely used for the stabilization of mudflats and marshland reclamation (Ranwell 1967). Efforts to establish the species in the tropics have so far failed; however, Ranwell (ibid.) considers that there is sufficient genetic variability in this taxon to warrant selection for heat-tolerant species.

Several spartinas produce high standing crop yields and have a moderately high protein content (Table 4). *Spartina townsendii*, *S. alterniflora*, and *S. patens* have been successfully used as hay for cattle and older sheep breeds; however, some modern European sheep breeds of the Clun type find *S. townsendii* of low palatibility. *S. townsendii* has been used to produce silage of satisfactory quality in Britain (Hubbard and Ranwell 1966). Although the crude protein content of several spartinas is relatively high, it should be noted that *S. alterniflora* was found to be low in amino acids, with the exception of lysine (Burkholder 1956). However, the concentration of amino acids can be increased considerably by mixed bacterial and protozoan fermentation of the hay (Taschdjian 1954). Vitamins of the B group have been found in adequate quantities in *S. alterniflora* and *S. patens* (Burkholder 1956, Udell et al. 1969).

Other promising halophytic forage species have received attention. These include *Distichlis spicata* (Udell et al. 1969), *Atropis distans* and *Nitraria retusa* (Tewari 1967), *Atriplex semibaccata* (Headdon 1929), *A. polycarpa* (Chatterton et al. 1970), *A. vesicaria*, *A. nummularia*, *Avicennia marina*, and *Prosopis juliflora* (Firmin 1968), *Kochia scoparia* (Erickson and Moxon 1947), and *Rhizophora mangle* (Morton 1965). Further investigation is required to determine the full potentials of these species as forage resources.

Several halophytes have been investigated as possible sources of cellulose for paper or rayon production. Promising early experiments with *Spartina townsendii* were not followed up due to difficulties of harvesting and cleaning the crop (Ranwell 1967). However, the methods successfully used to reap clean hay for silage (Hubbard and Ranwell 1966) could probably avoid most of these problems. Several North American spartinas may also be useful for paper production (Robinson 1947). In the Sundarbans, *Excoecaria agallocha* is the basis of a profitable newsprint industry and *Heritiera minor* has been investigated as a possible source of viscose grade rayon (Chapman 1972). In Australia, *Excoecaria agallocha*, *Camptostemon schultzii* and *Sonneratia acida*

Table 4. YIELDS AND CRUDE PROTEIN CONCENTRATION OF SOME HALOPHYTIC FORAGE CROPS

	Yield (dry wt., t/ha)	Crude Protein (% dry wt.)	Ref. Source
Spartina townsendii (s.l.)	4 - 12.3 $\bar{X} = 7.5$	8.2 - 21.5	Ranwell, 1967
S. alterniflora	$\bar{X} = 8.5$ max = 12.6	5.7 - 13.2	Ranwell, 1967
S. patens	8.96	5.4 - 12.7	Ranwell, 1967
S. cynosuroides		6.6 - 12.3	Ranwell, 1967
S. foliosa	8.0*	$\bar{X} = 4.4$ 12.4 - 17.9[+]	Phleger, 1971
Distichlis spicata	6 - 8.5 $\bar{X} = 6.6$	9.6	Udell et al., 1969
Cynodon dactylon		13.1	Udell et al., 1969

*Maudie 1970
[+] in hydroponic lab. cultures

have been found to yield a satisfactory pulp, comparable to that of most other hardwood pulp species (von Koeppen and Cohen 1955). Preliminary experiments with *Juncus maritimus* show that high cellulose yields (55% d. w.) can be obtained for production of paper with a quality similar to that of Esparto grass paper (Bloch et al. 1954).

Several species of mangroves are exploited for tannin production in subtropic and tropical regions (Chapman 1972). Species yielding a high percentage of soluble tannin include *Rhizophora mucronata* (36%), *Rhizophora harrisonii* (50-60% of the cutch), *Rhizophora mangle* (30%), *Bruguiera gymnorhiza* (36%), and *Xylocarpus obovata* (30%). Bark of *Ceriops tagal* and some ecotypes of *Laguncularia racemosa* is also fairly high in soluble tannin content, but this tannin seems to be of the pyrogallol type which produces an inferior color. Details of the methods of harvesting and extracting the tannin, and ratios of soluble to insoluble forms in various species are given by Chapman (ibid.).

Some mangrove species have local value for timber or fuel production. Details of the quality and uses of the most important species are summarized in Table 5. One of the main drawbacks to use of *Rhizophora mangle* timber is the tendency of the wood to crack and shrink in dry weather; however, it has been found that ringbarking prior to felling greatly reduces this problem (Chapman, ibid.). More widespread practice of this technique might greatly enhance the value of this timber. Controlled felling of economic mangrove species is carried out in the USA, India, Pakistan, Burma, Malaya, the Andaman Islands, and in parts of East Africa and Indonesia; in many regions, however, mangrove forests represent under-utilized and under-developed resources.

The present use of halophytes as ornamentals does not appear to offer significant economic opportunities. However, in view of the low salinity tolerance of many popular temperate-region ornamentals (see Table 6), it seems reasonable to look towards a variety of attractive halophytes for more widespread domestic use in water-short regions or saline soils. Other halophytic trees and shrubs could undoubtedly be added to the ornamental species identified in Tables 2 and 3.

It is beyond the scope of this review to evaluate the potential economic use of mangroves and salt marsh plants for medicine, food flavoring, or perfume production. Most of the uses listed in Tables 2 and 3 appear to be highly localized; other casual medicinal uses are listed by Chapman (1972) and Tewari (1967). However, one promising species deserves further consideration. *Matricaria chamomilla*, false chamomile, thrives in moderately saline or sodic soils and is cultivated in Europe and Russia for the production of flower heads from which a substitute for chamomile is extracted (Singh 1970a). In Yugoslavia alone, 230,000 ha of saline-sodic soils are under this crop. Recent trials in India indicate that at cultivation costs of $135/acre and current buying prices of $865-$362/acre for high quality produce, the typical

TABLE 5.

IMPORTANT MANGROVE TIMBER AND FUEL RESOURCES

(After Chapman 1972)

	Timber Use And Quality	Fuel Durability
Rhizophora mucronata and *R. apiculata*	Pit props (Nigeria); 2x as durable as teak	high
R. mangle (=R. racemosa)	poles, wharf piles, railway sleepers, planks; strong dense wood, resistant to fungi and termites but not toredo	ND
Heritiera littoralis	boat-building (E. Africa)	high
H. minor	boat-building, construction, furniture; hard, heavy and durable wood	ND
Bruguiera gymnorhiza	planks (E. Africa), poles (India); termite and toredo resistant	high
Ceriops tagal		high

Table 6: Classification of a Selection of Economic Plants by Salinity Class (after U. S. Salinity Laboratory, 1954; Bernstein, 1964; and Bernstein, et al. 1972.

ECONIMIC SPECIES

Salinity Class and EC_e	Field Crops	Truck Crops	Forage Crops	Fruit Crops	Ornamentals
2 2-4 mmhos/cm	field beans	radish, celery green beans	meadow foxtail, white dutch, alsike, red and ladino clover, burnet	citrus & stone fruits, almonds, berry fruits, avocado	Pittosporum (P. tobira, Algerian ivy, Grenoble rose, Burford holly, Pineapple, guava, star jasmine
3 4-8 mmhos/cm	corn, flax, sunflower, castorbean	cabbage, cauliflower, lettuce, sweet corn, potato, sweet potato, yam, bell pepper, carrot, onion, peas, cantaloupe, squash, cucumber	orchard grass, hay rye, wheat and oats, blue grama, meadow fescue, reed canary, big trefoil, smooth brome, tall meadow oatgrass cecer & sickle milkvetch, sour clover	pomegranate, fig, olive, grape	rosemary, Euonymus, Dracaena, oleander, bottle brush, spreading juniper, Pyracantha (P. graberi), silver-berry, arborvitae, Dodonea, Xylosma, boxwood, lantana, Texas privet, Viburnum, hibiscus cv/ Brillante, heavenly bamboo
8-16 mmhos/cm	safflower wheat, rice, rye, oats, sorghum, soybean, sesbania	table beets, kale, asparagus, spinach tomato, broccoli	tall fescue, hay barley, perennial rye, harding grass, birdsfoot trefoil, beardless wild rye, alfalfa, Rhodesgrass, rescue grass, Canadian wild rye, Western wheat grass, Hubam & strawberry clover, Sudan & Dallis grass, yellow & white sweet clover	date palm	Bougainvillea Natal plum
5 16 22 mmhos/cm	barley, sugar beets cotton		bermuda grass, tall & crested wheatgrass, saltgrass, alkali sacaton, Nuttall alkali-grass		

dry weight yields of 300 Kg/acre would support a profitable agricultural industry. Moreover, this halophyte accumulates fairly large amounts of Na$^+$ (up to 66 me/100 g dry wt.) and thus assists in the reclamation of saline soils. The cultivation of *Rosa damascena*, a source of perfume oil, has also been found profitable on sodic soils (Singh 1970b) but the salt tolerance range of this species is not known. Other economic species of purported salt or alkali tolerance are being tested at the Lucknow Botanic Gardens for their usefulness in saline or sodic soils; a list of approximately 50 potentially useful species is given by Singh (1970a).

There seem to be no obvious reasons (other than socio-political barriers) why a variety of halophytes should not be put to more widespread use to combat agricultural salinity problems or to extend agricultural potentials in both coastal and inland regions of the world. Selection for temperature tolerance and increased yield could undoubtedly be used to produce vigorous varieties or strains that were adapted to local climatic conditions, although it is unlikely that the boundaries between temperate and tropical climatic requirements could be bridged very far. It is also unlikely that mangroves could be profitably grown for timber purposes outside of the humid tropics since the climatic conditions characteristic of this region seem to be a prerequisite for optimal development (Macnae 1968).

Special cultivation techniques may be necessary for the propagation of some low-marsh halophytes that have a high soil moisture requirement. Several halophytes seem to demand relatively large amounts of NaCl for optimal growth; even though the minimal requirement seems to be quite low for most species (Brownell 1968) and is likely to be met by even moderately saline agricultural soils. Nonetheless, some species may require the addition of certain trace elements, e.g. *Spartina alterniflora* has a high Fe^{+++} requirement (Adams 1963) and *Rhizophora* may require relatively large amounts of Al (Hesse 1963). Little is known about the response of halophytes to fertilizers. Preliminary studies on the growth of *Salicornia*, *Suaeda* (Pigott 1968, Stewart et al. 1972) and Baltic shoreline vegetation (Tyler 1967) suggest that growth in some soils might be increased by the addition of nitrogen and phosphorous. Experimental trials on the cultivation requirements of promising genera, such as *Spartina* and *Rhizophora*, outside of their natural habitats are urgently needed.

The Comparative Salt Tolerance of Economic Plants And Halophytes

It is evident that many halophytes have characteristics which might make them suitable for more widespread economic use. In order to evaluate the gains that could be obtained from employing these halophytes for agricultural and other purposes, we should first carefully compare certain aspects of the salinity-related growth responses of conventional crops with those of a variety of halophytes.

The upper salinity tolerance limits of 116 agricultural and horticultural species are indicated in Table 6. These values may be roughly compared with those given for halophytes in Table 2 by assuming that % salt (TDS) = EC_e x .064, where EC_e is expressed in mmhos/cm at 25°C (Bernstein 1964). However, careful examination of the criteria used to evaluate salt tolerance is necessary. In Table 6, the salinity tolerance limit of a species is given as the EC_e at which a 50% decrease in yield, compared with a nonsaline control, is obtained. This relative growth criterion is commonly employed by agriculturalists and obviously has great economic significance in a competitive market. At the same time, however, it tends to obscure the fact that some crops may continue to grow, although more slowly, at much higher salinities. This criterion also accounts only for the total growth performance of the plant and fails to distinguish among differences in salt sensitivity at various growth stages.

In contrast to the agricultural emphasis on comparative yield or growth rate, ecologists have usually estimated the salt tolerance of a halophyte by the range of soil salinity with which a species is commonly associated. This criterion ignores the fact that the optimal growth of many halophytes occurs at a lower salinity than the maximum tolerated. Growth depression of halophytes at seawater concentrations above 20-50% has been reported for a variety of mangrove and salt marsh species by Clarke and Hannon (1970), for *Atriplex vesicaria* and *A. hastata* (Black 1960), and *Salicornia* and *Suaeda* (Tsuda 1961, Flowers 1972). These and other data reviewed by Chapman (1960), Strogonov (1964) and Barbour (1970) suggest that few halophytes grow optimally in undiluted seawater, at least under lab conditions, and only euhalophytic species maintain relatively high yields above 50% seawater. This apparent pattern of halophyte growth was confirmed by a recent study of selected Pacific Coast salt marsh species grown in a range of seawater concentrations. The relationship between salinity and relative yield of these salt marsh plants is shown in Figure 1, and can be directly compared with that of several economic crops that were grown under the same conditions (Mudie et al. 1972). It is evident that there is a fairly wide range in growth response to salinity among the halophytes which is not obviously correlated with the habitat salinity associations of the species. For example, both *Spartina foliosa* and *Frankenia grandifolia*, exhibited maximal growth rates at about 25% seawater (9 ppt). Both occupy habitats in Southern California that are frequently inundated by undiluted seawater (35-38 ppt) and they occur in soils with a saturation paste extract salinity of 40 ppt (Mudie 1970). It is also apparent that, under experimental conditions, the growth responses of some high marsh species, e.g. *Deschampsia caespitosa* ssp. *beringensis* and *Armeria maritima* (also, but not shown here, *Grindelia stricta* and *Hordeum hystrix*), were little different from those of the glycophytic crops, and were considerably poorer than that of the most salt-tolerant cultivar, namely table beet.

The persistence of miohalophytes in salt marshes is intriguing because it is not obviously related to growth potential under saline conditions. Perhaps the adaptive characteristics that permit survival of this group of facultative halophytes but exclude salt-tolerant glycophytes are determined by salt tolerance at some critical growth stage such as germination, emergence, or reproduction. However, it is also possible that many halophytes avoid salinity stresses through adaptive phenological responses such as rapid growth following dilution of the soil water by rain, through utilization of fresh water floating over a saline water table, or by association with highly porous substrates slightly elevated above tide level. In view of the growth depression shown by many halophytes at high salinities, it is suggested that much more attention should be directed towards the study of correlations between seasonal growth rates and soil water salinity fluctuations in the active root zone. It is particularly unfortunate that so few studies of these factors have been carried out in high marsh areas since most economically important halophytes appear to be miohalophytic, and are thus confined to the upland fringes of salt marshes.

Until more data is obtained on the salinity and related growth responses of halophytes, the assumption that the domestication of salt marsh and mangrove swamp species would greatly increase the acceptable range of agricultural soil and water salinity must be approached with caution. Clearly, the amount by which the permissible salinity may be increased varies among halophytic species (and probably ecotypes) and at present it can only be estimated with reasonable certainty that a significantly higher salinity can be tolerated by euhalophytic species.

On the other hand, the assumption that all economic crops are highly limited in salt tolerance should be closely scrutinized. For example, it is clear that under favorable conditions of aeration, nutrient availability and ionic balance, several crops can tolerate 50% seawater (TDS = 17g/1) and that some, e.g. table beet, produce acceptable yields up to almost 75% seawater (see Figure 1). These findings suggest that the relatively poor crop growth encountered in many saline soils may be attributable to factors other than total salt concentration. Confirmation of this hypothesis might lead to a significant increase in acceptable agricultural salinity by utilization of special cultivation conditions or substrates, e.g. porous, clay-free sands and gravels.

Similarly, the extent to which salinity tolerance can be enhanced by altering the ratio of beneficial to harmful ions needs to be thoroughly examined. For example, the growth of bush beans under saline conditions was remarkably improved by an increase in the ratio of Ca^{++} to Na^+ (La Haye and Epstein 1970) and an increase in the ratio of K^+ to Na^+ was reported to enhance the salt tolerance of salt-sensitive species such as citrus, groundnuts, and cowpeas (Heimann 1966, Heimann and Ratner 1966). Much remains to be learned about the specific action of ions on salt tolerance (Rains 1972) and it will be very interesting to see whether the manipulation of ion ratios has widespread application for increasing the salt tolerance of glycophytic species.

Increasing Salt Tolerance by Genetic Methods

In view of the widespread significance of salt accumulation in agricultural soils, it is surprising that relatively little effort has been devoted to genetic selection for salt tolerance and that virtually nothing is known about the inheritance of salt tolerance. Significant differences in salt tolerance have been found among cultivars of several economic plants including tomatoes (Tal 1971), alfalfa (Brown and Hayward 1956), barley (Greenway 1962), soybeans (Gates et al. 1966, 1970), creeping bent (Youngner et al. 1967) and among lines of intermediate wheatgrass (Hunt 1965). Marked differences in the accumulation of Na^+, Cl^- or B have also been shown to occur among rootstocks of varieties of grapes (Bernstein et al. 1969, Alexander and Woodham 1968), citrus species (Cooper and Edwards 1952, Cooper et al. 1951, 1952) and various stonefruit and almond varieties (Hansen 1948, 1955; Bernstein, Brown and Hayward 1956). Differences have often been correlated with scion performance under saline conditions. In general, then, it appears that salt tolerance is at least partially determined genetically and it should be possible to breed salt tolerance into crop species (Epstein 1972).

The occurrence of genetic variation in salt tolerance among cultivars, which have gene pools restricted by controlled breeding and selection for characters other than salt tolerance, strongly suggests that wide variations in salt tolerance should occur among wild halophyte populations. Unfortunately, little research has been done on this aspect of halophyte ecology, even though important clues might be derived regarding the inheritance or evolution of salt tolerance. Ecotypic differences in salt tolerance have been found among populations of *Typha* (McMillan 1959), *Spartina alterniflora* (Mooring et al. 1971), *Festuca rubra* and *Agrostis tenuis* (Hannon and Bradshaw 1968), *Lepidium perfoliatum* (Choudhuri 1968), *Rumex crispus* (Cavers and Harper 1967), and other species cited by Waisel (1972). Interpopulational differences in salinity tolerance are also implied in the studies of *Sporobolus texanus* (Ungar 1968) and *Atriplex polycarpa* (Chatterton et al. 1970). The studies of *Festuca*, *Agrostis* and *Rumex* species indicate, in these taxa at least, that differences in salinity tolerance between non-saline and saline soil ecotypes are not of the sharply defined type found between serpentine and non-serpentine races of some species (Kruckeberg 1951), or between lead-tolerant and intolerant populations of *Festuca ovina* (Gregory and Bradshaw 1965); rather, the upland (non-saline) ecotypes showed a fairly high tolerance of salinity, although of a lower degree than the saline soil ecotypes.

It has long been suggested (Isaacs 1964) that more attention should be paid to exploiting the genetic salt tolerance of wild halophytes through hybridization and graftage with glycophytic economic plants. Because of the paucity of information regarding the inheritance of salt tolerance, it is presently impossible to draw any firm conclusions regarding the magnitude of salt tolerance increase that can be expected from genetic manipulation.

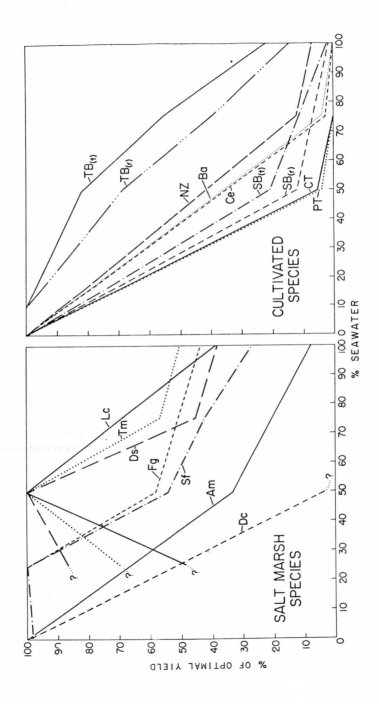

Figure 1.

Relative wet weight yields of a variety of Pacific Coast salt marsh species and a selection of crops grown hydroponically with diluted and undiluted seawater (34.5 g/l TDS). Yield is expressed as % of maximum yield.

Symbols: <u>Salt Marsh Species</u>

Am = *Armeria maritima*, Dc = *Deschampsia caespitosa* ssp. *beringensis*
Ds = *Distichlis spicata* ssp. *stricta*, Fg = *Frankenia grandifolia*
Lc = *Limonium californicum* var. *mexicanum*, Sf = *Spartina foliosa*
Tm = *Triglochin maritima.*

<u>Cultivated Species</u>

Ba = barley cv. 'Atlas '68', Ce = Celery cv. 'Utah Green',
CT = cherry tomato cv. 'Red Cherry', NZ = New Zealand spinach,
PT = paste tomato cv. 'San Marzano', SB = sugar beet, Union Sugar F_1 hybrid,
r = roots, t = tops.

However, it seems logical to look towards salt marsh or strand ecotypes of economic plants for sources of salt tolerant genes. It is also conceivable that the salt tolerance of wild halophytic species might be transferred to glycophytic economic plants by hybridization although there is a general reluctance among agricultural geneticists to attempt hybridization between taxa above the varietal level. It should be remembered that several important crops have been derived (naturally or artificially) through hybridization at the specific level (e.g. tobacco, potato, kapok, strawberry) or even at the generic level (e.g. wheat and possibly corn), and it has been pointed out by Harlan (1970) that much germ plasm is wasted because of mental blocks concerning taxonomic relationships. There is abundant evidence of natural and artificial interspecific hybridization in many plant genera (Knobloch 1972). Furthermore, about half of the 124 records of intergeneric hybrids investigated by Knobloch (ibid.) showed some degree of fertility which was not always low.

If a quantum increase in salt tolerance were the objective, it would be logical to focus on euhalophytic species as germ plasm sources. The predominant taxonomic pattern that seems to prevail, however, is that the euhalophytes are confined to small, almost exclusively halophytic families (e.g. Rhizophoraceae, Frankeniaceae, Avicenniaceae), to halophytic tribes or sections of families (e.g. Atriplicinae, Salicornioideae, Suaedeae, Halocnemoideae in the Chenopodiaceae, Staticeae in the Plumbaginaceae), or to small taxonomically isolated genera within larger families (e.g. *Spartina, Distichlis, Jaumea, Lumnitzera*). This taxonomic pattern would appear to reduce the possibility of using euhalophytes for hybridization or graftage with important economic glycophytes, but should be more thoroughly investigated.

There are many meso- or miohalophytes, however, that may be sufficiently salt-tolerant to warrant experiments on hybridization or graftage with relatively salt-sensitive economic glycophytes. Some of the more promising hybridization potentials include (halophytic parent listed first): *Agropyron junceum* x *Triticum aestivum, T. dicoccum* or *T. turgidum* (wheats); *Raphanus maritimus* x *R. sativus* (radish); *Crambe maritima* x *C. abyssinica* (a promising new source of vegetable oil); *Beta vulgaris* ssp. *maritima* x *B. vulgaris* varieties (sugar beet, table beet, chard). Relatively salt-tolerant rootstocks should be looked for in wild populations of *Persea borbonia* (redbay), *Bumelia celastrina* (saffron plum), and *Zanthoxylum fagara* (wild lime which occur on the hammocks within the Florida mangrove swamps), since these might be used to increase the salt tolerance of the highly salt-sensitive avocado, zapota, and citrus, respectively.

Insofar as genetic affinity reflects evolutionary pathways, it is interesting to speculate on the origin of halophytes in general. The apparent taxonomic isolation of the euhalophytes suggests that they diverted early from the mainstream of angiosperm evolution into the specialized halophytic niche.

Within the rather stable salt marsh niche (there have undoubtedly always been wave-protected intertidal areas although not always in the same places), they would have encountered little competition from upland plants, hence their frequently widespread occurrence and rather low morphological and taxonomic diversity (*Atriplex* excepted). Their dominance of the saline habitat and niche hyperspaces seems to have provided little opportunity for other taxa to enter except in the transition zone between highly saline and nonsaline upland soils. In this region, physiological flexibility to cope with both nonsaline and moderately saline conditions may be more important than the fixation of genes for extreme salt tolerance, hence the characteristic facultative halophytism of miohalophytes, namely, survival at high salinities but rapid growth only under mildly or nonsaline conditions.

At least one genus, however, appears to have taken the opportunity to enter a highly saline niche in relatively recent times. In the genus *Lasthenia*, several endemic Californian species have become adapted to the saline sinks of the Central Valley (Ornduff 1966), a region which only emerged in the Pleistocene, and at least one species, *Lasthenia glabrata*, appears to have euhalophytic characteristics (Mudie, unpublished data). Some of these hypotheses on halophyte evolution are being tested in our lab by means of a detailed study of the physiology and biosystematics of the closely-related group of species comprising *Lasthenia*. It would also be very interesting to test whether wild miohalophytes have the genetic potential for high salt tolerance under strong artificial selection pressure for this character.

The prospects for using halophytes to solve the widespread problems of soil salinity are presently very hazily defined and much remains to be determined about the relationship between salt tolerance and productivity, quality of useful product, effect on reproduction, and other important practical matters. It is my contention, however, that agriculturalists may well profit from seeking answers to the agricultural salinity problem from advances in the understanding of the physiology and ecology of halophytes. Likewise, halophyte ecologists may well derive a greater understanding of halophyte adaptations and evolution from focusing attention on the landward, rather than the seaward edge of the marsh.

Literature Cited

Adams, D. A. 1963. Factors influencing vascular plant zonation in North Carolina marshes. Ecology 44:445-456.

Adriani, E. D. 1945. Sur la phytosociologie, la synecologie et le bilan de halophytes de la region neelandaise meridionale, ainsi que de la Mediterranee francaise. S.I.G.M.A. 88:1-217.

Alexander, D. McE., and R. C. Woodham. 1968. Relative tolerance of rooted cuttings of four Vinifera varieties to sodium chloride. Aust. J. Exp. Agr. Anim. Husb. 8:461-465.

Allison, L. E. 1964. Salinity in relation to irrigation. Adv. in Agronomy 16:139-180.

Anon. 1966. Survey of saline and alkaline conditions in irrigated desert coastal valleys of South Peru. Arid Lands Res. Newsletter 21:2.

Aston, B. C. 1933. The Napier-Ahuriri lagoon lands. N. Z. J. Agric. 46:69-77, 260-266.

Atchley, S. C. 1938. Wild Flowers of Attica. Clarendon Press, Oxford. 60 pp.

Baker, H. G. 1965. Plants and Civilization. Wadsworth Publ. Co., Belmont, California. 183 pp.

Barbour, M. G. 1970. Is any Angiosperm an obligate halophyte? Amer. Midl. Natur. 84:105-120.

Bender, G. L. 1963. Native animals and plants as resources. Pages 309-337 in Aridity and Man. Edited by C. Hodge. AAAs Publ. 74, Washington.

Bernstein, L. 1962. Salt-affected soils and plants. Pages 139-174 in The Problems of the Arid Zone. Proceedings of the Paris Symposium. Arid Zone Research, 18, UNESCO.

Bernstein, L. 1964. Salt Tolerance of Plants. U.S.D.A. Agric. Info. Bull., No. 283. 23 pp.

Bernstein, L., J. W. Brown, and H. E. Hayward. 1956. The influence of rootstock on growth and salt accumulation in stone-fruit trees and almonds. Proc. Am. Soc. Hort. Sci. 68:86-95.

Bernstein, L., C. F. Ehlig, and R. A. Clark. 1969. Effect of grape rootstocks on chloride accumulation in leaves. J. Am. Soc. Hort. Sci. 94:584-590.

Bernstein, L., L. E. Francois, and R. A. Clark. 1972. Salt tolerance of ornamental shrubs and ground covers. J. Am. Soc. Hort. Sci. 97:550-556.

Bernstein, L., and H. E. Hayward. 1958. Physiology of salt tolerance. Ann. Rev. Plant Physiol. 9:25-46.

Black, R. F. 1960. Effects of NaCl on the ion uptake and growth of *Atriplex vesicaria* Heward. Australian J. Biol. Sci. 13:249-266.

Bloch, M. R., D. Kaplan, and J. Schnerb. 1954. *Juncus maritimus*, a raw material for cellulose. Bull. Res. Coun. Isr. 4:192-194.

Boyko, H. (ed.). 1966. Salinity and Aridity — New Approaches to Old Problems. Monographiae Biologicae. XVI. W. Junk. The Hague. 408 pp.

Brown, J. W., and H. E. Hayward. 1956. Salt tolerance of alfalfa varieties. Agronomy J. 48:18-20.

Brownell, P. F. 1968. Sodium as an essential micronutrient element for some higher plants. Plant and Soil 28:161-164.

Burkholder, P. R. 1956. Studies on the nutritive value of *Spartina* grass growing in the marsh areas of coastal Georgia. Bull. Torrey Bot. Club. 83:327-334.

Cavers, P. B., and J. L. Harper. 1967. The comparative biology of closely related species living in the same area. IX. *Rumex*: the nature of adaptation to a sea-shore habitat. J. Ecology. 55:73-82.

Chapman, V. J. 1960. Salt Marshes and Salt Deserts of the World. Interscience, N. Y. 392 pp.

Chapman, V. J. 1972. Mangroves of the World. Leonard Hill, London.

Chapman, V. J. Prof. of Botany, Auckland U., Auckland, New Zealand (personal communication).

Chatterton, N. J., C. M. McKell, F. T. Bingham, and W. J. Clawson. 1970. Absorption of Na, Cl and B by desert salt-bush in relation to composition of nutrient solution culture. Agron. J. 62:351-352.

Choudhuri, G. N. 1968. Effect of soil salinity on germination and survival of some steppe plants in Washington. Ecology 49:465-471.

Clarke, L. D., and N. J. Hannon. 1970. The mangrove swamp and salt marsh communities of the Sydney district. III. Plant growth in relation to salinity and waterlogging. J. Ecol. 58:351-369.

Clawson, A. B. 1937. Toxicity of arrowgrass for sheep and remedial treatment. USDA Tech. Bull. 580:1-16.

Cooper, W. C., and C. Edwards. 1952. Toxicity and accumulation of chloride in citrus on various rootstocks. Proc. Am. Soc. Hort. Sci. 59:143-146.

Cooper, W. C., C. Edwards, and E. O. Olson. 1952. Ionic accumulation in citrus as influenced by rootstock and scion and concentration of salts and boron in the substrate. Plant Physiol. 27:191-203.

Cooper, W. C., B. S. Gorton, and C. Edwards. 1951. Salt tolerance of various citrus rootstocks. Proc. Rio Grande Valley Hort. Inst. 5:46-52.

Darlington, C. D., and A. P. Wylie. 1955. Chromosome Atlas of Flowering Plants. Hafner Publ. Co., N. Y. 519 pp.

Davies, W. C. 1931. Tidal-flat and salt-marsh studies in Nelson Haven. N. Z. J. Sci. Tech. 12:338-360.

Dixey, F. 1966. Water supply, use and management. Pages 77-99 in Arid Lands. Edited by E. S. Hill. Methuen & Co. Ltd., London.

Doering, E. J., and L. C. Benz. 1972. Pumping an artesian source for water table control. Am. Soc. Civil Engineers Proc., J. Irrig. and Drainage Divn. 98, No. IR2. Pages 275-287.

Elzam, O. E., and E. Epstein. Salt relations of two grass species differing in salt tolerance. I. Growth and salt content at different salt concentrations. Agrochemica 13:187-195.

Epstein, E. 1972. Mineral Nutrition of Plants: Principles and Perspectives. John Wiley and Sons, Inc., N. Y. 411 pp.

Erickson, E. L., and A. L. Moxon. 1947. Forage from *Kochia*. S. Dak. Ag. Exp. Sta. Bull., 384.

Firmin, R. 1968. Forestry trials with highly saline or seawater in Kuwait. In Saline Irrigation for Agriculture and Forestry. Edited by H. Boyko. W. Junk Publ., The Hague. World Academy of Art and Sci. 4:107-132.

Flowers, T. J. 1972. Salt tolerance in *Suaeda maritima* (L.) Dum. J. exp. Bot. 23:310-21.

Fryxell, P. A. Research Geneticist, USDA Agricultural Research Divn., Texas A & M Univ., College Station, Texas (personal communication).

Furr, J. R., and C. L. Ream. 1968. Salinity effects on growth and salt uptake of seedlings of the date, *Phoenix dactylifera* L. Proc. Am. Soc. Hort. Sci. 92:268-273.

Gates, C. T., K. P. Haydock, and P. J. Claringbold. 1966. Response to salinity in Glycine. II. Differences in cultivars of *Glycine javanica* in dry weight, nitrogen and water content. Aust. J. Exp. Agr. Anim. Husb. 6:374-379.

Gates, C. T., K. P. Haydock, and M. F. Robins. 1970. Response to salinity in *Glycine*. IV. Salt concentration and the content of P, K, Na and Cl in cultivars of *Glycine wrightii (G. javanica)*. Aust. J. Exp. Agr. Anim. Husb. 10:99-110.

Gausman, H. W., W. R. Cowley, and J. H. Barton. 1954. Reaction of some grasses to artificial salinization. Agron. J. 46:414-416.

Gledhill, D. 1963. The ecology of the Aberdeen Creek mangrove swamp. J. Ecology. 51:693-703.

Greenway, H. 1962. Plant response to saline substrates. I. Growth and ion uptake of several varieties of *Hordeum vulgare* during and after NaCl treatment. Aust. J. Biol. Sci. 15:16-38.

Greenway, H. 1968. Growth stimulation by high chloride concentrations in halophytes. Israel J. Bot. 17:169-177.

Gregory, R. P. G., and A. D. Bradshaw. 1965. Heavy metal tolerance in populations of *Agrostis tenuis* Sibth. and other grasses. New Phytol. 64:131-143.

Gupta, R., and B. C. Gulati. 1967. Salt tolerance in *Cymbopogon nardus* (Linn.) Rendle. Pages 100-106 in Seminar on Sea, Salt and Plants. Edited by V. Krishnamurthy. Central Salts and Marine Chem. Res. Inst., Bhavnagar, India.

Gunn, C. R. 1972. Moonflowers, *Ipomoea* sect. *Calonyction* in temperate North America. Brittonia 24:150-168.

Hannon, N., and A. D. Bradshaw. 1968. Evolution of salt tolerance in two co-existing species of grass. Nature 220:1342-1343.

Hansen, C. J. 1948. Influence of the rootstock on injury from excess boron in French (Agen) prune and President plum. Proc. Am. Soc. Hort. Sci. 51:239-244.

Hansen, C. J. 1955. Influence of the rootstock on injury from excess boron in Nonpareil almond and Elberta peach. Proc. Am. Soc. Hort. Sci. 65:128-132.

Harlan, J. R. 1970. Evolution of cultivated plants. Pages 19-32 in Resources in Plants. Edited by O. H. Frankel and Bennet. Blackwell Sci. Publications, Oxford.

Hayward, H. E. 1954. Plant growth under saline conditions. Pages 37-41 in Arid Zone Research. IV. Reviews of Research Utilization of Salt Water.

Hayward, H. E., and L. Bernstein. 1958. Plant-growth relationships on salt-affected soils. Bot. Rev. 24:584-635.

Hayward, H. E., and C. H. Wadleigh. 1949. Plant growth on saline and alkali soils. Adv. in Agron. 1:1-38.

Headden, W. P. 1929. The Australian saltbush. Colorado Experiment. Station Bull. 345:27 pp.

Heimann, H. 1966. Plant growth under saline conditions and the balance of the ionic environment. In Salinity and Aridity. Edited by H. Boyko. Monographiae Biologicae 16:201-213.

Hiemann, R. H., and R. Ratner. 1966. Experiments on the basis of the principle of the "balance of ionic environment." In Salinity and Aridity. Edited by H. Boyko. Monographiae Biologicae 16:283-293.

Heizer, R. F., and L. K. Napton. 1969. Biological and cultural evidence from prehistoric human coprolites. Science 165:563-568.

Hesse, P. R. 1963. Phosphorus relationships in a mangrove swamp mud with particular reference to aluminum toxicity. Plant and Soil 19:205-218.

Hubbard, J. D. E., and D. S. Ranwell. 1966. Cropping *Spartina* salt marsh for silage. J. Br. Grassland. Soc. 21:214-217.

Hunt, O. J. 1965. Salt tolerance in intermediate wheatgrass. Crop Science 5:407-409.

Isaacs, J. D. 1964. The planetary water problem. Proc. First Internat. Conf. of Women Engineers and Scientists, June 15-21, N. Y. II. Pages 1-13.

Israelson, O. W., and V. E. Hansen. 1962. Irrigation Principles and Practices. Edn. 3. John Wiley and Sons, Inc., N. Y., 447 pp.

Kearney, T. H., and R. H. Peebles. 1960. Arizona Flora. Edn. 2 U. California Press, Berkeley. 1032 pp.

Kelley, W. P. 1951. Alkali Soils, Their Formation Properties and Reclamation. Reinhold Publ. Corp. 176 pp.

Kelley, W. P. 1963. Use of saline irrigation water. Soil Sci. 95:385-91.

Knobloch, I. W. 1972. Intergeneric hybridization in flowering plants. Taxon, 21:97-103.

Kruckeberg, A. R. 1951. Intraspecific variability in the response of certain native plant species to serpentine soil. Am. J. Bot. 38:408-419.

La Haye, P. A., and E. Epstein. 1970. Salt toleration by plants: enhancement with calcium. Science 166:395-396.

Leonard, J. 1969. A very promising source of proteins and vitamins: *Spirulina platensis*, an edible blue-green alga. Abstracts of papers presented at the 11th Internat. Bot. Congress, Aug. 24-Sept. 2, 1969:126. (abstract).

Lunt, O. R., V. B. Youngner, and J. J. Oertli. 1961. Salinity tolerance of five turfgrass varieties. Agron. J. 53:247-249.

Macnae, W. 1968. A general account of the fauna and flora of mangrove swamps and forests in the Indo-West Pacific region. Advances in Mar. Biol. 6:74-270.

Magistad, O. C. 1945. Plant growth relations on saline and alkaline soils. Bot. Rev. 11:181-230.

McMillan, C. 1959. Salt tolerance within a *Typha* population. Am. J. Bot. 46:521-526.

Meigs, P. 1966. Geography of Coastal Deserts. Arid Zones Res. Series, 28, UNESCO, UNESCO Publ. Center, N. Y. 140 pp.

Mobberley, D. G. 1956. Taxonomy and distribution of the genus *Spartina*. Iowa State College J. Sci. 30:471-574.

Mooring, M. T., A. W. Cooper, and E. D. Seneca. 1971. Seed germination response and evidence for height ecophenes in *Spartina alterniflora* from North Carolina. Amer. J. Bot. 58:48-55.

Morton, J. S. 1965. Can the red mangrove provide food, feed and fertilizer? Econ. Bot. 19:113-123.

Mudie, P. J. 1970. A Survey of the Coastal Wetland Vegetation of San Diego Bay. Unpublished report prepared for Calif. Dept. Fish and Game, Contract No. W26 D25-51.

Mudie, P. J., W. R. Schmitt, E. J. Luard, J. W. Rutherford, and F. H. Wolfson. 1972. Preliminary studies on seawater irrigation. Scripps Institution of Oceanography Reference Series 72-70.

Ornduff, R. 1960. A biosystematic survey of the goldfield genus *Lasthenia*. U. Calif. Publ. in Bot. 40:1-92.

Penfound, W. T., and E. S. Hathaway. 1938. Plant communities in the marshlands of southeastern Louisiana. Ecology Monogr. 8:1-56.

Phleger, C. F. 1971. Effect of salinity on growth of a salt marsh grass. Ecology 52:908-911.

Pigott, C. D. 1968. Influence of mineral nutrition on the zonation of flowering plants in coastal marshes. *In* Ecological Aspects of the Mineral Nutrition of Plants. Edited by I. H. Rorison. Symposium of the British Ecological Soc. 9:25-35.

Pirie, N. W. 1969. Food Resources, Conventional and Novel. Penguin Books, Inc., Baltimore, Maryland. 208 pp.

Raheja, P. C. 1966. Aridity and salinity. Pages 43-127 *in* Salinity and Aridity. Edited by H. Boyko. Monographiae Biologicae 16. W. Junk Publ., The Hague.

Rains, D. W. 1972. Salt transport by plants in relation to salinity. Ann. Rev. Plant Physiol. 23:367-388.

Rains, D. W., and E. Epstein. 1967. Preferential absorption of potassium by leaf tissue of the mangrove, *Avicennia marina*: an aspect of halophytic competence in coping with salt. Australian J. Biol. Sci. 20:847-857.

Ranwell, D. S. 1967. World resources of *Spartina townsendii* (s.l.) and economic use of *Spartina* marshland. J. Appl. Ecology 4:239-256.

Raven, P. H., and H. Curtis. 1970. Biology of Plants. Worth Publ., Inc., N. Y. 706 pp.

Robinson, B. B. 1947. Minor fiber industries. Econ. Bot. 1:47-56.

Saini, G. R. 1972. Seed germination and salt tolerance of crops in coastal alluvial soils of New Brunswick, Canada. Ecology 53:524-525.

Seneca, E. D. 1972. Seedling response to salinity in four dune grasses from the outer banks of North Carolina. Ecology 53:465-471.

Shreve, F. 1942. Desert vegetation of North America. Bot. Rev. 8:195-246.

Singh, L. B. 1970a. Utilization of saline-alkali soils for agro-industry without prior reclamation. Econ. Bot. 24:439-442.

Singh, L. B. 1970b. Utilization of saline-alkali soil without prior reclamation—*Rosa damascena*, its botany, cultivation and utilization. Econ. Bot. 24:175-179.

Stewart, G. R., J. A. Lee and T. O. Orebamjo. 1972. Nitrogen metabolism of halophytes. I. Nitrate reductase activity in *Suaeda maritima*. New Phytol. 71:167-263.

Strogonov, B. P. 1964. Physiological Basis of Salt Tolerance of Plants. English transl. by Israel Program for Sci. Transl. Jerusalem, 1964. 279 pp.

Tal, M. 1971. Salt tolerance in the wild relatives of the cultivated tomato: responses of *Lycopersicum esculentum, L. peruvianum,* and *L. esculentum minor* to sodium chloride solution. Aust. J. Agric. Res. 22:631-638.

Taschdjian, E. 1954. A note on *Spartina* protein. Econ. Bot. 8:164-165.

Tewari, A. 1967. Coastal vegetation and its utility. Pages 334-339 *in* Sea, Salt and Plants. Edited by V. Krishnamurthy. Central Salt and Marine Chem. Res. Inst., Bhavnagar, India.

Tsuda, M. 1961. Studies on the halophytic characters of the strand plants and of the halophytes in Japan. Jap. J. Bot. 17:332-370.

Turner, N. C., and M. A. M. Bell. 1971. The ethnobotany of the Coast Salish Indians of Vancouver Island. Econ. Bot. 25:63-104.

Tyler, G. 1967. On the effect of phosphorous and nitrogen, supplied to a Baltic shore-meadow vegetation. Bot. Notiser 120:433-447.

Udell, H. F., J. Zarudsky, T. E. Doheny, and P. R. Burkholder. 1969. Productivity and nutrient values of plants growing in the salt marshes of the town of Hempstead, Long Island. Bull. Torrey Bot. Club. 96:42-51.

Ungar, I. A. 1965. An ecological study of the vegetation of the Big Salt Marsh, Stafford County, Kansas. Univ. Kansas Sci. Bull. 46:1-99.

Ungar, I. A. 1966. Salt tolerance of plants growing in saline areas of Kansas and Oklahoma. Ecology 47:154-155.

Ungar, I. A. 1967. Vegetation-soil relationships on saline soils in northern Kansas. Am. Midland Naturalist. 78:98-120.

Ungar, I. A. 1968. *Sporobolus texanus* Vasey in Lincoln, Nebraska. Rhodora 70:450-451.

U. S. Salinity Laboratory. 1954. Diagnosis and improvement of saline and alkali soils. U.S.D.A. Agric. Handbook No. 60. 160 pp.

van Aart, R. 1971. Regional training course on drainage and land reclamation. Nature and Resources. 7:22-23.

von Koeppen, A., and W. E. Cohen. 1955. Pulping studies of five species of a mangrove association. Aust. J. Sci. 6:105.

Waisel, Y. 1972. Biology of Halophytes. Academic Press, N. Y. 395 pp.

Wheeler, W. A. 1950. Forage and Pasture Crops. D. van Nostraad Co., Inc., N. Y. 752 pp.

Youngner, V. B., O. R. Lunt, and F. Nudge. 1967. Salinity tolerance of 7 varieties of creeping bentgrass *Agrostis palustris* Huds. Agron. J. 59:335-336.

HALOPHYTES, ENERGETICS AND ECOSYSTEMS

Eugene P. Odum

Institute of Ecology
The University of Georgia
Athens, Georgia 30602

Halophytes are not only well adapted to stressful environments but in many situations they also exhibit a remarkable ability to exploit lack of competition and auxillary energy sources in such a way as to compensate for the energy cost of coping with the stress factors. Thus, the growth pattern and productivity of a stand of halophytes is as much determined by the energetics of its ecosystem as by physiological adaptations within the plant itself. In the past biologists who study estuarine and seashore organisms have been preoccupied with how such life adapts to temperature and salinity fluctuations, alternate flooding and exposure, and other obvious stresses: that some of these energy stresses might be converted to energy subsidies has only recently been fully appreciated. Water flow is a good example. Where the pattern of water flow is such that work beneficial to the organisms is accomplished to a degree that compensates for the energy lost in adapting to stress, the productivity of a stand of halophytes may equal or exceed that of an equivalent community growing in a more benign but less subsidized environment. Where water flow patterns are not favorable, the stress factor becomes limiting, and the stand is stunted in structure and low in productivity. Accordingly, assessment of systems energetics is an important part of the study of a particular species or community.

Since energy is an important common denominator in all ecosystems whether designed by nature or man, it provides the logical basis for a "first order" classification of ecosystem types. On this basis it is convenient to consider three major classes of ecosystems: (1) the unsubsidized solar-powered ecosystem, (2) the subsidized solar-powered ecosystem, and (3) the fuel-powered ecosystem. In the first category direct rays of the sun received on a unit-area basis is the principle, if not the sole, source of energy for the ecosystem; such ecosystems frequently suffer from shortages of nutrients, water or other resources since little auxillary energy is available for pumping in and recycling materials. Consequently productivity of unsubsidized solar-powered systems tends to be low, ranging from 1,000 to 10,000 kilo-calories per meter square per year, with an average of about 2,000. The open ocean and vast tracts of upland grasslands and forests are in this category, as are many halophyte communities, as already indicated. In contrast, where there are utilizable sources of energy other than the direct rays of the sun the ecosystem can be said to be "subsidized". Such auxiliary energy sources perform work of mineral cycling, food transport, waste removal and so on, and thereby enable the community to make more efficient use of the sun and food produced by plants. Subsidies can be natural, such as wind and rain in the rain forest or tides in the estuary. Or, of course, they can be provided by man, as in the case in agricultural ecosystems, where other work provides the subsidies. Productivity in both natural and fuel-subsidized solar-powered ecosystems ranges from 10,000 to 40,000 kcal $M^{-2}Yr^{-1}$, averaging about 20,000 — a good order of magnitude better than the unsubsidized solar-powered system. Halophyte communities such as salt marshes in tidal estuaries have a productivity in this range, and are clearly subsidized by the energy of the tides, waves and other water flows. The high productivity of such stands is in sharp contrast to halphyte communities growing in or around a salt lake where the same level of sun energy might be received, but where there is no mechanical energy subsidy to overcome the stress of the salty substrate.

The fuel-powered ecosystems, that is, man's cities and industrial developments, do not concern us here, except insofar as the waste from such systems have a major impact on estuaries and other halophyte habitats. Since the rate of energy flow is a a city is several orders of magnitude higher than that of natural systems (of order of 2,000 kcal $M^{-2}Yr^{-1}$) the impact of the one on the other is understandably very great. But some kinds of wastes could act as subsidies, depending on type and rate of input. There are several groups of investigators who are currently experimenting to determine to what extent salt marshes are able to assimilate sewage wastes at various stages of prior treatment. Because of their high rate of metabolism marsh systems may have considerable capacity as tertiary treatment plants for man, but it is too soon to establish a policy on this.

Now let us illustrate some of these concepts by specific examples. Grasses of the genus *Spartina* occur in a wide gradient of wetlands ranging from wet prairies and fresh water marshes to high salinity estuaries. Salt tolerant species also exhibit strong ecotypic adaptations that enable a single species to cover a wide range of edaphic conditions. For example, in the Georgia barrier island estuaries *Spartina alterniflora* occurs in almost pure stands over the major part of the intertidal gradient with distinctly different growth forms occupying successive zones. These growth forms which exhibit partial genetic fixation range from very tall (over 2 meters) slender plants growing on the soft tidal creek banks subjected to strong twice daily tidal inundations to short (less than 0.3 meters) stubby plants growing on peaty high marshes reached only by spring tides. Since we have found that the primary production of stands of *Spartina alterniflora* is roughly proportional to tidal amplitude within the moderate range of 1 to 10 feet, it is clear that the energy subsidy provided by tidal flow more than compensates for the energy drain of osmoregulation required by the high salinity environment. This halophyte also is a C4 plant (see Black, 1971) has a high net-to-gross production ratio (i.e., relatively low respiration) and is therefore well adapted to high temperatures and high light intensities. The deeply rooted growth form also enables this plant to recover from anerobic sediments nutrients that would be unavailable to any other autotroph in the estuaries. Since C4 plants also have low water requirements it is presumed that osmoregulatory energy requirements are reduced. These and other attributes combine to produce a highly productive community in what would generally be classed as a stressed environment for higher plant life.

Some figures from our Georgia studies will illustrate a marked increase in productivity associated with tidal irrigation. The following are round-figure annual net-production rates for the three major Spartina zones. Numbers are dry matter gm/m^2, lbs/acre, and Kcal/m^2, respectively (for more detail, see Odum and Fanning, 1973).

    Tall grass, low marsh
        (frequent and vigorous tidal irrigation)
        4,000; 17.8; 16,000
    Medium grass (gentle tidal irrigation)
        2,300; 10.2; 7,200
    Short grass, high marsh (infrequent tidal irrigation)
        750; 3.3; 3,000

Thus, the salt marsh stands do indeed use "tidal energy" and this factor can make a several-fold difference in production.

If we were to draw a performance curve to show these relationships by plotting tidal amplitude on X-axis and net production on the Y-axis we get a concave or "hump-backed" curve with an optimum in range of 5-8 feet tidal amplitude. Too high tides produce more stress than subsidy and too low or infrequent irrigation fails to compensate for the stresses.

Tides also greatly influence successional processes. The salt marsh estuary as a whole functions as a detritus ecosystem as does a young forest or a fresh water marsh, but unlike a stable water level marsh, or an early successional terrestrial community, litter does not accumulate in the estuary because tidal action hastens removal and decomposition of net production as fast as it is produced. Accordingly, we view the *Spartina* salt marsh in Georgia not as a successional stage leading to another type of vegetation but as a *pulse-stablized climax* where change over long periods of time is more a function of geological or "allogenic" forces (such as a changing sea level, for example) than autogenic processes within the ecosystem. This is not to say that all halophute communities are "climax"; clearly many are not stabilized and subject to autogenic change. For more on concept of pulse-stabilization see Odum, 1969.

Literature Cited

Black, C. C., 1971. "Ecological Implications of Dividing Plants into Groups With Distinct Photosynthetic Production Capacities". Advances in Ecological Research, 7:87-114.

Odum, E. P., 1969. "The Strategy of Ecosystem Development". Science, April 18, 1969, Vol. 164, pp. 262-270.

Odum, E. P. and Marsha E. Fanning, 1973. "Comparison of the Productivity of *Spartina alterniflora* and *Spartina cynosuroides* in Georgia Coastal Marshes". Georgia Academy of Science, Vol. 31, No. 1, pp. 1-12.

INDEX

A

Alaska salt marshes, 202-205
Alkali regions, 8-9
Allenrolfea, 262-265
Ammonia, 547-563
Amphipods, 451-452
Anderson, C. F., 307-344
Aster, 337-339
Atlantic coast halophytes, 23-50
Atlantic coast salt marxhes, 23-50, 407-427 441-447
Aquatic submerged plants, 379-390

B

Baja California salt marshes, 220-224
Barbour, M. G., 175-233
Barrier beaches, 407-427
Beach vegetation, 183-192
Birds, 463-474, 475-507
British Columbia salt marshes, 205-206
Burning marshes, 480, 492

C

Caldwell, M. M., 355-378
California salt marshes, 212-220
Cation exchange capacity, 444
Cattle, 463-474
Chapman, V. J., 3-19
Charophyta, 240-246
Cleantis, 429-440
Climate, 179-182
Coastal Halophytes, 23-233, 307-353 379-440, 449-508, 525-563
Cotnoir, L. J., 441-447
Crabs, 452

D

Daiber, F. C., 475-508
Davis, D. E., 531-545
Desert, salt, 3-19, 239-296
Desert, soils, 366-367
Detritus, 429-440
Diking, 480-493
Distichlis, 265-266, 321-327
Ditching, 480-487
Dredged material, 525-529
Duncan, W., 23-50

E

Ecology, 599-608
Economic considerations, 401-405, 565-597
Ecosystem considerations, 475-508
Edwards, A., 531-545
Energetics, 599-602
Eutrophication, 393-406, 547-563

F

Fertilizer, 547-563
Fungi, 429-440

G

Gallagher, J. L., 511-523
Gas exchange, 368-374
Gastropds, 451
Geology, 182-183
Godfrey, M. M., 407-427
Godfrey, P. J., 407-427
Gulf coast halophytes, 23-50

H

Halophytes, Atlantic coast, 23-50, 175-233
  307-353, 407-427, 525-529, 547-563
  desert, 355-378
  Gulf coast, 23-50, 379-390
  Inland, 235-305, 355-378
  Pacific coast, 175-233
  physiology, 345-378
Herbicides, 116-118, 531-545
Hordeum, 281-288
Humidity, 459
Hydroponics, 565-597

I

Imagery, 511-523
Impoundments, 487-493
Infrared photography, 511-523
Inland salt deserts, 4, 239-296
Inlet dynamics, 407-427
Insects, 452-455
Invertebrates, 449-462, 475-508, 531-556
Ion absorption, halophytes, 347-348
  356-357
Isopods, 451-452

J

Juncus, 333-337, 407-427, 511-523

K

Kraeuter, J. L., 449-462

L

Light, 457-458
Limonium, 316-321

M

Macdonald, K. B., 175-233
Mammals, 492-493
Management practices, 475-508

Mangrove, 51-174, 379-390
  adaptations, 87-113
  ecology, 63-65
  geographical distribution, 54-63
  herbicides, 116-118
  silviculture, 113-116
Marsh burning, 480, 493
Marsh soils, 441-447, 525-529
Marshes, inland, 8-9, 235-305, 354-378
Marshes, salt, 3-20, 175-233, 393-429
  449-508, 525-529, 547-563
May, M. S., 429-440
McMillan, C., 379-390
Modeling, 393-406, 547-563
Mosquitoes, 475-508
Mudie, P., 565-597

N

Nitrogen, 547-563
North Carolina barrier beaches, 407-427
North Carolina salt marshes, 307-344
Nutrient cycling, 393-406, 463-474,
  531-563

O

Odum, E. P., 599-602
Oregon salt marshes, 208-211
Organic soil classification, 442-444
Osmotic relationships, 356-357
Overwash, 407-427

P

Pacific coast halophytes, 175-233
Phosphorus movement, 393-406, 539-540
  547-563
Photography, 511-523
Potamogeton, 240-246
Productivity, 367-368, 393-406, 511-523
Puccinellia, 270-274

Q

Queen, W. H., 345-353

R

Raccoon, 463-508
Reimold, R. J., 393-406
Remote sensing, 511-523
Ruppia, 240-246

S

Salicornia, 246-252, 327-333
Salt deserts, 3-20, 239-296
Salt excretion, 360-362
Salt marsh invertebrates, 449-462
Salt marshes, 192-227, 407-427, 475-508, 599-602
Salt metabolish, 362-365
Salt tolerance, 357-360, 365-366, 379-390, 565-597
Scirpus, 288-293
Seagrasses, 380-383
Seneca, E., 525-529
Sesuvium, 252-254
Shanholtzer, R. S., 463-474
Silviculture, mangroves, 113-116
Soils, desert, 366-367
   marsh, 441-447
Spartina, 307-316, 393-440, 475-563, 599-602
Spectral reflectance, 511-523
Sporobolus, 277-281
Stabilization of salt marsh, 525-529

Suaeda, 254-262
Submerged aquatic plants, 379-390

T

Tamarix, 294-296
Teal, J., 547-563
Temperature tolerance, 374-375, 458-459
Thermal imagery, 511-523
Transplanting, 527-528
Triglochin, 274-277
Typhia, 475-508

U

Ungar, I. A., 235-305
Uniola, 407-427
Urea, 547-563

V

Valiela, I., 547-563
Vertebrates, 463-474, 502-508

W

Walsh, G. E., 51-174
Washington state salt marshes, 206-208
Wolf, P. L., 449-462

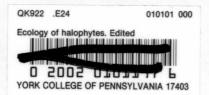